Resource Recovery from Water: Principles and Application

Resource Recovery from Water: Principles and Application

Edited by

Ilje Pikaar, Jeremy Guest, Ramon Ganigué, Paul Jensen, Korneel Rabaey, Thomas Seviour, John Trimmer, Olaf van der Kolk, Céline Vaneeckhaute and Willy Verstraete

CAPTURE

GHENT UNIVERSITY

IWA PUBLISHING

BioEngine

UNIVERSITÉ LAVAL

Published by **IWA Publishing**
Unit 104–105, Export Building
1 Clove Crescent
London E14 2BA, UK
Telephone: +44 (0)20 7654 5500
Fax: +44 (0)20 7654 5555
Email: publications@iwap.co.uk
Web: www.iwapublishing.com

First published 2022

British Library Cataloguing in Publication Data
A CIP catalogue record for this book is available from the British Library

ISBN: 9781789060317 (Paperback)
ISBN: 9781780409566 (eBook)
ISBN: 9781780409573 (ePUB)

This eBook was made Open Access in February 2022

Contents

Foreword

The water sector provides vital services to society that are essential for human health, sustaining economies, and protecting the environment. These roles will remain, but there are important changes underway that are calling for a new, changed paradigm for the way water is viewed and managed. With increasing global change pressures, cities of the future will experience difficulties in efficiently managing scarcer and less reliable water resources. In order to meet these challenges, there is a need for a fundamental change in the way we use and reuse water, based on a foundation of research, technology and innovation.

Climate change demands mitigation efforts to reduce greenhouse gas emissions. We live in a resource-constrained world in which the global population is expected to grow by another few billion. Such factors drive us towards a shift from a traditional approach to wastewater treatment to one aligned with a new paradigm in which the used water and materials and energy within it represent opportunities for resource recovery. The same applies for drinking water treatment, where the residuals from this treatment are often wasted. Minimising the movement of water, maximising reuse and redefining 'waste' as a resource can optimise productive use (matching quality to intended use) and reduce pollution. Sustainable (waste-) water management through resource recovery and reuse can be achieved with integrated business models that go beyond standard services and transform waste into valuable resources such as biofuels and fertiliser while also improving water management.

To make progress in this paradigm shift, there is a need to clarify underlying principles and to gather experiences of their application. This book contributes on both fronts, signalling opportunities in the water sector, industry, agriculture and the energy sector, in particular. These are opportunities that can be economically viable, with the prospect that further technological advances, growth in 'green' demand, and greater cost and scarcity of raw materials will add to that viability.

The water sector is poised to be able to help transform society. The relative predictability of potable water and wastewater flows through its infrastructure add to the sector's attractiveness as a partner in this transformation. Similarly, the relative long life of this infrastructure can provide a stable core around which implementation of resource recovery can be built.

However, because of this long life, change is needed now. Technology breakthroughs and innovative designs need to be coupled with comprehensive system changes to urban water processes, institutions, and regulations. We need to recognise the high-level relationships among water resources, energy, and land use in an urbanising world. More is needed than simply improving the performance and efficiency of the component parts of systems – change is needed at a system-wide level as well. This new book will help shape the actions that will be needed. It also points to the vibrant future we can expect in the sector – a vibrancy that stands to attract and inspire the next generation of young water professionals.

Dr. Kala Vairavamoorthy
Executive Director
International Water Association

doi: 10.2166/9781780409566

Chapter 1

Resource recovery from municipal wastewater: what and how much is there?

Pieter Ostermeyer[1,2], Gabriel Capson-Tojo[3,4], Tim Hülsen[3], Gilda Carvalho[3], Adrian Oehmen[5], Korneel Rabaey[1,2] and Ilje Pikaar[6]

[1]Center for Microbial Ecology and Technology (CMET), Ghent University, Coupure Links 653, 9000, Gent, Belgium
[2]Center for Advanced Process Technology for Urban Resource recovery, Frieda Saeysstraat 1, 9052 Gent, Belgium
[3]The Advanced Water Management Centre, The University of Queensland, St Lucia, QLD 4072, Australia
[4]Department of Chemical Engineering, CRETUS Institute, School of Engineering, Universidade de Santiago de Compostela, 15782, Santiago de Compostela, Spain
[5]The School of Chemical Engineering, The University of Queensland, St Lucia, QLD 4072, Australia
[6]The School of Civil Engineering, The University of Queensland, St Lucia, QLD 4072, Australia

1.1 INTRODUCTION

The introduction of sewer networks for the safe collection and subsequent treatment of municipal wastewater at down-stream wastewater treatment plants has had a major impact on public health and the protection of the aquatic environment. It is often identified as one of the ten great public health achievements of the 20th century [1]. To achieve this, management and engineering practice has always had a strong focus on treating wastewater by removing pollutants. The most well-known and most widespread approach to achieve effective pollutant removal is the conventional activated sludge process, which recently celebrated its 100-year anniversary [2]. Despite successfully protecting human and environmental health during the 20th century, traditional removal-focused approaches for domestic wastewater treatment will not suffice in the 21st century, as broader environmental and economic consequences of this mentality become increasingly well understood. The recovery of resources embedded within the wastewater matrix, without compromising human health and environment protection, represents an alternative fitting within a circular economy context. Municipal wastewater contains a wide range of resources in addition to water, albeit at relatively low concentrations. Nutrients (mainly nitrogen and phosphorus), dissolved organics, coagulants, cellulose and potentially other valuables such as metals, proteins and polysaccharides, could be recovered. The reuse of water has the potential to alleviate water scarcity, while recovered nutrients and organics can be valorized as fertilizer, feed, energy or higher value products for various industrial sectors. It is not the aim of this chapter to discuss these in detail, but to provide the reader with a general overview of the different resources contained in municipal wastewater. The current and potential recovery of these resources is also addressed. It should be noted that the composition of municipal wastewater differs to some extent in terms of concentrations of constituents, local availability and value of contained resources. The factors/characteristics causing this variability are briefly discussed in this chapter as well. As a final major remark, when considering resource recovery from municipal wastewater, the presence of pathogens, toxic compounds, waterborne viruses and other impurities cannot be ignored. The challenges arisen by the presence of these pollutants are, however, beyond the scope of this chapter.

1.2 LEARNING OBJECTIVES

At the completion of this chapter you should be able to:

- Describe the magnitude of municipal wastewater generation and treatment and understand its potential for resource recovery.
- Summarize the different factors affecting the composition and volume of municipal wastewater and its impact on resource recovery potential.
- Describe the key resources that can be recovered during the treatment of municipal wastewater that are *present* in or *added* during the treatment of municipal wastewater.
- Understand how centralized and decentralized municipal wastewater treatment systems impact the potential for resource recovery.
- Understand that resource recovery at municipal wastewater treatment plants is not a 'one size fits all' concept.

1.3 HOW DO WE DEFINE WASTEWATER?

Sewage, domestic or municipal wastewater all relate to wastewater originating from domestic use, and can be defined as '*the water supply of the community after it has been used in a variety of applications and which now contains constituents that render it unsuitable for most uses without treatment*' [3]. It mainly consists of a mixture of human excrements and wastewater originating from municipal toilet use, cleaning, washing and cooking. For simplicity, this mixture will be referred to as 'municipal wastewater' in this chapter. Other wastewaters, such as industrial wastewater, are discussed elsewhere in this book. Despite the fact that the composition can vary to some extent (see section 1.6), a key feature of municipal wastewater is that, except for the water itself, resources are present at low concentrations, which often complicates their efficient recovery.

1.4 HOW MUCH MUNICIPAL WASTEWATER IS PRODUCED?

It is estimated that 312 million megaliter (ML) of municipal wastewater is produced annually, of which 187 million ML is treated, equaling to about 60% of the total [4]. This is most likely an underestimation, as accurate data from highly populated and fast-growing developing countries is not available. It is not the aim of this chapter to provide detailed data on the amounts and differences in the daily production of municipal wastewater in different countries, but the following example is shown to illustrate that the amount of municipal wastewater produced per capita can differ substantially between countries. The average volume of wastewater produced per capita (arriving at the wastewater treatment plants (WWTPs)) in the USA is almost 690 L/day [5, 6], while in Belgium (Flemish region) this value is only about 378 L/day (data: Aquafin Inc., Aartselaar, Belgium). Interestingly, this relates to some 110 L/day of residential water consumption per capita in Belgium, hence the remaining 268 L originates from other sources such as run off, commercial and governmental buildings and industrial wastewater streams discharged into the sewer network (data: Aquafin Inc).

1.5 HOW IS MUNICIPAL WASTEWATER COLLECTED?

In developed countries and urbanized regions, wastewater treatment is generally centralized: municipal wastewater is transported via a complex network of underground sewers to a central WWTP for its

Figure 1.1 (a) Centralized wastewater treatment plant: The Stickney Water reclamation Plant, adapted from Metropolitan Water Reclamation District of Greater Chicago, 2018 [7]. (b) Decentralized wastewater treatment plant by constructed wetlands [8].

treatment prior to discharge into the receiving environment. In combined sewer systems, run-off and rainwater are collected together with domestic (and industrial) wastewater. In sanitary sewers, municipal wastewater is collected separately. Sanitary sewers are becoming increasingly dominant, as legislation is fading out the use of combined sewers. However, many cities still continue to use combined sewer systems. It is important to realize that in major urbanized areas, most of the municipal wastewater produced is treated in a few large WWTPs, with capacities often ranging between 100 000–1 000 000 population equivalents (PE), and even up to multiple million PEs. For example, the Stickney wastewater treatment plant in Chicago (USA, Figure 1.1(a)) has a maximum capacity of 2 300 000 PE, treating about 5451 ML/day under storm weather flow conditions [7]. One can imagine that the scale of the WWTP has a substantial impact on the economy of scale of resource recovery activities and the potential to recover resources at a practical relevant scale. In remote, less densely populated areas and developing countries, access to centralized sanitation infrastructure is often lacking, and smaller, decentralized, treatment concepts such as constructed wetlands (Figure 1.1(b)) are more common due to the lower investment costs when compared to centralized infrastructure. Smaller size WWTPs and decentralized systems have varying capacities, and can be used to treat wastewater from a single house or from several thousand PEs. Smaller scales often create additional burdens on the economics of resource recovery, given the inefficiencies and challenges of managing decentralized systems, and the low mass flows of resources that can be potentially recovered.

1.6 UNTREATED MUNICIPAL WASTEWATER – WHAT RESOURCES ARE IN THERE AND IN WHAT CONCENTRATION RANGE?

Water is the dominant component of wastewater with typically more than 99% of municipal wastewater comprised of water itself. Besides water, household municipal wastewater is primarily a mixture of feces, organic kitchen wastes, urine and water originating from cleaning, washing and cooking. At a household level, municipal wastewater can be identified as three separate streams, namely, black, grey and yellow water, which are typically collected together. Black water refers to a mixture of feces, urine, and flush water, and is characterized by high concentrations of organic matter, nitrogen and phosphorus. Grey water, which comprises water from cooking, washing and cleaning, is more dilute and represents most of the wastewater flow. Urine is occasionally collected separately as yellow water. Table 1.1 provides a summary of the typical relative contributions of waste streams generated at the household level to the total wastewater mass flow. The typical concentrations of various components found in each stream are also presented.

The varying flow rates and composition of the individual waste streams determine the characteristics of the raw wastewater as it enters the sewer system. The amount and characteristics

Table 1.1 Typical relative contributions of some of the major components embedded in domestic wastewater, expressed as percentage of total mass flow. Average concentrations of main components are also presented [9].

Parameter	Grey water	Urine	Feces
Typical daily flow (L/capita/day)	108	1.2[a]	0.2[a]
Contribution to overall Flow rate (%)	99	0.9	0.1
Contribution to overall COD (%)	50	11	31
Contribution to overall nitrogen (%)	8	79	13
Contribution to overall phosphorus (%)	28	47	25
Contribution to overall potassium (%)	7	71	22
Water content (%)	99	95	77
COD[b] (mg/L)	620	10 236	155 000
BOD_5[c] (mg/L)	279	4567	60 000
Total nitrogen (mg/L)	23	8661	7500
Total phosphorus (mg/L)	8.5	732	3000
Potassium (mg/L)	10	2047	4500

[a]*Mass flow, expressed in g/capita/day.*
[b]*COD refers to chemical oxygen demand and represents the total amount of oxidizable matter present in the wastewater.*
[c]*BOD_5 refers to biological oxygen demand and comprises the biodegradable fraction of the oxidizable matter (tested in 5 days at 20°C).*

of the grey water produced can substantially differ between countries, depending highly on local habits, and can be substantially higher than earlier mentioned numbers. A first factor affecting the volume and composition of wastewater is the volume of drinking water consumed. High per capita consumption of drinking water generally results in higher flow rates and more dilute wastewaters, except where other organics are introduced into the water (e.g., via kitchen grinders or in catchments where industrial trade waste enters the sewerage system). Additionally, the carbon and nutrient contents are influenced by local diets and caloric intake. Specifically, nearly all (99–100%) the nitrogen and phosphorus that is ingested is excreted in urine and feces, whereas a smaller fraction of caloric intake (2–10%) is ultimately excreted [10]. In general, in higher income countries the water consumption and nutrient excretion rates per capita are higher than in developing countries and communities. Even among the more developed countries, the differences can be substantial. Because of these differences, the concentrations of nitrogen in urine can differ significantly, with reported values ranging between 2.6 to 16 g total N/L, respectively [9]. It is important to note that in fresh urine most of the nitrogen is fixed as urea. Urea subsequently gets hydrolyzed into ammonium (NH_4^+) by enzymes produced by bacteria present in biofilms on the walls in the storage and collection systems. Therefore, the nitrogen load of municipal wastewater arriving at WWTPs typically comprises around 75–90% ammonium. Factors such as cleaning/washing habits, level of urbanization, and drinking water price also affect the wastewater composition and volume. As an example, in countries such as Japan or in regions like the Middle East, the use of toilet paper is either prohibited or not customarily used. As a consequence, the amount of organic matter in wastewater in these countries is substantially lower [11]. In addition to the above-described human aspects, climate and meteorological phenomena also affect the composition of the collected wastewater. In combined sewer systems (collecting run-off from precipitation) or when intrusion of sub-surface waters into sewer systems plays a significant role, wastewater volumes increase and concentrations decrease during precipitation events or wet weather seasons [12]. Rainfall patterns can thus have a substantial impact on the wastewater composition. However, it should be noted that in moderate

Table 1.2 Typical composition of raw municipal wastewater in terms of major compounds, adapted from Metcalf and Eddy [3].

Parameter	Low strength	Medium strength	High strength
COD total (mg/L)	500	750	1200
COD soluble (mg/L)	200	300	480
BOD_5 (mg/L)	230	350	560
Volatile fatty acid (VFA) (mg-acetate/L)	10	30	80
Total nitrogen (mg/L)	30	60	100
Total Kjeldahl nitrogen (Ammonia-N + organic-N) (mg/L)	30	60	100
Ammonia-N (mg/L)	20	45	75
Total phosphate (mg/L)	6	15	25
Total suspended solids (TSS) (mg/L)	250	400	600
Volatile suspended solids (VSS) (mg/L)	200	320	480

climates such as Europe and the USA, dry weather flow conditions typically prevail about 95% of the time. The effect of precipitation also interacts with the effect of urbanization due to, for example, impermeability of the soil in dense urban areas.

While the composition of municipal wastewater thus fluctuates between different locations in the world, municipal wastewater is a waste stream in which resources that potentially can be recovered are generally present in low concentrations, except the water itself (which is a recoverable resource i.e. often at a high concentration). Table 1.2 shows the typical composition and concentration range of the major components present in untreated municipal wastewater.

1.7 WHAT RESOURCES CAN BE RECOVERED DURING TREATMENT OF MUNICIPAL WASTEWATER?

As discussed in section 1.5, municipal wastewater is typically treated in a centralized manner. The majority of municipal wastewater is currently treated by the activated sludge systems such as nitrification/denitrification with chemical P removal and enhanced biological phosphorus removal (EBPR) [13–15]. It is important to highlight that configurations of these treatment systems may vary considerably. A simplified representation of a typical flow sheet of a centralized WWTP for biological nutrient removal by means of nitrification/denitrification coupled with chemical P removal is depicted in Figure 1.2. In the figure, the major fractions and mass flows of notable resources embedded in the wastewater matrix within the different compartments of the WWTP assuming typical influent concentrations for mid-strength sewage (see Table 1.2) are shown. The figure shows that a major fraction of the incoming reactive nitrogen is dissipated and emitted to the atmosphere as N_2 (g), with only about 30% being assimilated in sludge. Organics and other suspended solids are removed in a primary settler, resulting in primary sludge. In the activated sludge process part of the organic matter is oxidized to CO_2. The surplus biomass produced during the activated sludge process is removed in a secondary settler as secondary sludge. Biogas produced via anaerobic digestion can be used for the production of electricity and heat in a combined heat and power unit. The digestate leaving the reactor is dewatered through processes such as centrifugation and belt press where biosolids and reject water are produced. The reject water contains high(er) concentrations of NH_4-N and PO_4-P which opens up opportunities for their recovery.

There are more resources that can be recovered than those depicted in Figure 1.2. In this context, a summary of potential resources that can be recovered from municipal wastewater, including the product characteristics of resources recovered and examples of their potential end-use in various

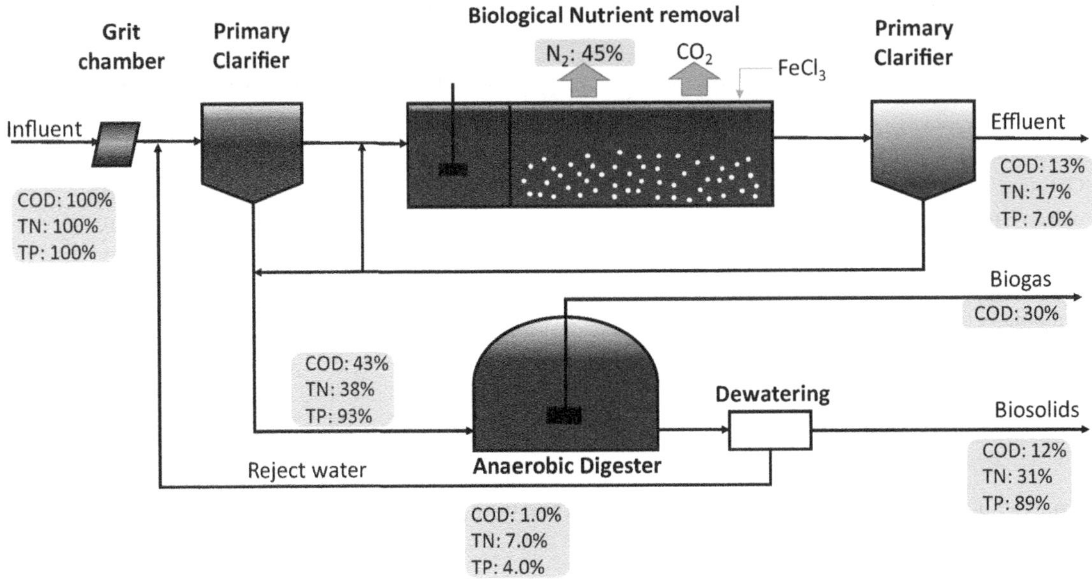

Figure 1.2 Example of a typical flowsheet of a centralized wastewater treatment plant for biological nutrient removal and the mass flows of COD, N and P. Note that the provided values should be considered as indicative only.

market segments, is provided in Table 1.3. An important emphasis to make is that the recoverable amounts depend on multiple aspects such as wastewater composition, the specific design and treatment capacity of the WWTP, the type, efficiency of the equipment used and mode of operation and so on. Equally important, some of the products listed in the table would require specific treatment steps that in fact would exclude the recovery of other components. For example, the recovery of cellulose through a micro-sieve step reduces the biogas production potential during anaerobic digestion (i.e., cellulose comprises a substantial fraction of the incoming COD load of a WWTP). Various other examples can be mentioned within this context with the key message being that the type and amount of resources that can be recovered depend on the treatment approach installed and priorities set forward by the wastewater utility regarding the preferred resources to be recovered. In the sub-sections below, we will discuss some of the key resources that can be recovered from municipal wastewater during its treatment in more detail.

1.7.1 Water reuse

First and foremost, one should appreciate the fact that over 99% of municipal wastewater comprises the water itself. This can be reused for various applications, depending on the local requirements and the quality of the produced recycled water. Water reuse practices can be classified into two main categories: non-potable and potable water reuse. The most common applications of non-potable recycled water include: agricultural irrigation, landscape irrigation, industrial reuse and groundwater recharge. Among them, agricultural and landscape irrigation are widely practiced worldwide, and have well-established health protection guidelines and agronomic practices [18, 19]. About 66% of the global population currently lives in water-stressed regions and 500 million people live in areas where water consumption exceeds local renewable water sources by a factor of two. Wastewater is increasingly being considered as an important water source for the production of drinking water [20]. This can be done indirectly through the recharge of water bodies used for drinking water production

Table 1.3 Overview of the potential resources that can be recovered from municipal wastewater and examples of their potential end-use in various market segments.

Resource category	Resource	Recovery potential	Examples of potential end-use/market segment
Water	Water	100–400 L/capita/day (daily water consumption substantially fluctuates depending on country/region)	Irrigation, non-potable domestic use, industrial use, potable domestic use. Injection to mitigate saltwater intrusion and so on.
Inerts	Sand	In the order of 0.1–3 kg/capita/year [16]	Construction industry
Organics	Cellulose	In the order of several kg/capita/year [11]	Biochemical industry, construction material
	Biosolids	It is nearly impossible to provide accurate numbers on the recovery potential of these compounds as the latter depends on a multitude of factors. 'Ball-park' figures that can be used are in the order of several kilograms per capita per year for each of these resources	Agriculture
	Alginate like substances		Pharmaceutical and food industry
	Biochar		Agriculture
	Volatile fatty acids		Biochemical industry
	PHA		Bioplastics/Agriculture
Energy	Biogas, as electricity[a]	In the order of 250 MJ/capita/year (theoretical) In the order of 33 MJ/ capita/year (practical)	Reuse onsite, local power grid
	Thermal energy (heat)[a]	In the order of 760 MJ/capita/year (theoretical) In the order of 291 MJ/capita/year (practical)	District heating/cooling
Nitrogen	Ammonia (NH_3)	1.6–7.4 kg N/capita/year [10]	Power generation (Denox)
	Ammonium sulfate		Agriculture
	Microbial protein		Agrifood, aquaculture
	Biosolids		Agriculture, landscaping
	Struvite		Agriculture
Phosphorus	Biosolids	0.4–1 kg P/capita/year [10]	Agriculture
	Struvite:		Agriculture
	Calcium phosphate		Agriculture
Metals	Large variety of metals in biosolids/ash	In the order of several grams/capita/year (for the sum of all metals)	Metallurgy
Coagulants	Predominantly Fe and Al based	In the order of 1 kg/capita/year	Soil amendment, construction, sulfide removal and odor control

[a]Assuming a water consumption per capita of 125 liter per day and data from [17].

(indirect potable reuse; e.g., aquifer recharge) or through the direct use of wastewater for the production of drinking water (direct potable reuse). Nevertheless, on a global scale, only a very small fraction of the municipal wastewater treated is currently being used for the production of drinking water. We would like to refer the reader to Chapter 4 of this book for a more detailed description on water reuse/reclamation and the most commonly used treatment methods.

1.7.2 Inerts

Depending on the size of the WWTP, sand is removed from wastewater in a dedicated treatment step to protect pumps and pipes from erosion and abrasion and reduce the amount of sludge processed in primary and secondary treatment. For example, in the Netherlands 88% of WWTPs remove sand as a separate treatment step [16]. Typical loads to a WWTP ranged between 0.1 and 3 kg capita^{-1} year^{-1}, albeit this can be substantially higher, especially during rain events in combined sewer systems. To give an illustration of the amount of sand that can enter a WWTP, it was found that with a capacity of 250 000 PE, some 840 tons of sand per year can be recovered [16]. About two-thirds of the recovered sand was reused in the construction industry, the remainder was landfilled. It is important to realize that it is a necessity to remove the sand from an operator's perspective in order to protect equipment and as such sand recovery can be seen as a low 'hanging fruit' in terms of its resource recovery potential.

1.7.3 Organic matter

Considering that the organics content (measured as COD) in municipal wastewater ranges between 500 and 1200 mg COD/L (see Table 1.1), and it is estimated that some 312 million megaliter (ML) of municipal wastewater is produced annually [4], the global theoretical mass flow of recoverable organics equals to a staggering 156–374 Mton COD/year. This organic matter comprises a mixture of organic compounds, including cellulose, extracellular polymeric substances (EPS), volatile fatty acids (VFAs), proteins, lipids and (complex) carbohydrates. Starting with cellulose, this compound originates primarily from toilet paper. In the Netherlands, for example, it is estimated that toilet paper accounts for about 125–360 mg COD/L, representing 25–30% of the total COD load in WWTPs. It was estimated that with a simple sieving step, 8–10 kg cellulose per capita per year could be recovered, amounting to a recovery potential of 4.1–6.1 Mton of cellulose in the EU-27 alone [11].

Nowadays, as also mentioned above, most of the organics in municipal wastewater (e.g., VFAs, proteins, lipids or carbohydrates) are removed via activated sludge-based processes such as conventional activated sludge, nitrogen/denitrification and enhanced biological phosphorus removal (EBPR) processes. Generally, these processes separate and transform the organic matter in the wastewater, generating two organic-concentrated streams, namely primary sludge and secondary sludge. These streams contain most of the organic matter originally present in the wastewater, thus facilitating its recovery. Primary sludge generally consists of easily biodegradable particulate organics, such as carbohydrates and lipids. Secondary sludge primarily comprises less to poorly degradable materials, especially at longer sludge ages. Table 1.4 provides an overview of typical compositions of primary and secondary sludge. Note that the values listed should be seen as a guideline only as the composition can deviate depending on factors such as the local sewer catchment, industry trade waste discharge into the sewer network, type of coagulant used and presence of a sand trap.

Primary and secondary sludges are processed via dewatering/stabilization, generating a concentrated product referred to as biosolids. Biosolids can be valorized in several ways. Anaerobic digestion is a common sludge stabilization method, with the benefit of concomitant biogas generation (as discussed in detail in Chapter 5). Current biosolids disposal methods vary, being land application in agriculture, landfilling or incineration the most commonly applied. Incineration is a more common practice in Europe, USA and Asia. Land application as fertilizer or soil amendment is often regarded as the preferred strategy for beneficial reuse of organics (as well as phosphorus and nitrogen). Despite the benefits, the intrinsic properties of the sludge also come with some environmental and human health risks due to the presence of persistent organic compounds, pathogens and emerging contaminants (e.g., Per- and polyfluoroalkyl substances (PFAS). Biosolids can also be used in the cement industry, where both the calorific value of the dried sludge and the ash resulting from its incineration can be used [21]). As an example, incinerated sludge rich in iron is often used in the cement industry to produce bricks with red color.

More recently, research and development is focusing on alternative resource recovery routes by transforming WWTPs into the so-called biorefineries, to produce higher-value organic compounds

Table 1.4 Typical composition of primary and secondary sludge, adapted from [3].

Parameter	Primary sludge	Secondary sludge
Total solids (TS) (%)	5–9	0.8–1.2
Volatile solids (VS) (as % of TS)	60–80	59–68
Nitrogen (%TS)	1.5–4	2.4–5.0
Phosphorus (%TS)	0.8–2.8	0.5–0.7
Potash (K_2O %TS)	0–1	0.5–0.7
Cellulose (%TS)	8–15	7–9.7
Iron (g Fe/kg)	2–4	–
Silica (SiO_2%)	15–20	–
pH	5.0–8.0	6.5–8.0
Grease and fats (%TS)	7–35	5–12
Protein (%TS)	20–30	32–41
Alkalinity (mg/L as $CaCO_3$)	500–1500	580–1100
Organic acids (mg/L as acetate)	200–2000	1100–1700
Energy content (MJ/kg TS)	23–29	19–23

such as carboxylates, including short and medium chain fatty acids, proteins, polyhydroxyalkanoates (PHA), extracellular polymeric substances (EPS) and alginate like compounds [22–24]. Despite the increasing interest, most of these emerging concepts are still at low(er) technological readiness levels (TRL levels) and have not been applied at full-scale. One very interesting approach that has shown strong practical and economic potential that has been recently implemented at full-scale is the recovery of the alginate like compound Kaumera Nereda® Gum [25]. This polysaccharide bio-based material is extracted from the aerobic sludge granules that are formed during the Nereda® wastewater treatment process [26].

1.7.4 Energy from wastewater

Energy in wastewater is present as chemical energy and thermal energy. Chemical energy is present in the form of organic matter, represented as COD. The theoretical energy content of a stream can be calculated assuming an ideal conversion of COD into methane (CH_4) via anaerobic digestion, followed by its transformation into electrical and/or thermal energy, generally via cogeneration (combined heat and power; CHP). Under these assumptions, the typical chemical energy content of one cubic meter of municipal wastewater has been estimated at 5.5–7.0 MJ/m^3. The latter implies that, assuming 312 million ML is being produced globally, around 2 184 000 TJ/year can theoretically be recovered [17, 27]. Nevertheless, it must be considered that, in practice, the presence of non-degradable COD, heat losses and non-ideal conversions reduce the chemical energy recovery to around 10–14% of the theoretical chemical energy (0.6–1.0 MJ/m^3) [17]. Practically, anaerobic technologies are able to recover 110–3300 kJ/m^3 from wastewater as biogas. This implies a global potential production of 34 000–1 030 000 TJ/year of biogas from municipal wastewater, of which 52% is usually recovered as heat and 30–40% as electricity via CHP. As an example of current chemical energy recovery practices, in the EU-27, 62 383 TJ_{biogas}/year is produced through anaerobic digestion of wastewater sludge [28]. The energy recovered is generally used to cover the energy demands of the WWTP (around 20–45 kWh per PE in state-of-the-art plants), covering usually around half of the total energy requirements [29].

Regarding thermal energy, this corresponds to the heat energy contained in the wastewater, largely coming from bathing, laundry cleaning, cooking and the difference between the ambient and the drinking water temperature [17]. Different commercial applications for thermal energy recovery from wastewater exist, mainly based on recovery from raw wastewater. However, challenges to this approach arise due to biofouling, scaling and corrosion issues (increasing maintenance costs and reducing heat transfer capacities). Therefore, treated wastewater (with temperatures around 10–16 and 20–25°C in temperate climates during winter and summer, respectively) is nowadays considered as the best place for heat recovery [17]. Applying water source heat pumps, the net energy recovery from treated wastewater has been estimated to be around 4–8 MJ/m^3, depending on the environmental conditions and the utilization of the stream (e.g., heating or cooling) [17, 30]. Compared to chemical energy, these values are 4–8 times larger, underlining the potential of thermal energy recovery from wastewater. To put this into perspective, a common size WWTP (i.e., 500 000 PEs and treating 125 000 m^3/d), recovering both chemical and thermal energy from wastewater, could produce enough energy to cover its demands and to supply the requirements of 8800 households (at optimal conditions of 9 MJ/m^3 recovered and energy consumption of 1.3 MJ/m^3 and assuming a consumption of 29 kWh/d per household (U.S. Energy Information Administration in 2017) [17, 31]). However, when the recovered energy is to be used to displace natural gas or fuel currently used for heating, challenges like the mismatch between supply and demand cannot be ignored and would need to be addressed [24].

1.7.5 Nitrogen

Municipal wastewater typically contains about 30–60 mg N/L. Globally, this equals to about 20 Mton of nitrogen that ends up at WWTPs [32]. Considering that yearly 119 Mton of reactive nitrogen used as fertilizer is produced via the Haber Bosch process, this means that complete recovery of nitrogen from wastewater would equal to about 17% of all nitrogen applied in agriculture [32]. Globally, an estimated 0–15% of nitrogen from human sanitation is reused on cropland [10]. Several nitrogen recovery approaches are available and are implemented at full scale, such as struvite precipitation and ammonia stripping. However, it is important to emphasize that these approaches are restricted to side-stream and/or decentralized processes such as digestate (which only comprises about 10–25% of the total nitrogen load to a WWTP) and urine which have much higher nitrogen levels. Recovery also occurs through reuse of biosolids in agriculture, as described above in section 1.7.3. Recently, there has been an increasing interest to further upgrade the recovered nitrogen through ammonia stripping into high value microbial proteins (see Chapter 12).

1.7.6 Phosphorus

Yearly, in the range of 15 Mton of mineral phosphorus is mined from apatite rock and converted into fertilizer (note that also about 8 Mton of P is recycled onto agricultural land by means of manure) [33]. About 3 Mton P is ultimately consumed by humans and ends up in urine and excrements, and thus municipal wastewater. Consequentially, if all P present in municipal wastewater could be recovered, the reliance on phosphate rock can be substantially reduced by about 20%. In this context, there are three main routes currently implemented at full-scale for recovery/reuse of P at centralized WWTPs, namely: (i) P reuse in the form of biosolids, (ii) struvite precipitation and (iii) P recovery from sludge incineration ash. In conventional centralized WWTPs, P removal is achieved by means of biological P removal, chemical precipitation using iron or alum-based coagulants, or a combination thereof. Biological P removal increases the fraction of P in the secondary sludge, which leads to increased phosphate concentration in the reject water of an anaerobic digester, hence increasing the recovery potentials. Chemical P removal results in poorly degradable metal bound precipitates that end up in the biosolids. Phosphorus from incinerated biosolids can be recovered from the ash through chemical leaching processes used in the metallurgic industry [34].

1.7.7 Heavy metals

Heavy metals in municipal wastewater originate from a variety of anthropogenic and natural sources, which are predominantly diffuse of nature. The latter makes it virtually impossible to develop source control strategies to avoid/minimize metals ending up in municipal wastewater. A wide variety of metals are present in municipal wastewater, including Cd, Cr, Cu, Ni, Pb, and Zn [35]. These metals ultimately end up in sludge and, despite the fact that they are typically present in relatively low concentrations (i.e., mg/kg range), their presence can significantly affect the potential of (repeated) land application of sludge, as stringent regulations are in place to avoid long-term environmental problems caused by the stability and persistence of metals in nature. For example, the U.S. Environmental Protection Agency (USEPA) has specified a 'ceiling concentration' and a limited cumulative pollutant loading rate for sewage sludge application on agricultural land, whereas in the EU, The Commission of the European Communities (CEC) Directive has set both a limit to the heavy metal concentrations in sludge and a total amount of heavy metals that can be added annually to agricultural land [36].

In addition to heavy metals, it has been recently found that sludge contains a wide variety of metals, including rare-earth elements and minor metals (e.g., Y, La, Ce, Pr, Nd, Sm, Eu, Gd, Tb, Dy, Ho, Er, Tm, Yb, Lu) and even platinum and gold [37], as depicted in Figure 1.3. According to the authors, the high value of some of these metals in combination with the high volume of municipal sludge could make their recovery an interesting option, with a total estimated 'metal' value of $460/ton sludge (as dry solids). In the case of sludge incineration, these metals would be concentrated in the ash. By including the ash into the feedstock of metallurgical refiners, these metals could be recovered via already employed pyro- and hydrometallurgical processes. It should be noted that due to the complex mixture of a wide variety of metals in combination with their very low concentrations (Figure 1.3), their recovery remains highly challenging from a technical and economic perspective. Note that the concentrations presented in Figure 3 can differ depending on the geographical location and catchment characteristics.

1.7.8 Coagulants

Aluminum and iron-based coagulants are commonly used in wastewater systems for phosphorus and/or sulfide removal. To illustrate the reliance of urban water management on chemical dosage, according to a recent market analysis, the global chemical market for the water industry was estimated to be 6 billion USD, with a further estimate increase to about 8.5 billion USD by the year 2023 [38]. Iron salts (mostly ferrous and/or ferric chloride) are often dosed to sewer networks to combat hydrogen sulfide induced corrosion of concrete sewer pipes, a notorious and costly problem for wastewater utilities

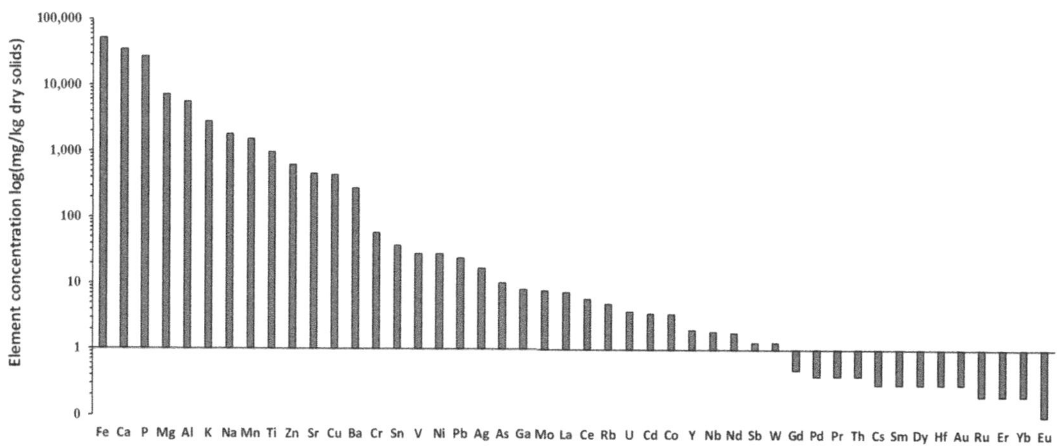

Figure 1.3 Example of the wide variety of metals (and other elements) that can be present in sludge. The figure shows data of sludge from a WWTP in arizona, as described in detail elsewhere [36].

worldwide [39]. Most WWTP still rely heavily on the addition of either iron (Fe) or aluminum (Al) based salts (mainly $FeCl_2$ or $FeCl_3$ and $Al_2(SO_4)_3$) for chemical phosphate removal. Lastly, iron salts are often dosed to anaerobic digesters for sulfide control. It is expected that these chemicals will keep playing an important role in the coming decades. This makes their recovery interesting. Typical dosing concentrations are in the order of 5–50 mg/L Fe or Al based coagulants, respectively. Depending on the dosing rate and WWTP configuration, Fe and Al contents in the sludge are typically in the order of ~0–50 g/kg dry solids.

1.8 CHAPTER SUMMARY

Municipal wastewater can be defined as a waste stream consisting of a mixture of human excrements from toilet use and water originating from household activities such as cleaning, washing and cooking. In this chapter, the typical production rates and composition of the various streams (i.e., yellow, grey and black water) comprising municipal wastewater and how they are typically collected has been discussed. Moreover, the chapter describes how much municipal wastewater is generated per capita and, globally, how this wastewater is typically collected and treated. The key resources that are present (or added during the treatment process) in municipal wastewater that potentially can be recovered during its treatment have also been outlined. Moreover, the typical concentrations ranges at which these different resources are present, as well as in what form these resources can be recovered, have been discussed. After reading this chapter, the reader should have gained an appreciation of the order of magnitude of municipal wastewater generation and the concentration and mass flows of the plethora of resources embedded in municipal wastewater. In this context, the importance of centralized systems for treating municipal wastewater on the potential for resource recovery in terms of economy of scale should not be ignored. Many resources can potentially be recovered from municipal wastewater, depending on the technology implemented. However, it is evident that there is no 'one size fits all' combination of technologies/configurations that allows to recover each and every resource present in the wastewater and choices need to be made which product to prioritize. Finally, one should never forget that with >99% of the total mass flow, the major recoverable resource in municipal wastewater is water. All other resources are present in diluted form at relatively low concentrations.

1.9 EXERCISES

Exercise 1.1: A wastewater treatment plant services a community of 50 000 people residing in a catchment of 30 km^2. Assume that 90% of the inhabitants are connected to the combined sewer network and that 10% of the people are living too remotely to be connected to the centralized sewer network and have household based septic tanks. In order to improve the sustainability of urban wastewater management of the community, resource recovery of nitrogen (in the form of ammonium) has been raised as a potential priority by the city council. In this context, recent developments aimed to recover nitrogen from urine (urine has a high N content, see Table 1.1) have gained special interest in recent years. Let us assume that indeed ammonium recovery from urine at a household level can be achieved, and based on this assumption, calculate the following:

(a) Calculate the amount of nitrogen that can be recovered on a daily and annual basis per household (assuming an average of four people per household) and an N recovery efficiency of 90%.
(b) Calculate the amount of nitrogen that can be recovered at the WWTP assuming a recovery potential of 25% of the incoming N load at the WWTP.
(c) Considering the values found in (a) and (b), describe the practical limitations of resource recovery at a household level.

(d) In addition to the practical issues raised in (c), it is important to realize that ammonium is considered a bulk product produced at low cost and at very large industrial scale through the Haber–Bosch process. Explain why this could further reduce the economic potential of decentralized N recovery.

Exercise 1.2: A community in Indonesia must build a centralized wastewater treatment system. The system is to service the 60 000 people of the district. An anaerobic pond is selected as a low-cost treatment system in order to treat the water to achieve a safe discharge level of BOD_5 of 50 mg/L and capture the biogas for powering gas burning stoves in the community, or burning it to provide a local and renewable source of power for the community. The typical per capita wastewater characteristics highlighted in Table 1.1 apply.

(a) Estimate the following influent parameters for the sewage stream (mg/L): BOD_5, COD, TN, TP.
(b) Calculate how much methane (m^3/d and GJ/d) could be recovered from the anaerobic pond (assume 0.18 $m^3_{methane\ produced}/kgBOD5_{removed}$ and that the energy content of methane is approximately 37 MJ/m^3) and determine the number of households that would be supported through the use of this gas for the case of:
   (i) Cooking gas: where an average household of 5–6 uses 70 MJ/d of methane to power their cooking stoves, or
   (ii) Electricity: assume the energy conversion from methane to electricity is 0.3 and a typical family uses 10 kWh of electricity per day (1 kWh = 3.6 MJ). Comment on the advantages/disadvantages of each proposed plan for methane recovery.

Exercise 1.3: The Canadian city of Halifax is building an activated sludge system at its main wastewater treatment plant that services 170 000 people. Two resource recovery strategies are considered for the sludge produced from this treatment facility, the first consists of one-stage anaerobic digestion for methane recovery from biogas and the other is polyhydroxyalkanoate (PHA) recovery through a 3-stage process. The 3-stage PHA process consists of: (1) anaerobic sludge fermentation to volatile fatty acids (VFA), (2) aerobic culture selection through a feast/famine process fed with the effluent of (1), and (3) PHA accumulation from the selected culture in (2) using the effluent of (1). Assume standard (i.e., Table 1.1) wastewater flows and characteristics and that 80 gVSS of sludge are produced per m^3 of wastewater treated by the facility. Considering that the value of PHA is \$3.5/kg and that the value of methane is \$0.4/m^3, where 1 m^3 of biogas (with a methane content of 65%) is produced per kg of VSS, while 0.1 kg of PHA are produced per kg of VSS in the 3-stage process:

(a) Determine the relative value of the sludge stream using either the PHA recovery or biogas recovery strategy
(b) Discuss the key factors impacting the capital and operational expenditures associated with biogas or PHA production. Which process is likely to incur higher production costs? Why? What would you consider to be the key points impacting your decision on the process to be implemented?

Exercise 1.4: A utility is required to establish wastewater treatment strategies for remote communities at the household level (average of four inhabitants per household) and is considering source separation of the urine from the influent wastewater for nitrogen recovery as a fertilizer to be applied agriculturally. Considering the N loading per capita of Table 1.1 and that the N load per hectare required for fertilization is 14 kg N/ha, how many households would be required to meet the fertilizer demand for 12 ha in 30 days?

Exercise 1.5: An apartment building in Beijing houses approximately 2500 inhabitants, where source separation of urine is considered to be implemented for both nitrogen and phosphorus recovery that will be transported to a neighboring region for fertilization purposes. Consider that the N and P loading per capita of Table 1.1 applies and that the N and P loads per hectare required for fertilization are 21 kg N/ha and 9 kg P/ha, respectively. How many hectares could be fertilized from this building per year? What would be the limiting nutrient?

Exercise 1.6: A music festival in Lisbon will have approximately 12 000 people in attendance during the event. The water utility of the region plans to institute source separation of urine from the wastewater at the event to provide a fertilizer for a nearby agricultural region where nitrogen is the limiting nutrient. Considering the N loading per capita of Table 1.1 and that the N load per hectare required for fertilization is 18 kg N/ha, how many hectares could be fertilized per day?

Exercise 1.7: The Singapore airport services approximately 227 000 passengers per day, where source separation of urine will be implemented for both nitrogen and phosphorus recovery for fertilization. Considering the N and P loading per capita of Table 1.1 and that the N and P loads per hectare required for fertilization are 31 kg N/ha and 2.5 kg P/ha, respectively, how many hectares could be fertilized from the airport per day? Will N or P be the limiting nutrient?

Exercise 1.8: Titanium dioxide nanoparticles are increasingly used in a variety of commercial products such as textiles, paints and personal care products, and the fate of a great part of this titanium is wastewater. Indeed, up to 4% of the TiO_2 applied onto textiles can wash off in a single wash [40], and they are part of numerous daily life products such as sunscreen and toothpaste, resulting in relevant concentrations of TiO_2 nanoparticles being washed into municipal wastewater systems (Figure 1.3). TiO_2 must be removed in WWTP since it has been demonstrated that, if left untreated, it can be harmful to aquatic life. Up to 85% of titanium typically ends up in the biosolids due to the low solubility of TiO_2. The state of Arizona conducted a feasibility study regarding viability of recovering TiO_2 in WWTP. It was found that the titanium concentrations in raw wastewater ranged from 181 to 1233 μg/L (median of 26 samples was 321 μg/L) [41]. Consider for this study the WWTP of Phoenix 91st Avenue, which serves a population of 2.5 million people, treating a flow rate of approximately 870 ML/day. Two processes of TiO_2 recovery are proposed, one from liquid influent, and another from the biosolids, which are produced at a rate of 100 kg DW/ML of influent wastewater treated. Estimate the minimum concentration of TiO_2 in the influent or in the biosolids to make recovery economically viable for either scenario, assuming recovery efficiencies of 95 and 90% for the liquid effluent and the biosolids, respectively. Assume process costs of $30/ML of influent wastewater and $58/tonDW of sludge produced, and a value of TiO_2 of $60 per kg.

Exercise 1.9: Given that a small town with 3000 inhabitants is interested in investing in the implementation of an anaerobic digester at their centralized WWTP:

(a) Calculate the practical calorific biogas production potential of the small town using the data provided in Table 1.3.
(b) In (a) you have calculated the potential biogas production. How does this compare to natural gas given that the annual per capita natural gas consumption equals to 2361 Nm^3 (average consumption in the US? Assume that 1 Nm^3 of natural gas has a caloric value of 40 MJ/m^3.

Exercise 1.10: In this chapter, the different waste streams comprising municipal wastewater have been discussed, without the help of the information provided in this chapter, fill in the table below.

	Grey water (%)	Urine (%)	Feces (%)
Flow rate			
COD			
Nitrogen			
Phosphorus			
Potassium			

Exercise 1.11: Considering (i) the metal concentrations depicted in Figure 1.3, (ii) typical coagulant dosing rates in the form of either alum or iron-based coagulants and (iii) sludge production rates of 8 000 000 and 9 253 000 ton DM sludge/year for the US and EU, respectively, estimate the total mass flows by filling in the table below.

Metal	Concentration in sludge (mg/kg dry weight)	Total estimated amount in the EU (ton/year)	Total estimated amount in the USA (ton/year)
Fe			
Al			
Ti			
Zn			
Cu			
Ag			
Yt			
Nd			
Au			
Pd			

1.10 DISCUSSION QUESTIONS

Question 1.1 (technology and economy): As the innovation manager of a large water utility, you are in charge of reorganizing the existing water infrastructure from its current situation to a more circular approach within a timeframe of 15 years. The current wastewater treatment infrastructure comprises two very large-scale wastewater treatment plants, with a capacity of 400 000 and 500 000 PE, both using conventional activated sludge plants of 400 000 and 350 000 PE. In addition, it comprises more than 50 small size WWTPs with a capacity between 500 and 1500 PE. You are asked to give a presentation to the board of directors in which you evaluate the current status and justify your masterplan. Where would you focus on in terms chosen technology, location (i.e., which WWTP to focus on) and which resources would you target? What are your key considerations/motivations?

Question 1.2 (drivers and market analysis): A wastewater utility operating a WWTP with a capacity of 500 000 PE is evaluating the feasibility of upgrading their treatment processes using advanced treatment processes in order to generate an effluent that exceed drinking water quality. As the business development manager of the wastewater utility, you are asked to give a presentation to the board of directors in which you evaluate the economic potential, technical constraints as well as social implications/considerations that need to be taken into account. What are the key criteria and considerations that should be considered?

Question 1.3 (drivers, risks and social acceptance): As discussed in this chapter, the ultimate disposal (or reuse) route of biosolids differs amongst the different regions in the world with beneficial reuse in agricultural application, landfill and incineration the most commonly applied approaches. Evaluate these three options and provide a list of potential advantages and disadvantages for each of these options.

Question 1.4 (drivers, targets, policy and sustainability): With an ever-growing population and increasing living standards, the world needs increasing amounts of inorganic fertilizer. Important ingredients of fertilizer include ammonium and phosphorus. Ammonium is produced in large amounts (i.e., over 100 million tons per year) via the energy-intensive Haber–Bosch process. The process involves the reaction of N_2 (which comprise 80% of our atmosphere!) with hydrogen under high temperature and pressure in the presence of an iron catalyst. Phosphorus on the other hand, is a non-renewable resource that is mostly extracted from mineral phosphorus-rock deposits. An important aspect is that only a few countries have significant deposits, with Morocco, China, Algeria and Syria being the top four. Considering the above-described information, discuss the necessity and key drivers for nitrogen and phosphorus recovery and also evaluate the potential environmental benefits and social impact.

Question 1.5 (market analysis, risks and economy-of-scale): In developed countries, the wastewater infrastructure is ageing and would require major upgrades in the coming decades. In less developed countries and emerging economies a significant amount of new infrastructure will need to be realized in the coming decades. Considering the above, there is an ongoing debate regarding how our future urban waster infrastructure will look, that is, more centralized or more modular and decentralized. Evaluate the general characteristics of centralized and decentralized systems and provide advantages and disadvantages for each of these scenarios in the context of resource recovery, taking into account economy of scale, practical feasibility and maintenance, monitoring and quality control and market requirements of recovered resources.

Question 1.6 (rare earth metals, economics): As depicted in Figure 1.3, a wide variety of elements can be found in sludge. In fact, even gold and platinum can be found in sludge. Due to the high value of gold and platinum and other elements, it was estimated that when adding up the values based on the individual prices of these elements, sludge can be valued at $460/ton sludge. Some people argue this is a too simplistic economic assessment. Discuss potential economic and technical challenges that would substantially reduce the overall process economics of recovery of high-value elements from sludge.

Question 1.7 (rare earth metals, environmental impact): In the questions above, you have highlighted several economic and technical challenges with respect to recovery of high value elements from sludge. Let us assume a situation where all economic and technical constraints have been solved for the situation where the sludge is incinerated and the metals can be recovered from the sludge ash. Even in this scenario one could provide arguments as to why other approaches such as beneficial land use as a fertilizer would be better. Examine the various sludge management strategies and discuss what could be considered important aspects/drivers that would need to be assessed in detail.

Question 1.8 (design considerations, process stability, down-stream processing): Various approaches aim to produce and recovery high-value products as a means to improve the overall economics of resource recovery from wastewater. In this context, a very interesting approach is Kaumera Nereda® Gum. Interestingly, and an important aspect to consider, is that the Nereda® technology was not specifically developed and designed with the purpose to recover Kaumera. In fact, the primary objective was to reduce the costs of municipal wastewater treatment and the observation that the polysaccharide bio-based material Kaumera could be recovered from excess aerobic granular sludge in a completely independent side-stream process was made at a later stage. Discuss the potential benefits Kaumera may have from a process stability/robustness and wastewater operator confidence and willingness to introduce this recovery method in relation to other resource recovery methods such as struvite precipitation and ammonia stripping.

Question 1.9 (economic considerations, market demands, decision-making): For exercises 1.4–1.7, comment on the potential for N and P recovery and the process challenges that would be encountered in each case. Comment on the impact of economies of scale on the economic and practical potential of recovery of N and P through source separation.

Question 1.10 (economic considerations, market demands, decision-making): You are the Innovation Manager of a large wastewater utility that needs to prioritize the resources to be recovered from your largest wastewater treatment facilities. Not all resources that are of potential interest can be recovered simultaneously, while the recovery of some components may impact your capacity to recover other components. Outline your top five resource priorities to be recovered and justify well each of your choices.

Question 1.11 (integrated management, multiple reuse, asset management): Sewer networks are million-dollar assets that are exposed to severe corrosion problems arising from sulfate present in wastewater. Large amounts of iron salts are continuously applied to precipitate sulfate thereby reducing the hydrogen sulfide generated in sewers in order to control corrosion and odor. Knowing that coagulants are also used in drinking water treatment systems and for the removal of phosphorus in WWTP, discuss an integrated approach for water management in a perspective of achieving multiple objectives from a single coagulant application. What possible obstacles could arise is reusing iron salts within the urban water cycle?

Question 1.12 (design considerations, enhanced energy generation, unwanted side-effects) In various cities around the world, the idea has been proposed to use the existing underground sewer network to transport other waste rather than just sewage. In fact, in several parts of the world the installation of grinders in kitchen sinks are a common way to deal with organic kitchen waste. For example, according to the USA Census Bureau (2011), almost half of the kitchens in the US were equipped with these grinders. As organic kitchen waste increases the degradable organic load in the sewage that could be recovered as biogas during anaerobic digestion, one could see the synergy – reducing the amount of solid waste landfilled coupled with enhanced biogas production. While principally this holds true, the approach also comes with important disadvantages that need to be considered such as unwanted methane formation in the sewer network itself. Discuss whether the increased methane production outweighs the disadvantages and risks.

FURTHER READING MATERIALS

Lema, J. M. and Suarez Martinez, S. (2017). Innovative Wastewater Treatment & Resource Recovery Technologies: Impacts on Energy, Economy and Environment. IWA Publishing, London.

Tchobanoglous, G., Stensel, H. D., Tsuchihashi, R., Burton, F. L., Abu-Orf, M., Bowden, G. and Pfrang, W. (2014). Wastewater Engineering: Treatment and Resource Recovery/Metcalf & Eddy, AECOM. McGraw-Hill Education, New York, NY.

Von Sperling, M. (2007). Wastewater Characteristics, Treatment and Disposal. IWA Publishing, London, UK.

REFERENCES

1. Centers for Disease Control and Prevention (1999). Ten Great Public Health Achievements – United States, 1900–1999. Centers for Disease Control and Prevention, Atlanta, Georgia, USA.
2. Jenkins D. and Wanner J. (2014). Activated Sludge – 100 Years and Counting. IWA Publishing, London, UK.
3. Tchobanoglous G., Burton F. L. and Stensel H. D. (2013). Wastewater Engineering Treatment and Reuse, 5th edn. Metcalf & Eddy Inc. (ed). McGraw-Hill Education, New York.
4. FAO Aquastat (2018). AQUASTAT Database Query Results. Available from: http://www.fao.org/nr/water/aquastat/data/query/results.html (accessed: January 25, 2022).
5. USGS (2016). How Much Water Does the Average Person Use at Home Per Day? Available from: https://water.usgs.gov/edu/qa-home-percapita.html. (accessed: January 25, 2022).
6. Water-link (2019). Gemiddeld huishoudelijk waterverbruik in Vlaanderen. Available from: https://water-link.be/sites/default/files/2018-03/gemiddeld-waterverbruik_4.pdf (accessed: January 25, 2022).
7. Metropolitan Water Reclamation District of Greater Chicago (2018). Stickney Water Reclamation Plant Fact Sheet. Chicago. Metropolitan Water Reclamation District of Great Chicago, Chicago, Illinois, USA.
8. Srikanth M., Vanwonterghem I. and Rabaey K. (2019). "Empowered Septic Tank as Decentralized Wastewater Treatment System" Under "Reinvent the Toilet Challenge". DBT-BIRAC and Bill & Melinda Gates Foundation. BIRAC, New Delhi, India.
9. Larsen T. A., Udert K. M. and Lienert J. (2013). Source Separation and Decentralization for Wastewater Management. IWA Publishing, London, UK.

10. Trimmer J. T., Cusick R. D. and Guest J. S. (2017). Amplifying progress toward multiple development goals through resource recovery from sanitation. *Environmental Science and Technology*, **51**, 10765–10776.
11. Ruiken C. J., Breuer G., Klaversma E., Santiago T. and van Loosdrecht M. C. M. (2013). Sieving wastewater – cellulose recovery, economic and energy evaluation. *Water Research*, **47**(1), 43–48.
12. Pons M. N., Spanjers H., Baetens D., Nowak O., Gillot S., Nouwen J. and Schuttinga N. (2004). European Water Management Online Wastewater Characteristics in Europe – A Survey. European Water Association (EWA), Hennef (Sieg), Germany.
13. Henze M., Gujer W., Mino T. and van Loosedrecht M. (2006). Activated Sludge Models ASM1, ASM2, ASM2d and ASM3. IWA Publishing, London, UK.
14. Grady C. P. L., Daigger G. T., Love N. G. and Filipe C. D. M. (2011). Biological Wastewater Treatment, 3rd edn. CRC Press, Boca Raton, FL, USA.
15. van Haandel A. and van der Lubbe J. (2012). Handbook of Biological Wastewater Treatment: Design and Optimisation of Activated Sludge Systems. IWA Publishing, London, UK.
16. Jong C. D. and Veldhoen A. (2017). Sand Removal at WWTPs in the Netherlands. STOWA, Amersfoort, The Netherlands.
17. Hao X., Li J., van Loosdrecht M. C. M., Jiang H. and Liu R. (2019). Energy recovery from wastewater: heat over organics. *Water Research*, **161**, 74–77.
18. Asano A., Burton F. L., Leverenz H. L., Tsuchihashi R. and Tchobanoglous G. (2007). Water Reuse: Issues, Technologies, and Applications. McGraw-Hill, New York.
19. Lazarova V. and Asano T. (2013). Milestones on Water Reuse: The Best Success Stories. IWA Publishing, London, UK.
20. United Nations World Water Assessment Programme (WWAP) (2017). The United Nations World Water Development Report 2017. Wastewater: The Untapped Resource. UNESCO, Paris.
21. De Meyer A., Cattrysse D., Ostermeyer P. and Van Orshoven J. (2016). Implementation of OPTIMASS to optimise municipal wastewater sludge processing chains: proof of concept. *Resources, Conservation and Recycling*, **114**, 168–178. https://doi.org/10.1016/j.resconrec.2016.03.012
22. Seviour T., Yuan Z., van Loosdrecht M. C. M. and Lin Y. (2012). Aerobic sludge granulation: A tale of two polysaccharides? *Water Research*, **46**(15), 4803–4813.
23. Alloul A., Ganigué R., Spiller R., Meerburg F., Cagnetta C., Rabaey K. and Vlaeminck S. E. (2018). Capture–ferment–upgrade: A three-step approach for the valorization of sewage organics as commodities. *Environmental Science and Technology*, **52**(12), 6729–6742.
24. van der Hoek J. P., de Fooij H. and Struker A. (2016). Wastewater as a resource: strategies to recover resources from Amsterdam's wastewater. *Resources. Conservation and Recycling*, **113**, 53–64.
25. https://kaumera.com/english/kaumera/ (accessed: January 25, 2022).
26. Lin Y., de Kreuk M., van Loosdrecht M. C. M. and Adin A. (2010). Characterization of alginate-like exopolysaccharides isolated from aerobic granular sludge in pilot-plant. *Water Research*, **44**(11), 3355–3364.
27. Scherson Y. D. and Criddle C. S. (2014). Recovery of freshwater from wastewater: upgrading process configurations to maximize energy recovery and minimize residuals. *Environmental Science and Technology*, **48**, 8420–8432.
28. Kampman B., Leguijt C., Scholten T., Tallat-Kelpsaite J., Brückmann R., Maroulis G., Lesschen J. P., Meesters K., Sikirica N. and Elbersen B. (2017). Optimal Use of Biogas From Waste Streams – An Assessment of the Potential of Biogas From Digestion in the EU Beyond 2020. European Commission, Brussels, Belgium.
29. Capodaglio A. G. and Olsson G. (2020). Energy issues in sustainable urban wastewater management: Use, demand reduction and recovery in the urban water cycle. *Sustainability*, **12**(1), 1–17.
30. Verstraete W. and Vlaeminck S. E. (2011). ZeroWasteWater: Short-cycling of wastewater resources for sustainable cities of the future. 2nd Xiamen Int. Forum Urban Environment, Institute of Urban Environment, Chinese Academy of Sciences, Xiamen, China, pp. 253–264.
31. Mininni G., Braguglia C. M., Ramadori R. and Tomei M. C. (2004). An innovative sludge management system based on separation of primary and secondary sludge treatment. *Water Science and Technology*, **50**(9), 145–153.
32. Bodirsky B. L., Popp A., Lotze-Campen H., Dietrich J. P., Rolinski S., Weindl I., Schmitz C., Müller C., Bonsch M., Humpenöder F., Biewald A. and Stevanovic M. (2014). Reactive nitrogen requirements to feed the world in 2050 and potential to mitigate nitrogen pollution. *Nature Communications*, **5**(1), 1–17.

33. Cordell D., Drangert J.-O. and White S. (2009). The story of phosphorus: global food security and food for thought. *Global Environmental Change*, **19**, 292–305.
34. Desmidt E., Ghyselbrecht K., Zhang Y., Pinoy L., Van Der Bruggen B., Verstraete W., Rabaey K. and Meesschaert B. (2015). Global phosphorus scarcity and full-scale P-recovery techniques: A review. *Crit. Rev. Environmental Science and Technology*, **45**, 336–384.
35. Du F., Freguia S., Yuan Z., Keller J. and Pikaar I. (2015). Enhancing toxic metal removal from acidified sludge with nitrite addition. *Environmental Science and Technology*, **49**(10), 6257–6263.
36. Collivignarelli C. M., Abbà A., Frattarola A., Carnevale Miino M., Padovani S., Katsoyiannis I. and Torretta V. (2019). Legislation for the reuse of biosolids on agricultural land in Europe: overview. *Sustainability*, **11**(21), 1–22.
37. Westerhoff P., Lee S., Yang Y., Gordon G. W., Hristovski K., Halden R. U. and Herckes P. (2015). Characterization, recovery opportunities, and valuation of metals in municipal sludges from U.S. wastewater treatment plants nationwide. *Environmental Science and Technology*, **49**(16), 9479–9488.
38. Research B. (2018). Specialty Water Treatment Chemicals: Technologies and Global Markets. BCC Research, Wellesley, Massachusetts, USA.
39. Pikaar I., Sharma K. R., Hu S., Gernjak W., Keller J. and Yuan Z. (2014). Reducing sewer corrosion through integrated urban water management. *Science (New York, N.Y.)*, **345**(6198), 812–814.
40. Windler L., Lorenz C., von Goetz N., Hungerbühler K., Amberg M., Heuberger M. and Nowack B. (2012). Release of titanium dioxide from textiles during washing. *Environmental Science and Technology*, **46**(15), 8181–8188.
41. Westerhoff P., Song G., Hristovski K. and Kiser M. A. (2011). Occurrence and removal of titanium at full scale wastewater treatment plants: implications for TiO_2 nanomaterials. *Journal of Environmental Monitoring*, **13**(5), 1195–1203.

doi: 10.2166/9781780409566

Chapter 2

Resource recovery from industrial wastewater: what and how much is there?

Eric Walling[1], *Wim Moerman*[2], *Willy Verstraete*[3,4], *Denys K. Villa Gomez*[5] *and Céline Vaneeckhaute*[1]

[1]*Research Team on Green Process Engineering and Biorefineries (BioEngine), Chemical Engineering Department, Université Laval, 1065 avenue de la Médecine, Québec, QC, Canada*
[2]*NuReSys, Hoekstraat 3, B-8540 Deerlijk, Belgium*
[3]*Center for Microbial Ecology and Technology (CMET), Ghent University, Coupure Links 653, B-9000 Gent, Belgium*
[4]*Avecom NV, Industrieweg 122P, 9032 Wondelgem, Belgium*
[5]*The School of Civil Engineering, The University of Queensland, St Lucia, QLD 4072, Australia*

2.1 INTRODUCTION

Industrial production of goods and services is at the heart of our modern economy and society, and accounts for approximately 20% of global water consumption (FAO, 2020). This number increases significantly in developed and industrialized nations, as highlighted in Figure 2.1, with industry in Europe and Canada accounting for 54 and 80% of total water use, respectively (FAO, 2020; Statistics Canada, 2014). This water is key to the production of a variety of goods, such as processed food, clothing, chemicals, materials, and energy. Through its use in industry, a portion of this water will inevitably come into direct contact with a wide variety of raw resources, contaminants, or intermediate products, leading to its contamination. This contaminated water is defined as industrial wastewater, and it does not include contributions from agricultural and municipal wastewater discharge. Economic, environmental and safety regulations have driven industrial sectors to systematically manage and treat their wastewater, which can carry undesirable by-products, chemical residues, organics, pesticides, heavy metals, nutrients and minerals, to meet certain standards for process water, rendering it safe to be recycled and/or reused in processes, or returned to the environment. This treatment, which can take the form of mechanical, biological, physical, chemical, and thermal processes, often applied in series, represents a significant cost to industry. For example, in Canada, industrial wastewater treatment and discharge costs represent between 28% and 63% of total industrial water costs, averaging out at about 37% for all industries (Statistics Canada, 2014). Furthermore, in addition to continuous and ongoing growth in high-income countries, rapid industrialization and population growth in developing nations are expected to boost water demand by the manufacturing sector by 400% by 2050 (Marchal *et al.*, 2011). These treatment processes, which amount to annual costs of tens of billions of dollars at a global scale, can, however, present an alluring opportunity to recover resources, reducing the economic burden associated with industrial wastewater treatment.

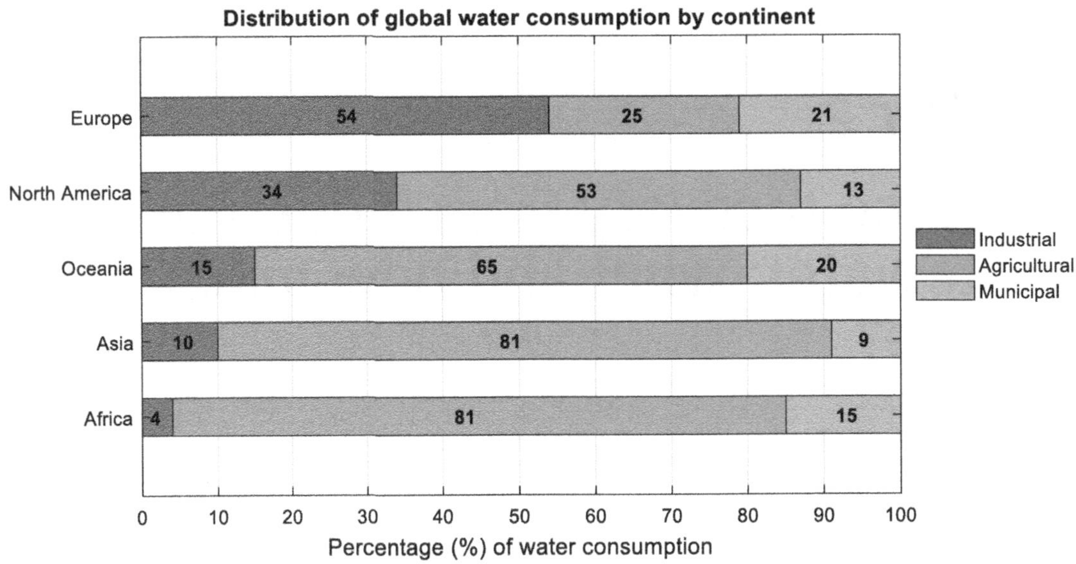

Figure 2.1 Distribution of global water consumption by continent (data from FAO, 2020). The industrial segment is generally made up of a variety of major sectors, each encompassing specific industries. Generally, regarding wastewater, we can consider major industrial sectors such as manufacturing, mining, and energy, each of which can then be further subdivided. For example, manufacturing encompasses industries such as textiles and leather, pulp and paper, chemicals, among many others.

Indeed, though often approached as waste streams, these industrial wastewaters can provide a rich source of water, nutrients, minerals, and compounds that can be recovered into valuable products. Besides the recovery of water as process water, there are two main fields of resource recovery, that is inorganic/organic compound recovery and energy recovery. In most industries, the largest driver is the water recovery, though recovery of high cost/value compounds in the wastewaters of certain processes, such as spent catalysts, can overtake this. The aim of this chapter is to outline these industrial wastewater resource recovery opportunities. The following sections will discuss major industries that contribute to the production of wastewater and their general characteristics, current practices in industrial wastewater treatment, and resources that can be recovered from industrial wastewater treatment processes.

2.2 LEARNING OBJECTIVES

At the completion of this chapter, you should be able to:

- Describe the sources and magnitudes of industrial wastewater generation and treatment, and understand its potential for resource recovery.
- Describe the key resources that can be recovered during the treatment of industrial wastewater.
- Identify recovery potential(s) in different types of industries.
- Understand that the added value from resource recovery processes can be evaluated not only by considering product value, but also through additional benefits for the overall treatment of the wastewater.

2.3 THE MAJOR INDUSTRIES THAT PRODUCE WASTEWATER AND THEIR CHARACTERISTICS

Industrial wastewaters stem from a variety of processes and industries, all being characterized by very different compositions, and physical and chemical characteristics. These waste flows are as unique as the processes that produce them and, therefore, require specific consideration when developing and implementing resource recovery processes. Examples of industries that produce large volumes of wastewater include: power generation, pulp and paper production, mining and refining processes, oil and gas production and refining, industrial food production and processing, chemical manufacturing, and textile production.

Indeed, as presented in Figure 2.2, the magnitude and share of industrial wastewater discharges from the major industrial sectors (each encompassing a variety of specific industries) vary greatly by region and depend on a variety of local factors. Though Figure 2.1 presents this distribution by continent and Figure 2.2 highlights how variable this can be between countries, the same kind of variability can be found between municipalities and regions within a country. This variability is a representation of local realities, reflecting the specific types of industries and sectors present in a region. For example, Figure 2.2 shows a stark contrast between the distribution of wastewater generation in Canada and the United States compared with the average across 20 countries (mostly based on countries in Europe), as well as with the three other example nations. However, these differences reflect and can be explained by national behavior. Indeed, on a per capita basis, we find that both Canada and the United States consume more than three times the amount of energy than the other countries in Figure 2.2; around 12 000–14 600 $kWh \cdot person^{-1} \cdot year^{-1}$ for the United States and Canada, compared to approximately 4000–4600 $kWh \cdot person^{-1} year^{-1}$ for Poland and China (EIA, 2019). Similarly, Figure 2.2 also reflects the central role mining plays in Poland's economy, being one of the world's main producers of coal, copper, silver, and rhenium (USGS, 2020). In contrast, China's dominance of global manufacturing, representing 28% of global output (World Bank, 2019), is mirrored in the distribution of the country's wastewater discharge.

Furthermore, all of the different industries contained within the industrial sectors presented in Figure 2.1 have unique characteristics based on the specific industrial processes and wastewater treatment scenarios applied. This can be seen in both Figures 2.3 and 2.4, where wastewater discharge characteristics for China and Europe are presented, respectively. In Figure 2.3, which presents the wastewater and chemical oxygen demand (COD) discharges from 22 industries in China, we can see just how variable these can be. This variability is observed both in the amount discharged, as well as the composition of the discharged wastewater, measured through COD. For example, based on Figure 2.3, the textile goods segment is responsible for the third largest quantity of wastewater discharged, while it represents the highest COD discharge in China, reflecting the high organic matter content of the textile industry's waste stream. Conversely, wastewater discharges from electricity and energy production (expressed as both 'Electricity, steam' and 'gas production') are large but contain a small COD load relative to the other segments.

This variability in industrial wastewater characteristics continues when looking at more specific compounds/pollutants, such as chlorinated organic compounds, heavy metals, and other organic and inorganic pollutants, such as COD, hydrocarbons, and phenols (organic), and chlorides, cyanides, fluorides, and nutrients (inorganic), as shown in Figure 2.4. As mentioned, these flows are extremely variable based on sectors, processes, and treatment scenarios, with a deeper dive into these sectors being provided throughout the following section. Generally, we can note that chlorinated organics mainly stem from the pulp and paper and chemical industries, while heavy metals are present across all industries, being particularly present in general manufacturing and metal and mining related industries. Other organic compounds are prevalent in most sectors (with the exception of metal and mining related industries), while inorganic compounds are the main pollutants discharged from the chemical industry.

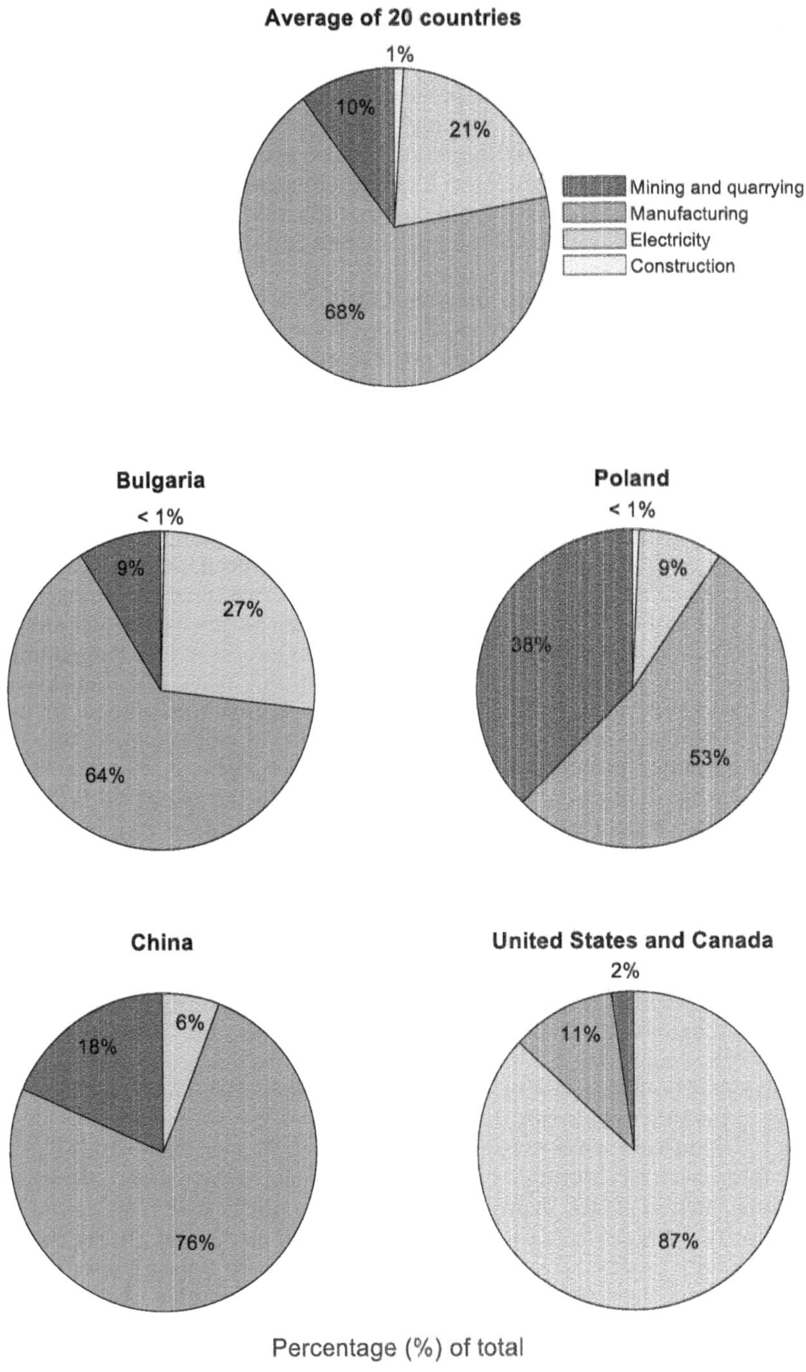

Figure 2.2 Breakdown of wastewater generation by major industrial sectors for various countries. Note that no data was available for the contribution of construction to the totals given by Canada, China, and the United States. Based on data and estimates from the United Nations World Water Assessment Program (WWAP) (2017), Dieter *et al.* (2018), Statistics Canada (2014), and Guo *et al.* (2018).

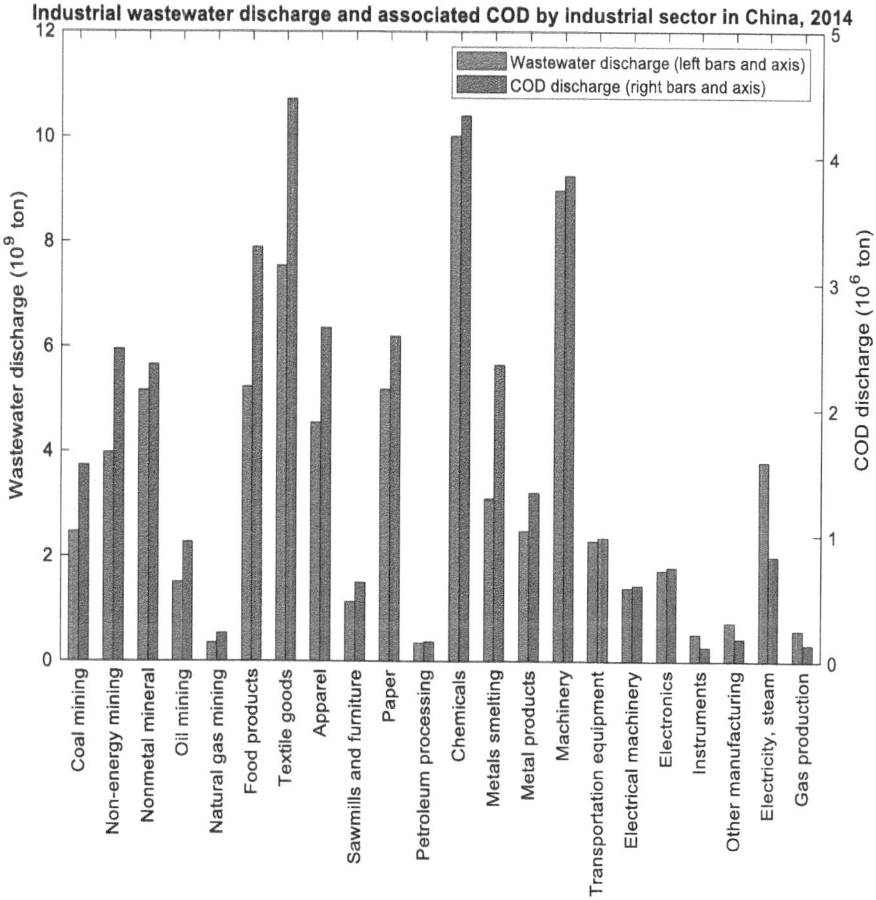

Figure 2.3 Industrial discharge by sector in China for the year 2014. Based on the estimates of Guo *et al.* (2018).

The figures presented thus far have provided a good picture of the different industrial segments that generate wastewater and some of the main pollutants in those wastewater streams. However, resource recovery processes are often aimed at specific compounds, and not general pollutant classes. Table 2.1 presents a list of the major types of contaminants found in industrial wastewater and the industries with which they are generally associated. These include substances with significant impacts to public and/or environmental health, such as heavy metals, synthetic organic substances, inorganic substances, nutrients (such as N, P, K), organic matter and emerging contaminants. Some of the contaminants in Table 2.1 can be recovered, recycled and/or reused, as described in Section 2.5 and later chapters of this book. Other contaminants in Table 2.1 are not recovered but may still impact the selection of treatment technologies and the value of recovered resources.

Currently, on-site industrial wastewater treatment often focuses on managing organics (COD) or toxic compounds (e.g., heavy metals and chlorinated organics), while other inorganic substances (including nitrogen and phosphorus) are important discharges from industrial wastewater treatment. Indeed, in 2011, the European Environment Agency reported that inorganic substances made up more than 98% in mass of the directly released (i.e., from the industrial plant directly into waterbodies)

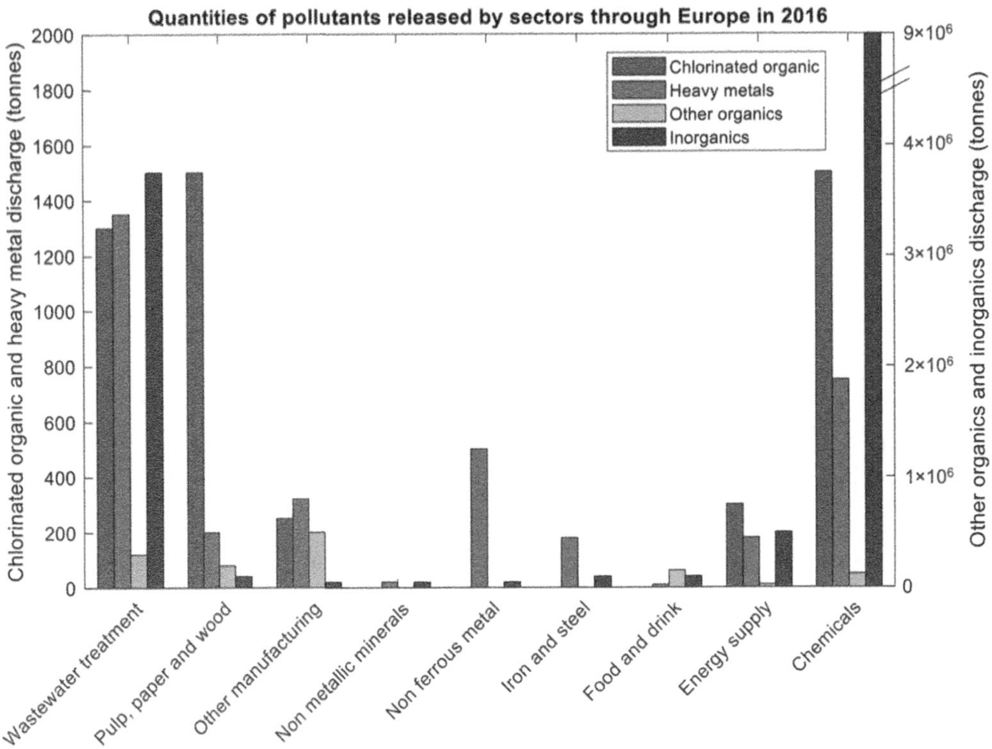

Figure 2.4 Pollutant discharge by sector throughout Europe in 2016, adapted from EEA (2019).

substances (EEA, 2019), while Canada's 2009 inventory showed the importance of nutrients, with nitrogen representing 90.4% of industrial discharge (47.1% nitrate and 43.3% total ammonia), followed by phosphorus at 5.1% (Statistics Canada, 2011). Organic substances account for 2% of total release, with all other contaminants making up the remainder. Indeed, despite higher discharge limits for COD than most other contaminants, the massive amount of inorganics used, produced, and released during industrial processes generally outweighs the mass of discharged organics. This is highlighted in Figure 2.4, where the release of inorganic substances through Europe is in the order of 10^7 tons·year^{-1}, compared to about 10^6 tons·year^{-1} of organic compounds, mostly stemming from the chemical manufacturing industry. Furthermore, chemical manufacturing is one of the primary industrial sectors in many countries (e.g., China; see Figure 2.3) and is thus an important source of industrial wastewater discharge. Another factor leading to the prominence of inorganics release is that regulations on the nutrient loads of discharged waters vary widely, and often only focus on limiting discharge in areas susceptible to eutrophication. However, legislation throughout the world is shifting to stricter guidelines (Preisner *et al.*, 2020). Despite the relatively low presence (by mass) of other substances, such as heavy metals and emerging contaminants in industrial wastewater, their potency and significant impacts on human health and the environment still render them very problematic.

The following section will highlight the main characteristics of a few major industries, focusing on what differentiates them from one another, given the great variability found in industrial wastewater composition.

Table 2.1 Major contaminants found in industrial wastewaters and their primary industrial sources.

Substance Type	Substance	Major Sources
Nutrients	Nitrogen (N)	Most sectors
	Phosphorus (P)	
	Sulfur (S)	
	Potassium (K)	
Heavy metals	Arsenic (Ar)	Metal and glass manufacturing, energy, car repair, tanneries
	Cadmium (Cd)	
	Chromium (Cr)	
	Copper (Cu)	
	Lead (Pb)	
	Mercury (Hg)	
	Nickel (Ni)	
	Zinc (Zn)	
Organic matter	Biochemical/chemical oxygen demand (BOD/COD)	Food industry, pulp and paper, wood preservation
	Suspended solids (SS)	
Organic substances	Chlorinated organics	Chemical manufacturing, pulp and paper, wood preservation
	Benzene	Most sectors
	Di(2-ethylhexyl)phthalate (DEHP)	
	Halogenated organics (AOX)	
	Linear alkylbenzene sulfonates (LAS)	
	Naphthalene	
	Nonylphenol and ethoxylates (NPE)	
	Organotin compounds	
	Phenols	
	Polycyclic aromatic hydrocarbons (PAH)	
	Toluene	
	Xylenes	
Inorganic substances	Chlorides	Energy, iron and steel production, chemical manufacturing
	Cyanides	
	Fluorides	
	Brominated diphenyl ethers (PBDES)	Flame retardants (textile, furnishing and electrical insulation manufacturing)
	Chlorinated paraffin	Plastic manufacturing, flame retardant, sealants, paints
	Pharmaceutical compounds	Pharmaceutical industries
	Polydimethylsiloxanes (PDMS)	Lubricants, electrical insulators and antifoam production
Emerging contaminants	Per – and Polyfluoroalkyl substances (PFAS)	Flame retardants, cleaning products, stain-resistant products, paints, automotive, construction, electronics

2.3.1 Food and beverage industries

The food production and processing sector encompasses a large variety of industries, including animal production, animal processing, fruit and vegetable washing, general food processing (fruit, vegetable, meat, dairy), bottling and packaging, and so on. A common characteristic of wastewater generated in these industries is high organic matter content, notably high biochemical oxygen demand (BOD) and high total suspended solids (TSS). There are also additional contaminants that are more industry and process specific. For example, wastewater from animal industries may contain blood, tissue/body residues, and often high amounts of oil and grease, while wastewater from fruit and vegetable industries may contain pesticides. Salts, surfactants, flavoring compounds, pathogens, and nutrients are also present in variable concentrations.

Given the role of organic compounds in the food and beverage sector – such as organic acids used to regulate pH, act as preservatives, and enhance flavors (Quitmann *et al.*, 2013; Valta *et al.*, 2015) – they are one of the primary contributors to the release of non-chlorinated organic substances (EEA, 2019). An overview of wastewater production and associated BOD and total nitrogen (TN) concentrations from food and beverage production and processing is presented in Table 2.2.

The largest wastewater producer in the food and beverage sector is the dairy industry, which is characterized by wastewaters with high BOD, COD, and salt contents. Between 50 and 80% of

Table 2.2 Wastewater characteristics from select sectors of the food and drink industries.

Sector	Water Production (m^3/Unit)	BOD (mg/L)	BOD/TN (–)[a]	Unit of Production
Fruits and vegetables				
Sugar cane	0.5–10	250–5000	25–60	1 t produced
Canning (fruit/vegetables)	4–50	600–7500	27–37	1 t processed
Pea processing	13–18	300–1350		1 t processed
Tomato processing	4–8	450–1600		1 t processed
Carrot processing	11	800–1900		1 t processed
Potato processing	7.5–16	1300–3300	5–14	1 t processed
Citrus processing	9	320	27–37	1 t processed
Animal industries				
Fish farming[b]	0.5–1.5	0.4–300	1–20	1 t produced
Dairy (without cheese)	1–10	300–5000	3–14	1000 L milk
Dairy (with cheese)	2–10	500–8000		1000 L milk
Chicken processing	15–60	100–2400	3–10	1 t produced
Beef processing	10–16	200–6000		1 t processed
Fish processing	5–35	2700–3500	7–43	1 t processed
Slaughterhouse	0.5–3	150–8500	–	1 cattle/2.5 pigs
Beverage and confectionary				
Sweets/candies	5–25	200–1000	–	1 t produced
Margarine	20	1500	–	1 t processed
Yeast production	150	7500	–	1 t produced
Alcohol distillation	60	3500	30–100	1 t cane processed
Brewery	5–20	500–4000	7–43	1 m^3 produced
Soft drinks	2–5	600–2000	5–200	1 m^3 produced
Wine	5	–	–	1 m^3 produced

Various sources from practice.

Table 2.3 Wastewater characteristics from some dairy industries effluents.

Process Effluent	BOD (mg/L)	COD (mg/L)	TSS (mg/L)	TN (mg/L)	TP (mg/L)	pH
Butter	220–2650	8900	700–5070	–	–	12.1
Cheese	590–5000	1000–63500	190–3400	18–830	5–280	3.3–9.5
Fluid milk	500–1300	950–2400	90–450	–	–	5–9.5
Ice cream	2500	5200	3100	–	14	5–7
Mixed dairy	240–5900	500–10400	60–5800	10–660	0–600	4–11
Whey	9500–60 000	50 000–100 000	1250–22 150	200–2000	120–530	3.9–6.5
Whey processing	590–1200	1070–2180	80–440	–	–	5–9

Adapted from Kolev Slavov (2017).

the water used by the dairy industry becomes contaminated, with estimates placing the volume of wastewater generated at 2.5 times the volume of processed milk (Kolev Slavov, 2017), which would result in over 2 billion m^3 of wastewater generated yearly. The most polluted waters stem from equipment cleaning, sanitary waters, and manufacturing by-products. These wastewaters are high in organic matter and solids, with whey being one of the primary pollutants, while sanitary waters are an important source of nitrogen (Kolev Slavov, 2017). Currently, multiple processes are used to treat dairy wastewaters, including mechanical, physicochemical, chemical, and biological treatment approaches, as will be discussed in section 2.4.

Table 2.3 provides an overview of the composition of wastewaters from dairy industries, adapted from the work of Kolev Slavov (2017). As noted, these wastewater streams remain highly variable, even for the same process effluent, given the great variability in processes applied throughout industries. We will explore the potential to recover resources from wastewater in section 2.5, but we can already note that the high level of organic contaminants (BOD and COD) can be targeted, while nitrogen and phosphorus could potentially be recovered from certain waste flows (notably those associated with cheese and whey). Indeed, the use of on-site anaerobic digesters has been growing quickly in recent years as a strategy to recover energy for many dairy industries.

2.3.2 Textile industries and leather production

Textile manufacturing and leather production are two other primary generators of industrial wastewaters. As highlighted in Figure 2.3, in China, the textile industry is responsible for the third highest industrial wastewater discharge and the highest COD discharge. Indeed, wastewaters from textile processes are high in volume and generally contain very high amounts of BOD, COD, and TSS, though lower amounts of nutrients. The pH is a key parameter that varies between industries. This is significant as pH corrections may be required to enable some treatment or resource recovery technologies. The general characteristics from various textile manufacturing processes are presented in Table 2.4.

Beyond these general properties, textile industries are also characterized by the abundance of certain compounds (Bisschops & Spanjers, 2003; Yaseen and Scholz, 2019). For example, due to the frequent and heavy use of dyes within many of these industries, flows often have highly variable tinctorial characteristics. Additionally, reactive dyeing processes use large quantities of sodium chloride, leading to wastewaters reaching up to 6000 mg Cl^-/L and 7000 mg Na^+/L (Yaseen & Scholz, 2019). Oil and grease (lipids) are also very common, often being by-products or reagents for certain treatment/manufacturing steps. For example, wool scouring (washing) produces high amounts of oil and grease, while oils and fats are important additives for fiber spinning and fabric manufacturing. Metals can also be present throughout these wastewaters, notably from dyes, many of which contain chromium, cadmium, lead, zinc, copper, or cobalt (Bisschops & Spaniers, 2003; Halimoon and Yin,

Table 2.4 Wastewater characteristics from textile manufacturing processes (bisschops & spanjers, 2003; bond & straub, 1974; von Sperling, 2007).

Process	Water Production (m^3/unit)	BOD (mg/L)	COD (mg/L)	TSS (mg/L)	pH	Unit of Production
Cotton	120–750	200–1500	400–1800	200	8–12	1 t produced
Wool	500–600	500–600	–	–	–	1 t produced
Rayon	25–60	500–1200	–	–	–	1 t produced
Nylon	100–150	350	–	–	–	1 t produced
Polyester	60–130	1500–3000	–	–	–	1 t produced
Cotton desizing	–	–	950–20 000	1000–26 200	8.8–9.2	–
Wool scouring	20–70	2000–60 000	2000–90 000	1000–30 000	7.6–11	1 t produced
Cotton scouring	–	100–2900	8000	184–17 00	7.2–13	–
Synthetic scouring	–	–	500–2800	600–3300	8–10	–
Wool dyeing	20–60	400–5000	620–7920	900	4.6–8	1 t produced
Cotton dyeing	–	970–1460	1115–4585	130–25 000	9.2–10.1	–
Textile bleaching	–	250–300	288–13500		6–13.5	–
Tanning[a]	20–40	600–4000	2000–11200	600–3000	11–12	1 t hide processed
Laundry	–	1600	2700	250–500	8–9	–
Leather[b]	–	2000	2340–7180	–	4.1–4.7	–
Shoes	5	3000	–	–	–	1000 pairs produced

[a]Kannaujiya *et al.* (2019); [b]Ramasamy (2019).

2010). Indeed, in dyeing wastewater, heavy metal concentrations can be as high as 12.1 mg Cu/L, 2.7 mg Cr/L, 7.5 mg Cd/L, and 3.4 mg Zn/L (Bisschops & Spaniers, 2003), while the effluent from tanneries can reach up to 391 mg/L of Cr and 268 mg/L of Mg (Sahinkaya *et al.*, 2017). Another class of substance used in high quantities in textile manufacturing that ends up in wastewaters is surfactants, notably anionic and non-ionic surfactants, with concentrations ranging from 15 mg/L in silk and lycra printing plants up to 2000 mg/L for desizing wastewaters (Bisschops & Spaniers, 2003). Nutrients such as nitrogen, phosphorus, and sulphates are also present, as in most wastewaters, often ranging between 70 and 80 mg/L for total Kjeldahl nitrogen (TKN), between 0 and 300 mg/L of phosphate, and between 0 and 2250 mg/L of sulphates, depending on the process (Yaseen & Scholz, 2019). Regarding nutrients, sulfur and/or phosphorus recovery could be of particular interest for some of these textile waste streams, given the potential for high concentrations.

Wastewaters from the leather industry are also characterized by very high COD and BOD, as observed in Table 2.4 (leather; tanning). Beyond high organic loads, tanneries discharge a significant amount of chromium, due to the prevalence of the chrome-tanning process throughout the world. Wastewaters from plants using the chrome process have been reported to contain chromium (Cr) concentrations of over 500 mg/L (Kannaujiya *et al.*, 2019). Sulfides are also prevalent, ranging between 50 and 900 mg/L (Kannaujiya *et al.*, 2019). A variety of resources can be targeted for recovery from leather and tanning wastewaters, including sodium sulfide and fats (Sawalha *et al.*, 2020).

2.3.3 Wood-related industries

Pulp and paper industries generally produce waste effluents with high levels of organic material, such as BOD, COD, and suspended solids, while generating significant volumes of wastewater (around 70 m^3 per ton of paper produced) (Hubbe *et al.*, 2016), alongside important chlorinated organics, methanol, and sulfur concentrations. Indeed, processes that involve further wood treatment, such

Table 2.5 Wastewater characteristics from processes used throughout pulp and paper industries.

Process	BOD (mg/L)	COD (mg/L)	TSS (mg/L)	N (mg/L)	P (mg/L)	S (mg/L)	MeOH (mg/L)	pH
Thermomechanical pulping whitewater	1541	2713	127	7	–	–	–	4.6
Thermomechanical pulping	2800	5600–7210	383–810	12	2.3	72	25	4.2
Chemi-thermomechanical pulping	3000–4000	6000–9000	500	–	–	167	1500	6.2
Kraft mill	2000–10 700	4000–16 000	3620	306–600	1–2	6–375	421–8500	8.2
Bleach Kraft mill	128–184	1124–1738	37–74	2	–	–	40–76	10.1
Sulfite mill	2000–5110	4000–27 100	–	–	–	800–1270	–	2.5–5.9
Bleached pulp mill	1566	2572	1133	–	–	–	–	7.5
Wood preparation	250	–	600	–	–	–	–	–
Paper making	330–27 000	610–5020	760–800	11	0.6	97	9	7.8
Newsprint mill	–	3500	250	–				–
Chip wash	12 000	20 000	6095	86	36	315	70	–
Digester house	13 088	38 588	23 319	–				11.6
Spent liquor	13 300	39 800	253	55	10	868	90	–

Adapted from Bajpai (2000) and Ashrafi *et al.* (2015).

as bleaching, can generate organic substances, notably chlorinated organics like trichloromethane (chloroform). Wood processing and preserving industries also tend to have waste flows with high levels of organic materials, while also using many toxic compounds such as heavy metals (arsenic, copper, chromium) and organic substances (phenols, oils). Overall, wood-related industries are the second most important contributors to the release of organic substances in Europe, as well as the most important source of chlorinated organics (EEA, 2019). General wastewater characteristics from pulp and paper processes are presented in Table 2.5.

The pulp and paper industry already implements significant resource recovery to recover many of the chemicals used throughout the various processes. One notable example is the recovery boiler, applied with most Kraft processes, alongside some sulfide processes. These boilers are employed to recover the chemicals used during the pulping process from the wastewater, known as black liquor, through a variety of steps, alongside energy through combustion (Vakkilainen, 2005). From this process, black liquors are used to generate heat, reduce emissions of inorganic sulfur, produce sodium carbonate and sodium sulfide (both of which are used as reagents during pulping), and recover inorganic chemicals. While recovery boilers have been a staple of the pulp and paper industry since the 1930s, significant progress has been made in recent years regarding resource recovery, leading to a constant reevaluation of recovery alternatives in the field. Generally, energy is the primary resource targeted for recovery, due to the vast quantities of organic residues arising from the processes. Though alternatives such as incineration, gasification, and pyrolysis can be of interest for the liquors from pulping processes (sulfite and kraft mills), as will be discussed in sections 2.4 and 2.5, some recovery processes can face significant challenges. For example, pulp and paper wastewaters typically contain very low nutrient concentrations, with the exception of sulfur. Therefore, if biological recovery processes such as anaerobic digestion are sought, additional nutrients are required to enable the biological pathways. Furthermore, compounds such as chlorine and its reactive byproducts are plentiful, leading to hundreds of different chlorinated hydrocarbons, including chlorinated lignosulfonic acids, chlorinated resin acids, chlorinated phenols, guaiacols, catechols, benzaldehydes, vanillins, syringovanillins, and chloropropioguaiacols, as well as products of lignin degradation, notably chlorolignins (Hubbe *et al.*, 2016), many of which can be detrimental to biological processes through toxicity (Chen *et al.*, 2014; Yin *et al.*, 2001).

As mentioned, most nutrients in pulp and paper wastewaters tend to be low. The exception to this is sulfur, which can be very high in wastewaters from certain pulp and paper processes. This is notably the case for chemical pulping, which is either undertaken through sulfite or sulfate (Kraft) pathways. Both processes are pulping processes that seek to extract almost pure cellulose from wood. The first (sulfite) process achieves this extraction through a reaction of aqueous sulfur dioxide (SO_2) with a base (calcium, sodium, magnesium or ammonium). The second, and more popular, (Kraft) process generally uses sodium hydroxide (NaOH) and sodium sulfide (Na_2S) to produce wood pulp. Consequently, a significant amount of sulfur ends up in various process effluents, as shown in Table 2.5, with concentrations potentially ranging above 1000 mg L^{-1}. As mentioned, recovery boilers already seek to recover a portion of this sulfur, but a significant portion of sulfur still leaves pulp and paper mills as waste and could be valorized.

Methanol is another by-product of pulp mills that could be targeted for resource recovery, as it is generated in massive amounts and is generally considered as a waste product. Indeed, the Kraft process typically produces around 5 kg of methanol per ton of dry pulp, ranging upwards to 15 kg per dry ton (Joyce, 1979; Zhu *et al.*, 2000). Given its volatile nature, the condensate from Kraft evaporators is particularly rich in methanol, with concentrations ranging between 1000 and 46 000 mg/L, representing between 80 and 96% of total COD for this effluent (Badshah *et al.*, 2012). Another area of complexity that must be addressed when treating wastewaters from pulp and paper plants are the variable wood extracts that are hydrophobic and soluble in neutral solvents. These compounds include resin acids, fatty acids, sterols, diterpene alcohols, and tannins. They are among the main contributors to pulp mill effluent toxicity but are also resistant to chemical degradation (Hubbe *et al.*, 2016).

2.3.4 Metal and mining industries

As noted previously, the mining industry is among the main contributors to wastewater use and discharge throughout the world. Many processes involved in mining and refining of metals and ore lead to significant wastewater generation, with mining alone being responsible for 800 million m^3 of wastewater annually throughout Europe (WWAP, 2017). These waters can stem from a variety of sources, including water intensive processes such as electroplating, flotation, and cooling. However, given the 'open' nature of many of these processes, rain waters are also an important source of wastewater, washing exposed surfaces, transporting, and leaching various substances, both from the rock surfaces and from industrial equipment (oils and grease, hydraulic fluids, etc.). Indeed, the recovery and refining of metals and ores lead to the contamination of process water with the metals, any other elements found alongside the desired product, and the chemicals used in these processes. For example, a significant amount of arsenic tends to be found in gold deposits (Straskraba & Moran, 1990), leading to the release of arsenic, cyanide, and very fine particles of gold following the commonly applied cyanide leaching (cyanidation) process (Straskraba & Moran, 1990). Similarly, gold also tends to naturally form with sulfide-type ores (Welham, 2001). Following extraction of the desired elements, the sulfur inevitably ends up forming sulfuric acid in acid mine drainage, that is acid solutions rich in heavy metals (Druschel *et al.*, 2004). Another common extraction and recovery process is flotation. Flotation is the most popular mineral processing method and therefore the greatest consumer of water in mineral processing plants (Li *et al.*, 2019). These waters tend to be heavily contaminated with (non-organic) suspended solids, organic flotation reagents, and heavy metal ions (Li *et al.*, 2019).

Coolant water also tends to be a major source of contaminated water. Indeed, during refining, notably of iron and steel, water is used as a coolant for the high temperature reduction reactions used in these processes. As such, the water comes in contact with a variety of contaminants, notably ammonia, sulfides, and cyanide, alongside a variety of organic compounds (Biswas, 2013). There is also

Table 2.6 Wastewater characteristics from processes used throughout the metal and mining industry.

Compound	Blast Furnace Gas Cleaning	Slag Crushing	Rolling Mills	Cooling of Pig Iron	Pickling Wastewater	Flotation and Enrichment of Lead and Zinc Ores
TSS (mg/L)	330–670	500–600	1000–1500	500–3500	–	20 000–140 000
TDS (mg/L)	800–4000	450–550	400–500	500–2000	–	–
Cyanide (mg/L)	0.6–1.3	Negligible	–	–	–	2–5
Thiocyamates (mg/L)	0–17	3–4	–	–	–	2–5
Iron (mg/L)	140–1180	–	–	–	80–600	–
Chloride (mg/L)	–	–	–	30–300	–	–
Sulfate (mg/L)	–	100–150	100–150	20–650	200–2000	–
Calcium (mg/L)	–	–	–	–	50–200	–
Aluminium (mg/L)	–	–	–	–	0–50	–
Lead (mg/L)	–	–	–	–	–	5–10
Zinc (mg/L)	–	–	–	–	–	0.1–10
Copper (mg/L)	–	–	–	–	–	0.4–8
pH	7–9	–	–	7–8	1.5–4.5	–

Adapted from Jørgensen (1979).

a major source of acidic wastewater from the conversion of iron to steel that is further contaminated by ferrous salts. Given the role of metals in this sector, mining and refining industries are the most important direct source of heavy metals and are nearly on-par with the chemical manufacturing industry for overall (direct and indirect) release of these substances (EEA, 2019). Table 2.6 presents a general overview of wastewater composition of some of the major processes involved in metal and mining industries, while Table 2.7 highlights some of the metals that can be present in various metal and mining waters. It is important to note that the compositions in Table 2.7 are extracted from a limited sample of works and are only intended to highlight the presence of the different metals and their variability.

2.3.5 Oil and gas production and refining

Another major contributor to global wastewater generation, the oil and gas sector generates more than 3 billion m^3 of wastewater annually (Saunders, 2017). Oil and gas production generates two main kinds of wastewaters: process water and produced water. Process water is the result of the various chemical processes used to produce petrochemicals, often leading to waters rich in hydrocarbons, organic acids and compounds, and inorganic substances (Allen, 2008). Produced water arises from salt water trapped in underground formations that surfaces during extraction. This water contains high quantities of hydrocarbons and metals, as well as spent caustics and salt that can be a burden to treatment equipment (Wei *et al.*, 2019), and is the largest source of wastewater (by volume) of oil and gas extraction, with ratios of produced water to oil potentially ranging upwards of 100:1 (EPA, 2019). An overview of a variety of these wastewaters is provided in Table 2.8, while Table 2.9 provides general characteristics from petroleum refineries and petrochemical plant wastewaters.

2.3.6 Chemical industry

Chemical manufacturing is the main source of inorganic substances in industrial wastewaters, as well as an important contributor to heavy metal release (EEA, 2019). The substances found in these wastewaters are dependent on the chemicals being used and manufactured, but are generally comprised

Table 2.7 Some examples of heavy metal loads of various flows emanating from the metal and mining industries.

Type of Metal	Electroplating[a,b,c]	Basic Mine Water[d]	Acid Mine Drainage[e]	Mine-impacted Groundwater[f]	Smelting Wastewater[f]	Scrap Leachate[g]
Magnesium (mg/L)	–	74.6	342	1950	–	–
Aluminium (mg/L)	–	0.022–0.024	54.3	–	–	166
Vanadium (mg/L)	–	–	–	–	–	–
Chromium (mg/L)	34–225	–	0.12	–	2.3	44
Manganese (mg/L)	–	0.001–0.003	6.05	–	–	–
Iron (mg/L)	2.6–8.55	–	391	675	88	244
Cobalt (mg/L)	–	–	8.9	–	0.04	–
Nickel (mg/L)	28–190	–	3.78	–	12	642
Copper (mg/L)	0.92–36.5	–	44.9	60	164.48	12 293
Zinc (mg/L)	24–239	–	5.9	65	455.6	4375
Palladium (mg/L)	–	–	–	–	–	183.1
Silver (mg/L)	–	–	–	–	–	10.9
Cadmium (mg/L)	–	–	0.01	–	76.05	–
Platinum (mg/L)	–	–	–	–	–	20.9
Gold (mg/L)	–	–	–	–	–	11
Lead (mg/L)	2.5	–	6.9	–	4.6	111
Bismuth (mg/L)	–	–	–	–	85	–
Indium (mg/L)	–	–	–	–	–	1008
Tin (mg/L)	–	–	–	–	–	122

[a]Chang and Kim (2007); [b]Sankararamakrishnan *et al.* (2008); [c]Kumar *et al.* (2011); [d]Nordstrom *et al.* (2015); [e]Sahinkaya *et al.* (2017); [f]Weijma *et al.* (2002); [g]Umeda *et al.* (2011).

of unreacted reagents, co or intermediary products, end products, acids or bases used to control pH, as well as heavy metals that are often employed in a variety of processes. A range of sectors fall within this industry, such as chemicals, petrochemicals, polymers, fertilizers, pharmaceuticals, cosmetics, and many other consumer products. Wastewater characteristics from some of these industries are presented in the tables below, highlighting how variable these streams can be between industries. Indeed, Table 2.10 presents wastewater characteristics from the pharmaceutical industry, in which we note extremely variable ranges for compounds and properties such as TSS, total dissolved solids (TDS), COD, BOD, and pH, while having relatively high levels of heavy metals, chlorides, sodium, and oil and grease. Table 2.11 presents the interesting case of wastewater generated from the plastic recycling industry during pre-washing and washing, demonstrating a wide variety of contaminants due to the nature of their use, being in contact with an array of different compounds.

2.4 CURRENT PRACTICE IN INDUSTRIAL WASTEWATER TREATMENT

Given the wide range of substances found in industrial wastewaters, as highlighted in section 2.3, treatment processes are also as variable. Table 2.12 presents an overview of conventional treatment processes used to treat industrial wastewaters. These processes are divided into mechanical, physical, chemical, biological, and thermal processes. The first two are mainly separation processes, with mechanical and physical treatments using physical properties, such as particle size, solubility, density, boiling point, and so on., to drive separation. Chemical treatments can be either focused on separation

Table 2.8 Wastewater characteristics from oil and gas production wastewater (EPA, 2018).

Compound	Drilling Wastewater	Produced Water	Process Wastewater	Flue Gas Desulfurization[a,b]
TSS (mg/L)	168–47 300	57–353	–	–
TDS (mg/L)	557–39 500	2861–226 733	191.9–151 713	–
BOD (mg/L)	79.8–1119	244–2120	–	–
COD (mg/L)	153–9270	1360–3070	–	–
Aluminium (mg/L)	1.7–6916	–	–	0.147
Ammonia (mg/L)	0.98–34.98	–	0.147–2.86	–
Barium (mg/L)	2.55–471	0.963–787	0.05–6.86	–
Benzene (mg/L)	–	0.0015–1.7	–	–
Bromide (mg/L)	–	270–798	14.71	–
Chloride (mg/L)	158–23 469	698–141 200	5–74 975	–
Mercury (mg/L)	–	–	–	0.066–0.289
Potassium (mg/L)	–	0–2190	–	–
Selenium (mg/L)	–	–	–	1.24–3.13
Sodium (mg/L)	167–15 726	733–63 284	–	–
Strontium (mg/L)	1.8–663	nd–4370	0.11–149.35	–
Sulfate (mg/L)	nd–1568	nd–3350	13.06–16.32	–

[a]Staicu *et al.* (2017); [b]Gingerich *et al.* (2018).

Table 2.9 General wastewater characteristics from petroleum refineries and petrochemical plants.

Compound	Range	Average
TSS (mg/L)	20–930	120
TDS (mg/L)	270–87 810	2900
BOD (mg/L)	1–14 230	110
COD (mg/L)	75–265 100	610
Ammonia (mg/L)	20–65	30
BTEX (mg/L)	0.005–1290	10
Heavy metals (mg/L)	0.01–100	6
Oil and grease (mg/L)	3.5–2990	60
Phenol (mg/L)	0.2–210	50
Sulfate (mg/L)	1.5–100	25
Sulfide (mg/L)	18–870	25

Adapted from Jain *et al.* (2020).

or transformation, while potentially allowing for both simultaneously. Regardless of whether the goal is transformation or separation, chemical reactions are the driving factor. The last two treatment types (biological and thermal), tend to be used to breakdown and transform problematic contaminants into more manageable or less hazardous forms. Biological treatments involve the use of microorganisms to undertake biochemical reactions, notably to treat organic matter and some organic and inorganic substances. In some cases, especially when there are significant quantities of heavy metals and

Table 2.10 General wastewater characteristics from pharmaceutical industry wastes.

Compound	Range	Metal	Range
TSS (mg/L)	30–1200	Cadmium (mg/L)	0.036–0.56
TDS (mg/L)	135–4000	Chromium (mg/L)	0.01–1.11
BOD (mg/L)	20–15 660	Copper (mg/L)	0.02–1.67
COD (mg/L)	130–38 640	Iron (mg/L)	0.02–2.35
TN (mg/L)	80–500	Lead (mg/L)	0.03–6.53
TAN (mg/L)	74–116	Nickel (mg/L)	8.5–10.8
TP (mg/L)	18–47	Manganese (mg/L)	6.41–8.47
Chloride (mg/L)	200–2800	Zinc (mg/L)	0.2–1.3
Sodium (mg/L)	155–2000		
Potassium (mg/L	128–140		
Sulfate (mg/L)	80–360		
Sulfide (mg/L)	40–100		
Oil and grease (mg/L)	0.5–3965		
pH	3.9–8.5		

Adapted from Rana *et al.* (2017).

emerging contaminants, thermal processes can be applied to ensure the destruction (or at least concentration to facilitate safe disposal) of these compounds at high temperature.

2.5 WHICH RESOURCES CAN BE RECOVERED FROM INDUSTRIAL WASTEWATER TREATMENT?

Though some of the substances contaminating industrial wastewaters are generally undesirable, many of them have some value and can be reused in other circumstances. Of the contaminants highlighted in section 2.3, the following can be of interest to recover and reuse: inorganic/organic compounds (nutrients, metals, chemical compounds, stabilized organic biosolids), water and energy. Recovery pathways are presented in Figure 2.5 and briefly described below.

Table 2.11 General wastewater characteristics from plastics recycling.

	Polyethylene Terephthalate		High Density Polyethylene: Polypropylene	
Compound	Pre-washing	Washing	Pre-washing	Washing
TS (mg/L)	660–1310	5180–5455	650–1090	7810–8090
TSS (mg/L)	345–634	732–1136	450–876	211–466
COD (mg/L)	315–684	750–983	340–581	100–267
Oil and grease (mg/L)	23–122	20–96	156–266	70–130
Cadmium (mg/L)	–	–	–	0.04–0.06
Iron (mg/L)	4.75–8.06	6.73–7.86	7.13–12.90	3.03–6.06
Lead (mg/L)	0.06–0.14	0.8–1.84	–	0.80–1.35
Manganese (mg/L)	0.08–0.11	0.07–0.08	0.09–0.15	0.06–0.11
pH	7	11.9	7	12.4

Adapted from Santos *et al.* (2005).

Table 2.12 Conventional processes applied to industrial wastewater treatment and the main potential recoverable resources.

Process Type	Process	Description	Main Potential Recoverable Resources
Mechanical treatment	Centrifugation	Separation of particles through centrifugal force based on their size, density and viscosity	Organic solids, nutrients
	Filtration	Separation of particles based on size. Undertaken with membranes (reversed osmosis, ultra- and microfiltration)	Organic solids, high-quality water (if reversed osmosis)
	Flotation	Separation of particles based on their hydrophobic/philic nature (can be controlled by chemical additives)	Organic solids, nutrients, metals
	Sedimentation	Separation of particles due to gravity	Organic solids, nutrients, metals
Physical treatment	Adsorption	Separation of certain substances by physically immobilizing them onto an adsorbent	Nutrients, chemical compounds, metals
	Coagulation and flocculation	Agglomeration of particles into flocs to facilitate their separation	Organic solids, nutrients, metals
	Distillation	Separation of substances based on boiling points	Chemical compounds
	Evaporation	Separation by surface vaporization of liquids below their boiling point	Organic solids, nutrients
	Extraction	Separation of substances based on their relative solubility in different immiscible liquids	Nutrients, metals, chemical compounds
	Irradiation	Disinfection method to kill off pathogens using ultraviolet rays	Potential high-quality water
	Precipitation	Formation of a solid precipitate due to a change in physical conditions (temperature, concentration)	Nutrients, metals, chemical compounds
	Freeze concentration	Freezing (crystallizing) water to facilitate the removal of other compounds	Organic solids, nutrients
	Stripping	Separation of volatile components from a liquid stream to a vapor stream	Nutrients, chemical compounds
	Absorption	Separation of certain compounds by uptake into another	Nutrients, metals, chemical compounds
Chemical treatment	Electrolysis	Separation of ions using membranes and direct electric current	Nutrients, metals, chemical compounds
	Ion-exchange	Removal of dissolved ions from a solution by replacing them with ions of similar charge	Nutrients, metals, chemical compounds
	Oxidation/reduction	Removal of organic and inorganic substances through oxidation reactions	Potential high-quality water
	Precipitation	Formation of a solid precipitate due to chemical reactions	Nutrients, metals, chemical compounds
	Scrubbing	Removal of substances from a gas stream using a liquid solution. Often applied following stripping	Nutrients, chemical compounds

(Continued)

Table 2.12 Conventional processes applied to industrial wastewater treatment and the main potential recoverable resources. *(Continued)*

Process Type	Process	Description	Main Potential Recoverable Resources
Biological treatment	Aerobic treatment	Breaking down of organic contaminants and conversion of nutrients by aerobic microorganisms	Organic solids, nutrients
	Anaerobic treatment	Breaking down of organic contaminants and conversion of nutrients by anaerobic microorganisms	Energy, organic solids, nutrients
	Nitrification/ denitrification	Conversion of ammonia-nitrogen and nitrate-nitrogen to dinitrogen gas using anoxic and aerobic systems operated in series	Organic solids
Thermal treatment	Gasification	Thermal degradation with a controlled amount of oxygen. No combustion	Energy, chemical compounds, organic solids, nutrients
	Incineration	Combustion of organic substances	Energy, inorganic solids (ash), nutrients, metals
	Pyrolysis	Thermal degradation in the absence of oxygen. No combustion	Energy, chemical compounds, organic solids, nutrients
	Thermal hydrolysis	Thermal treatment at high temperature and pressure to improve organic matter for anaerobic digestion	Organic solids, nutrients

2.5.1 Nutrients

As noted in section 2.3, inorganic substances, notably nutrients, are among the main contaminants of industrial wastewaters. These nutrients, such as nitrogen (N), phosphorus (P), and potassium (K), are the major plant nutrients used as fertilizers throughout agriculture to stimulate plant growth. Given the vast amount of the nutrients used as fertilizer, it is evident that recovery of this type of products is absolutely necessary to ensure global food production without further disrupting the worldwide ecosystem. Benefits are not only derived from alternative production of N and P, diminishing the impact associated to fossil resource use on climate change, but also from lessening dependence on the limited supply of certain commodities and increasing food security, all of which warrant maximum resource recovery. Indeed, the global supply of some nutrients such as potassium and phosphorus is mainly produced in a few countries (China, Morocco, USA, Canada, Russia, Belarus) (USGS, 2020), requiring extensive exportation to international markets. The ability to recover these nutrients at a local level can therefore be of extreme benefit, both to the environment and in creating a local and sustainable nutrient supply chain.

Beyond the three major macronutrients (N, P and K), industrial wastewaters can also be rich in three other important macronutrients that can serve as fertilizers: sulfur, calcium and magnesium. A specific hot spot is regeneration of water softening or reverse osmosis brines. Common processes used to recover nutrients include absorption, adsorption, electrolysis, extraction, ion-exchange, precipitation, and scrubbing (Table 2.12). Furthermore, processes such as anaerobic digestion, composting, gasification, pyrolysis, and incineration can be used to recycle certain nutrients, allowing for their recovery in downstream processes (Walling *et al.*, 2019). For example, phosphorus can be

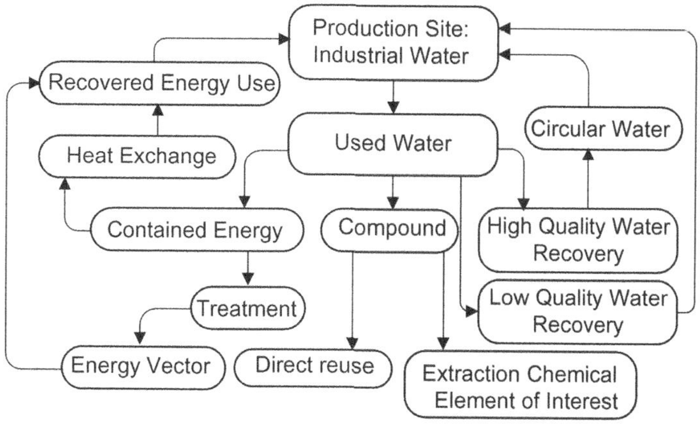

Figure 2.5 Three main recovery applications for industrial wastewater.

recovered from incineration ash through chemical extraction, while nitrogen can be recovered from the remaining liquid (digestate) after anaerobic digestion through stripping and scrubbing (Walling *et al.*, 2019).

2.5.2 Metals

A number of used water flows from the metal processing or metals finishing industries will contain an array of different metals and heavy metals that, up until now, are processed towards disposal of the metals after a chemical treatment converting the soluble metals to a sludge separated from the treated used water. Alternative techniques such as extraction, adsorption, or ion exchange, could be applied to recover these metals.

Indeed, as is the case for phosphorus, some metals can be recovered from incineration ash (Šyc *et al.*, 2020). There is also a growing interest in metal recovery from various industrial wastewaters, especially mining and metal processing wastewaters (section 2.3.4). This has mainly been explored through a combination of biological, membrane (filtration), and precipitation processes, with particular focus on the biological sulfate-reduction process (Gómez & Lens, 2017; Huisman *et al.*, 2006; Kumar & Pakshirajan, 2021; Weijma *et al.*, 2002). This process uses microorganisms to reduce sulfate to sulfide, producing sulfidic metal precipitates along the way, and has currently been applied at full scale to recover metals such as copper (Cu), nickel (Ni), and Zinc (Zn) (Gómez & Lens, 2017). Further processes that have been implemented or are being developed for metal recovery include: hybrid membrane technologies, electrocoagulation, ion exchange, and solvent extraction (Brooks, 2018; Parga *et al.*, 2009). Recently, interest in recovering lithium (Li) has also been growing, especially with regard to its recovery from the brine in wastewaters, either through sorption or evaporation. Oil and gas and mining wastewaters are of particular interest, as well as wastewaters from battery recycling (Kim *et al.*, 2018; Kumar *et al.*, 2019; Park *et al.*, 2015).

2.5.3 Chemical compounds

Depending on the industry, there can be an interest in extracting molecules such as hydrocarbons and organic acids from wastewater, often using extractive or transformative processes. A classic example of a transformative pathway is fermentation (anaerobic biological process), which is commonly used to obtain acids or alcohols from organic matter, notably ethanol and lactic acid. The process can also be used to obtain lipids, proteins, and/or biopolymers. Ethanol can serve as a chemical precursor or reagent for many processes, or as a source of clean energy, while lactic acid is used throughout food, cosmetic, and pharmaceutical industries in a wide variety of roles (Walling *et al.*, 2019). To generate good yields of ethanol or lactic acid, substrates with high sugar or starch concentrations are needed. Therefore, food processing industries can be of particular interest, given the sugars and starch found in fruit and vegetable residues. Corn and potato starch wastewaters are good examples of flows that have been targeted for fermentation, while other products, such as bioplastics and lipids, can also be targeted (Xue *et al.*, 2010). Pulp and paper wastewaters can also be rich in starch, but the presence of complex molecules such as cellulose, hemicellulose, and lignin can require pretreatment to render these complex sugars accessible before fermentation.

If fermentation is not viable, either for technical or economic reasons, processes such as gasification and pyrolysis can be used to obtain a variety of compounds. Beyond applying these technologies to recover energy through hydrogen (H_2) and methane (CH_4) generation (alongside a large variety of energetic co/by-products), organic compounds can be formed through the thermal transformation pathways. One such example is the use of gasification to produce dimethyl ether and/or methanol from black liquor (Kraft wastewater, discussed in section 2.3.3) (Ekbom *et al.*, 2003; Naqvi *et al.*, 2010b). Furthermore, a portion of the significant sulfur found in black liquors can be converted to hydrogen sulfide (H_2S), which can then be sent to a Claus process for desulfurization and recovery of elemental sulfur (and carbon dioxide). The sulfur can then be reused in the pulping process. However, work still remains before gasification can be used as a reliable recovery technique for sulfur recovery from pulp and paper liquors (Hruška *et al.*, 2020; Naqvi *et al.*, 2010a).

Non-transformative pathways can also be used to recover compounds already present in wastewaters through a variety of chemical or physical processes (Table 2.12). As discussed in section 2.3.5, wastewaters from the oil and gas sector contain high levels of organics and sodium, making them strong targets for compound recovery through processes such as evaporation for sodium, or extraction, filtration, and distillation for hydrocarbon recovery. Chemical recovery from black liquor is also of interest. We previously discussed the potential to recover energy and methanol from these pulping wastewaters (section 2.3.3), and many other recovery alternatives exist. One such example is the recovery of lignin from pulp and paper liquors. The lignin can then be available to use as an energy source, while current research is also exploring the prospects of using lignin to produce carbon fiber (Akpan, 2019).

2.5.4 Stabilized organic biosolids

To maintain a healthy and arable soil, maintaining soil organic matter content is crucial. Intensive agriculture tends to deprive arable soils of this essential component. Processed organics, such as biosolids from anaerobic digestion and composting, have a high added value in providing the necessary stable carbon to build up soil organic matter. These processed organics are obtained following all mechanical processes, given that these generally aim at separating solid and liquid phases, as well as certain physical processes, such as evaporation and freeze concentration, which leave behind solid organic fractions. However, to be able to apply these solids to land as soil amendments and organic fertilizers, these biosolids must meet specific criteria regarding pathogen elimination and heavy metal and toxic compound limitations. As such, the solids obtained from the above-mentioned processes are often further processed through biological or thermal pathways to meet regulatory restrictions. These treatments can involve aerobic processes (such as activated sludge treatments and composting), anaerobic processes (such as fermentation and anaerobic digestion), and/or thermal processes, as detailed in Table 2.12.

2.5.5 Water

Water is an extremely valuable resource, representing a major cost for both acquisition and proper disposal. Therefore, water recovery is an important part to any industrial process design. Water can be reused at different levels depending on the required level of quality. A major distinction can be made between reuse after partial purification by either degrading or extracting pollutants from the used water, which would account for a low-grade quality reuse. High grade quality reuse typically uses reverse osmosis as a final treatment step to generate circular water that can be reintroduced in the main production process.

2.5.6 Energy

The final major resource is energy. Energy recovery form industrial wastewaters can take two main forms: (1) indirect recovery of the calorific potential from the substances found in the waste streams, and (2) direct recovery of energy from the flow through heat transfer. Indirect energy recovery from the substances in the waste is related to the transfer of energy rich chemically reduced equivalents, contained in the organic substances, towards a uniform energy vector with different possible applications. Multiple processes can be used to recover this energy, including incineration, anaerobic digestion, fermentation, gasification and pyrolysis, while emerging technologies are continually being investigated and deployed to further recover energy rich compounds. One such example is the recovery of fat from sludges through air flotation, providing a great substrate for heating or as a co-substrate for anaerobic digestion (Fields *et al.*, 2020). The alternative of direct energy recovery by heat transfer (e.g. using heat exchangers) is a well-known field with technology and design considerations that are common knowledge.

2.5.7 Symbiotic resource recovery

Finally, it must be remarked that, with regard to resource recovery, the operational spectrum should not focus strictly on reuse for a given individual company. The exchange of recovered resources between different companies within each other's vicinity is also a viable model, which can constantly be adapted to the alternating needs of the companies involved. One such example is the Kalundborg Symbiosis, which is a major project in Kalundborg, Denmark, involving 12 different companies constantly exchanging resources, including energy, chemicals, and waters of various qualities.

This more holistic approach to resource recovery is extremely important. It requires going beyond a focus on a specific part of a process or even the whole process chain itself. Instead, it looks further to work toward optimizing resource flows across the local surroundings. A solution that is optimal for one plant can be completely unfeasible for another, despite apparent similarities. For example, ammonium sulfate, which is often recovered via stripping and absorption with sulfuric acid with the intention to be used as a fertilizer, could possibly be of more interest locally as a chemical reagent, depending on the situation. Indeed, ammonium sulfate is used in a variety of industries, including the manufacture of chemicals, dyes, flame retardants, and laboratory chemicals. Therefore, one could imagine an industrial park where one plant is seeking to recover its nitrogen. Agricultural lands are far away, meaning that transporting ammonium sulfate as a fertilizer would be costly, as it would either require inefficiently transporting a large amount of liquid over long distances or the crystallization of the ammonium sulfate. However, a nearby chemical manufacturer requires large quantities of ammonium sulfate, presenting a fantastic avenue for local recovery and valorization. Further symbiosis and local integration could be achieved if another local industry generates sulfuric acid, given that it is a common industrial byproduct. Indeed, flue-gas desulfurization is often used to generate high quality sulfuric acid from gas emanating from the combustion of fossil fuels or wastes. Furthermore, recovering nitrogen through the pathway of stripping and scrubbing is not only limited to the production of ammonium sulfate, but can be used to produce a variety of ammonium salts, depending on the acid used in the scrubbing process. Therefore, there can be flexibility in developing a resource recovery process. This flexibility was also highlighted earlier when discussing the opportunities of transformative technologies such as fermentation and gasification.

2.6 CHAPTER SUMMARY

In this chapter, we have explored the sources of industrial wastewater, the contaminants that are found within these waters, and the value and potential of recovering some of these contaminants as resources. Particular detail was given to select industries, highlighting how process and sector dependent these waste flows are. Examples include high organic matter contents generated by the food and beverage sector, the presence of tinctorial compounds and associated heavy metals in the textile industry, the massive wastewater generation and dissolved solid and ions from oil and gas production, and the dominance of chemical manufacturing in inorganics release, among many others. Following this, a plethora of wastewater treatment processes were presented, with notable emphasis on the recovery of resources, including nutrients, energy, chemical compounds, organic matter, metals, and water itself.

Overall, the takeaway message of this chapter is that industrial processes are a major contributor to global water consumption and wastewater generation, and their proper management is therefore critical. These wastewater flows represent a rich and valuable source of resources that should be targeted for recovery. It is important to remember, however, that the compositions of these flows are as unique as the processes that make them, as highlighted throughout this chapter. Therefore, when exploring, developing, and implementing industrial resource recovery processes, it is important to know the characteristics of the flows being assessed, given that they can render certain alternatives completely infeasible, while making others more attractive. Resource recovery from industrial wastewaters is a paradigm: resources can be found in any waste flow, and the key to the development and implementation of appropriate recovery approaches relies on openness, creativity, and awareness of the tools and techniques (processes) that are available. Indeed, what can seem farfetched to someone in one field or industry can be commonly known and applied in another, strengthening the need to learn and share to build a more sustainable future.

Following this chapter, you should have developed an understanding of the interest in recovering resources in industry, both from an economic and environmental perspective, and you should be able to answer the following:

- How important is industrial wastewater generation throughout the world?
- What industries primarily contribute to this generation?
- How and why is the generation of industrial wastewater variable from region to region?
- What contaminants are found in industrial wastewaters?
- How do wastewaters from various industries compare to one another?
- What is the reason for differences in wastewater composition within and between industries?
- What resources can be recovered from industrial wastewaters?
- Why recover resources from industrial wastewaters?
- What processes can be used to recover resources from industrial wastewaters?

2.7 EXERCISES

Exercise 2.1: A dairy producing industry fabricating 2000 L of milk per week (no cheese) is looking for ways to improve its wastewater management. How much wastewater is produced on average by the company over a year? Assuming that half of this wastewater flow comes from fluid milk processing, how much BOD, COD, and TSS is generated at this site over a year?

Exercise 2.2: Compare and discuss the mass loads of BOD, COD, TN, TP, TS, and pH across the following industries:

(1) Dairy industry
(2) Wood-related industries

(3) Textile and leather
(4) Oil and gas production and refining

Exercise 2.3: What resources could be recovered from pulp and paper industrial sludge? What conventional technologies could be applied that would allow to recover these resources, all while treating the sludge? What compounds could hinder recovery?

Exercise 2.4: Which of the following steps involved in the manufacturing of bleached pulp are responsible for the highest and lowest generation of BOD and TSS: assuming that wood preparation generates 300 L/s of wastewater, pulping produces 150 L/s of wastewater, and bleaching produces 550 L/s (assuming average values for BOD and TSS concentrations).

Exercise 2.5: A citrus processing plant processes 10 tons of citrus per day, how much wastewater would this process generate annually? What would the yearly BOD production be? How much nitrogen could be expected (low and high, yearly, estimates)? Is this process of interest for energy and/or nutrient recovery?

Exercise 2.6: Maximum sulfate levels for discharge of wastewaters vary, ranging up to 1000 mg/L, while the standard for drinking water recommended by the World Health Organization is 250 mg/L. Treatment and recovery processes for sulfate include both chemical and biological pathways, such as precipitation and reduction by sulfate-reducing bacteria. Of the industrial wastewaters presented throughout this chapter, which ones would require consideration of sulfates during their treatment?

Exercise 2.7: Wastewater from an industrial process is characterized by a BOD and COD in the orders of 10^2 and 10^3 mg/L, with a pH ranging from low acidic to slightly alkaline. This wastewater is also noted as having a significant load in sodium, chloride and heavy metals, notably copper and chromium, while nutrient content (TKN and TP) is low. What process does this water stem from?

Exercise 2.8: The major macronutrients recovered from wastewaters for reuse in agriculture are nitrogen, phosphorus, and potassium. What are three other macronutrients that can be recovered from industrial wastewaters and which of the industrial sources and processes presented throughout this chapter could prove interesting for recovery of some of these nutrients?

Exercise 2.9: What compounds, if any, could be of interest to recover in produced water emanating from oil and gas wastewater?

Exercise 2.10: What type(s) (mechanical, physical, chemical, biological, thermal) and process(es) do the following descriptions relate to:

(1) Conversion of COD into energy by microorganisms in conditions without oxygen.
(2) Recovery of water and removal of contaminants by vaporization at ambient temperatures.
(3) Conversion of COD into energy at high temperatures and in the presence of oxygen.
(4) Recovery of particulate matter, such as organic solids, based on size using membranes.
(5) Immobilization of contaminants onto a surface based on their ionic charge.
(6) Separation of compounds based on differences in freezing/boiling points.
(7) Recovery of phosphate and ammonium by addition of magnesium to form struvite ($NH_4MgPO_4 \cdot 6H_2O$).
(8) Recovery of energy, biosolids and biooil at high temperatures without oxygen.
(9) Agglomeration of particles into flocs and separation based on size or density.
(10) Recovery of stripped ammonia (gas) using sulfuric acid (H_2SO_4) to form ammonium sulfate, $(NH_4)_2SO_4$.

2.8 DISCUSSION QUESTIONS

Question 2.1: An industrial park is willing to improve its resource management. The park is composed of 10 industries (four food and drink industry, two mining industry, four power generation), each producing various types of effluents that are currently treated by conventional methods. You are recruited as an engineer to improve the overall resource management of the industrial park, all while reducing conventional treatment costs. What data or information will you collect from each industry? What management approach will you apply? What criteria will your choices be based on?

Question 2.2: A progressive food processing industry driven by environmental regulations wants to improve its overall sustainability. It currently produces significant amounts of solids (sludge) through conventional treatment of its wastewaters. These sludges are currently being landfilled. Nutrients are biologically treated through nitrification-denitrification to transform nitrogen into nitrogen gas that is released into the atmosphere, all while producing biological sludge. Phosphorus is being precipitated as an aluminium phosphate sludge. You are asked to provide a presentation to the board of directors, indicating the current practice and the flows that could potentially be recovered. You are also asked to provide motivations for why resources should be recovered, as well as general considerations that will define the technico-economic feasibility of resource recovery.

Question 2.3: A wood-related industry is willing to install a production site in a dense city with about 900 000 inhabitants. You work as a wastewater engineer for the city and are asked to evaluate the health and environmental risks associated to the installment of such industry in the city. Provide a table with the potential risks and potential solutions to circumvent these risks (if any).

Question 2.4: All of the wastewater characterizations shown throughout this chapter have been variable, sometimes ranging by multiple orders of magnitude for a single compound within a process. As an engineer potentially tasked with conceiving processes around such variability, how would you circumvent or take into account this uncertainty?

Question 2.5: The recycling and recovery of industrial wastewaters within industry has been undertaken for decades now and is an integral part of process design. However, as our society transitions towards circular economies and resource recovery becomes ever more present, recovered resources such as nutrients are now making their way into agricultural fields and residual biomass is being used to produce energy. This can lead to potential issues with adoption of these products due to the lack of social awareness or societal acceptance. What reasons or risks, be they real or perceived, can lead to someone being opposed to the recovery and reuse of products derived from industrial wastewaters? Are these reasons justified?

Question 2.6: Beyond the technical feasibility of recovering resources, what other considerations must be taken into account?

Question 2.7: What industries that you are aware of could potentially present an interesting avenue for resource valorization? What resources could be recovered, how feasible would this be, and how would you go about it?

Question 2.8: Based on Table 2.12, how would you describe, in your own words, the differences between the various types of treatment and recovery processes (mechanical, physical, ...)?

Question 2.9: An industrial wastewater treatment process is applying anaerobic digestion to treat their wastewater. During this process, a significant amount of nitrogen in the wastewater is converted to ammonia, which is then sent to an ammonia stripping and scrubbing section to recover ammonium sulfate (H_2SO_4). The ammonium sulfate can either be crystallized and distributed as a solid, or left in solution; what benefits could each alternative offer and how could this ammonium sulfate be valorized?

Question 2.10: You are tasked with conceiving the wastewater treatment process for a mine in a very isolated area. Knowing that access to resources such as chemical reagents can be very costly, given large shipping and storage fees, how would your design take this reality into consideration? Are there any treatment and recovery processes that you might initially be drawn to consider? Why?

REFERENCES

Akpan E. I. (2019). Bio-sourced lignin: recovery techniques and principles. In: Sustainable Lignin for Carbon Fibers: Principles, Techniques, and Applications, E. I. Akpan and S. O. Adeosun (eds), Springer, Cham, pp. 65–150.

Allen E. W. (2008). Process water treatment in Canada's oil sands industry: I. Target pollutants and treatment objectives. *Journal of Environmental Engineering and Science*, **7**(2), 123–138.

Ashrafi O., Yerushalmi L. and Haghighat F. (2015). Wastewater treatment in the pulp-and-paper industry: a review of treatment processes and the associated greenhouse gas emission. *Journal of Environmental Management*, **158**, 146–157.

Badshah M., Parawira W. and Mattiasson B. (2012). Anaerobic treatment of methanol condensate from pulp mill compared with anaerobic treatment of methanol using mesophilic UASB reactors. *Bioresource Technology*, **125**, 318–327.

Bajpai P. (2000). Treatment of pulp and paper mill effluents with anaerobic technology. Randalls Road, Leatherhead, UK.

Biswas J. (2013). Evaluation of various method and efficiencies for treatment of effluent from iron and steel industry – a review. *International Journal of Mechanical. Engineering and Robotics Research*, **2**, 67–73.

Bisschops I. and Spanjers H. (2003). Literature review on textile wastewater characterisation. *Environmental Technology*, **24**(11), 1399–1411.

Bond R. and Straub C. (1974). CRC Handbook of environmental control: Vol. 4, Wastewater: Treatment and disposal. CRC Press, Cleveland, Ohio.

Brooks C. S. (2018). Metal Recovery From Industrial Waste. CRC Press, Boca Raton, Florida, USA.

Chang I. S. and Kim B. H. (2007). Effect of sulfate reduction activity on biological treatment of hexavalent chromium [Cr (VI)] contaminated electroplating wastewater under sulfate-rich condition. *Chemosphere*, **68**(2), 218–226.

Chen J. L., Ortiz R., Steele T. W. and Stuckey D. C. (2014). Toxicants inhibiting anaerobic digestion: a review. *Biotechnology Advances*, **32**(8), 1523–1534.

Dieter C. A., Maupin M. A., Caldwell R. R., Harris M. A., Ivahnenko T. I., Lovelace J. K., Barber N. L. and Linsey K. S. (2018). Estimated Use of Water in the United States in 2015 (No. 1441). US Geological Survey, Reston, Virginia. Available at: https://pubs.usgs.gov/circ/1441/circ1441.pdf (accessed: January 26, 2022).

Druschel G. K., Baker B. J., Gihring T. M. and Banfield J. F. 2004 Acid mine drainage biogeochemistry at iron mountain, California. *Geochemical Transactions*, **5**(2), 13–32.

EEA (European Environment Agency). (2019). Industrial waste water treatment – pressures on Europe's environment, Report No 23/2018, European Environment Agency, Luxembourg.

EIA (U.S. Energy Information Administration). (2019). International: Electricity – Electricity Net Consumption, Washington, DC. Available at: https://www.eia.gov/international/data/world/electricity/electricity-consumption (accessed 16 April 2021).

Ekbom T., Lindblom M., Berglin N. and Ahlvik P. (2003). Technical and Commercial Feasibility Study of Black Liquor Gasification with Methanol/DME Production as Motor Fuels for Automotive Uses – BLGMF. Nykomb Synergetics AB, Stockholm.

EPA. (2018). Detailed Study of the Centralized Waste Treatment Point Source Category for Facilities Managing Oil and Gas Extraction Wastes. U.S. Environmental Protection Agency, Washington, DC, EPA 821-R-18-004. Available at: https://www.epa.gov/sites/production/files/2018-05/documents/cwt-study_may-2018.pdf (accessed: January 26, 2022)

EPA. (2019). Study of Oil and Gas Extraction Wastewater Management Under the Clean Water Act. U.S. Environmental Protection Agency, Washington, DC, EPA-821-R19-001. Available at: https://www.epa.gov/sites/production/files/2019-05/documents/oil-and-gas-study_draft_05-2019.pdf (accessed: January 26, 2022)

FAO. (2020). FAO's Global Information System on Water and Agriculture. Food and Agriculture Organization of the United Nations, Rome, Italy. Available at: http://www.fao.org/aquastat/en/overview/methodology/water-use (accessed: January 26, 2022)

Fields W. M., Yuanchun X. U. and Zhang G. (2020). Water Environmental Technology and Water Environmental Technology. System for recovering fat, oil and grease from wastewater. U.S. Patent Application 16/629,066.

Gingerich D. B., Grol E. and Mauter M. S. (2018). Fundamental challenges and engineering opportunities in flue gas desulfurization wastewater treatment at coal fired power plants. *Environmental Science: Water Research & Technology*, **4**(7), 909–925.

Gómez D. K. V. and Lens P. N. (2017). Metal recovery from industrial and mining wastewaters. In: Sustainable Heavy Metal Remediation, E. E. Rene, E. Sahinkaya, A. Lewis and P. L. N. Lens (eds), Springer Nature, New York, USA, pp. 81–114.

Guo X., Ho M. S., You L., Cao J., Fang Y., Tu T. and Hong Y. (2018). Industrial water pollution discharge taxes in China: a multi-sector dynamic analysis. *Water*, **10**(12), 1–21.

Halimoon N. and Yin R. G. S. (2010). Removal of heavy metals from textile wastewater using zeolite. *Environment Asia*, **3**, 124–130.

Hruška M., Variny M., Haydary J. and Janošovský J. (2020). Sulfur recovery from syngas in pulp mills with integrated black liquor gasification. *Forests*, **11**(11), 1–21.

Hubbe M. A., Metts J. R., Hermosilla D., Blanco M. A., Yerushalmi L., Haghighat F., Lindholm-Lehto P., Khodaparast Z., Kamali M. and Elliott A. (2016). Wastewater treatment and reclamation: a review of pulp and paper industry practices and opportunities. *BioResources*, **11**(3), 7953–8091.

Huisman J. L., Schouten G. and Schultz C. (2006). Biologically produced sulphide for purification of process streams, effluent treatment and recovery of metals in the metal and mining industry. *Hydrometallurgy*, **83**(1–4), 106–113.

Jain M., Majumder A., Ghosal P. S. and Gupta A. K. (2020). A review on treatment of petroleum refinery and petrochemical plant wastewater: a special emphasis on constructed wetlands. *Journal of Environmental Management*, **272**, 1–21.

Jørgensen S. E. (1979). Waste water from the iron and steel industry and mining. *Environmental Science*, **5**, 217–227.

Joyce T. W. (1979). Significance of Methanol Recovery to Kraft Mill Total Energy Budget. IPS technical paper series, 71. The Institute of Paper Chemistry, Appleton, Wisconsin, USA.

Kannaujiya M. C., Mandal T., Mandal D. D. and Mondal M. K. (2019). Treatment of leather industry wastewater and recovery of valuable substances to solve waste management problem in environment. In: Environmental Contaminants: Ecological Implications and Management. Microorganisms for Sustainability, R. Bharagava (ed.), Springer, Singapore, Vol. **14**, 311–340.

Kim S., Kim J., Kim S., Lee J. and Yoon J. (2018). Electrochemical lithium recovery and organic pollutant removal from industrial wastewater of a battery recycling plant. *Environmental Science: Water Research & Technology*, **4**(2), 175–182.

Kolev Slavov A. (2017). General characteristics and treatment possibilities of dairy wastewater – a review. *Food Technology and Biotechnology*, **55**(1), 14–28.

Kumar M. and Pakshirajan K. (2021) Continuous removal and recovery of metals from wastewater using inverse fluidized bed sulfidogenic bioreactor. *Journal of Cleaner Production*, **284**, 1–11.

Kumar R., Bhatia D., Singh R., Rani S. and Bishnoi N. R. (2011). Sorption of heavy metals from electroplating effluent using immobilized biomass Trichoderma viride in a continuous packed-bed column. *International Biodeterioration & Biodegradation*, **65**(8), 1133–1139.

Kumar A., Fukuda H., Hatton T. A. and Lienhard J. H. (2019). Lithium recovery from oil and gas produced water: a need for a growing energy industry. *ACS Energy Letters*, **4**(6), 1471–1474.

Li Y., Xie S., Zhao Y., Ling X., Li H. and Song S. (2019). The life cycle of water used in flotation: a review. *Mining, Metallurgy & Exploration*, **36**, 385–397.

Marchal V., Dellink R., Van Vuuren D., Clapp C., Chateau J., Magné B. and Van Vliet J. (2011). OECD Environmental outlook to 2050. *Organization for Economic Co-Operation and Development*, **8**, 397–413.

Naqvi M., Yan J. and Dahlquist E. (2010a). Black liquor gasification integrated in pulp and paper mills: a critical review. *Bioresource Technology*, **101**(21), 8001–8015.

Naqvi M., Yan J. and Fröling M. (2010b). Bio-refinery system of DME or CH4 production from black liquor gasification in pulp mills. *Bioresource Technology*, **101**(3), 937–944.

Nordstrom D. K., Blowes D. W. and Ptacek C. J. (2015). Hydrogeochemistry and microbiology of mine drainage: an update. *Applied Geochemistry*, **57**, 3–16.

Parga J. R., Vázquez V., Casillas H. M. and Valenzuela J. L. (2009). Cyanide detoxification of mining wastewaters with TiO_2 nanoparticles and its recovery by electrocoagulation. *Chemical Engineering & Technology*, **32**(12), 1901–1908.

Park H., Singhal N. and Jho E. H. (2015). Lithium sorption properties of HMnO in seawater and wastewater. *Water Research*, **87**, 320–327.

Preisner M., Neverova-Dziopak E. and Kowalewski Z. (2020). An analytical review of different approaches to wastewater discharge standards with particular emphasis on nutrients. *Environmental Management*, **66**(4), 694–708.

Quitmann H., Fan R. and Czermak P. (2013). Acidic organic compounds in beverage, food, and feed production. In: Biotechnology of Food and Feed Additives, H. Zorn and P. Czermak (eds), Springer, Berlin, Heidelberg, pp. 91–141.

Ramasamy B. (2019). Short review of salt recovery from reverse osmosis rejects. In: Salt in the Earth, M. Cengiz and S. Karabulut (eds), IntechOpen, London, UK, pp. 63–77.

Rana R. S., Singh P., Kandari V., Singh R., Dobhal R. and Gupta S. (2017). A review on characterization and bioremediation of pharmaceutical industries' wastewater: an Indian perspective. *Applied Water Science*, **7**(1), 1–12.

Sahinkaya E., Uçar P. and Kaksonen A. H. (2017). Bioprecipitation of metals and metalloids. In: Sustainable Heavy Metal Remediation, E. R. Rene, E. Sahinkaya, A. Lewis and P. N. L. Lens (eds), Principles and Processes. Springer International Publishing, Cham, Vol. **1**, pp. 199–231.

Sankararamakrishnan N., Kumar P. and Chauhan V. S. (2008). Modeling fixed bed column for cadmium removal from electroplating wastewater. *Separation and Purification Technology*, **63**(1), 213–219.

Santos A. S. F., Teixeira B. A. N., Agnelli J. A. M. and Manrich S. (2005). Characterization of effluents through a typical plastic recycling process: an evaluation of cleaning performance and environmental pollution. *Resources, Conservation and Recycling*, **45**(2), 159–171.

Saunders N. (2017). Getting Dangerously Creative with Oil and Gas Wastewater. Environmental Defence Fund, New York, Vol. 2019. Available at: http://blogs.edf.org/energyexchange/2017/10/12/getting-dangerously-creative-with-oil-and-gas-wastewater/ (accessed 11 December 2021).

Sawalha H., Al-Jabari M., Elhamouz A., Abusafa A. and Rene E. R. (2020). Tannery wastewater treatment and resource recovery options. In: Waste Biorefinery, T. Bhaskar, A. Pandey, E. R. Rene and D. C. W. Tsang (eds), Elsevier, Amsterdam, The Netherlands, pp. 679–705.

Staicu L. C., Morin-Crini N. and Crini G. (2017). Desulfurization: critical step towards enhanced selenium removal from industrial effluents. *Chemosphere*, **172**, 111–119.

Statistics Canada. (2011). Human Activity and the Environment: Wastewater Discharges, Statistics Canada, Ottawa, Ontario. Available at: https://www150.statcan.gc.ca/n1/pub/16-201-x/2012000/part-partie4-eng.htm (accessed: January 26, 2022)

Statistics Canada. (2014). Industrial Water Use. Statistics Canada, Ottawa, Ontario. Available at: https://www150.statcan.gc.ca/n1/pub/16-401-x/16-401-x2014001-eng.pdf (accessed: January 26, 2022)

Straskraba V. and Moran R. E. (1990). Environmental occurrence and impacts of arsenic at gold mining sites in the western United States. *International Journal of Mine Water*, **9**(1), 181–191.

Šyc M., Simon F. G., Hykš J., Braga R., Biganzoli L., Costa G., Funari V. and Grosso M. (2020). Metal recovery from incineration bottom ash: state-of-the-art and recent developments. *Journal of Hazardous Materials*, **393**, 1–17.

Umeda H., Sasaki A., Takahashi K., Haga K., Takasaki Y. and Shibayama A. (2011). Recovery and concentration of precious metals from strong acidic wastewater. *Materials Transactions*, **52**(7), 1462–1470.

USGS (United States Geological Survey). (2020). Mineral Commodity Summaries 2020. Reston, Virginia. Available at: https://pubs.usgs.gov/periodicals/mcs2020/mcs2020.pdf (accessed: January 26, 2022)

Vakkilainen E. (2005). Kraft Recovery Boilers – Principles and Practice. Suomen Soodakattilayhdistys r.y., Helsinki, Finland.

Valta K., Kosanovic T., Malamis D., Moustakas K. and Loizidou M. (2015). Overview of water usage and wastewater management in the food and beverage industry. *Desalination and Water Treatment*, **53**(12), 3335–3347.

von Sperling M. (2007). Wastewater Characteristics, Treatment and Disposal. IWA Publishing, London, UK.

Walling E., Babin A. and Vaneeckhaute C. (2019). Nutrient and carbon recovery from organic wastes. In: Biorefinery, J. R. Bastidas-Oyanedel and J. Schmidt (eds), Springer, Cham, pp. 351–373.

Wei X., Zhang S., Han Y. and Wolfe F. A. (2019). Treatment of petrochemical wastewater and produced water from oil and gas. *Water Environment Research*, **91**(10), 1025–1033.

Weijma J., Copini C. F. M., Buisman C. J. N. and Schultz C. E. (2002). Biological recovery of metals, sulfur and water in the mining and metallurgical industry. In: Water Recycling and Recovery in Industry, P. Lens, L. Hulshoff Pol, P. Wilderer and T. Asano (eds), IWA Publishing, London, UK, pp. 605–622.

Welham N. J. (2001). Mechanochemical processing of gold-bearing sulphides. *Minerals Engineering*, **14**(3), 341–347.

World Bank. (2019). Data: Manufacturing, Value Added (Current US$). Washington, D.C. Available at: https://data.worldbank.org/indicator/NV.IND.MANF.CD?most_recent_value_desc=true (accessed: April 16th 2021).

WWAP (United Nations World Water Assessment Programme). (2017). The United Nations World Water Development Report 2017. Wastewater: The Untapped Resource. UNESCO, Paris. Available at: http://www.unesco.org/new/en/natural-sciences/environment/water/wwap/wwdr/2017-wastewater-the-untapped-resource/ (accessed: January 26, 2022)

Xue F., Gao B., Zhu Y., Zhang X., Feng W. and Tan T. (2010). Pilot-scale production of microbial lipid using starch wastewater as raw material. *Bioresource Technology*, **101**(15), 6092–6095.

Yaseen D. A. and Scholz M. (2019). Textile dye wastewater characteristics and constituents of synthetic effluents: a critical review. *International Journal of Environmental Science and Technology*, **16**(2), 1193–1226.

Yeo S. E., Binkowski F. P. and Morris J. E. (2004). Aquaculture Effluents and Waste by-Products Characteristics, Potential Recovery, and Beneficial Reuse. NCRAC Technical Bulletins. NCRAC Publications Office, North Central Regional Aquaculture Center, Ames, Iowa, USA. Available at: https://dr.lib.iastate.edu/server/api/core/bitstreams/18004648-f522-469d-b0e8-83f0bc25f8ab/content (accessed: January 26, 2022)

Yin C. R., Seo D. I., Baek S. H., Ohlen K. and Lee S. T. (2001). Inhibitory effect of chlorinated guaiacols on methanogenic activity of anaerobic digester sludge. *Biotechnology Letters*, **23**(17), 1379–1383.

Zhu J. Y., Yoon S. H., Liu P. H. and Chai X. S. (2000). Methanol formation during alkaline wood pulping. *Tappi Journal*, **83**(7), 1–13.

doi: 10.2166/9781780409566

Chapter 3

Resource recovery from drinking water production facilities: what and how much is there?

Ilje Pikaar[1], Katrin Doederer[2], Tessa van den Brand[3], Olaf van der Kolk[4] and Wolfgang Gernjak[5,6]

[1]*School of Civil Engineering, The University of Queensland, St Lucia, QLD 4072, Australia*
[2]*The Advanced Water Management Centre, The University of Queensland, St Lucia, QLD 4072, Australia*
[3]*KWR Water Research Institute, Groningenhaven 7, 3433 PE Nieuwegein, The Netherlands*
[4]*AquaMinerals, Groningenhaven 7, 3433 PE Nieuwegein, The Netherlands*
[5]*Catalan Institute for Water Research (ICRA), Emili Grahit 101, 17003 Girona, Spain*
[6]*Catalan Institute for Research and Advanced Studies (ICREA), Passeig Lluis Companys 23, 08010 Barcelona, Spain*

3.1 INTRODUCTION

Cost-effective, reliable production and safe transportation of high-quality drinking water through a comprehensive distribution network is one of the pillars of modern society that has enabled mankind to live in densely populated cities. The greatest acute health risk when consuming contaminated water are waterborne pathogens that can lead to outbreaks of diseases which in extreme cases can result in death and can affect a large proportion of the community (Baldursson & Karanis, 2011; https://www.un.org/sustainabledevelopment/water-and-sanitation/). In addition to the safety of drinking water from a human health perspective, drinking water needs to be of a high 'aesthetic' quality in order to gain full consumer acceptance. Important aesthetics aspects include appearance, that is off-tastes, odors and staining (Hargesheimer & Watson, 1996). Considering the vital role of providing safe drinking water to our society, it is not surprising that significant efforts have been made to develop and design regulations and guidelines that offer an authoritative reference on: (i) what defines 'safe and good quality' water, (ii) how it can be achieved, and (iii) how this can be ascertained. The difference between regulations and guidelines is that the first are maintained directives while guidelines are non-mandatory standards. It should be noted that the nature and form of regulations and guidelines varies among countries and regions. Similarly, and somewhat surprisingly, there is no recognized standard that jurisdictions follow in adapting existing knowledge to their own regional context. Often this results in different safe values proposed for water quality parameters in different jurisdictions.

To ensure drinking water of good quality, the treatment process generally consists of a multi-barrier system. Its function must continuously be maintained as well as that of the clear water storage and reticulation mains. Distribution mains can be located underground or above the surface. While colder climates try to avoid freezing by burying the mains, warmer climates lay water mains above the ground. It is generally assumed that warmer waters increase the necessity for maintaining a disinfectant residual during distribution. While the primary goal of water treatment is first and foremost to produce high-quality drinking water, we will discuss here whether this can be achieved with concomitant recovery of resources.

While resource recovery from water has gained a lot of momentum in recent years, one often associates this with recovery of materials and resources during the treatment of domestic and/or industrial wastewater. In comparison, resource recovery from drinking water treatment plants has attracted less attention. However, there are plenty of resources to be recovered during the production of drinking water and/or during water recycling operations. In fact, resource recovery of residuals from drinking water treatment has some advantages compared to recovery of resources from domestic wastewater. The presence of pathogens, toxic metals, and microbial contamination in domestic wastewater often comes with regulatory hurdles and issues around public/customer perception and market acceptance. Certainly, technological advancement in the last decades have made it possible to overcome these quality and regulatory issues and have allowed us to recover resources from wastewater at such qualities that in some cases the quality exceeds that of virgin raw materials. Nevertheless, the importance of public/customer perception and market acceptance is something that cannot be underestimated. One can imagine that recovering products from drinking water seems something that could more easily gain full consumer acceptance.

3.2 LEARNING OBJECTIVES

At the completion of this chapter you should be able to:

- Describe water treatment and drinking water consumption and explain its potential opportunities for resource recovery.
- Describe key resources present in various sources for drinking water production that can be recovered; that is surface water, ground water, sea water and reclaimed water.
- Describe the key resources that can be recovered during the production of drinking water that are added during the water purification process.
- Understand the differences between resource recovery from wastewater and drinking water production in relation to product quality, regulatory hurdles and technology requirements.
- Understand the importance of the non-technological aspects such as creating a strong stakeholder community, effective communication and market/customer acceptance.

3.3 MAJOR SOURCES FOR THE PRODUCTION OF DRINKING WATER

There are four major sources used for drinking water production, namely surface water, ground water, seawater, and reclaimed effluent from wastewater treatment plants. Surface water sources can be lakes, reservoirs, rivers, streams or ponds. Treated effluent from wastewater treatment plants which would normally be discharged to the receiving environment can be further treated using advanced technologies to produce water at (or even) exceeding drinking water quality (see Chapters 4 and 18 for more detail). High quality (recycled) water exceeding drinking water quality is often used in industrial applications such as ultra-pure water for cooling towers in power plants or in high tech industry.

Depending on the geological location of the water treatment plants the predominant source of water used will be different (Figure 3.1). For example, due to its dry climate and limited supply of fresh water, about 40% of the drinking water produced in Israel comprises desalination of seawater and saline groundwater (Feitelson & Rosenthal, 2012). On the other hand, European countries like France, Netherlands and Germany produce as much as 70% of their drinking water from groundwater while Australia and Canada mainly rely on surface water from rivers, reservoirs and lakes. In addition, substantial regional differences within countries exist. A good example is China, where 70% of the national groundwater reserves are located in its southern part, generating a distinct regional imbalance (Liu & Zheng, 2016).

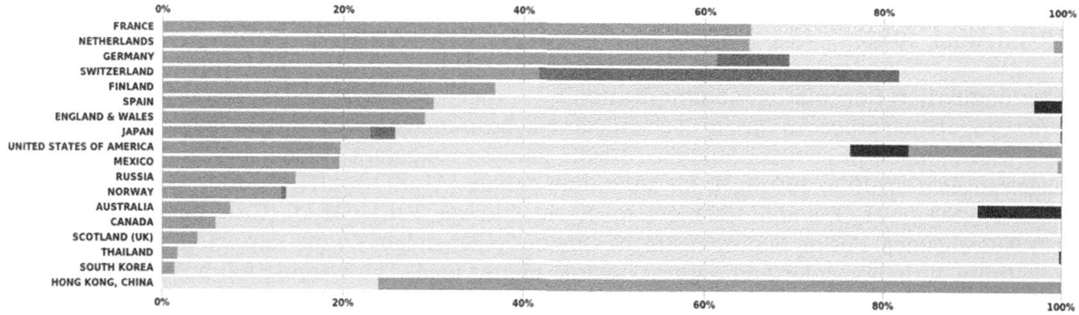

Figure 3.1 Distribution of the different water sources used for the production of drinking water (2014) (http://waterstatistics.iwa-network.org/).

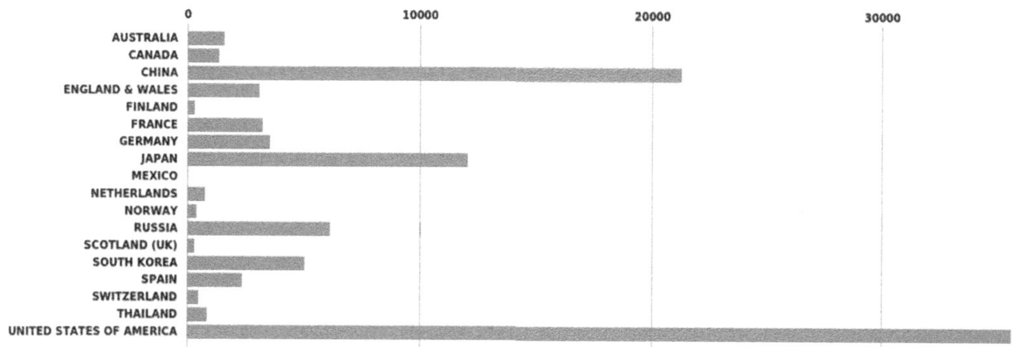

Figure 3.2 Water delivered to households and small businesses in 2014 in million m³ per year, adapted from (http://waterstatistics.iwa-network.org/).

In general, centralized water treatment is applied in major urban areas and in most parts of the developed world. This means source water is treated at a centralized treatment plant and then delivered to consumers through a pressurized distribution system. Centralized treatment plants in major urban areas produce up to several hundred thousand m^3 of drinking water per day. One could imagine that such a large scale and centralized production of drinking water in general has a beneficial impact on the potential of resource recovery due to the economy of scale. Figure 3.2 shows the volumes of water delivered in different countries per year (note: this should give a good indication on the national resource recovery potential of the countries listed). In 2016, the USA delivered the most drinking water to households and small businesses (no data available on water consumption of the industrial sector) with ~35 billion m^3. China, with a four times greater population, only produced 21 billion m^3, while European countries, including Germany and France, delivered between 3.2 and 3.5 billion m^3.

3.4 CURRENT PRACTICE IN WATER TREATMENT

In this paragraph, the most commonly used treatment processes for drinking water production are described. It is evident that the type of process selected depends on the desired production capacity and the characteristics of the water source used. It is not the aim of this chapter to discuss each process in detail, but rather to provide the reader with a general overview of these processes in relation to their resource recovery potential. To obtain in-depth process understanding, we refer the reader to textbooks where these treatment processes are discussed in great detail (Crittenden & Borchardt, 2012; Edzwald, 2010).

3.4.1 Coagulation-flocculation-sedimentation

For the coagulation process a variety of chemicals can be used. However, the vast majority of treatment plants uses either iron or aluminum-based salts as base coagulant (Pikaar *et al.*, 2014). A typical dosage ranges from 10 to 150 mg/L for aluminum sulfate ($Al_2(SO_4)_3{\cdot}14H_2O$), 10–250 mg/L for ferric sulfate ($Fe_2(SO_4)_3{\cdot}9H_2O$) and 5–150 mg/L for ferric chloride ($FeCl_3{\cdot}6H_2O$) (Crittenden & Borchardt, 2012). The chemical dosing requirements are obviously dependent on the raw water quality in terms of alkalinity, turbidity, color and natural organic matter load. The coagulation process contributes the major load of the solid water treatment residuals containing either aluminum or iron oxides 'polluted' with organic matter (i.e., mainly humic and fulvic acids), clay, silt, colloidal material, and microorganisms with other chemicals employed in the process such as a diversity of flocculants (Crittenden & Borchardt, 2012). The vast majority of these solids are then separated in a clarifier and gathered by the sludge removal system. Sometimes powdered activated carbon (PAC) is applied for the control of taste and odor compounds as well as for the sorption of emerging contaminants, which then ends up in the solid waste as well. Another source of solids is from the regular backwash of rapid media filters that further treat the effluent of the clarifier. Besides carry-over from the clarifier, solids generated by the addition of filter aid polymers and/or the oxidation of manganese and iron to insoluble forms accumulate on the filter and result in a relatively large volume of residual produced in a short time.

3.4.2 Lime-soda ash softening

Lime-soda ash and pellet softening are common methods used to reduce the hardness of source waters by precipitation that also produces a solid residual stream. These processes are preferred in waters high in calcium and magnesium where also heavy metals and radioactive elements like radon may be removed. Lime-soda ash is a precipitative process that typically includes coagulation, flocculation, and sedimentation applying quicklime (i.e., CaO(s)), or hydrated lime (i.e., $Ca(OH)_2$). Sodium hydroxide is used when at least half of the calcium hardness is carbonate hardness. An alternative to coagulation-flocculation-sedimentation is pellet softening. At the core of this technology is a column type reactor filled with fine sand. In this process, the water is fed through the column filled with sand at fairly high up-flow velocities, normally ranging between 60 and 100 meter/hour, creating a fluidized bed reactor operation. Caustic soda is added (and rapidly mixed) at the bottom of the column in order to almost instantaneously elevate the pH of the water to $\sim$8.5, causing crystallization of the calcium (and magnesium) in the form of calcium carbonate (calcite) onto the sand, making the grains grow into pellets of about 1 mm, which is when they become obviously heavier, are less efficiently fluidized and subsequently removed and substituted with fresh sand. Note that the softened water often requires re-carbonation in order to restore the total carbonate concentration as well as to lower the pH.

3.4.3 Ion exchange

During drinking water treatment ion-exchange resins are primarily used for water softening (i.e., removal of calcium and magnesium and enhanced removal of organic matter to reduce disinfection by-product formation and lower their coagulant and total oxidant demand. A saline solution (i.e., normally NaCl as it is a cheap and readily available bulk product) is used to regenerate the resins and produces highly concentrated brine containing the target pollutant, organic acids, sulfate, bicarbonate and chloride (McAdam & Judd, 2008).

3.4.4 Membrane filtration

Membranes are a physical barrier to remove dissolved solids and particulates. There are different classes of pressure driven membranes, classified by their pore size, including microfiltration (MF), ultrafiltration (UF), nanofiltration (NF) and reverse osmosis (RO), as described in detail elsewhere (Crittenden & Borchardt, 2012). MF and UF are typically used to remove pathogens and solids from their feedwater. Sometimes they also provide pre-treatment before NF or RO processes. MF and UF are generally operated in a dead-end configuration and periodic backwashing, normally every 30–60 minutes, is applied in order to minimize the build-up of particles/organics onto the membrane surface. The latter produces a residual stream containing chlorine and organics. On the other hand, RO and NF membranes, which are characterized by much smaller pore sizes, should rather be imagined as dense polymer layers with material-free void spaces in the separation layer of the thin-film composite membrane. They are often used for brackish and seawater desalination, resulting in a highly saline residual stream that is continuously produced, often referred to as the concentrate. Concentrate streams typically range from 10% to 30% for brackish and from 45% to 60% for seawater desalination of the feed volume. Those membranes also find application in water reuse to provide a barrier for viruses in addition to reducing salinity or micropollutants.

To protect the membranes from biofouling, commonly hypochlorite/chlorine or chloramines are added to the feed. Since chlorine in solution can damage the polymer membranes, sodium bisulfite is typically used to reduce chlorine in larger RO systems as it readily reacts with any residual chlorine to form sodium sulfate that ends up in the concentrate stream. Moreover, depending on the pH and hardness level of the water source used, pH correction with acids and added anti-scalants can be employed in order to maintain operation at a constant membrane pressure. Anti-scalants are a family of chemicals designed to inhibit the formation and precipitation of crystallized mineral salts that form scale. In addition, potentially added dispersants are polymers designed to inhibit the agglomeration and deposition of foulants onto the membrane surface. Foulants include inorganic metal oxides and hydroxides, polymerized silica, organic/inorganic colloids, and organic and biological matter. RO and NF membranes typically also require regular acid and base cleaning, which results in the generation of residual acidic or alkaline streams.

3.5 WHICH RESOURCES CAN BE RECOVERED?

Table 3.1 provides a summary of resources that can be recovered at drinking water treatment plants using commonly applied treatment processes. Note that a clear distinction can be made between the resources naturally present in the water sources and the resources (i.e., mostly chemicals) that are added during the treatment processes. As shown in Table 3.1, depending on the drinking water treatment process applied, there are various waste streams being produced, which are generally referred to as water treatment residuals (Cornwell & Roth, 2000). In a conventional water treatment plant as much as 3–5% of the plant influent can end up as solid, semisolid, and liquid residuals (Crittenden & Borchardt, 2012).

A schematic representation that provides a clear and holistic overview of the different material flows and their relative importance in terms of input and output is a so-called Sankey diagram (see Figure 3.3). The widths of the arrows are proportional to the mass flow of the different inputs and outputs added and produced during the production of drinking water. The example in Figure 3.3 depicts the average material flow of 20 drinking water production facilities using ground water as their source water for the year 2018 (personal communication, AquaMinerals 2018). Note that in the diagram the input in terms of assets like the distribution network are also included, often not considered in the context of recovery/recycling. The difference in input and output (i.e., output exceeds input) may be somewhat confusing but can be attributed to, for example, replacement of heavy old concrete/cast iron pipes to PVC as well as a temporary peak in final disposal of 'end-of-service life' infrastructure.

As depicted in Table 3.1, a wide variety of residuals is produced during the production of drinking water. It is important to realize that both the quantity of these residuals generated as well as the quality in terms of their purity and product strengths depends on both the treatment process applied and water source used. This is of crucial importance when trying to link the recovered resources with the market

Table 3.1 Commonly applied treatment processes during the production of drinking water using different water sources and the potential resources that can be recovered.

Treatment Process	Main Treatment Objective	Water Source	Chemicals Added During Treatment	Residual Stream	Potential Resources That can be Recovered	Load/Characteristics of the Produced Residual Stream
Aeration of ground water	Removal of ferrous iron ions	GW	N.A.	Iron sludge	Iron(hydr)oxides	Brown-red sludge, odorless Up to 90% of dry weight
Coagulation-flocculation-sedimentation	Removal of organic matter[a], pathogens,	GW, SW, SE	Aluminum sulfate	Alum sludge	Aluminum Organic acids	Grey sludge, often difficult to dewater, may smell depending on organics
	Ca, Mg and bicarbonate		Iron chloride Iron sulfate	Iron sludge	Iron, organic acids	Brown-red sludge, may smell depending on organics
			$Ca(OH)_2$ Sodium carbonate, caustic soda	Softening sludge	$CaCO_3$, $Mg(OH)_2$ Organic acids	2–5% of plant flow Natural colored (white-blonde-gray-brown) sludge or pellets; odorless
Media filtration (sand, activated carbon, anthracite)	Removal of solids post sedimentation	GW, SW	NA	Spent filter backwash	Aluminum, iron Manganese, organics Media filter itself (e.g., activated carbon)	A mix of Al, Fe, Mn, organics at mg/L levels
Low-pressure membranes (MF/UF)	Solids, pathogens	GW, SW, SE	NA	Spent filter backwash	Organic acids	Load highly depends on plant recovery; typically <5% of plant feed flow
	Chemical cleaning		Citric acid Caustic soda Chlorine	Cleaning solution		
High-pressure membranes (NF/RO)	Salts, pathogens, organics	GW, SEA, SE	Antiscalant Antifoulant Sodium bisulfite	Concentrate	Sodium bisulfate Phosphonates	Load highly depends on plant recovery; typically 10–25% of plant feed flow
	Chemical cleaning		Citric acid Caustic soda	Cleaning solution		
Ion-exchange (IEX)	Organic matter, nitrate, hardness	GW, SW	Salt, sodium-bicarbonate Caustic soda	Regenerate solution	Organic acids (humic and fulvic acids)Also (Na^+, NO_3^-, Cl^-, Ca^{2+}, Mg^{2+} and HCO_3^-)	0.05–10% of treated product flow depending on IEX system; low percentages can be achieved by the MIEX® process. A mix of NaCl, SO_4^-, HCO_3^- and humic and fulvic acids at g/L levels

GW = ground water, SW = surface water, SEA = sea water, SE = secondary effluent of waste water treatment plant, NA = not applicable. [a]Organic matter is the main cause of color and turbidity.

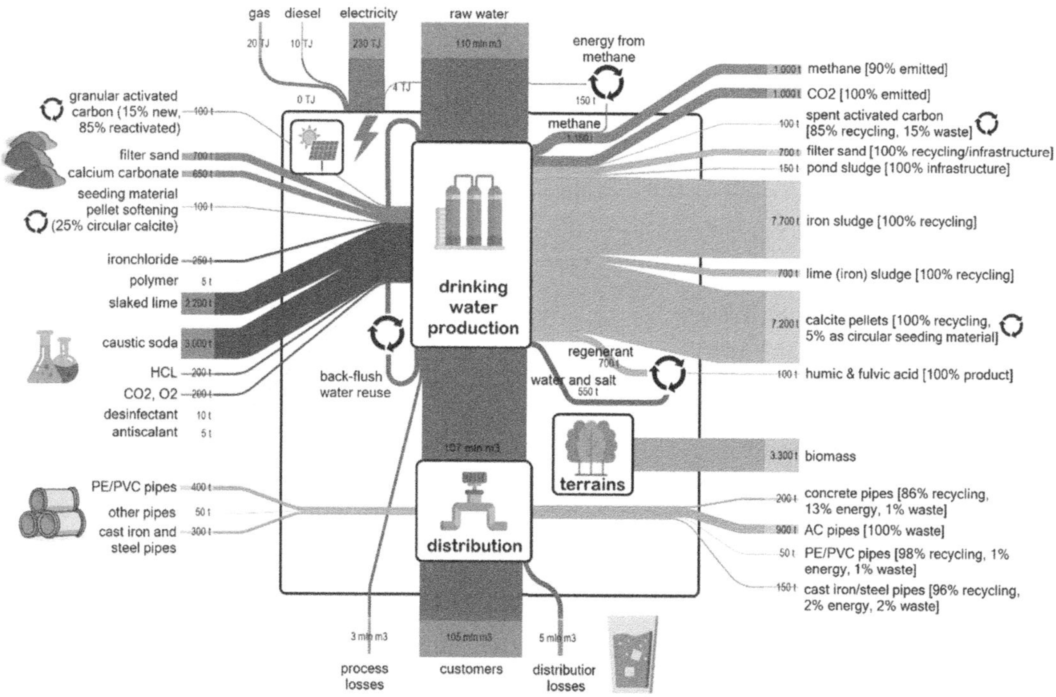

Figure 3.3 Sankey diagram of the material flows for drinking water production from ground water. The figure describes the average from 20 drinking water production facilities in the Netherlands for the year 2018 (photo courtesy: Aquaminerals).

demands and requirements. Moreover, especially in more moderate climates like part of the USA and Europe, there are also seasonal effects that affect the composition of a water source. In the warmer summer months, surface water often contains higher concentrations of natural organic matter, which results in a higher chemical coagulant dosing and thus increased sludge production. The difference in source water quality is important as there is evidently a strong correlation between the type, concentration and overall volume of residuals produced during the treatment processes and the quality of the source water. When the source water has a high load of contaminants, removing them leads in general to a larger quantity of residuals; mainly due to an increase in chemical dosing requirements to meet the required drinking water quality (i.e., drinking water production still heavily depends on the addition of chemicals). A good example to illustrate this is to compare the residual production rate using ground water and surface water as water source. The almost universal presence of suspended solids (and natural organic matter) in surface water, which is almost negligible in groundwater, induces higher coagulant dosage leading to rather large volumes of coagulation sludge. In a similar context, one could imagine that the amount of residuals produced during treatment of municipal wastewater is substantially higher due to the higher load of pollutants in the untreated sewage. Indeed, for example, for the Dutch situation, approximately 0.68 kg wet residual is generated per treated m^3 of municipal wastewater (http://statline.cbs.nl/) versus on average only 0.17 kg wet residual per m^3 drinking water produced from ground and surface water (personal communication, AquaMinerals 2018).

Regarding potential resources that can be recovered (Table 3.1), it is very important to realize that these resources often contain other impurities that in some cases need to be removed prior to beneficial reuse of the target resource. This depends on both the type and concentration of these

Table 3.2 Properties and product characteristics of resources recovered from water residuals and their potential end-use.

Residual	Color	Morphology/ Structure	Main Elemental Component	Other Substances/ Contaminants	End-use/Market Segment
Calcite pellets	White-Brown	Round pellets	$CaCO_3$	Fe, organic matter, Mn, Si	Glass, carpet, agriculture, feed, cosmetics, reuse in drinking water production, concrete (tiles), steel production
Lime sludge	White-Brown	Dewatered/ dried sludge	$CaCO_3$	Fe	Soil improvement, construction
Iron-rich sludge (produced through ground water aeration)	Brown-Red	Dewatered/ dried sludge	$Fe(OH)_3/Fe_2O_3$	Mn, As	Soil amendment, construction industry, sulfide control in sewers and digesters, constructed wetlands, P-removal surface water, As-removal drinking water
Iron-rich coagulation sludge	Brown	Dewatered/ dried sludge	Fe-complexes Organic matter Clay	Various	Soil amendment, construction industry, constructed wetlands, raw material for clay brick and cement production
Alum-rich coagulation sludge	(dark-) Grey	Dewatered/ dried sludge	Al-complexes Organic matter Clay	Various	Soil amendment, construction industry, phosphate removal, constructed wetlands cement production, alum-industry (melters)
Brine	Colorless-Yellow	Liquid	H_2OSalt	N, S	Reuse of the salt for regeneration of the resin
Humic-/ Fulvic Acids	Yellow-Black	Liquid-Powder	C-complexes	Salt	Biostimulant for crop growth, additive in feed

Note that also the filter sand/gravel (from sand filter units) as well as activated carbon (from activated carbon filtration) can be considered resources that can be recovered once they need replacement, but they are not listed in this table.

impurities and the envisaged end-use and the targeted market segment for the recovered product. Table 3.2 provides a summary of some existing end-use routes for recovery and reuse of water treatment residuals implemented by the water industry at full scale.

3.6 CHAPTER SUMMARY

In this chapter, we have discussed the major water sources used for the production of drinking water, namely, surface water, ground water, seawater and reclaimed effluent from wastewater treatment plants. The most common water purification methods currently implemented by the water industry have also been described. Furthermore, we have explained the different resources that are naturally present in these water sources themselves as well as resources that are present in the water treatment, residuals added during drinking water production and how and in which form (and concentration) these resources can be recovered. Lastly, the properties and product characteristics of resources recovered from water residuals and their potential end-use in different market segments have been described. Finally, we have also highlighted the importance of the volumes, concentrations and presence of other compounds/contaminants of the recovered resources on the resource recovery and valorization potential in various industry sectors.

3.7 EXERCISES

Exercise 3.1: In the Netherlands, some 0.17 ton (wet) residual per mega liter (ML) drinking water produced is generated. Calculate the amount of residuals produced per year in the Netherlands assuming a water usage per capita per day of 130 liters.

Exercise 3.2: Use the same residual production rate as in Exercise 3.1, but now calculate for the total amount of residuals produced on a yearly basis for the USA. Note that the average water consumption in the USA is over 400 liters per capita per day.

Exercise 3.3: Assuming an average coagulant dosing of either aluminum sulfate or iron chloride, what would the yearly production of coagulation sludge (in dry matter) be for the Netherlands, Australia, USA, India and China?

Exercise 3.4: Assuming an average coagulant dosing of either aluminum sulfate or iron chloride (i.e., $Al_2(SO_4)_3{\cdot}14H_2O$ and $FeCl_3{\cdot}6H_2O$), what would the yearly production of coagulation sludge similarly to Exercise 3.3 (in dry matter) be for two typical water treatment plant (WTP), for example 30 000 m^3/day (small to medium city) and 300 000 m^3/day (typical large plant).

Exercise 3.5: What are the key resources that can be recovered at drinking water treatment plants that are either naturally present in the various sources (i.e., ground water, surface and sea water) used for the production of drinking water as well as the resourced added to the drinking water production process? More specifically, what are the major resources that can be recovered from the following processes?

(a) Aeration of ground water;
(b) Coagulation-flocculation-sedimentation;
(c) Low-pressure membrane filtration;
(d) High-pressure membrane filtration;
(e) Ion-exchange.

Exercise 3.6: What are the most critical impurities in relation to the major resources that can be recovered identified in Exercise 3.4.?

Exercise 3.7: Provide a qualitative overview of the major resources identified in Exercise 3.4 that can be recovered from:

(a) Aeration of ground water;
(b) Coagulation-flocculation-sedimentation;
(c) Low-pressure membrane filtration;
(d) High-pressure membrane filtration;
(e) Ion-exchange.

Exercise 3.8: As discussed in this chapter, there is a clear distinction between the resources naturally present in the water sources and the resources added during the treatment processes. Design a scheme that provides an overview of resources added in the treatment processes and naturally present in different water sources used.

Exercise 3.9: A clay brick manufacturer produces 25 million bricks per year, each of them weighing about 2 kg. The manufacturer uses 10% m/m coagulation sludge from drinking water production in the brick production process. Let us assume that an average drinking water production site produces 7500 tonnes of sludge with a dry weight of 33.3% annually. How many drinking water production locations need to work together to meet the demands of this manufacturer?

3.8 DISCUSSION QUESTIONS

Question 3.1 (*drivers and market analysis*): A city of 500 000 inhabitants is evaluating the feasibility of providing iron-rich coagulation sludge from the its drinking water treatment plant for phosphate removal to its wastewater treatment plant. The WWTP currently uses iron chloride for the removal of phosphate. What data or information do you need as a basis for the feasibility analysis? What are the key criteria that could favour the use of coagulation sludge? What are the criteria that could favor the use of iron chloride coagulant?

Question 3.2 (*regulations, health protection and economic viability*): A progressive city council has firmly anchored its ambition to implement 'circular economy' by the year 2030. It is considering to apply alum-rich drinking water sludge as a soil amendment in its various parks and green areas. Its current practice is landfilling these materials as there is no clear interest from local industry to reuse this material and because of the uncertainty of the long-term impact on the soil quality (i.e. the soils are quite acidic and as such there is a potential for leaching of alum in the surrounding aquifers). You are asked to give a presentation in which you evaluate the risks and opportunities of using alum-rich sludge as soil amendment by the city council.

Question 3.3 (*technology, economy*): As the innovation manager of a large water utility, you are in charge of reorganizing the existing water infrastructure from its current situation to a more circular approach within a timeframe of 15 years. The current drinking water treatment infrastructure comprises two very large-scale drinking water treatment plants, with a capacity of 200 000 and 120 000 m^3/day for a total of 1 300 000 million inhabitants and local industry, both of which are using conventional coagulation-flocculation, sedimentation, filtration, and disinfection treatment processes. In addition, it comprises about 50 smaller drinking water treatment plants with a total capacity that equals that of the two very large-scale treatment plants. You are asked to give a presentation to the board of directors in which you evaluate the current status and justify your masterplan. Where would you focus on in terms of choice of resources to recover, location, and market segments/ end-uses? What are your key considerations/motivations?

Question 3.4 (*technology, economy*): As discussed in this chapter, in general, centralized water treatment using large drinking water production facilities is prevalent in major urban areas. Discuss the implications of this in terms of the economic potential of recovery of resources with subsequent market uptake and replication. What are the most critical non-technological aspects in order to make a successful link with the market?

Question 3.5 (*overall sustainability, social considerations*): Discuss the relative importance and impact of non-technological aspects in order to transform our drinking water infrastructure and overall management strategy as such that it perfectly fits within the circular economy. In the same context, discuss the importance of human behavior in relation to achieving sustainable water management in the 21st century, for example not using a resource may be even better than recovering resources. How can global change and resource scarcity drive the implementation of the circular economy?

Question 3.6 (*economy, product quality and volumes*): As discussed in this chapter, there are various sources for the production of drinking water including surface waters (i.e. rivers, lakes, and the sea), ground water and spring water. Which of these sources will, according to you, provide the best resources to recycle. In your answer discuss the purity of the recovered resources, economy of scale and total volumes.

Question 3.7 (*technology, regulation and social considerations*): Discuss the similarities and differences between resources that can be recovered from drinking water treatment plants and municipal wastewater treatment plants. In your answer, discuss the following aspects: (semi-)public owned, mid-to long term planning, quality and quantity, public perception and hygiene.

Question 3.8 (*supply chain considerations and robustness*): The supply chain of resources recovery from drinking water production is considered to be stable and reliable in cases where the transport/delivery of these resource is undisrupted and reliable to the end-user or processor over a prolonged period of time. However, in practice situations can take place that cause disruption to the supply chain. Discuss and analyze the following situations/scenarios: (i) failure or maintenance at the production site, (ii) change of treatment steps changing the volume and/or quality of the residual, (iii) scaling the production up or down (e.g. during a dry season or when (dis-)connecting a neighborhood or changing the source of the water.

Question 3.9 (*supply chain considerations, circular resources and end-user*): Calcite can be recovered during drinking water production in cases where a softening step is required. Calcite can be ground, sieved and sterilized and can be re-used within the treatment process as seeding material, calcinated calcite can be use as new quicklime for the softening process. Discuss the terms supply chain considerations, end-user and process stability in relation to the benefits of reusing the calcite.

Question 3.10 (*product quality, impurities end-user*): A benefit of using ion-exchange resins for the removal of natural organic matter (NOM) is that upon regeneration of the ion-exchange resins, a concentrated NOM solution is generated that can be beneficially used as a biostimulant in agricultural applications for plant growth promotion. Discuss the terms supply chain considerations, end-user and product quality in relation to the economic benefits and practical feasibility. With respect to the product quality, in particular discuss the potential implications of the fact that regeneration of the ion-exchange resins is typically achieved by using a concentrated salt solution.

FURTHER READING MATERIALS

AWWA & ASCE. (2012). Water Treatment Plant Design, 5th edn. McGraw-Hill, New York.

Babatunde A. O. and Zhao Y. Q. (2007). Constructive approaches toward water treatment works sludge management: an international review of beneficial reuses. *Critical Reviews in Environmental Science and Technology*, **37**(2), 129–164.

Bratby J. (2006). Coagulation and Flocculation in Water and Wastewater Treatment. IWA Publishing, London, UK.

Matilainen A., Vepsäläinen M. and Sillanpää M. (2010). Natural organic matter removal by coagulation during drinking water treatment: a review. *Advances in Colloid and Interface Science*, **159**(2), 189–197.

Smeets P. W. M. H., Medema G. J. and Van Dijk J. C. (2009). The Dutch secret: How to provide safe drinking water without chlorine in the Netherlands. *Drinking Water Engineering and Science*, **2**(1), 1–14.

WHO. (2017). Guidelines for Drinking-water Quality, 4th edn, incorporating the 1st addendum. World Health Organization, Geneva.

REFERENCES

Baldursson S. and Karanis P. (2011). Waterborne transmission of protozoan parasites: review of worldwide outbreaks – an update 2004–2010. *Water Research*, **45**(20), 6603–6614.

Cornwell D. A. (1999). Water treatment plants residuals management. In: Water Quality and Treatment, R. D. Letterman (ed.), McGraw Hill, New York, pp. 16.1–16.51.

Crittenden J. C. and Borchardt J. H. (2012). MWH's Water Treatment Principles and Design. John Wiley & Sons, Hoboken, N.J.

Edzwald J. (2010). Water Quality and Treatment. McGraw-Hill Professional Publishing, New York.

Feitelson E. and Rosenthal G. (2012). Desalination, space and power: the ramifications of Israel's changing water geography. *Geoforum*, **43**(2), 272–284.

Hargesheimer E. E. and Watson S. B. (1996). Drinking water treatment options for taste and odor control. *Water Research*, **30**(6), 1423–1430.

Liu J. and Zheng C. (2016). Towards integrated groundwater management in China. In: Integrated Groundwater Management, A. J. Jakeman, O. Barreteau, R. J. Hunt, J. D. Rinaudo and A. Ross (eds), Springer, Cham, Switzerland, pp. 455–475.

McAdam E. J. and Judd S. J. (2008). Biological treatment of ion-exchange brine regenerant for re-use: a review. *Separation and Purification Technology*, **62**(2), 264–272.

Pikaar I., Sharma K. R., Hu S., Gernjak W., Keller J. and Yuan Z. (2014). Reducing sewer corrosion through integrated urban water management. *Science*, **345**(6198), 812–814.

doi: 10.2166/9781780409566

Chapter 4

Water reuse: a pillar of the circular water economy

Valentina Lazarova

Water Globe Consulting, 36 rue du président Wilson A56, 78230 Le Pecq, France

4.1 INTRODUCTION

The human civilization of the 21st century is changing the Earth in ways that threaten its ability to operate within the planetary boundaries within which humanity can continue to develop and thrive for generations to come [1]. Water resources, water cycle and water security are challenged by climate change, increased demand and polluted water supplies. Other global changes (e.g., urbanisation, changing consumption patterns, deforestation, and intensification of agriculture) add to these challenges. According to a recent report from the United Nations [2], almost half of the world's population currently live in areas vulnerable to water scarcity and more than six billion people could suffer water shortages by 2050. It is suggested that this number may be an underestimation, and scarcity of clean water may be worse, in particular in developing countries.

To secure and preserve fresh water, a new era of water management is emerging – the era of 'One-Water' – where the old distinct barriers of water and wastewater are fading and where water in all of its states is looked upon as a valuable commodity and precious resource that has to be closely monitored, accounted for, and recycled/reused whenever possible, rather than being considered as a linear product, a waste after use [3]. Nowadays, the water supply planning paradigm is evolving from solely relying on conventional fresh water resources (i.e., mainly groundwater and surface water from rivers and lakes) towards building an environmentally sustainable diversified water portfolio where low-cost, conventional water sources are balanced with more costly but also more reliable and sustainable water supply alternatives such as recycled or desalinated water [4]. This builds on centuries of experience in reusing water for agriculture, often without much treatment but still adding considerable value to the water cycle [5].

Indeed, water reuse is becoming a cornerstone of sustainable water management and urban planning and is considered to be a key chain-link of the emerging circular economy [6]. Circular economy and rational, responsible, renewable and sustainable use of water resources are closely intertwined. Looking beyond the current take-make-dispose extractive linear approach, a circular economy aims to redefine growth, focusing on positive society-wide benefits by recovering and reusing resources, thereby creating a closed loop. This is also considered one of the most efficient ways to solve another important challenge of the 21st century – the water-energy nexus – better preserving our environment and maximizing energy and resource recovery from wastewater [7].

It is important to highlight the enormous advances in science and technology in the last decade(s) that have greatly contributed to the implementation of new more efficient and reliable wastewater treatment trains. The latter has resulted in the adoption of different methods and approaches that have created diversified safe water reuse practices [8]. Indeed, producing water from wastewater (i.e., often referred to as reclaimed water) of a specified (fit-for-purpose) quality to fulfil multiple water use objectives has become a reality due to the progressive evolution of water reclamation technologies, regulations, and environmental and health risk protection measures. Today, technically proven water reclamation and purification technologies are producing pure water of almost any quality desired, including purified water of quality equal to or higher than that of drinking water. A diverse portfolio of technologies for water scarcity management is available, and such technologies range from basic reclamation with natural systems for lower quality uses to multi-barrier systems incorporating membrane filtration, reverse osmosis and advanced oxidation when higher quality is required.

While the focus in the past has been on technology advancement, other crucial enabling factors have been given less attention [8]. Such enabling factors align with the three key pillars of sustainable development: social, economic and environmental aspects of water management and reuse. For water reuse projects, key barriers and opportunities include diverse aspects that link the technological challenges and these three pillars. Issues include economic aspects such as water price and affordability, social aspects related to public education and acceptance, corporate social responsibility and strategy, policy and regulation frameworks, and environmental aspects, which are often the main drivers for water reuse.

This chapter presents the basic principles and technologies enabling water recycling, identifies the main lessons emerging from large projects and full-scale water reuse experiences around the globe and illustrates how to resolve the major technical, economic, social and economic challenges for the successful implementation of water reuse practices. Water reuse is an essential element of water resources management, which is strongly supported by the availability of efficient and reliable technologies, but it requires the political decision to implement it and adequate management strategies to address the social, economic and environmental issues associated with water reuse. A specific emphasis is given on water reuse drivers, water reuse planning, selection of 'fit-to purpose' water reclamation technologies, safety (management of health, environmental, agronomic and technology risks) and economic viability.

4.2 LEARNING OBJECTIVES

At the completion of this chapter the reader should be able to:

- Understand what water reuse means and define its role in the context of sustainable water management and circular (water) economy.
- Describe the main drivers, challenges and benefits of water reuse.
- Explain the main steps of water reuse project development and explain how to select relevant water reuse applications.
- Identify and characterise potential water reuse schemes.
- Select 'fit to purpose' treatment technologies, key design criteria and water quality parameters to ensure safe water reuse practices.
- Evaluate key specific challenges related to economic viability, social acceptance and health or environmental risks.

4.3 WATER REUSE AS A KEY PILLAR OF THE CIRCULAR (WATER) ECONOMY

First of all, it is important to understand the water reuse vocabulary. Several different terms are used as synonyms to describe the beneficial use of treated wastewater such as water reuse, water reclamation and water recycling [9, 10]. As a rule, wastewater reclamation was used in the past to indicate the preliminary treatment of wastewater, compared to some reuse schemes where untreated sewage or poor-quality wastewater was used for land application [5]. Because today wastewater treatment is becoming mandatory and the general treatment requirements are determined by the intended use and related health risk, the produced 'design water' is no longer considered to be wastewater and the term of reclamation (i.e., treating and processing wastewater in order to make it suitable for beneficial reuse) is losing its key importance. Water recycling is selected as the most appropriate term by regulations in some regions of the world (e.g., Australia, California) in order to reflect the notion of treatment and circularity, that is water reuse. Moreover, other terms and concepts reflecting the water value are emerging, such as NeWater in Singapore and One Water in the USA [3, 8].

In the context of increasing water stress and scarcity, the most difficult challenge for long-term sustainable development is to find alternative ways to manage water resources. The conventional linear approach of water management is replaced by the 'One Water' concept of integrated management of systems for water recovery and recycling (Figure 4.1) [7, 8]. The large investment needed for traditional engineering practice and management of the anthropogenic water cycle (i.e., dams, wells, water transportation, treatment and supply, wastewater collection and treatment, etc.) could be easily repaid by the benefits from the recovery of purified water.

The long-term experience in water reuse – with recent great successes of new water reuse schemes and applications – demonstrates that water recycling provides an important and economically and technologically viable opportunity to augment traditional water sources [8, 10]. As a multi-disciplinary component of water resources management, water reuse can help to close the loop between water supply and wastewater disposal. The successful development of this alternative drought-proof water resource depends upon close coordination of water policies and regulation, infrastructure and facilities planning, water quality management and wastewater treatment process selection and reliability.

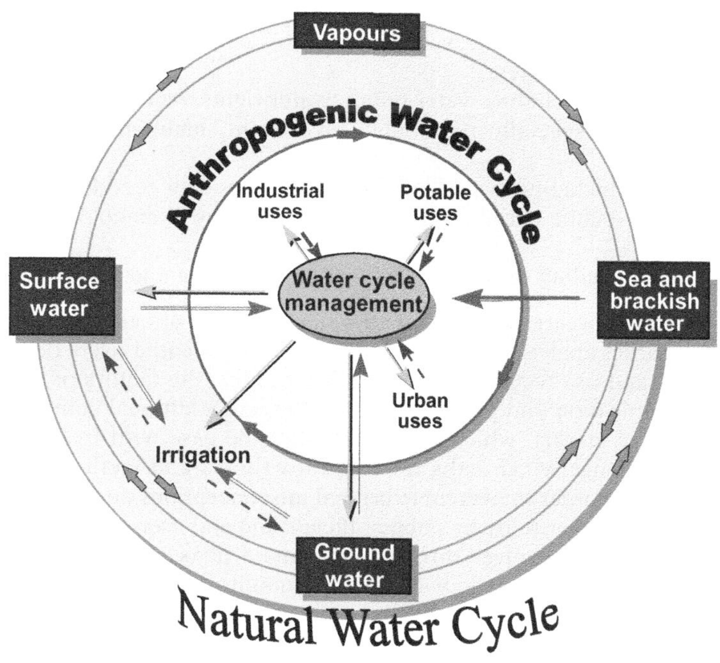

Figure 4.1 Sustainable water cycle management based on water recycling, energy and nutrient recovery, and efficient environmental protection.

4.4 WATER REUSE PLANNING

Water reuse planning is an iterative process following the step-by step approach of water resource management [10]:

(1) Identifying the key drivers and objectives;
(2) Gathering background data in terms of available water resources, water demand and supply, wastewater management, potential end users of recycled water, policies and regulations, economics, and so on.;
(3) Defining alternatives, including water reuse;
(4) Evaluating and ranking alternatives, including risks, costs and benefits analysis;
(5) Selecting alternatives for implementation.

The key components of successful water reuse planning include not only the technical know-how and good engineering design/practice, but also a rigorous market analysis and economic, environmental and social considerations.

The main element to be considered during water reuse planning is a clear and comprehensive definition of project drivers and the ability of water recycling to resolve existing problems. On the basis of the analysis of local conditions favouring water reuse (e.g., water scarcity, water demand projections and water supply deficit, wastewater treatment and disposal, new sewer systems, new tourist, agricultural or economic developments, etc.) and water reuse market assessment, different alternative scenarios will be formulated and the best solutions will be chosen by means of a multi-criteria analysis.

The multi-criteria screening and evaluation of alternative reuse and non-reuse options of water management typically include the following feasibility criteria [8]:

(1) Engineering feasibility, particularly the possibility for implementation of wastewater treatment, storage and distribution systems;
(2) Health and environmental impacts, such as potential health risks and negative effects on soils, groundwater, crops or ecosystems;
(3) Institutional feasibility, including water policy, water rights, regulations, and enforcement;
(4) Economic feasibility, especially reasonable investment, maintenance and operation costs (Capex and Opex);
(5) Financial feasibility, including available funding and subsidies, recycled water rates;
(6) Social impacts and public acceptance, mainly support by stakeholders, end users and the public at large;
(7) Market feasibility, including who will be using recycled water and at what conditions.

The first six criteria on the list are common for all water resources projects. The new crucial element for water reuse is the market analysis and the identification of potential users of recycled water and their willingness to pay and use recycled water. In this respect, the factors of public health, water quality, adequate communication and public acceptance create additional complexity in identifying and securing a water reuse market, which is generally not the case with freshwater supply. Social acceptance is a great challenge which calls for innovative tools and methodologies for enhanced public education/outreach in order to overcome critical misconceptions and concerns of health risks and liability associated with water reuse by public officials and end users.

Water reuse planning usually evolves through three main phases: (1) project conceptualisation and feasibility study, which could include water quality monitoring and pilot studies for treatment train selection and economic evaluation, (2) advanced project solutions (APS) and scoping, with detailed process specifications, flow diagrams and technical-economic evaluation, and (3) facility planning and construction followed by commissioning with process performance and water quality validation.

4.4.1 Key water reuse drivers

Water reuse is a global trend, which requires local solutions. In fact, water reuse drivers are strongly influenced by local climate conditions, water demand and availability, demographic and economic growth, and political and economic context. The typical drivers of water reuse projects are as follows:

(1) Water scarcity, such as chronic water deficits and/or more frequent and prolonged droughts due to climate change, as well as declining quality of available water resources, leading to imbalances between water availability and water demand;
(2) Increasing water demand and water deficits due to population growth, intensive agriculture, accelerated urbanisation and economic development, which ultimately result in unsustainable water supply and impeded business development;
(3) Environmental protection, such as the improvement of river water quality, bathing zones, sensitive habitats and wetlands;
(4) Political and policy decisions in response to a water crisis, as a solution to create climate change resilience, or the ambition of governments to transition into the emerging circular economy.

4.4.2 Key water reuse challenges

The key issues related to the implementation of water reuse projects, their ranking and some of the foreseeable impediments is highly dependent on location and specific local conditions. The major water reuse challenges are [8]:

- economic viability;
- social acceptance: public perception and support by users and local authorities;
- policy and regulations;
- technical feasibility and energy efficiency;
- innovation and fast implementation of new tools, technologies and good engineering practices.

4.4.2.1 Economic viability

Securing economic viability is an important challenge for most water reuse projects [8, 10]. Unfortunately, water reuse feasibility is often suppressed by the use of undervalued and/or subsidised conventional water resources (as a rule, water tariffs do not include the full cost of water resource and supply management, as well as wastewater collection, treatment and discharge). Full-cost recovery of water reuse is a desirable objective but depends on ability to pay. The cost-benefit analysis of water reuse projects must include other management objectives and socio-environmental criteria, based on a holistic approach and catchment scale (e.g., multi-criteria analysis including climate change impacts, environmental impacts, water quality and resource depletion, economic development, etc.).

The implementation of wastewater reclamation facilities generally requires a substantial capital investment. While water reuse is considered a sustainable approach that can be cost-effective in the long run, the additional treatment and monitoring requirements, as well as the construction of recycled water distribution systems, could be costly as compared to water supply alternatives such as imported water or groundwater. As such, government grants or subsidies may be required to implement water reuse. Unfortunately, institutional barriers, as well as varying government and community priorities, can make it difficult to implement large scale water reuse projects in some cases.

4.4.2.2 Social acceptance

Independent of the type of reuse application and country, the knowledge and understanding of the public with respect to the safety and suitability of recycled water is a key factor for the success of any water reuse programme [8, 10–12]. Consistent communication and easy to understand messages need to be developed for the public, with politicians explaining the benefits of water reuse for long-term water security and sustainable water resources management, as well as the efficient control of adverse health, agronomic and environmental impacts.

To date, the major emphasis of water reuse has been on non-potable applications such as agricultural and landscape irrigation, industrial applications (e.g., use as cooling and boiler make-up water), and on residential or commercial building applications such as toilet flushing in large buildings. Understandably, potable reuse raises more public concerns. In any case, the value of water reuse is weighed within the context of greater public issues. The implementation of water reuse projects continues to be influenced by diverse debates surrounding drought and availability of water, growth vs. no growth, urban sprawl, traffic noise and air pollution, perception of recycled water safety, and public policy governing sustainable water resources management.

4.4.2.3 Policy and regulation

The development and enforcement of water reuse standards is an essential step for the social acceptance of water recycling. However, in some cases, regulations could be a challenge and burden for water reuse, as in the case of very stringent requirements based on the precautionary principle. For example, the health risk-based regulations for irrigation, such as those developed in Australia and used as the basis for new WHO and European regulations, require an additional health risk assessment (Quantitative Microbial health Risk Assessment, QMRA) and validation of log removal of treatment technologies, in addition to water quality monitoring. These new requirements lead to significantly higher permit and operation costs, without any guarantee for lower health risk or better process reliability. A recent review [13], performed by a large group of leading experts, demonstrated that the treatment-based approach used for years in the United States, in particular in California, does not increase public health risk and that modifying the standards to make them more restrictive will not improve public health. Water reuse standards must be adapted to the country's specific conditions (administrative infrastructure, economy, climate, etc.), should be economically viable and should be coordinated with the country's water conservation strategy.

4.4.2.4 Technical issues and energy efficiency

Technical challenges facing water reuse are not yet completely resolved. In particular, for industrial, urban and potable water reuse applications, it is extremely important to improve performance, efficiency, reliability and cost-effectiveness of treatment technologies. Water recycling facilities are facing tremendous challenges related to high variations in raw water quality, salinity spikes due to seawater or brackish water intrusion into sewers, and variations in water quantity caused by extreme conditions of very limited water demand, flooding or the need for alternative disposal of recycled water [11]. It is not the objective of this chapter to discuss each of these challenges in detail. We would like to refer the reader to the following excellent literature in which these technological challenges are discussed in detail [10].

Energy efficiency, carbon and environmental footprints have gained importance in recent years [7, 14]. The ambitious goals of sustainable development and of achieving zero net carbon and pollution emission footprints call for a new holistic approach to the management of the water cycle with an increased role for water reuse. With increasing urbanization in many parts of the world and further growth of megacities and increasing efforts to optimise energy efficiency, water recycling is of growing interest and is expected to become an important pillar of the circular economy. Decentralised or semi-centralised water distribution systems are more efficient for future cities when water reuse is inevitably considered [15]. Water supply can be tailored to match water demand more closely in centralised water infrastructures, adapting water quality to the given use.

The energy consumption for water reclamation is significantly lower than that for seawater desalination and represents only a fraction of the energy consumed for water supply, treatment, and distribution [7]. Nevertheless, energy-intensive processes such as reverse osmosis should be limited to high quality applications, whereas alternative solutions may be more energy efficient for irrigation use.

4.4.2.5 Innovation

Demonstration studies and scientific advances will play a key role in realizing widespread acceptance and adoption of water reuse practices as a key water management strategy. The roadmap of technology innovation in water reuse should be focused on the development of reliable 'practical' solutions, in order to unlock the regulatory, economic and social barriers for building a cost competitive water reuse market. The major actions included in such a roadmap that would be easy to implement over the short term include [16]:

(1) Improvement of reliability, performance, flexibility and robustness of existing technologies;
(2) Development of new cost effective and energy efficient technologies;
(3) Development of new tools and methods for improved water quality and process performance monitoring;
(4) Advancement and implementation of 'soft science' innovation to resolve the socio-economic challenges of water reuse.

A good example of an innovative solution is the implementation of direct potable reuse [17, 18]. While there has been a clear preference for non-potable and indirect potable reuse applications, a number of factors make it less feasible to further increase water reuse in these applications. Most of the economically viable non-potable reuse opportunities have been exploited. For example, the typical cost for parallel distribution of tertiary-treated recycled water is 0.3–1.7 USD/m^3, whereas the typical cost for highly-treated purified water, which could be delivered directly into the distribution system, is 0.6–1.0 USD/m^3 [17]. It is inevitable that purified water will be used as a source of potable water supply in the future – as now 'de facto' reuse is a common practice.

4.5 WATER REUSE APPLICATIONS

Water reuse practices can be classified into two main categories: *non-potable* and *potable* water reuse. The most important characteristics, key issues and lessons learned for water reuse practices are summarised in Table 4.1. The most common applications of recycled water include agricultural irrigation, landscape irrigation, industrial reuse and groundwater recharge. Among these, agricultural and landscape irrigation are widely practiced throughout the world and have well-established health protection guidelines and agronomic practices [8, 10].

Case study: Producing designer recycled water tailored to customer needs in West Basin, California, United States [19]

The Edward C. Little Water Recycling Facility in El Segundo, located near the Los Angeles International Airport, is one of the largest reclamation facilities in the world and the only one which is producing five distinct types of recycled water for various applications. The plant was expanded step by step, enabling it to consistently demonstrate economic viability, treatment efficiency and water quality compliance. A number of fundamental and applied research projects contributed to the rapid implementation of innovative tools, methods and technologies. A comprehensive public outreach and education programme enabled it to achieve large public support and acceptance and increase the number of end users. This experience is one of the best examples of successful integrated resource management in urban and coastal areas with water reuse.

Key figures: Start-up in 1995, 160-km network, five distinct types of 'designer' recycled water
Treatment capacity: 132 500 m^3/d

Key to success: Reliable drought-proof local resource, creative pricing strategy, cost competitiveness, step by step implementation, public outreach and education

Benefits: Increased water supply security, potable water produced by an advanced multiple barrier approach, secured social and economic development

Tertiary multi-barrier treatment trains: (1) conventional tertiary Title 22 treatment of secondary effluents for urban uses by means of flocculation, multi-media filtration and chlorine disinfection, (2) additional treatment of this conventional recycled water by nitrification/chlorination for cooling water production, (3) additional polishing of conventional recycled water by microfiltration (MF) and two stages of reverse osmosis (RO/RO) for boiler make-up water, (4) advanced treatment of secondary effluents by MF/RO/RO for boiler make-up water in industrial reuse, and (5) barrier product water for aquifer recharge against salt intrusion where the MF/RO is supplemented with advanced oxidation (AOP) by high dose UV with hydrogen peroxide addition.

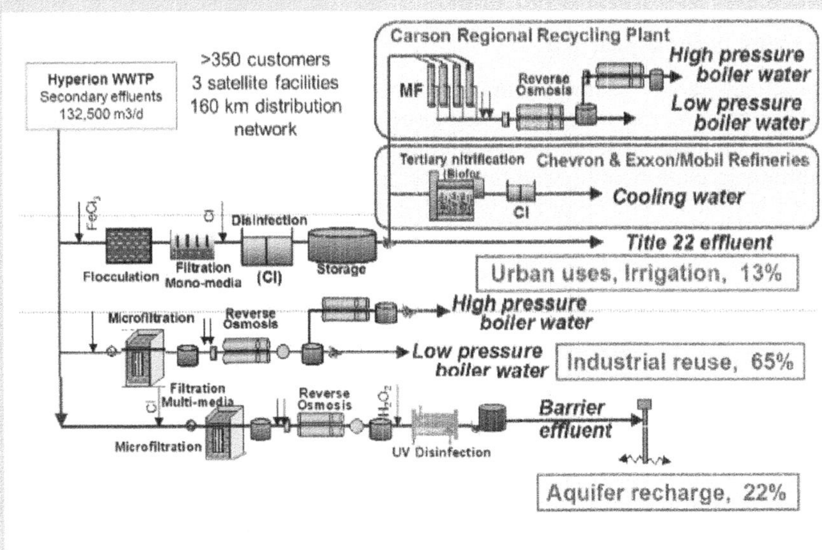

Energy saving: Approximately 10% of the energy demand is covered by solar energy. During each step of expansion, new more competitive technologies have been implemented, such as the replacement of RO pre-treatment from pressure-driven MF by vacuum-driven MF.

Major challenges: Developing separate facilities and infrastructure, meeting water quality requirements for industrial customers, declining secondary effluent quality (increasing salinity, TOC, turbidity, etc.)

Table 4.1 Categories of municipal wastewater reuse applications and related issues or constraints (adapted from [8, 10]).

Category		Potential application	Issues/constraints	Lessons learned
Non-potable water reuse				
Agricultural irrigation	Unrestricted or restricted	Food crop eaten raw Food crop processed or cooked Pastures for milk production Orchards, vineyards with or without contact with edible fruits Fodder and industrial crops Ornamental plant nurseries	Water quality impacts on soils, crops, and groundwater Runoff and aerosol control Health concerns Farmers' acceptance and marketing of crops Storage requirements	Good practices available to mitigate adverse health and agronomic impacts (salinity and sodicity) Storage design and irrigation technique are important elements Numerous reported benefits
Landscape irrigation		Golf courses and landscape Public parks, school yards, playgrounds, private gardens Roadway medians, roadside plantings, greenbelts, cemeteries	Water quality impacts on ornamental plants Runoff and aerosol control Health concerns Public acceptance Water quality control in distribution systems	Successful long-term experience Good agronomic practices On-line water quality control can ensure health safety Numerous benefits
Urban uses		In-building recycling for toilet flushing Landscaping (see irrigation) Air conditioning, Fire protection Commercial car/truck washing Sewer flushing Driveway and tennis court washdown Snow melting	Health concerns Control of water quality and biological growth in distribution systems Cross-connection control with potable water Cost of distribution systems	Dual distribution systems require efficient maintenance and cross-connection control No health problems reported even in the case of cross-connections (for tertiary disinfected reclaimed water)
Environmental/ Recreation uses	Unrestricted or restricted	Recreational impoundments Environmental enhancement (freshwater or seawater protection) Wetlands restoration Fisheries Artificial lakes and ponds Snowmaking	Health concerns Eutrophication (algae growth) due to nutrients Toxicity to aquatic life	Emerging application with numerous benefits for the cities of the future: improving living environment, human wellbeing, biodiversity, and so on. On-line water quality control can ensure health safety

(Continued)

Table 4.1 Categories of municipal wastewater reuse applications and related issues or constraints (adapted from [8, 10]). (*Continued*)

Category		Potential application	Issues/constraints	Lessons learned
Industrial reuse		Cooling water Boiler feed water Process water Heavy construction (dust control, concrete curing, fill compaction, and clean-up)	Scaling, corrosion and fouling Biological growth Cooling tower aerosols Blowdown disposal	Water quality to be adapted to the specific requirements of each industry/process Request for high reliability of operation, cost and energy efficiency
Potable water reuse				
Indirect potable reuse with replenishment of:	Aquifers	Groundwater replenishment by means of infiltration basins or direct recharge by injection wells Barrier against brackish or seawater intrusion (direct recharge) Ground subsidence control	Health concerns Groundwater contamination Toxicological effects of organic chemicals Salt and mineral build-up Public acceptance	Successful practice since 1970s Multiple barrier treatment ensures safe potable water production Efficient control by means of advanced modelling tools
	Reservoirs	Surface reservoir augmentation Blending in public water supply reservoirs before further water treatment	Health concerns Public acceptance Eutrophication (algae growth) due to nutrients	Successful practice since 1970s Multiple barrier treatment ensures safe potable water production Improvement of water quality
Direct potable reuse		Pipe-to-pipe blending of purified water and potable water Purified water is a source of drinking water supply blended with source water for further water treatment	Health concerns and issues of unknown chemicals Public acceptance Economically attractive in large scale reuse and chronic water scarcity Environmental buffers	Multiple barrier treatment ensures safe potable water production No health problems related to recycled water in Namibia since 1968 Cost efficient compared to indirect potable reuse

Indirect potable reuse, and in particular aquifer recharge, has been implemented in many countries as an efficient response to the need to increase water supply. One of the key advantages of aquifer recharge is the complementary polishing and storage of recycled water in an environmental buffer, thus improving not only water quality but also public acceptance. Finally, direct potable reuse is emerging as a solution for regions subjected to water scarcity. As mentioned previously [2], water demand to 2050 will probably grow even more than the recent forecasts, in addition to the reduction of water quality of natural resources. Local patterns will be more critical than global patterns, making the problem more difficult to solve. In this context, direct potable reuse could be a relevant, safe and cost competitive solution, as demonstrated by the long-term (over 50-year) experience in Namibia [18].

4.6 WATER REUSE TREATMENT AND DESIGN

Wastewater treatment is the most critical element of any water reuse scheme because it is the most cost-effective way to reduce health and environmental risks associated with the use of recycled water. The terms of primary (including preliminary screening), secondary and tertiary treatments are commonly used to indicate the main wastewater treatment steps for the removal of grit and suspended solids followed by biological treatment for the removal of organics and nutrients (N and P) and additional polishing (Figure 4.2). As a rule, tertiary treatment includes disinfection and, if necessary, the removal of residual suspended solids and/or nutrients when very stringent discharge limits for nitrogen and phosphorus are in place. These conventional tertiary treatment trains are commonly used for the majority of non-potable reuse applications, such as agricultural and landscape irrigation, environmental enhancement and for some other urban uses.

In comparison to conventional wastewater treatment, more stringent requirements are applied with respect to water source characterisation/monitoring, treatment process selection and design of water recycling schemes [10, 20]. Secondary treatment is mandatory for the majority of water reuse regulations, with the exception in cases where preserving the fertilisation capacity of reclaimed water is of utmost importance, for example in Mexico for agricultural irrigation [21].

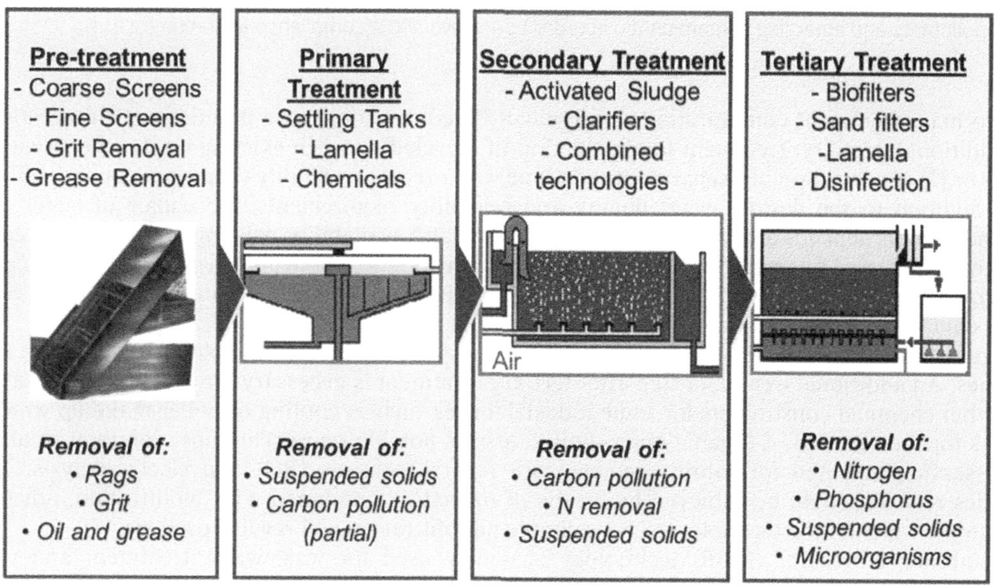

Figure 4.2 Wastewater treatment steps and typical processes used to removal various pollutants.

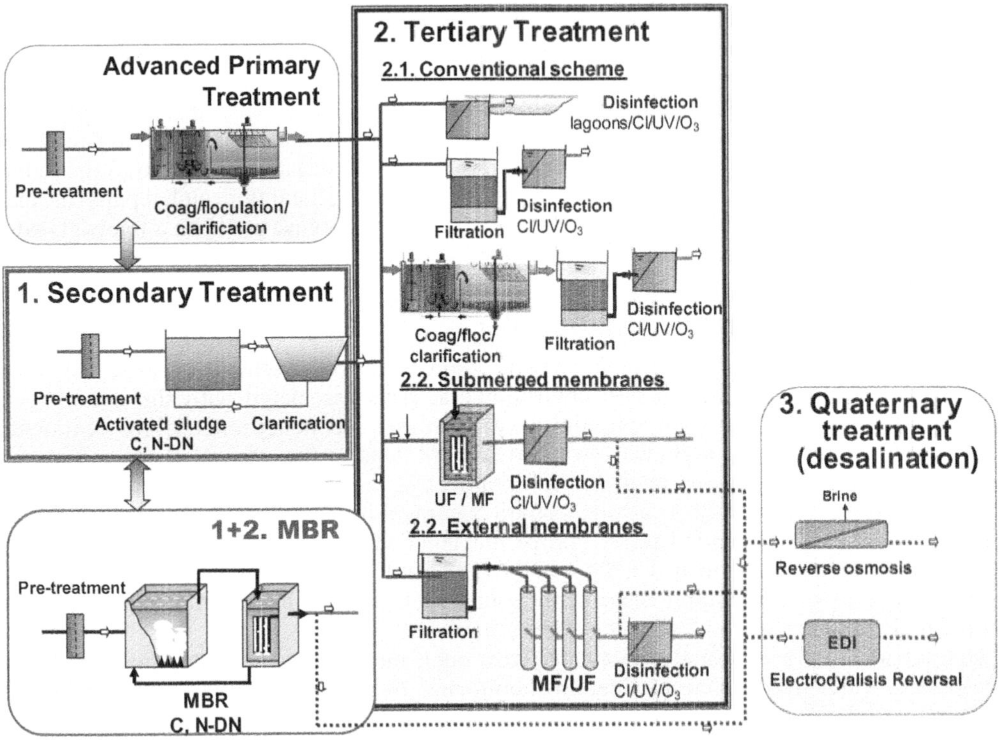

Figure 4.3 Typical combinations of wastewater treatment processes for water reuse. In the case of potable reuse, it is of utmost importance to implement multi-barrier treatment approaches, for example the use of several treatment processes for the removal of each contaminant class of concern, including pathogens, waterborne viruses, organic micro-pollutants and emerging contaminants, in order to achieve 100% compliance all the time.

Two main treatment configurations are typically used in water reuse practices: (1) integration of the additional (tertiary) treatment for production of recycled water in existing wastewater treatment plants or (2) construction of a separate stand-alone water recycling facility treating secondary effluent.

In addition to the desired water quality and reliability requirements, the choice of water reuse treatment trains depends on size of the treatment plant, land availability, existing infrastructure, energy and labour costs and financial resources. In selecting appropriate treatment for a given reuse application, the provision of multi-barriers is a very important consideration, in particular in the presence of risk of direct contact with recycled water as well as for creating a social license to operate [10].

Figure 4.3 summarises some typical treatment processes that have been implemented in water reuse schemes. An additional treatment step after tertiary treatment is necessary for the removal of salinity and other chemical constituents for some industrial uses such as cooling or boiler make-up water, as well as for the removal of organic micro-pollutants for potable reuse. The most common treatment processes implemented for salinity removal are reverse osmosis (RO) and electrodialysis. Trace organics removal could be achieved by means of ozonation combined with biofiltration, advanced oxidation, or membrane technologies, for example nanofiltration and reverse osmosis.

Membrane bioreactor (MBR) technology is widely used for wastewater treatment and reuse, providing the double advantage of secondary and tertiary treatment for production of high-quality recycled water, improving the environmental footprint of wastewater treatment [22].

4.6.1 Typical treatment trains for agricultural and landscape irrigation

Depending on the type of irrigated crops, irrigation method and public access, additional treatment of secondary effluent can include a disinfection step with or without preliminary filtration (Figure 4.4a). Enhanced and almost total removal of suspended solids is needed to consistently achieve very stringent water quality standards with near-complete removal of microbial pollution (i.e., indicator microorganisms below the detection limits).

Depending on the quality of the secondary effluent, chemically enhanced filtration or a combination of high rate clarification (i.e., coagulation, flocculation and lamella settling) and filtration could be needed [10, 20]. Disinfection is the final polishing step, with chlorination and UV irradiation as typical processes. For this application, some extensive treatment technologies based on natural purification processes such as maturation ponds have the important advantage of providing storage, in addition to their ability to remove pathogens (disinfection by UV light, e.g.).

Good disinfection without carbon and nitrogen removal can be achieved by advanced primary treatment using high rate sedimentation followed by filtration and disinfection (see Figure 4.3).

Membrane treatment is not economically viable for agricultural irrigation, but it is increasingly applied for landscape irrigation and other urban uses [12, 23].

The most common tertiary treatments for agricultural reuse are direct disinfection of secondary effluents by means of chlorination or UV irradiation, with a preliminary filtration in the case of stringent standards.

4.6.2 Typical treatment trains for urban reuse

Urban uses are characterised by relatively high risk of exposure (i.e., direct contacts or cross connections with recycled water) compared to agricultural irrigation. For this reason, the commonly applied treatment trains aim to achieve a quasi-total removal of suspended solids and microorganisms (similar to irrigation of crops for raw consumption). Interest in using ozonation for this application has increased in recent years because of its high effectiveness in removing viruses and bacteria, combined with colour removal and oxidation of some organic micropollutants.

As mentioned previously, MBRs and membrane filtration are commonly used technologies enabling the production of high-quality recycled water. An additional advantage of the use of membranes is the 'physical barrier' of membranes, thereby enhancing public acceptance. The typical reuse applications using membrane technologies are in-building water recycling (Figure 4.4b), car washing, air-conditioning and recreational uses [10, 12, 24].

Figure 4.4 (a) Typical treatment levels for agricultural water reuse and (b) View of in-building recycling plant in Battery Park, New York City.

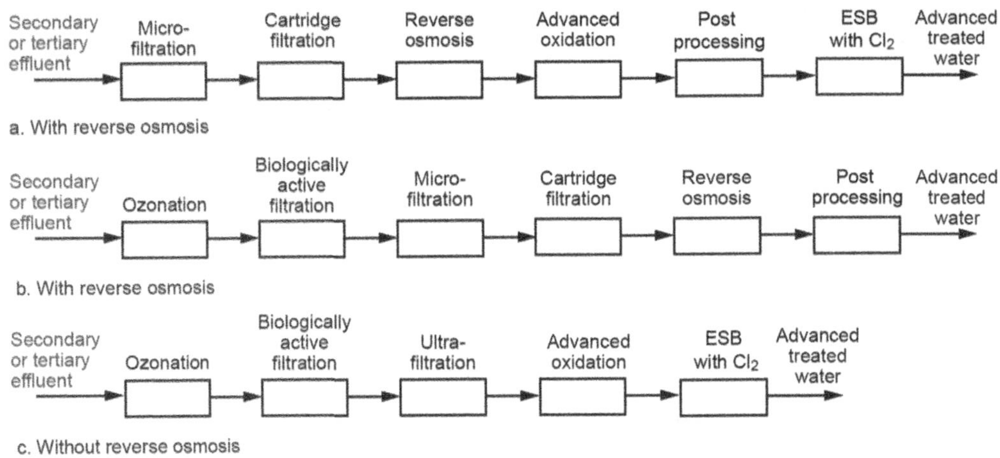

Figure 4.5 Example of technologies used for indirect and direct potable reuse.

4.6.3 Typical treatment schemes for potable reuse

As mentioned previously, potable reuse treatment trains are typically designed based upon the so-called multi-barrier concept. However, it should be noted that some indirect potable reuse projects, such as aquifer recharge using surface spreading and replenishment of storage reservoirs or dams, require less complex tertiary treatment (filtration plus disinfection) because of the additional treatment capacity of natural soil aquifer treatment (SAT).

The treatment trains for indirect potable reuse via direct aquifer recharge and direct potable reuse include at least 2–3 additional barriers for pathogen and trace organic removal to consistently achieve drinking water quality (Figure 4.5, adapted from [17]). Dual membrane treatment by low pressure membranes (microfiltration or ultrafiltration) and reverse osmosis, followed by advanced oxidation (e.g., UV plus hydrogen peroxide) is becoming very popular and is being considered as the best available technology in California. The management of brine from reverse osmosis is the main problem for such schemes, in particular in inland locations. For this reason, an increasing interest is reported in conventional advanced treatment trains for trace organics removal through a combination of ozonation, biological activated carbon, ultrafiltration or nanofiltration, and advanced oxidation [17].

4.6.4 Typical process design

Process design of water reuse treatment is determined based on both the source water quality and the target water quality [10, 20]. In addition to water quality specifications, the process performance is largely influenced by the robustness and reliability of the selected treatment technologies, as well as by the operation and maintenance.

Table 4.2 illustrates some typical water quality and process design parameters of the most commonly used treatment processes. It can be seen that the design parameters change depending on water reuse applications [adapted from 10, 20].

Physical-chemical disinfection is strongly influenced by inlet water quality, and in particular by suspended solids concentration [25, 26]. The impact of water quality on disinfection performance is highest for UV irradiation, followed by chlorination (mainly due to by-product generation) and ozonation, which was identified as the least sensitive process.

The most viable design of ozone contactors for wastewater treatment should be based on short contact times and enhanced mass transfer [27, 28]. Good mixing and dose control are the key design

Table 4.2 Indicative design parameters and microbial log removal of selected treatment processes.

Treatment process	Design for target water quality			Indicative microbial log removal		
	≤10 *E. coli*/100 mL	≤200 *E. coli*/100 mL	≤1000 *E. coli*/100 mL	*E. coli* and enteric bacteria	Viruses	Protozoa
Secondary treatment	NA			1–3	0.5–2	0.5–1.5
Membrane bioreactors	Achievable by MF/UF		NA	6	4–6	6
Non-potable water recycling trains						
Chemically enhanced dual media filtration	<5 mg TSS/L, turbidity 2–5 NTU Polyelectrolyte dosage 0.5–1.5 mg/L Filtration rates 10 ± 5 $m^3/m^2 \cdot h$ (up to 25 for pressure filters)		NA	0–1	0.5–3	1–3
Maturation ponds (typical depth of 0.5–1.5 m)	NA	10–30 mg TSS/L (30–50% of algae) Residence time 15–30 days		2–5	1–4	3–4
Chlorination[a] (chlorine dose/CT 15 mg min/L)	1–5	2–4	2–9 mg/L	2–6	1–3	0.5–1.5
UV irradiation (UV dose)	<2.5 mg TSS/L, <2 NTU, 100–120 mJ/cm^2	<5 mg TSS/L, 60 mJ/cm^2	<10 mg TSS/L, 30–40 mJ/cm^2	2–6	1–6	3–6
Ozonation, transferred ozone dose	2–5 mg/L	4–8 mg/L	6–10 mg/L	2–6	3–6	1–4
Potable water reuse recycling trains						
Reverse osmosis	Below the detection limit	NA	NA	6	6	6
Advanced oxidation (UV + H_2O_2)		NA	NA	6	6	6
Ozone plus biological activated carbon		NA	NA	4–6	4–6	2–4
Soil-aquifer treatment		NA	NA	6	6	6

NA, not applicable.
[a]chlorination of nitrified effluents.

parameters of wastewater disinfection systems, with hydraulic residence time being less important compared to drinking water treatment.

An indicative range of microbial removal is also provided in Table 4.2 for reference, according to recent guidelines [29, 30]. It is important to underline that the demonstration of log removal by operational monitoring is very difficult and even impossible in some cases due to low initial concentrations, low accuracy of available analytical methods and the specificity of treatment technologies.

Challenge testing at bench and pilot scale runs allow high log removal levels to be demonstrated by spiking influent with higher than normal microorganism concentrations to allow detection in the effluent. Nevertheless, this method is limited to selected processes (e.g., membranes, disinfection) and could be very expensive or inapplicable at full-scale.

4.7 KEY CHARACTERISTICS AND MILESTONES OF AGRICULTURAL WATER REUSE

The majority of water reuse projects worldwide are implemented for agricultural irrigation and are driven by increasing water scarcity and ever-growing agricultural water demand. The need for alternative water resources has been accelerated over the past few years by severe droughts, which have occurred and continue to occur not only in traditionally arid areas of the United States, the Mediterranean region, the Middle East, Australia and South Asia, but also in a number of temperate-climate states and countries in Europe and North America. For example, according to an FAO study [31], the drought of 1999 in the Near East and North Africa (NENA) region resulted in a relative decline of food production by 51%. Another good example of the economic repercussions is the 2003 drought in Europe, with estimated economic losses exceeding €13 billion [32]. In this context, water reuse is becoming more and more valued as a viable alternative water resource. Indeed, various countries are already using a large portion of their treated wastewater for irrigation. Examples include, but are not limited to, Argentina, China, Cyprus, Egypt, Israel, Jordan, Kuwait, Libya, Mexico, Saudi Arabia, Spain, Syria, Tunisia and United Arab Emirates.

The basic principle of beneficial water reuse in agriculture is that municipal wastewater may be used for all kinds of irrigation applications as long as the water has been previously treated to appropriate levels that meet specific water quality requirements [10, 20, 33]. It is worth noting that besides the well-recognized benefits of water reuse, the use of recycled water for irrigation may have adverse impacts on public health and the environment, depending on treatment level, local conditions and irrigation practices. In all cases, the existing scientific knowledge and practical experience can be used to lower the risks associated with water reuse by implementation of sound planning and effective management of irrigation practices with recycled water [34, 35].

The milestones in the development of safe water reuse practices for agricultural irrigation are illustrated in Figure 4.6 [8]. The proposed three categories of irrigation projects are based on the

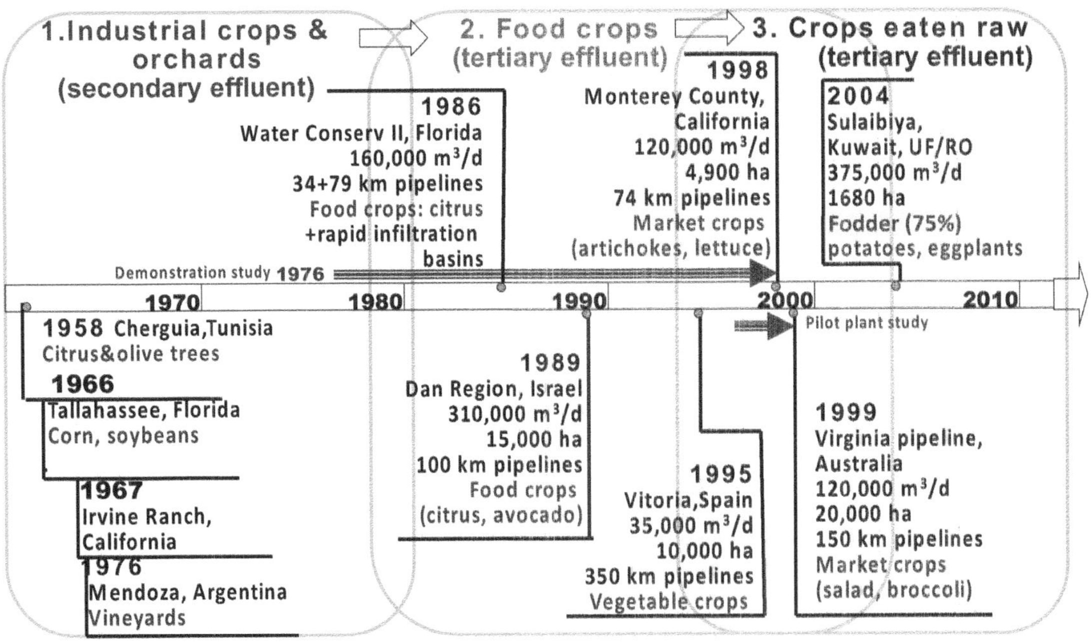

Figure 4.6 Milestones in water reuse for agriculture with selected cornerstone projects.

history of agricultural water reuse and reflect the evolution of regulations and advances in wastewater treatment technologies and monitoring:

(1) Irrigation of industrial crops, fodder and seed crops, orchards, forests, and so on., irrigated with secondary effluent often after storage and polishing in open lagoons such as maturation ponds. Implementation of large projects in the Unites States and Tunisia in the 1960s, in Argentina in the 1970s, and in Spain and in France in the 1980s.
(2) Irrigation of food crops (eaten cooked or processed) with tertiary effluents. In the 1980s, two large projects in Florida (e.g., Water Conserv II) and in Israel (Dan Region) demonstrated the safety and benefits of water reuse by means of extensive scientific studies.
(3) Unrestricted irrigation of crops consumed raw with highly-treated and disinfected recycled water (tertiary filtered and disinfected effluent or ultrafiltration-treated secondary effluent). Long-term extensive research, completed as part of the implementation of water reuse projects in Monterey, California [35] and Virginia pipeline, Australia [36], has demonstrated the safety of recycled water and has convinced stakeholders of the benefits of water reuse.

Many successful water reuse projects for agricultural irrigation have been developed worldwide. However, the projects presented in Figure 4.6 are very important for the success of water recycling because they are used as benchmark examples.

Case study: Virginia Pipeline, Adelaide (Australia) – Irrigation of crops eaten raw [36]

The Virginia pipeline scheme is one of the largest projects for agricultural reuse in the world and an excellent example of mitigation of water scarcity impacts for the sustainable development of local economy and food security. Lessons learned from this project demonstrate the possibility to consistently achieve high quality standards by means of conventional treatment technologies, as well as the economic viability of water reuse.

Key figures: start-up in 1999, irrigation of 20 000 ha (root and salad crops, cauliflower, broccoli, cabbage, wine grapes and olives), 250 supply agreements, 150 km of pipes + 20 km extension in 2009

Tertiary treatment (110 000 m^3/d): maturation pond, coagulation/flocculation, dissolved air flotation, filtration and chlorination (<10 *E. coli*/100 mL)

Key to success: city initiative and national program to assist growers to produce more with less water; guidance manuals for farmers

Economic viability: long term 'take or pay' contracts, recycled water charges are covering O&M costs, +50% increase of land value, AU$120 million vegetable production

4.7.1 Water quality requirements

Water quality requirements for agricultural reuse depend on the type of irrigated crops (food or industrial crops, eaten raw or after processing, potential contact with recycled water, etc.) and the risk of contact with workers and the general public (restricted or unrestricted irrigation depending on the access control to irrigated areas). The most stringent standards apply for irrigation of crops eaten raw, requiring almost full removal of suspended solids and disinfection (from <1 to <10 *E. coli*/100 mL to 100–200 *E. coli*/100 mL).

4.7.2 Health risk management

Despite the numerous success practices, preventing health risks (human or animal) remains a critical issue [10]. Efficient health risk management in the context of water reuse in agriculture includes several types of actions that can be classified in four main groups:

(1) Policy, regulations and institutional initiatives, including set-up of water quality criteria and their enforcement;
(2) Engineering actions such as wastewater treatment and storage, adequate operation and water quality monitoring, control of recycled water application, and in particular, the selection of irrigation method and technology;
(3) Additional health protection measures and barriers such as crop restrictions, human exposure control and immunisation of field workers;
(4) Good agronomic practices, including crop selection, control of the timing and frequency of irrigation, crop harvesting measures, and so on.

It is very important to underline that the safe and beneficial implementation of water reuse schemes are significantly facilitated by the development of appropriate codes of good agronomic and irrigation practices, which are as important for farmers and operators as water quality and monitoring requirements [34].

4.7.3 Major findings and lessons learned

Based upon the large amount of water reuse projects for unrestricted agricultural irrigation implemented around the globe, the key lessons learned can be summarised as follows [8]:

(1) Keys to success:
  - local and national incentives, subsidies, creative economics and water recycling pricing, demonstration of economic benefits;
  - efficient and reliable treatment methods and good engineering practices;
  - communication, education and training of farmers.

(2) Major challenges:
  - food safety and public perception are very important issues for farmers (influence of *E. coli* outbreaks such as the fresh spinach issue in the USA in September 2006 resulting in a revenue loss of over \$74 million and the 'cucumber' crisis in Europe in May 2011 resulting in a revenue loss for farmers of over €600 million);
  - agronomic aspects: salinity, sodicity and toxic ion management should be taken into account [33];
  - operation and maintenance (O&M) of irrigation networks (e.g., biofouling, bacterial recontamination or regrowth control).

4.8 KEY CHARACTERISTICS AND MILESTONES OF URBAN WATER REUSE

Water reuse in urban areas includes a wide variety of applications and schemes, with a common denominator that all of these purposes do not require potable water quality. The main categories of urban water reuse are as follows:

(1) Landscape irrigation, which is the primary use and includes irrigation of public parks, sport fields, green belts, golf courses, as well as private residential areas and gardens.
(2) Other non-irrigation urban uses such as street cleaning, car washing, fire protection, air conditioning, toilet flushing and some commercial applications.
(3) In-building recycling, which refers mostly to water recycling in high-rise buildings, including office buildings, commercial malls and private residential buildings.
(4) Environmental enhancement and recreational uses for replenishment of water bodies, lakes and urban streams including those used for swimming (with or no body contact), leisure or fishing purposes.

As a general rule, urban water reuse requires adequate infrastructure, and in particular dual distribution (or reticulation) and dual plumbing systems. Dual distribution and plumbing systems are relatively easy to install in new urban areas or buildings with relatively low initial investment. Because of the risk for the community of direct contact with recycled water, the water reuse requirements, as well as the rules for cross-connection control, are very stringent (i.e., total disinfection and on-line control) [10].

4.8.1 Landscape irrigation

Landscape irrigation includes applications of recycled water to restricted and unrestricted areas, the latter requiring the highest water quality. Agronomic water requirements are also similar to those for agricultural irrigation and depend on a crop's sensitivity. For example, irrigation of ornamental plants requires careful analysis of agronomic water quality parameters, while the majority of turfgrass species are very tolerant to salinity and sodicity [37]. Typically, the recycling scheme components for landscape irrigation are more sophisticated than for agricultural irrigation because of the use of water efficient irrigation techniques (e.g., sprinklers, micro-sprinklers and drip irrigation), which may require additional filtration after storage, backflow prevention devices, pressure control and automation [10]. Similar devices and efficient operation and cross-connection control practices are required for dual distribution networks.

Relatively few cross-connection incidents with backflow from recycled water systems have been reported with no reported illness. The major causes of cross connections between recycled and potable water systems reported in Australia, the United States and the UK are:

- illegal connection of private residences;
- inadequate construction, records and pipe identification; and
- higher pressure in recycled water system.

4.8.2 Other urban uses

Water recycling in urban areas, including toilet flushing, is becoming an acceptable practice, with confirmed operating experience and recognised economic viability for: (1) large municipal scale toilet flushing and irrigation, (2) large office buildings (e.g., Japan, California-United States) and (3) large residential areas (e.g., Florida-United States, Australia) [38]. Because of the high importance of this type of reuse for urban water cycle management and for cities of the future, the major milestones in urban reuse are classified not only by application or technologies, but also by the size and scheme of implementation (Figure 4.7):

- recycling in high-rise buildings;
- large-scale dual distribution systems;
- new concepts for eco-cities with low water and energy footprint.

4.8.3 Main steps and milestones in the development of urban reuse

Selected cornerstone water reuse projects of urban water reuse are presented in Figure 4.7 [8].

In 1977, the city of St. Petersburg in Florida implemented one of the largest urban dual water systems of its kind in the world. Even though the initial objective was to avoid wastewater discharge

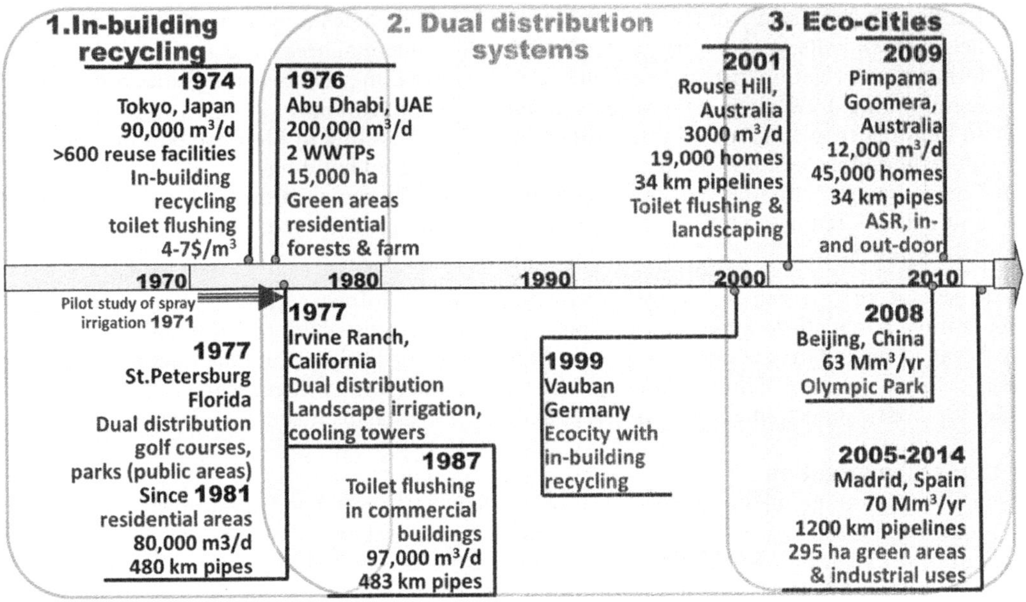

Figure 4.7 Milestones in urban water reuse with selected cornerstone projects.

in surface waters, the water recycling systems have become an essential element of urban water cycle management [10]. The operational experience and research and development (R&D) studies contributed to the development of good practices for optimum watering and selection of ornamental plants, as well as for the increase in cost efficiency by using a looped distribution network with pipes of small diameter.

The history of recycled water use in individual buildings in Japan dates back to 1965 and earlier, although the number of installations was small [39]. A severe drought in 1978 prompted the use of recycled water in buildings in Fukuoka in southern Japan. There have been about 100 new installations annually in Japan since 1980, with more than 1475 individual office buildings and apartment complexes equipped with on-site water recycling systems in 1997. In 2009, 9% of the total volume of treated wastewater was recycled.

New concepts of sustainable water management and eco-cities with reduced potable water demand by means of dual distribution systems are under development in other countries, and particularly in Australia. On the basis of the operational experience of the first large residential project in Rouse Hill in Sydney, a new project in Pimpama Goomera in Queensland was aiming to cover 45% of the water demand of 45 000 residential homes using recycled water, lowering the drinking water demand to about 16% of its typical baseline level. However, in 2013 the city voted to stop supply of A^+ recycled water to households over a three-year period due to the high operation and monitoring cost, which has far outweighed the revenue.

Case study: Water Recycling in nine high-rise buildings in Shinjuku Area, Tokyo [39]

Japan has one of the best and oldest experiences in high-rise building recycling in the world. Few technical and economic constraints favoured the development of large semi-decentralised water recycling schemes, whose first milestone is the project in Shinjuku Area in Tokyo. High quality recycled water is supplied for toilet flushing in 30 high-rise office buildings.

The project success is based on very efficient control and management of water storage and supply, meeting large hourly variations of water demand 100% of the time, an adequate pricing strategy and public education.

Since 1984, 1400 m^3/d recycled water, expanded to 3000 m^3/d

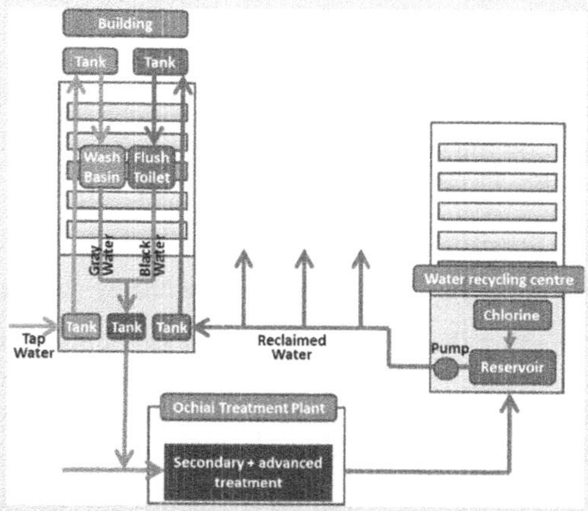

- **Semi-decentralised** (urban district) tertiary treatment of secondary effluents: sand filtration followed by chlorination
- **Water quality**: no *E. coli* detected and <3 total coliforms/100 mL
- **Key to success**: early planning, low transportation distance, political support, cost savings for owners
- **Major challenges**: reliable recycled water supply despite high variation of hourly water demand, aesthetic quality without colour and odour

4.8.4 Major findings and lessons learned

Based upon the extensive knowledge gained from the various projects of urban water reuse implemented around the globe, the key lessons learned can be summarised as follows [8]:

(1) Keys to success:
- governmental and regional incentives;
- consistent water quality and supply reliability;
- community education, communication and market assessment.

(2) Major challenges:
- water quality: health safety (total disinfection), aesthetic parameters, others (nutrients, dissolved salts, etc.);
- reliability of supply: interruptible in many cases (fire protection, toilet flushing, cooling) with backup water system and storage capacity for irrigation purposes to meet peak demand;
- costs: very high cost of dual distribution systems (prohibitive for >15–25 km), dual plumbing viable for new buildings, additional expenses for potable water when a substitution is needed;
- O&M of dual systems: cross-connection control, management of leaks, corrosion, scaling, and bacterial regrowth.

4.9 KEY CHARACTERISTICS AND MILESTONES OF INDUSTRIAL WATER REUSE

Industry is the second largest market for water supply after agriculture (70%) with around 25% of global freshwater demand. Industrial reuse and internal recycling schemes within factories have become standard practice in many countries and industries. Moreover, many industries have set forward ambitions to achieve closed loop cycles and zero liquid discharge as a long-term strategic goal, whereas inter-sector water reuse, and in particular the use of recycled urban wastewater for industrial purposes, is growing rapidly. The potential for industrial reuse of treated municipal wastewater is expected to increase even further in the future as raw potable water supplies become more limited, the cost of potable water increases due to more demanding standards and the discharge regulations become more stringent.

The major factors that influence the potential for industrial water reuse include availability of water, the industry's discharge requirements, the industrial process requirements, input water quality, volume, economics and reliability. Although there is a wide range of industrial water uses, the major uses are:

- cooling system make-up water;
- boiler feedwater (boiler make-up water);
- process water;
- washdown water;
- miscellaneous other uses, including site irrigation, fire protection, road cleaning, and so on.

The first three categories are of particular interest because they are high-volume and high-quality applications with excellent prospects for using recycled municipal wastewater. In general, the closing of the industrial water cycle can be divided into three main strategies around water saving and wastewater minimisation initiatives:

(1) Cascading reuse, involving direct reuse with no or little treatment;
(2) Wastewater recycling after appropriate treatment;
(3) Source reduction by decreasing the need for water in a given industrial process.

4.9.1 Water quality requirements

The water quality requirements and type of application of water recycling is highly affected by the type of industry (e.g., power generation, food and beverage, chemical, paper, textile industry), particular industrial process, and target performance (e.g., impacts on product quality, on installations, etc.). For these reasons, it is not possible to generalise water quality requirements for industrial applications, which in some cases can be more stringent than drinking water standards.

4.9.2 Management of adverse water quality impacts

The greatest concern of the use of recycled water in cooling towers is the risk posed by inhalation of pathogens entrained in the aerosols. In this case, biocides can be added, but they cause risks for corrosion and inclusion in aerosols. Moreover, corrosion, scaling and biofouling problems of equipment and distribution systems are a common issue. As a rule, for water to be suitable to be used as feedwater for boilers or stream generators, it must be of a higher quality than water used for cooling purposes [11]. The quality requirements increase as operating pressure and temperature increase. The control or removal of hardness is required. Insoluble salts of calcium and magnesium are the main contributors to scale formation in boilers and are removed by processes such as ion exchange and reverse osmosis. The development of an adequate treatment line from the initially available water to the process water is crucial to keep costs under control without affecting the process.

4.9.3 Milestones in industrial reuse

Successful water reuse practices are implemented in the petroleum industry, oil refineries, thermoelectric power generation plants, pulp and paper facilities, the textile industry and even microprocessor, electronic and food industries. The milestones in industrial water reuse can be categorised in two groups, similar to indirect potable reuse, with the cornerstones being the implementation of membrane technologies, and in particular reverse osmosis.

The West Basin Municipal Water District in California is a pioneer in the production of 'designer' recycled water from municipal wastewater, from which three qualities are produced for industrial purposes. Since the first start-up of combined MF/RO treatment of municipal wastewater at this plant in 1995, the long-term operational experience has demonstrated that recycled water provides a reliable source of water for industry when potable water supply is uncertain at lower cost [40]. Several other similar recycling facilities have been constructed in Singapore, Mexico, India, Northern California and Hawaii [8]. The German experience in industrial water recycling has demonstrated that there are technical possibilities for water recycling for almost any application [41]. Key factors for the success of industrial water recycling are treatment efficiency and reliability, as well as technical and economic feasibility. As a result, water-recycling schemes have reduced wastewater volume by 78–92% in textile, paper and food and beverage industries.

Case study: Water recycling in refineries and naphtha cracker industry using advanced multi-barrier system [42]

This project is a very relevant example showing how advanced water recycling technologies are contributing to Indian industrial development, and in general, to emerging economies securing water supply. The production of ultra-pure recycled water for boiler make up is enabling an expansion of refinery capacity with the objective to achieve zero liquid discharge in the mid-term. Effective management of a number of technical and water quality control challenges have been developed to ensure reliable provision of process water and environmental protection.

Start-up in 2006, Paniprat refinery recycling plant, India
Key figures: treatment capacity 42 500 m^3/d; 9 Mm^3/yr
Additional treatment: UF + RO
Water quality: <10 μg/L silica, <0.1 mg/L TDS
Key to success: economic growth in India, environmental awareness, advanced technologies, experienced plant operator
Major challenges: fluctuating water quality, stringent water quality requirements, zero liquid discharge to be achieved

4.9.4 Major findings and lessons learned

Based upon the extensive knowledge gained from the various industrial reuse projects successfully implemented worldwide, the major conclusions can be summarised as follows [8]:

(1) Keys to success:
- consistent water quality, efficient water quality control and reliable uninterrupted water supply;
- implementation of proven advanced technologies, in particular membrane bioreactors, membrane filtration (micro- and ultrafiltration, reverse osmosis), and ozonation;
- availability of funding, economic success of the clients/investors;
- cost efficiency and heat, energy and resource recovery;
- availability of an adjacent municipal wastewater plant to avoid an extensive and expensive reclaimed water distribution network.

(2) Major challenges:
- fluctuations of raw water quality and quantity;control of water quality, in particular salinity, silica and hardness;
- zero liquid discharge target in the long-term;
- disposal and treatment of RO brines and other concentrates.

4.10 KEY CHARACTERISTICS AND MILESTONES OF POTABLE WATER REUSE

4.10.1 Key points in development of potable reuse

The success story of potable water reuse started in the early 1960s with water augmentation by means of groundwater (aquifer) recharge in Montebello Forebay, California (1962) and direct potable reuse in Windhoek, Namibia (1968). The more than 50-year operational experience of direct 'pipe-to-pipe' reuse in Windhoek demonstrated the practical feasibility and lack of adverse health effects of this practice [18]. Public opposition and concerns of unknown micropollutants were the main constraints for the development of this water reuse practice.

The most common practices of planned indirect potable reuse include aquifer recharge and reservoir replenishment. Indirect potable reuse occurs when some fraction of the raw water used for drinking purposes is of wastewater origin. Unplanned indirect potable reuse is a common situation in which inherent water quality issues are not fully addressed because of the divided responsibilities for wastewater discharge and downstream water supplies [42].

Because of health concerns, a 'multiple barriers' treatment concept is applied in order to achieve an extremely high degree of reliability, much higher than that of conventional wastewater treatment and reuse schemes. The feedback from full-scale operation (as well as scientific studies) has demonstrated that human health can be safeguarded, and, moreover, long-term robust and reliable operation can be achieved using adequate treatment, water quality control and good practices of managed aquifer recharge [10, 43, 44].

4.10.2 Milestones in potable water reuse

The key milestones in the development of safe water reuse practices for indirect potable reuse are illustrated in Figure 4.8, adapted from [8]. The two proposed categories are based on the advance in wastewater treatment coming from the technological breakthrough of membrane filtration in the late 1990s.

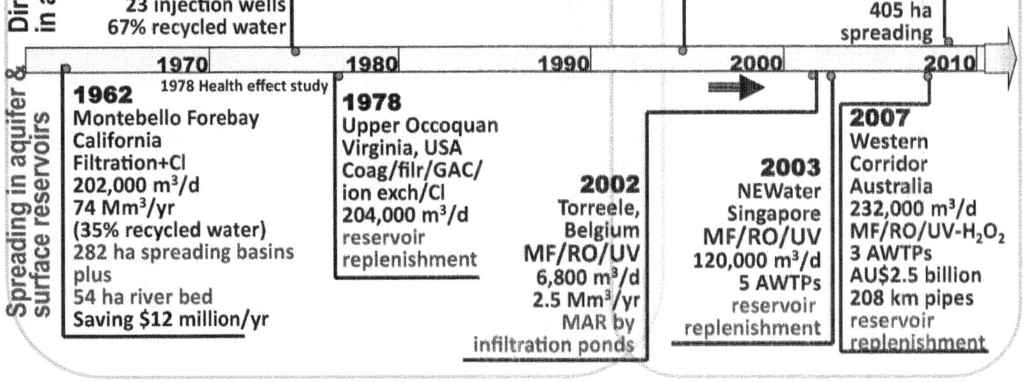

Figure 4.8 Milestones in indirect potable water reuse with selected cornerstone projects.

Two subcategories, covering both aquifer recharge and reservoir replenishment, are dissociated according to the water quality requirement:

(1) surface spreading in aquifers and surface reservoir replenishment; and
(2) direct injection in aquifers, mostly as a seawater intrusion barrier.

In fact, the operational experience demonstrated that, for all indirect water reuse applications, the advancement in membrane efficiency with decreasing cost favoured the implementation of advanced multi-barrier treatment schemes, including the combination of micro- or ultrafiltration, reverse osmosis, and advanced oxidation by high UV dose and hydrogen peroxide.

The significance and the lessons learned from recharge of the unconfined aquifer of Montebello Forebay in the eastern Los Angeles County are of crucial importance for the success of this practice. In fact, the first experience started in 1962 with the recharge of spreading basins with disinfected secondary effluent from the Whittier Narrows Water Recycling Plant, which demonstrated the economic viability and health safety of this water reuse practice [45]. A five-year health effect epidemiological and toxicological study (1978–1983) and a follow-up epidemiological study (1996–1999) did not

Case Study: Comparison of over 40–50 years of experience with potable water reuse

The two selected projects are excellent examples of successful implementation of potable water reuse by means of conventional treatment technologies and multi-barrier approaches. Comprehensive water quality monitoring programmes are consistently demonstrating the robustness and reliability of treatment trains and compliance with drinking water standards. During such long-term operation, no reclaimed water-related health problems were experienced. These projects illustrate well the human dimensions of water reuse and the role of public education, appropriate communication, adequate pricing strategy, good evaluation of water reuse benefits and the decisive role of decision makers and government support.

Case study 1: Indirect potable reuse for the Occoquan reservoir replenishment near Washington D.C., United States [46]

Start-up in 1978

Key figures: treatment capacity 204 400 m^3/d, 44 Mm^3/yr (8% of the annual average reservoir inflow)

Multi-barrier conventional treatment, including biological nitrogen removal, lime clarification, multimedia filtration, granular activated carbon adsorption and chlorination/dechlorination

Drivers: need to ensure long-term ability to meet regional water supply needs, federal construction grant funding

Key to success: state and local government support and collaboration, independent monitoring, continued demonstration of improvements to water quality over project life

Benefits: high quality reclaimed water has enhanced both watershed and reservoir water quality, reduced local capital cost of initial project

Major challenges: increased emphasis on controlling pollution from urban runoff, developing local support for more expensive reuse infrastructure, future expansions will all be locally funded

Case study 2: Direct potable reuse of the New Goreangab Reclamation Plant, Windhoek, Namibia [18]

Start-up in 1968

Key figures: treatment capacity 21 000 m^3/d, 6.4 Mm^3/yr

Advanced multi-barrier treatment, including pre-ozonation, coagulation, flotation, dual media filtration, ozonation, BAC, GAC, ultrafiltration and chlorination

Drivers: severe water deficit, polluted surface water, other alternatives that are too expensive

Key to success: vision and great dedication of the potable reclamation pioneers, excellent information policy and education practice, no reclaimed water-related health problems experienced, reliable operation and on-line process and water quality control, public-private partnership

Benefits: increased water supply security, savings from produced potable water, secured social and economic development

Major challenges: optimum management of the multiple-barrier approach (non-treatment, treatment and operational barriers), increasing salinity, concern of emerging micro-pollutants, increasing water demand

demonstrate any measurable adverse effects on groundwater quality or the health of the population drinking this water.

4.10.3 Major findings and lessons learned

Based upon the extensive knowledge gained from the various potable reuse projects successfully implemented worldwide, the major findings can be summarised as follows [8]:

(1) Keys to success:
  - pilot plant testing and selection of efficient and reliable treatment technologies based on a multiple-barrier approach;
  - government and stakeholder collaboration on adopting water reuse projects;
  - continued demonstration of consistent water quality, independent water quality monitoring, on-line process and water quality control, and further water quality improvement over the project life;
  - effective public outreach, community education, communication and use of positive terminology (e.g., 'purified water', NEWater, 'eco-water');
  - governmental grant funding, subsidies, phasing of expansions and public-private partnerships.

(2) Major challenges:
  - public support and regulatory approval;
  - high capital and operational costs, costly monitoring for emerging micropollutants and other constituents;
  - increased emphasis on controlling pollution from urban runoff for both aquifers and surface reservoirs;
  - management of the complex membrane facilities during extreme conditions such as low production rates or flooding;
  - increased salinity and pollution of wastewater, in particular in coastal areas and aged wastewater treatment plants.

4.11 THE COST-RISK NEXUS

Health risk analysis in water reuse is based on drinking water regulations. The enormous progress in research on process engineering and water quality monitoring, as well as the feedback from long-term operation of water reuse projects, represent an excellent background for the development of safe water reuse practices. Nevertheless, the increasing knowledge in water quality monitoring and analytical equipment may adversely impede water reuse development due to very stringent regulatory requirements that are impossible to implement in practice.

Risk analysis consists of three principal elements: risk assessment, risk management, and risk communication [10]. Risk assessment is the qualitative or quantitative characterisation and estimation of potential adverse health effects associated with human exposure to recycled water. Qualitative risk assessment is based on a combined evaluation of the magnitude of consequences and the likelihood that those consequences can occur. For non-potable reuse projects, qualitative risk assessment is the most appropriate and economically feasible methodology [47]. Risk management is the process of evaluating, and if necessary, controlling sources of exposure and risk in order to ensure that water of a safe quality is provided to end-users.

Chemical and microbial risk assessment is used with success to set drinking and water reuse standards. Nevertheless, a number of serious limitations exist when applying risk assessment to water reuse [48]. Several important simplifying assumptions are needed, including the selected model, the limited availability of exposure and dose-response data, the health outcomes investigated and treatment of the variability and uncertainty of the data used to calibrate the selected model. In addition, it is important to underline that wastewater microbial content varies considerably from country to country and even for a given wastewater reclamation facility. Moreover, efficiency and reliability of treatment processes strongly depend on the selected technology and operation and maintenance frequency.

The current experience with water reuse regulations based on water quality limits and treatment requirements has demonstrated the lack of adverse health impacts, including for direct potable reuse [43]. The feedback from the operation of plants with risk-based regulations in Australia has demonstrated the limits of this approach and the excessive costs required, which have caused some water reuse projects to stop. In the future, it would be relevant to consider risk-based water reuse regulations, but the key challenge is to define the acceptable or tolerable risk that is economically affordable. The selection of a benchmark level of risk can be a complex process that involves technical, political, regulatory and social factors and should take into account increasing health risks related to other sources.

4.12 INNOVATION AND RESEARCH NEEDS

The safe implementation of water reuse within the context of an emerging circular economy opens up a plethora of opportunities for innovation. It is virtually impossible to list all opportunities and research needs because they highly depend on the specific water reuse application and local context. We summarise the general innovation needs below, categorised by the specific aspects that have to be developed.

(1) Technology research and innovation:
- Improvement of performance, reliability, energy efficiency and robustness of existing treatment processes.
- Development of new, more efficient treatment technologies with improved performance, lower carbon footprint and competitive costs. Specific focus is needed for the scale-up of new technologies.
- Development of innovative, efficient, robust and low-cost tertiary treatment (filtration and disinfection) for water reuse, allowing seasonal operation for irrigation and other uses with intermittent water demand.
- New tools and methods for monitoring of chemical and microbial pollutants, and development of on-line (real-time) monitoring of water quality and process performance. A specific challenge is the monitoring of pathogens in raw wastewater and complex matrixes (sludge and soil), as well as new pollutants (nanoparticles, microplastics, antibiotic resistance).
- Investigate and develop databases for a better understanding of pathogen removal efficiencies and the variability of performance of various unit processes in multi-barrier wastewater reclamation trains.

(2) Health, social, and environmental issues:
- Address critical gaps in the understanding of health impacts concerning human exposure to constituents in recycled water in comparison to other well-defined health risks (e.g., dose response, relevant pathogen selection, target population, etc.).
- Extensive scientific studies may be needed to demonstrate health safety associated with emerging parameters (e.g., antibiotic resistance, pharmaceuticals, endocrine disruptors, personal care products, nanomaterials, perfluorinated substances, etc.) for agricultural irrigation.
- Assessment of the potential impacts of environmental applications of recycled water in sensitive ecological communities.
- Quantitative assessment of the non-monetized costs and benefits of potable and non-potable water reuse compared with other water supply sources to enhance water management decision making.
- Development of easy to use tools for assessment of public acceptance and public education.

4.13 CHAPTER SUMMARY

As demonstrated by the milestones in water reuse development worldwide, with many regions or communities approaching the limits of their available water supplies, wastewater reclamation has become an attractive option for conserving and extending available water supplies by:

(1) Using recycled water for applications that do not require high-quality drinking water;
(2) Augmenting water sources and providing an alternative source of supply to assist in meeting both present and future water needs;
(3) Protecting aquatic ecosystems by decreasing the diversion of freshwater, reducing the quantity of nutrients and other toxic contaminants entering waterways;
(4) Reducing the need for water control structures such as dams and reservoirs;
(5) Complying with environmental regulations by better managing water consumption and wastewater discharges.

Producing reclaimed water of a specified quality to fulfil multiple water use objectives is a reality due to the progressive evolution of water reclamation technologies, regulations, and environmental and health risk protection throughout the last decades. However, the ultimate decision to promote water reuse is dependent on economic, regulatory, public policy, and, more importantly, public acceptance factors reflecting the water demand, safety, and need for reliable water supply under local conditions.

A recent market forecast of Global Water Intelligence (Desalination Markets 2016) shows that water reuse will likely grow faster than desalination over the next five years. Technology is playing a critical role as an enabler of water reuse and diversification of water reuse practices, as demonstrated by several cornerstone projects described within this chapter. Increasing concerns of water scarcity, climate change impacts and promotion of circular economy are becoming major drivers for the increasing use of recycled water in the years ahead.

With the fast growth of water reuse, the major challenges for its successful implementation are changing. In the early 2000s, for example, economic and financial challenges were reported as the main barriers to water reuse. Technical challenges were, and still are, considered to be the less important obstacle. Nevertheless, with new, more stringent regulations and requirements – which are impossible or too expensive to implement in practice – technical challenges are becoming a major issue, providing large opportunities for innovation.

4.14 EXERCISES

4.14.1 Multiple choice test

Instructions: Please read each question carefully and then select the letter of the correct answer! Only one answer is correct!

(1) Water reuse could be the most competitive option to mitigate water stress in the following conditions:
    (a) Existence of water reuse regulations
    (b) Needs for water transfers for neighbouring river basins
    (c) Avoiding water conservation measures
    (d) Expansion of agricultural lands and food production
(2) What are the most important factors/components for the success of water reuse projects?
    (a) Water conservation policies, water reuse regulations and standards
    (b) Availability of wastewater treatment technologies
    (c) Financial and political incentives
    (d) Public acceptance and education
(3) Which water reuse application requires the least treatment for reuse?Toilet flushing
    (a) Aquifer recharge

   (b) Agricultural irrigation
   (c) Industrial reuse
(4) Which of the following contaminants are not critical for water reuse safety in terms of health, environmental, agronomic and technical risks?Suspended solids
   (a) Nitrates
   (b) Salinity
   (c) Redox potential
(5) Which of the following contaminants cannot be removed by conventional tertiary treatment?
   (a) Turbidity
   (b) Dissolved solids
   (c) Viruses
   (d) Bacteria
(6) Why do membrane technologies have a key role in high-quality water recycling around the world?
   (a) Affordable costs and easier operation
   (b) Possibility for easy to verify virus log removal
   (c) Physical barrier to microorganisms (viruses, bacteria and protozoa) and dissolved matter
   (d) High public acceptance and trust in reliability

Exercise 4.1: Estimate the UV dose needed to disinfect filtered secondary effluent according to the regulatory requirements of 4 log removal of MS2 phages (an indicator of the most resistant viruses and microorganisms) on the basis of the bench (collimated beam tests) and pilot scale shown in Figure 4.9.

What are the main factors influencing the difference between bench (collimated beam tests) and pilot scale results? What UV technology will you recommend for this application?

Exercise 4.2: Define the transferred ozone dose needed to disinfect secondary effluents to comply with the Spanish standards for non-restricted irrigation with recycled water of <200 *E. coli*/100 mL, in the case when the inlet coliform concentration is 6 logs (10^6 E. *coli*/100 mL). The results of the

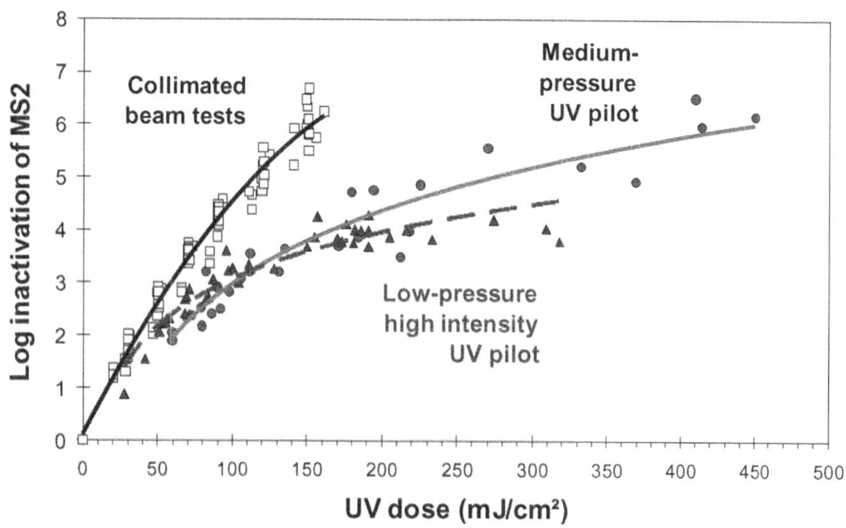

Figure 4.9 Bench and pilot-scale results of inactivation of MS2 phages.

Table 4.3 Pilot plant results of ozonation of filtered secondary effluents.

Ozone transferred dose, mg/L	Concentration of *E. coli*/100 mL		Log removal $-\log_{10}$ (N/N_0)
	Secondary effluents	Inlet of the ozone bubble column	
1	32 000	500	1.81
2	360 000	900	2.60
3	530 000	300	3.25
4	1 500 000	410	3.56
5	280 000	50	3.75
10	250 000	2	5.10

pilot plant study are shown in Table 4.3. What should the full-scale ozone dose be, assuming an ozone transfer efficiency of 85%?

Exercise 4.3: The withdrawal permit for irrigation of the agricultural area in the Green Valley will expire in three years. Local authorities informed the irrigation consortium that the permit for irrigation will be not renewed due to the severe overexploitation of the local aquifer. The irrigation areas and the main categories of cultivated crops are shown in Figure 4.10.

Two kinds of irrigation systems are implemented (recent rehabilitation): (1) sprinkler irrigation for the maize and potatoes, and (2) drip irrigation for the other crops (fruit trees and vegetables). Irrigated fields are located at a distance of 9 km from an existent wastewater treatment plant with conventional

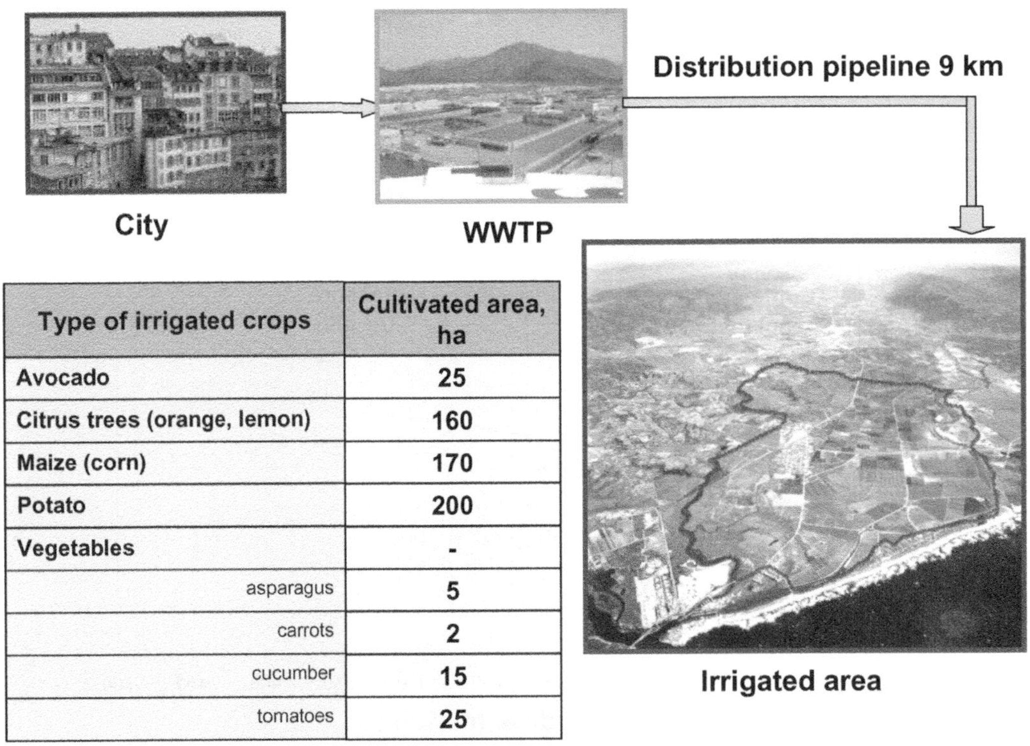

Type of irrigated crops	Cultivated area, ha
Avocado	25
Citrus trees (orange, lemon)	160
Maize (corn)	170
Potato	200
Vegetables	-
asparagus	5
carrots	2
cucumber	15
tomatoes	25

Figure 4.10 Scheme of the water recycling project and the cultivated areas (total irrigated area of 602 ha).

activated sludge treatment. The plant's average daily capacity is 100 000 m^3/d and the discharge consent (95%ile) is 30 mg/L suspended solids and 25 mg/L BOD_5. The water quality requirements for the use of recycled water for agricultural irrigation are as follows:

- Crops eaten raw: <0.1 helminth/L, <100 *E. coli*/100 mL, turbidity <10 NTU, suspended solids <20 mg SS/L
- No contact, cooking, pasture: <0.1 helminth/L, 1000 *E. coli*/100 mL, <35 mg SS/L
- Drip irrigation, fodder, nurseries: <0.1 helminth/L, 10^4 *E. coli*/100 mL, <35 mg SS/L

The main task of the exercise is to evaluate the costs and benefits of the water reuse option by means of a multi-criteria analysis and select the most cost competitive and easy to operate tertiary treatment for the production of recycled water for irrigation. The sub-tasks are as follows:

(a) Calculate the *hydraulic loading rate* for irrigation and the associated daily water demand:
  (1) Calculate the net irrigation requirements (depending on the type of crops and leaching requirements)
  (2) Calculate the irrigation hydraulic loading rate, taking into account water efficiency of irrigation methods
  (3) Calculate the daily water volume required for irrigation on the basis of the area of the irrigated fields
(b) Define the *level of disinfection* of recycled water for irrigation of cultivated crops and propose the appropriate tertiary treatment
  (1) Analyse water reuse regulations
  (2) Identify the required level of disinfection for each type of cultivated crop
  (3) Define the recycled water quality to be provided for irrigation, and explain the main criteria for the choice and why only one treatment should be applied
  (4) Select an appropriate tertiary treatment scheme (one-two alternative options)
(c) Evaluate economic and social dimensions of water reuse (multi-criteria analysis)
  (1) Identify the potential benefits of the use of recycled water for irrigation, including avoided direct and indirect costs and environmental impacts
  (2) Discuss the economic viability of the project
  (3) Discuss social dimensions of water reuse, including public acceptance, public education, market strategies, and so on. When should public communication start?

The characteristics of the irrigated fields are given in Table 4.4. The growth period is assumed to be the same for all crops (six months), and the agronomic parameters of wastewater are in an acceptable range. Irrigation efficiency is 65% for sprinkler irrigation and 85% for drip irrigation.

The hydraulic loading rate for irrigation L_w is the water that is consumed through evapotranspiration (ET_c), leaching and water losses in distribution systems. Evapotranspiration (ET_c) depends on the type of crops and is calculated from the reference evapotranspitation (mm/unit of time), which is influenced by climate conditions and the crop coefficient $K_c \cdot K_c$ varies from 0.04 to 1.27 as a function of crops.

Hydraulic loading rate is calculated using the following equation:

$$L_w = \mathrm{NR}/(E_i/100) = (\mathrm{ET}_c - P)(1 + \mathrm{LF}/100)(100/E_i) \quad (4.1)$$

where NR is net irrigation requirement (mm/unit of time) NR = ($ET_c - P$) (1 + LF/100); ET_c – crop evapotranspiration, mm/unit time; P – precipitation, mm/unit time; E_i – irrigation efficiency, %; LF – leaching fraction defined previously, %.

To simplify the calculation, it is recommended to use the water requirements ($ET_c - P$) from Table 4.4, which take into account crop requirements, precipitation and evapotranspiration. The repartition of water demand is assumed to be uniform during the growth period. In fact, water demand increases

Table 4.4 List of crops grown by the farmers and their water requirements.

Type of crop	Water requirements, mm/growing period (ET_c-P)	Sensitivity to water supply	Leaching fraction, %
Avocado	1000	Medium high	40
Citrus trees (orange, lemon)	1200	Low to medium-high	
Maize (corn)	800	High	
Potato	700	Medium high	
Vegetables:			
Asparagus	1000[a]	High[a]	0
Carrots			80
Cucumber			20
Tomatoes			20

[a]To facilitate the exercise, the water requirements are assumed to be the same during the irrigation period.

with plant growth to reach its maximum at mid-season. On the basis of the calculated hydraulic loading rates in mm/day, the quantity of irrigation water must be calculated in m^3/d in order to design the hydraulic capacity of tertiary treatment.

The most common treatment schemes enabling the achievement of various levels of disinfection of secondary effluents are illustrated in Figure 4.11.

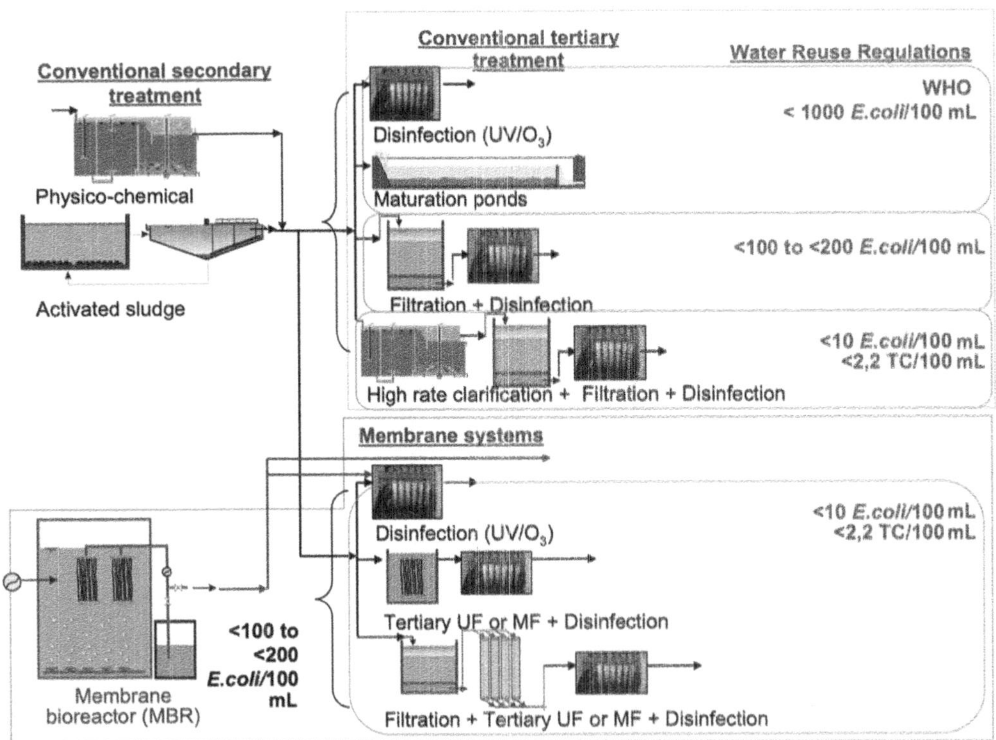

Figure 4.11 Selected tertiary treatment trains enabling to achieve various disinfection levels.

Direct cost comparison of wastewater treatment capital and operations costs is characterised by great uncertainty and imprecision due to the strong influence of technical and local conditions, various patterns of fund granting, different treatment targets and large variations of plant size. The most important consideration regarding wastewater treatment costs is that over the operational lifetime of the given sewage treatment plant, the operation and maintenance costs (O&M costs) may be as important as or even higher than constructions costs, in particular for advanced tertiary treatment.

For this exercise, it is proposed to evaluate the economic viability on the basis of the life cycle costs of additional treatment and pipelines from the treatment plant to the irrigated areas. It can be assumed that the Capex of the tertiary treatment by filtration and UV for such a high treatment capacity is 3.7 million €, plus the Capex of the distribution system of 7.2 million € (at the basis of the cost of pipelines in a rural area of 800 €/mL, linear meter). Life cycle costs were estimated at 0.05 €/m^3. The costs of pumping stations and storage reservoirs are not included. As a rule, the cost of pumping stations is negligible compared to the cost of the distribution network and the most common storage reservoirs for agricultural irrigation are constructed using only soil excavation and impermeabilisation (open reservoirs similar to lagoons, in many cases receiving also rain and surface water).

One of the most important parts of this exercise is to evaluate the potential benefits of the use of recycled water for irrigation, including avoided direct and indirect costs and environmental impacts. It would also be relevant to discuss the pricing strategy of the project, assuming that farmers did not pay for irrigation water or had fresh water fees of 0.02 €/m^3. The economic benefits could be related to the revenue effects on agriculture of water shortage and/or water supply cut that can occur during the next years. In several cases, only water reuse can provide a drought proof alternative resource for irrigation, enabling farmers to maintain agricultural activities. For example, each crop is characterised by a given income multiplier from 1.2 to 2.5, meaning that a 1€ change in farm-gate revenue results in a 1.2–2.5€ revenue change in the overall regional economy. As a rule, fruits have the highest multiplier among all crop categories.

Exercise 4.4: Two methods of tertiary treatment are being considered for golf course irrigation using conventional high load activated sludge and activated sludge with nitrogen removal, filtration and disinfection. Nutrient level ranges are 25 ± 10 and 7 ± 5 mg N_{tot}/L for the first and second option, respectively. In general, turf grass requires up to 16 nutrients, from which nine are needed in much larger quantities (macronutrients, including carbon, hydrogen, oxygen, nitrogen, phosphorus, potassium, calcium, magnesium, and sulphur). Nitrogen applications occur more frequently and in larger quantities than any other fertiliser-supplied nutrient because plants require more nitrogen and it is typically the most yield-limiting nutrient. Treated wastewater could provide a great part of the needed nitrogen and a part of some of the other macronutrients. For this exercise, it is necessary to estimate the fertilisation capacity of recycled water for the irrigation of 10 ha for each option in kg N/ha, assuming a water requirement for irrigation of 700 mm per year for an irrigation period of four months and recommended fertilisation rate in the range of 40–50 kg N/ha per month. Evaluate and discuss the feasibility, advantages and disadvantages of each option.

Exercise 4.5: The use of disinfected tertiary effluents is planned for a golf course with a total area of 24 ha. One pond on the golf course will be used for the storage and supply of recycled water to alleviate the effect of peak water demand and minimise the cost of the distribution system. Calculate the required volume of operational storage to cover the peak irrigation demand for 3 days without recycled water supply. The allowed timing for irrigation is only during the night, between 22h00 and 4h00. During the peak daily demand, 10 mm of water is applied in a 2 h duration every day for irrigation of the total area, which is equally divided into three areas. Evaporation and irrigation efficiency will be not taken into account. Discuss the result and feasibility of such storage.

Exercise 4.6: A large municipality has upgraded its wastewater treatment plant with a conventional activated sludge system for nitrogen removal, followed by sand filtration and UV disinfection to meet a very stringent standard of <10 *E. coli*/100 mL for unrestricted agricultural irrigation. The total capacity of the plant is one million person equivalents (p.e.) and dry weather flow of 350 000 m^3/d, with the annual capacity for the production of high-quality recycled water of 124 million m^3/yr. The amount of water required for the irrigation of 15 000 ha of rice and corn is 500 mm for three months of growing period. The pumping cost of recycled water for three months of operation is 30 000 €/yr. The fertilising value of recycled water can be assumed to be 0.04 €/m^3. In the past, farmers were using surface water for irrigation, free of charge. Calculate the right sale price of recycled water for farmers, if the total O&M cost of 0.12 €/m^3 of the plant should be covered only by the agricultural use, as well as the recycled water pumping cost. Discuss the willingness and ability of farmers to cover this cost and propose other strategies to ensure the economic viability of this project.

4.15 DISCUSSION QUESTIONS

4.15.1 Planning and evaluation of the feasibility of water reuse

Question 4.1 (*drivers and market analysis*): A city of 300 000 p.e. is evaluating the feasibility of providing recycled water to a golf course, which currently has its own wells for irrigation. What data or information do you need as a basis for feasibility analysis? What are the key criteria that could favour the replacement of groundwater by recycled water?

Question 4.2 (*regulations, health protection and economic viability*): A new regulation is under discussion to design water recycling systems for agricultural irrigation based on tolerable drinking water health risk with no detectable *E. coli*/100 mL and demonstration of 6 log removal of adenovirus, one of the most resistant pathogens to UV irradiation. The average total annualised cost of treatment and monitoring to consistently achieve such stringent standards compared to WHO (2006) guidelines is estimated to average 0.5 €/m^3 for large projects (10 000 m^3/d) versus 0.05–0.1 €/m^3. The additional log removal credit of recycled water storage in open reservoirs and other health protection measures during recycled water application are not taken into account. Discuss the pros and cons of such stringent regulation considering various factors involved in implementing a water reuse project, as well as the economic viability (who should cover the extra cost).

Question 4.3 (*technology, economy*): As the innovation manager of a large water utility, you are in charge of reorganising the existing water infrastructure from its current situation to a more circular approach within a timeframe of 15 years. The current wastewater treatment infrastructure comprises two very large-scale wastewater treatment plants, with capacities of 400 000 and 500 000 p.e., both of which use conventional activated sludge plants of 400 000 and 350 000 p.e. In addition, it comprises more than 50 small size WWTPs with capacities between 500 and 1500 PE. You are asked to give a presentation to the board of directors in which you evaluate the current status and justify your master plan. Where would you focus in terms of choice of technology, location (i.e., which WWTP to focus on) and which resources to target? What are your key considerations/motivations?

4.15.2 Water reuse technologies

Question 4.4 (*centralised versus decentralised treatment, technical-economic evaluation*): A middle size municipality is proposing to construct a dual water recycling system to secure the increasing water demand and climate change impacts. The city is located on the Mediterranean coast and the total population served in summer is over two times higher compared to the low tourist season in winter. Collected wastewater is treated in one centralised wastewater treatment plant with a total capacity of 200 000 p.e., designed for only carbon removal using activated sludge. The plant is also receiving sewage from several decentralised on-site septic tanks of large hotel and tourist complexes. What water quality and level of treatment will you recommend to the city council? What would be the main options to consider: dual distribution system or decentralised treatment? Discuss the constraints and benefits of each option.

Question 4.5 (*selection of water recycling treatment options*): Recycled water will be used for park irrigation and toilet flushing. Based on the information in Figure 4.3, what types of processes should be considered for tertiary treatment of secondary effluents? What would be the best option if an MBR is used for secondary treatment? Explain and discuss the solutions you are proposing.

Question 4.6 (*evaluation of water quality and selection of complementary treatment for water reuse*): A new centralised wastewater treatment plant is constructed employing MBR treatment with nitrogen removal. The municipality is planning to make the high investment profitable by implementing a water reuse project for direct aquifer recharge to control seawater intrusion. What should the recycled water quality be, and what additional treatment and complementary studies are needed to develop this project?

4.15.3 Water reuse applications

Question 4.7 (*irrigation versus in-building recycling*): Compare the similarities and differences between the two case studies of agricultural irrigation in Adelaide, Australia and in-building water reuse in Tokyo, Japan in terms of water quality and treatment requirements, the key to success, as well as major challenges.

Question 4.8 (*industrial uses*): A refinery is considering the use of municipal recycled water for boiler feedwater. A water reclamation plant is located in close proximity, and currently it is producing tertiary treated recycled water for landscape irrigation. Discuss what additional treatment should be provided and what key water quality parameters should be considered.

Question 4.9 (*environmental and recreational uses*): Environmental and recreational uses are quite similar as their key function is the discharge of treated wastewater to the receiving water body. What factors differentiate this type of reuse from a wastewater discharge? What are the major benefits and constraints for wide implementation of this kind of reuse? How do you see the growth of this type of reuse in the future and in what regions?

Question 4.10 (*indirect potable reuse*): Many communities currently use surface water sources that contain significant amounts of wastewater (de facto indirect potable reuse). Discuss and compare the health risks between planned indirect potable reuse and incidental (de facto) potable reuse? If you are a decision maker, what are your recommendations to manage the increase of the de facto indirect potable reuse in the future due to population growth and climate change?

FURTHER READING MATERIALS

Asano, T., Burton, F. L., Leverenz, H., Tsuchihashi, R. and Tchobanoglous, G. (2007) Water Reuse: Issues, Technologies, and Applications. McGraw-Hill, New York.

Lazarova, V., Choo, K. H. and Cornel, P. (eds.), (2012) Water-Energy Interactions in Water Reuse. IWA Publishing, London, UK.

Lazarova, V., Asano, T., Bahri, A. and Anderson, J. (eds.), (2013) Milestones on Water Reuse: The Best Success Stories. IWA Publishing, London, UK.

Lazarova, V. & Bahri, A. (eds.), (2005) Irrigation with Recycled Water: Agriculture, Turfgrass and Landscape. CRC Press, Boca Raton, FL, USA.

REFERENCES

1. Barnosky A. D., Hadly E. A., Bascompte J., Berlow E. L., Brown J. H., Fortelius M., Getz W. M., Harte J., Hastings A., Marquet P. A., Martinez N. D., Mooers A., Roopnarine P., Vermeij G., Williams J. W., Gillespie R., Kitzes J., Marshall C., Matzke N., Mindell D. P., Revilla E. and Smith A. B. (2012). Approaching a state shift in Earth's biosphere. Review. *Nature*, **486**, 52–58.

2. United Nations (2018). The United Nations World Water Development Report 2018. World Water Assessment Programme, New York, USA. Available at: www.unwater.org/publications/world-water-development-report-2018/ (accessed 20 July 2018).
3. US Water Alliance (2016). One Water Roadmap: The Sustainable Management of Life's Most Essential Resource. Available at: http://uswateralliance.org/sites/uswateralliance.org/files/publications/Roadmap%20FINAL.pdf (accessed 18 July 2018).
4. Lazarova V. (2013). Global milestones in water reuse: keys to success and trends in development. *Water* **21**, 12–14.
5. Angelakis A. N., Asano T., Bahri A., Jimenez B. E. and Tchobanoglous G. (2018). Water reuse: from ancient to modern times and the future. *Frontiers in Environmental Science*, **11**, 1–17. Available at: https://www.frontiersin.org/articles/10.3389/fenvs.2018.00026/full (accessed 7 December 2018). https://doi.org/10.3389/fenvs.2018.00026
6. IWA (2016). Water Utility Pathways in a Circular Economy. Available at: http://www.iwa-network.org/wp-content/uploads/2016/07/IWA_Circular_Economy_screen.pdf (accessed 15 June 2018).
7. Lazarova V., Peregrina C. and Dauthuille P. (2012). Toward energy self sufficiency of wastewater treatment. In: Water-Energy Interactions in Water Reuse, V. Lazarova, K. H. Choo and P. Cornel (eds.), IWA Publishing, London, UK, pp. 87–126.
8. Lazarova V. and Asano T. (2013). Milestones in water reuse: main challenges, keys to success and trends of development. An overview. In: Milestones on Water Reuse: The Best Success Stories, V. Lazarova, T. Asano, A. Bahri and J. Anderson (eds.), IWA Publishing, London, UK, pp. 1–22.
9. ISO 20670 (2018). Water Reuse – Vocabulary. International Organization for Standardization, Geneva, Switzerland (under publication). Available at: https://www.iso.org/standard/68785.html (accessed: January 22, 2022).
10. Asano T., Burton F. L., Leverenz H., Tsuchihashi R. and Tchobanoglous G. (2007). Water Reuse: Issues, Technologies, and Applications. McGraw-Hill, New York.
11. Lazarova V., Oelker G. and Won W. (2012). Production of custom-made recycled water for various reuse purposes: lessons learned from one of the world's largest recycling facilities. *Water Practice & Technology*, **7**(3), 2012052.
12. Lazarova V., Sturny V. and Tong Sang G. (2012). The role of the community engagement towards the success of water reuse in isolated islands and tourist areas. *Journal of Water Reuse and Desalination*, **2**(1), 1–12.
13. Olivieri A. W., Seto E., Cooper R. C., Cahn M. D., Colford J., Crook J., Debroux J. F., Mandrell R., Suslow T., Tchobanoglous G., Hultquist R. A., Spath D. P. and Mosher J. J. (2014). Risk-based review of california's water-recycling criteria for agricultural irrigation. *Journal of Environmental Engineering*, **140**(6), 04014015. https://doi.org/10.1061/(ASCE)EE.1943-7870.0000833
14. Meda A., Lensch D., Schaum C. and Cornel P. (2012). Energy and water: relations and recovery potential. In: Water-Energy Interactions in Water Reuse, V. Lazarova, K. H. Choo and P. Cornel (eds.), IWA Publishing, London, UK, pp. 21–35.
15. Bieker S., Zeig C. and Cornel P. (2012). Semizentral Germany: energy self-sufficient infrastructure systems for livable cities of the future. In: Water-Energy Interactions in Water Reuse, V. Lazarova, K. H. Choo and P. Cornel (eds.), IWA Publishing, London, UK, pp. 171–182.
16. Lazarova V. (2017). Roadmap on Development and Technology Innovation in Water Reuse. IWA Regional Water Reuse Conference, 11–15 June 2017, Murcia, Spain.
17. Tchobanoglous G., Leverenz H. and Asano T. (2011). Direct potable reuse: A future imperative. *Journal of Water Reuse and Desalination*, **1**(1), 2–10.
18. Lahnsteiner J., du Pisani P., Menge J. and Esterhuizen J. (2013). More than 40 years of direct potable reuse experience in Windhoek. In: Milestones on Water Reuse: The Best Success Stories, V. Lazarova, T. Asano, A. Bahri and J. Anderson (eds.), IWA Publishing, London, UK, pp. 351–364.
19. Walter J., Oelker G. and Lazarova V. (2013). Producing designer recycled water tailored to customer needs. In: V. Lazarova, T. Asano, A. Bahri and J. Anderson (eds.), Milestones on Water Reuse: The Best Success Stories. IWA Publishing, London, UK, pp. 37–52.
20. Lazarova V. (2005). Wastewater treatment for water recycling. In: Irrigation With Recycled Water: Agriculture, Turfgrass and Landscape, V. Lazarova and A. Bahri (eds.), CRC Press, Boca Raton, FL, USA, pp. 164–234.
21. Jimenez B. and Asano T. (2008). Water Reuse: An International Survey of Current Practice, Issues, and Needs. Scientific and Technical Report No. 20. IWA Publishing, London, UK.

22. Lazarova V., Martin Ruel S., Barillon B. and Dauthuille P. (2012). The role of MBR technology for the improvement of environmental footprint of wastewater treatment. *Water Science & Technology*, **66**(10), 2056–2064.
23. Jefferson B., Judd S. J. and Stephenson T. (2000). Membrane bioreactors and their role in wastewater reuse. *Water Science & Technology*, **41**(1), 197–204.
24. Liu Y., Giraldo E. and LeChevallier M. W. (2013). Water reuse in the america's first green high-rise residential building – the solaire. In: Milestones on Water Reuse: The Best Success Stories, V. Lazarova, T. Asano, A. Bahri and J. Anderson (eds.), IWA Publishing, London, UK, pp. 162–168.
25. Savoye P., Janex M. L. and Lazarova V. (2001). Wastewater disinfection by low-pressure UV and ozone: a design approach based on water quality. *Water Science & Techoloy*, **43**(10), 163–171.
26. Lazarova V. and Savoye P. (2004). Technical and sanitary aspects of wastewater disinfection by UV irradiation for landscape irrigation. *Water Science & Technology*, **50**(2), 203–209.
27. Xu P., Janex M.-L., Savoye P., Cockx A. and Lazarova V. (2002). Wastewater disinfection by ozone: main parameters for process design. *Water Research*, **36**(4), 1043–1055.
28. Lazarova V., Liechti P. A., Savoye Ph. and Hausler R. (2013). Ozone disinfection: main parameters for process design in wastewater treatment and reuse. *Journal of Water Reuse and Desalination*, **3**(4), 337–345.
29. USEPA (2012). Guidelines for Water Reuse. U.S. Environmental Protection Agency, EPA/600/R-12/618, Washington, DC, USA.
30. WHO (2017). Potable Reuse – Guidance for Producing Safe Drinking-Water. World Health Organization, Geneva.
31. FAO (Food and Agriculture Organization) (1999). Special Report: Drought Causes Extensive Crop Damage in the Near East Raising Concerns for Food Supply Difficulties in Some Parts. Available at: http://www.fao.org/docrep/004/x2596e/x2596e00.htm (accessed 24 July 2018).
32. UNEP (2004). Impacts of Summer 2003 Heat Wave in Europe. Available at: http://www.grid.unep.ch/products/3_Reports/ew_heat_wave.en.pdf (accessed 24 July 2018).
33. FAO (Food and Agriculture Organization) (1985). Water Quality for Irrigation. Irrigation and Drainage Paper n°29, by Ayers, R. S. & Wescot, D. W. FAO, Rome, Italy.
34. Lazarova V. and Bahri A. (2005). Code of practices for health protection. In: Irrigation With Recycled Water: Agriculture, Turfgrass and Landscape, V. Lazarova and A. Bahri (eds.), CRC Press, Boca Raton, FL, USA, pp. 83–102.
35. Sheikh B. (2005). Monterey county water recycling projects: a case study in irrigation water supply for food crop irrigation. In: Irrigation With Recycled Water: Agriculture, Turfgrass and Landscape, V. Lazarova and A. Bahri (eds.), CRC Press, Boca Raton, FL, USA, pp. 374–379.
36. Stevens D. and Anderson J. (2013). Irrigation of food crops in Australia. In: Milestones on Water Reuse: The Best Success Stories, V. Lazarova, T. Asano, A. Bahri and J. Anderson (eds.), IWA Publishing, London, UK, pp. 201–210.
37. Sheikh B. (2005). Codes of practices for landscape and golf course irrigation. In: Irrigation With Recycled Water: Agriculture, Turfgrass and Landscape, V. Lazarova and A. Bahri (eds.), CRC Press, Boca Raton, FL, USA, pp. 152–163.
38. Lazarova V., Hills S. and Birks R. (2003). Using recycled water for non potable urban uses : a review with particular reference to toilet flushing. *Water Supply*, **3**(4), 69–77.
39. Kitamura K., Saeki K. and Funamizu N. (2013). Semi-decentralized water recycling in megacities: The example of Tokyo shinjuku area. In: Milestones on Water Reuse: The Best Success Stories, V. Lazarova, T. Asano, A. Bahri and J. Anderson (eds.), IWA Publishing, London, UK, pp. 151–159.
40. Rosenwinkel K. H., Borchmann A., Engelhart M., Eppers R., Jung H., Marzinkowski J. and Kipp S. (2013). Closing loops – industrial water management in Germany. In: Milestones on Water Reuse: The Best Success Stories, V. Lazarova, T. Asano, A. Bahri and J. Anderson (eds.), IWA Publishing, London, UK, pp. 243–258.
41. Lahnsteiner J. and Mittal R. (2010). Reuse and recycling of secondary effluents in refineries employing advanced multiple barrier systems. *Water Science & Technology*, **62**, 1813–1820.
42. Wells M. J. M., Mullins G. A., Bell K. Y., Da Silva A. K. and Navarrete E. M. (2017). Fluorescence and quenching assessment (EEM-PARAFAC) of de facto potable reuse in the Neuse River, North Carolina, United States. *Environmental Science & Technology*, **51**, 13592–13602.
43. Lahnsteiner J., van Rensburg P. and Esterhuizen J. (2018). Direct potable reuse – a feasible water management option. *Journal of Water Reuse and Desalination*, **8**(1), 14–28.
44. Lazarova V., Emsellem Y., Paille J., Glucina K. and Gislette Ph. (2011). Water quality management of aquifer recharge using advanced tools. *Water Science & Technology*, **64**(5), 1161–1168.

45. Crook J. (2004). Innovative Applications in Water Reuse: Ten Case Studies. WateReuse Association, Alexandria, VA, USA.
46. Angelotti R. W. and Grizzard T. J. (2013). 34 Years of experience with potable water reuse in the Occoquan Reservoir. In: Milestones on Water Reuse: The Best Success Stories, V. Lazarova, T. Asano, A. Bahri and J. Anderson (eds.), IWA Publishing, London, UK, pp. 323–337.
47. ISO 20426 (2018). Guidelines for Health Risk Assessment and Management for Non-Potable Water Reuse. International Organization for Standardization, Geneva, Switzerland. Available at: https://www.iso.org/obp/ui/#iso:std:iso:20426:ed-1:v1:en (accessed: January 22, 2022).
48. Olivieri A. W., Seto E. and Soller J. (2007). Application of Microbial Risk Assessment Techniques to Estimate Risk Due to Exposure to Reclaimed Waters (WRF-04-001). WateReuse Research Foundation, Alexandria. Available at: https://www.waterboards.ca.gov/water_issues/programs/grants_loans/water_recycling/research/04_011_01.pdf (accessed: January 22, 2022).

doi: 10.2166/9781780409566

Chapter 5

Established full-scale applications for energy recovery from water: anaerobic digestion

Paul D. Jensen[1], Sergi Astals[2], Xue Bai[1], Ludwika Nieradzik[1], Peter Wardrop[1], Damien J. Batstone[1] and William P. Clarke[3]

[1]*The Advanced Water Management Centre, The University of Queensland, St Lucia, QLD 4072, Australia*
[2]*Department of Chemical Engineering, University of Barcelona, Barcelona, Spain*
[3]*The School of Civil Engineering, The University of Queensland, St Lucia, QLD 4072, Australia*

GLOSSARY

AD	Anaerobic digestion
AL	Anaerobic lagoon
AnMBR	Anaerobic membrane bioreactor
BMP	Biochemical methane potential
BOD	Biological oxygen demand
CAL	Covered anaerobic lagoon
CHP	Combined heat and power
COD	Chemical oxygen demand
EGSB	Expanded granular sludge bed reactor
FOG	Fat, oil and grease
HRT	Hydraulic retention time
IC	Internal circulation reactor
LNG	Liquefied natural gas
OLR	Organic loading rate
PHB	Polyhydroxybutyrate (renewable plastic)
PFR	Plug flow reactor
UASB	Upflow anaerobic sludge blanket reactor
UV – Vis	Ultraviolet – Visible
TAN	Total ammonical nitrogen

TKN	Total Kjeldahl nitrogen
TS	Total solids
TSS	Total suspended solids
VFA	Volatile fatty acids
VS	Volatile solids
VSS	Volatile suspended solids
WAS	Waste activated sludge

5.1 INTRODUCTION

Anaerobic digestion (AD) is a complex biochemical process where organic matter decomposes in the absence of oxygen to produce biogas (methane and carbon dioxide) and residual digestate (stabilized organic matter). AD is currently the most successful application of resource recovery technology used in urban water management and particularly wastewater treatment. From a waste treatment perspective AD has multiple advantages over aerobic treatment and composting processes in that AD does not require aeration energy and does not generate large volumes of microbial sludge (Batstone & Jensen, 2011). AD also has advantages over common thermal processing technologies such as incineration in that AD can be applied to both wet and dry wastes without large energy inputs. The methane content of biogas has the potential to be an important source of renewable energy. Captured biogas can be burned in boilers, engines or combined heat and power (CHP) units to generate heat and/or electricity, therefore AD processes are generally considered as net-energy producing rather than net-energy consuming; and this has driven the expansion of AD applications beyond waste treatment technology into a dedicated renewable energy technology where energy crops are grown specifically as feedstocks for biogas plants.

Anaerobic digestion is a highly flexible and broadly applicable technology. There are successful full-scale installations treating sewage, manure, agricultural residues, mixed municipal wastes, food processing residues, beverage wastes, paper and textile industry wastes, energy crops and others. Anaerobic digestion can be applied at a single household scale or at large industry scales up to and exceeding millions of EP and generating hundreds of GWh electricity per year.

The design, operation and interpretation of engineered anaerobic processes has greatly advanced over the last 30 years. This improvement is based on improved understanding of underlying concepts such as the microbial functions and dynamics; and physico-chemical interactions. As such, AD technology is often implemented to operate stably and reliably without intervention. The process itself has: (i) multiple microbial steps, mediated by different organisms; (ii) different metabolic steps that can be rate limiting under different feedstock loads or process conditions; (iii) interaction with the physicochemical system, particularly weak acid and base inhibition of microbial processes, and (iv) nonlinear behaviour, particularly with respect to pH regulation and inhibition.

Feedstocks for anaerobic digestion can be highly variable in terms of biological characteristics, including biogas potential and speed of degradation; and in terms of physical characteristics including strength and materials handling properties. This variability occurs both within wastewater applications where upstream processes have a significant impact on waste sludge and more broadly where AD is applied for co-treatment or organic wastes from other industries. Similarly, anaerobic digestion can be implemented using a broad range of reactor configurations and process complexity. Therefore, application of anaerobic technology needs careful thought to: (i) determine and understand the viability of the process; (ii) select the most appropriate technology for the application; and (iii) achieve an optimally engineered process for a specific application. This chapter provides an introduction into the fundamentals of anaerobic digestion, feedstock characterisation and presents some guiding principles for the selection, design and operation of anaerobic processes.

5.2 LEARNING OBJECTIVES

At the completion of this chapter you should be able to:

- Describe the four main biological steps of the anaerobic digestion process and the sequence in which they occur.
- Estimate the theoretical biogas yields for anaerobic digestion feedstocks with different compositions.
- Explain the parameters used to estimate practical methane potential of complex substrates and outline procedures to determine these parameters.
- Describe commonly applied forms of anaerobic digestion technology, including the advantages, disadvantages and typical application areas of each technology.
- Use waste characteristics, selection diagrams and site-based constraints to select appropriate forms of anaerobic digestion technology for a described application.
- Set a design basis for anaerobic digestion technologies, conduct preliminary reactor sizing and calculate performance predictions.
- List ancillary equipment required to support the AD process, including common options for materials handling, materials transport and dewatering.
- Design a monitoring program for an AD processes. Use process data to calculate key performance measures and assess against performance expectations.
- Identify when a process is not operating at optimal capacity and use decision trees to troubleshoot the process.

5.3 CONCEPTUAL OVERVIEW OF ENERGY RECOVERY THROUGH ANAEROBIC DIGESTION

5.3.1 Fundamental principles of anaerobic digestion

Anaerobic digestion proceeds through a series of parallel and sequential interdependent reactions, carried out by a complex and diverse microbial community (Campanaro *et al.*, 2018; Kirkegaard *et al.*, 2017; Vanwonterghem *et al.*, 2016). All microbial growth requires energy, which is generated by the flow of electrons between molecules undergoing oxidation (electron donating) and reduction (electron accepting) reactions. As opposed to aerobic wastewater treatment technologies, the anaerobic digestion process occurs in absence of an external electron acceptor (e.g., O_2, NO_3^-, SO_4^{2-}) (Batstone & Jensen, 2011). The provision of appropriate internal electron acceptors relies on the syntrophic relationships between the electron producer and the methanogenic electron consumer. The reactions occurring during the anaerobic digestion process have been typically grouped into four steps: (i) hydrolysis, (ii) acidogenesis, (iii) acetogenesis and (iv) methanogenesis (Figure 5.1).

Hydrolysis is an enzyme-mediated process, which decomposes particulates and polymeric substrates that cannot be directly utilized by the anaerobic microorganisms. Polymers, such as carbohydrates, proteins, and lipids are broken down into their monomers (i.e., sugars, amino acids, and long chain fatty acids) using a wide range of extracellular enzymes (e.g., cellulases, amylases, peptidases, and lipases).

Acidogenesis, also known as primary fermentation, converts hydrolysis products into intermediate compounds, such as volatile fatty acids (VFA), alcohols, lactate, formate and hydrogen. Typically, sugars and amino acids undergo fermentation, while long chain fatty acids can only be oxidized anaerobically (mainly β-oxidation).

Acetogenesis, or secondary fermentation, is the degradation of fermentation products to acetate using an external electron acceptor, such as the hydrogen ion or bicarbonate.

Methanogenesis is the production of methane from acetate, hydrogen, formate and carbon dioxide, generated during previous fermentation processes. Methanogenesis is carried out through two distinct pathways: acetoclastic methanogenesis, where acetate is cleaved into methane and carbon dioxide ($CH_3COOH \rightarrow CH_4 + CO_2$) and hydrogenotrophic methanogenesis where hydrogen (electron donor) and carbon dioxide (electron acceptor) produce methane ($4H_2 + CO_2 \rightarrow CH_4 + 2H_2O$).

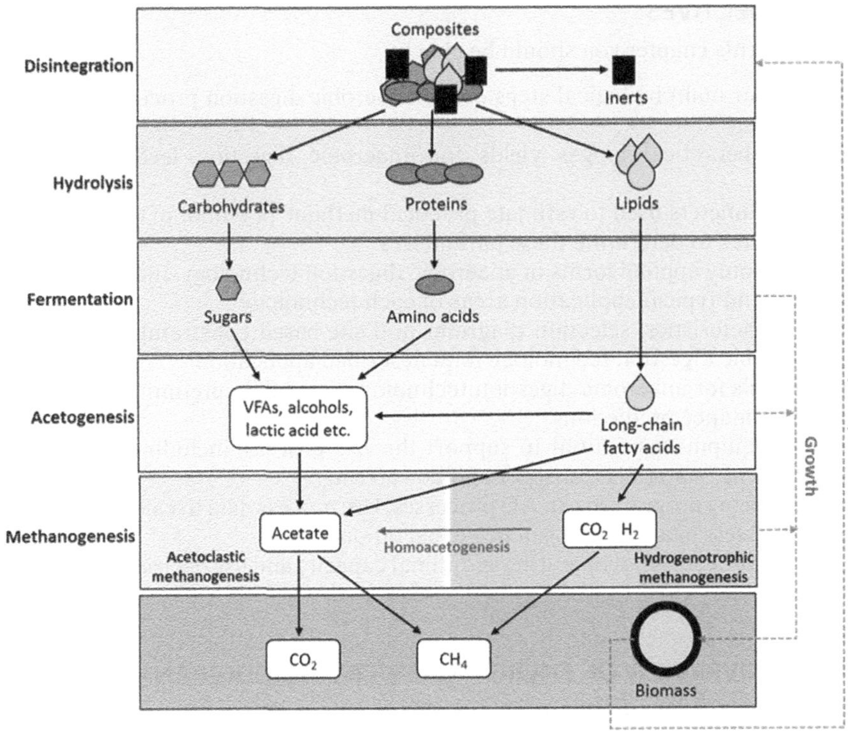

Figure 5.1 A schematic of key anaerobic digestion reactions, progressively breaking down composite particulate wastes into soluble intermediate products and ultimately into methane (adapted from Van Lier *et al.* 2008).

AD processes are categorized into three temperature ranges: psychrophilic (up to 15°C); mesophilic (25–40°C) and thermophilic (50–70°C) (Batstone *et al.*, 2002). Temperature has a strong impact on the biological, chemical and physical processes that occur during AD. In particular, temperature impacts the growth rate of AD microorganisms and the reaction rates of metabolic processes. Importantly, temperature can impact the different stages of AD in different proportions and can therefore impact the balance between acid-producing steps and acid consuming steps, thus changing the risks associated with process overload and inhibition. Mesophilic AD processes support the largest range of microorganisms and therefore have the most diverse microbial communities. This diversity allows a higher degree of microbial redundancy that can improve process robustness. Thermophilic AD reaction rates are typically faster than mesophilic reaction rates, however thermophilic processes have less diverse microbial communities, can be less stable and are more susceptible to ammonia inhibition (see section 5.3.5). Psychrophilic AD reaction rates are generally slower than mesophilic reaction rates, methanogenesis in particular is slower under psychrophilic conditions and this can create difficulties in managing AD of rapidly degrading materials where organic acids can readily accumulate. More detailed information on the impacts of temperature on AD are presented by Batstone and Jensen (2011).

5.3.2 Characterizing feedstock and methane potential

The feasibility, design and operation of anaerobic digestion systems requires the measurement of a range of physical, chemical and biological parameters. Feedstock composition is a key parameter that impacts both the metabolic pathways required for anaerobic degradation and the degradation end products. During anaerobic degradation of organic material, electron transfer occurs between carbon atoms such that organic carbon is both oxidized to CO_2 and reduced to CH_4. The ratio of CH_4 to CO_2, and therefore the theoretical biogas yield of a feedstock, is related to the chemical structure and oxidation state of the carbon.

Two common analytical parameters used to describe the concentration of wastewater and other organic wastes are the chemical oxygen demand (COD) and volatile solids (VS). The COD describes the amount of oxygen needed to completely chemically oxidize the material and is expressed as g $O_2 \cdot kg^{-1}$ (or $gO_2 \cdot L^{-1}$). Stoichiometrically, 350 L (0°C, 1 atm) of methane is possible for each kilogram of COD in the feedstock. The VS is an estimate of the amount of organic material in the waste, defined as the amount of matter in a dried sample that is lost at a temperature of approximately 550°C in air (APHA *et al.*, 2005), and is expressed as gVS kg^{-1} (gVSL^{-1}).

There is no direct universal stoichiometric relationship between VS and methane potential. The relationship between COD and VS of a sample is used to assess the energy density (oxidation state) of a feedstock and therefore the theoretical methane potential and biogas composition. Table 5.1 shows the theoretical methane yields and biogas compositions for a range of model organic substrates. Wastewater sludge can be described as a mixture of the macro-components in Table 5.1. Nopens *et al.* (2009) fractionates sludge as 33% carbohydrates, 33% proteins and 33% lipids, however this ratio will vary depending on the intake catchment and the upstream processes.

In practice, biogas production is lower than the theoretical yields shown in Table 5.1. This is due to multiple factors including:

- Sludge and other complex feedstocks contain compounds which are not anaerobically degradable, such as lignin or some forms of endogenous cell debris. Non-degradable materials do not degrade and depending on the physical structure of the material, the presence of non-degradable material can also prevent microorganisms contacting and degrading other compounds;
- Insufficient reaction time to allow for complete degradation of biodegradable organic matter. Typically, 80–90% of biogas potential is recovered in well-designed processes;
- Growth of anaerobic microorganisms. Typically, 5–10% of the organic material degraded is converted into anaerobic biomass;
- Unfavourable environmental conditions in the reactor.

Practical estimates of methane potential are based on experimental anaerobic biodegradability assays, such as biomethane potential (BMP) tests. These assays are used to determine the ultimate methane potential (B_0), which represents the maximum volume of methane that can be produced from a specific sample. BMP tests also provide the following important information about the material

Table 5.1 Theoretical characteristics of model substrate components (van Lier *et al.*, 2008).

Substrate	Formula	Carbon Content (g C/g VS)	COD/VS (g COD/g VS)	Theoretical Methane Yield (mL CH_4/g VS)	Theoretical Biogas Composition[a] (% CH_4)
Carbohydrate	$(C_6H_{10}O_5)_n$	0.444	1.19	415	50
Protein[b]	$C_5H_7NO_2$	0.531	1.42	496	50
Lipids[b]	$C_{51}H_{98}O_6$	0.756	2.88	1007	71
Ethanol	C_2H_6O	0.522	2.09	730	75
Propionic acid	$C_3H_6O_2$	0486	1.51	530	58
Acetic acid	$C_2H_4O_2$	0.400	1.07	373	50

[a]Estimated based on stoichiometry and not considering any physico-chemical reactions in the process.
[b]Standard average formula has been applied. There are multiple compounds in this category with different formula.

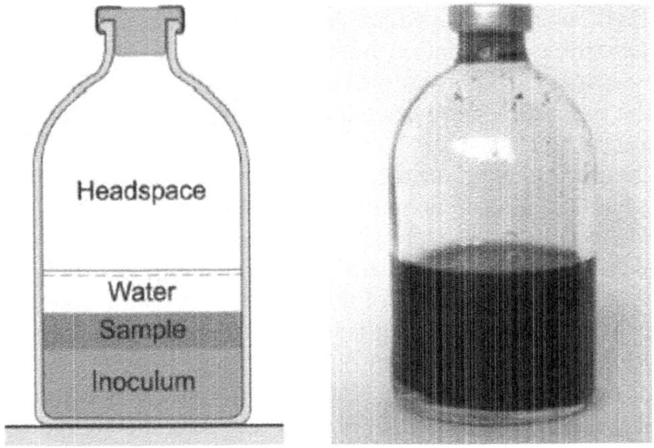

Figure 5.2 Example of batch biochemical methane potential assay (Angelidaki *et al.*, 2009).

under study: (i) anaerobic biodegradability (f_d), which is the fraction of COD that can be converted to methane and (ii) the apparent first order hydrolysis rate coefficient (k_{hyd}), which represents the speed of degradation (day^{-1}) (Angelidaki *et al.*, 2009; Holliger *et al.*, 2016; Lesteur *et al.*, 2010). BMP tests are batch experimental assays (Figure 5.2) where a known mass of test sample is combined with anaerobic inoculum containing active anaerobic biomass. There are multiple variations of the BMP assay where the selection of inoculum, the inoculum-to-substrate ratio, the addition of nutrient medium, mixing regimes and storage temperature are key experimental design parameters. In all tests, the volumetric biogas production and biogas composition are measured at regular intervals to generate a methane production curve (Da Silva *et al.*, 2018; Holliger *et al.*, 2016). Sewage sludge BMP tests generally take 30–40 days to complete at 35°C, however timing is dependent on the particular sewage sludge and experimental conditions used, therefore monitoring should continue until biogas production is exhausted. An example result from a BMP test is shown in Figure 5.3, while Figure 5.4 shows several examples of ultimate methane potential (B_0) and biodegradability (f_d) for sewage sludge compared to different organic wastes. Data from BMP testing together with the COD and VS measurements are the most frequently used parameters to design and monitor full-scale anaerobic digesters, as described in more detail in section 5.3.4.

In BMP tests examining the properties of particulate wastes such as sewage sludge, hydrolysis is expected to be the rate limiting step. Hydrolysis is a first order process, therefore, data obtained from BMP assays is typically fitted using a first order kinetic model, as shown in Equation (5.1). For waste or wastewater streams where a high fraction of the organic material is soluble, hydrolysis is no longer the key limiting step and Monod kinetics are commonly applied:

$$S_{CH4} = B_0\left(1 - e^{-k_{\text{hyd}}t}\right) \tag{5.1}$$

where B_0 is the ultimate methane potential of the substrate, S is the concentration of substrate, k_{hyd} is the apparent hydrolysis rate coefficient (for solid or particulate substrates).

As a microbial process, AD can be impacted by a range of unfavourable environmental conditions, or compounds with inhibitory or toxic effects. Where required, more complex modelling may be applied to determine specific inhibition characteristics of a waste material, such as the inhibition constant (K_I). There are a broad range of inhibitory compounds, examples include ammonia, organic acids, heavy metals, and sulphide. The inhibition concentrations for these compounds vary, as does the mechanism through which the microbial cells are impacted. Therefore, different mathematical

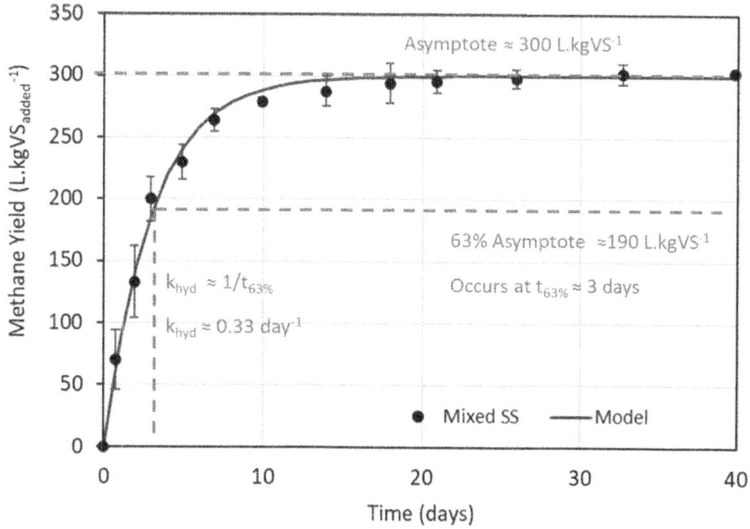

Figure 5.3 Example output from BMP test. Error bars on the data indicate 95% confidence interval from triplicate batches. A graphical method for estimating B_0 and k_{hyd} is shown on the figure in red.

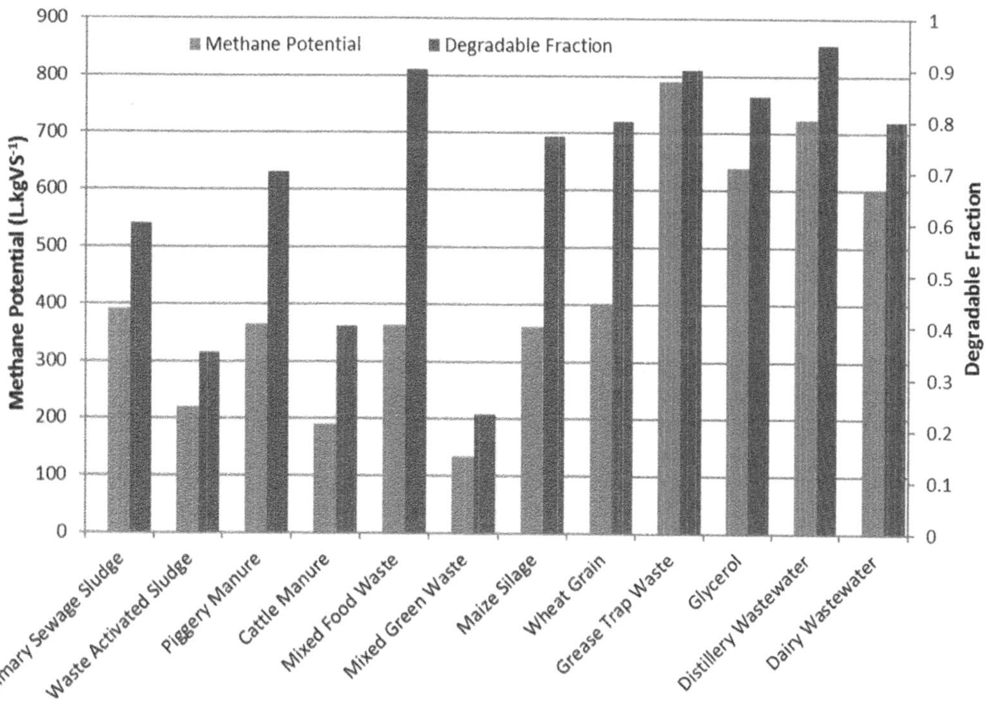

Figure 5.4 Examples of ultimate methane potential for different feedstocks.

equations must be used to represent the inhibition impact. Equation (5.2) is used for non-competitive inhibition and is most commonly applied to AD. A first order model with inhibition function is shown as Equation (5.3). For soluble compounds, Monod kinetics would be an alternative approach for modelling BMP results. Monod kinetics with an inhibition function is shown in Equation (5.4):

$$I = \frac{K_I}{K_I + S_I} \tag{5.2}$$

$$\frac{dS_{CH4}}{dt} = k_{\text{hyd}} SI \tag{5.3}$$

$$\frac{dS_{CH4}}{dt} = B_o k_{m,} \frac{1}{1 + K_S/S} I \tag{5.4}$$

where k_{hyd} is the apparent hydrolysis rate coefficient (for solid substrates), k_m is the maximum substrate uptake rate (for soluble substrates) and K_s is the half saturation constant (i.e., the concentration where substrate is uptake is 50% of the maximum). I is the inhibition function and represents the impact on microbial activity. K_I is the inhibition coefficient and represents the concentration of the inhibitory compound where activity is 50% of the maximum and S_I is the concentration of the inhibitory compound.

5.3.3 Overview of anaerobic digestion technologies

AD is a highly flexible and broadly applicable technology with installations treating sewage, manure, agricultural residues, mixed municipal wastes, food processing residues, beverage wastes, paper and textile industry wastes, energy crops and others. These different waste streams have different material handling properties and different degradation characteristics. AD is applied in household, industrial, rural and urban settings, each with different space considerations and operating requirements. A broad range of reactor designs have been developed to meet the different challenges associated with different AD applications.

5.3.3.1 Lagoon-based anaerobic technologies

Anaerobic lagoons (ALs) are large engineered earthen basins that represent a relatively low-cost treatment option with respect to both capital cost and operating expenditure. Typical anaerobic lagoons have a depth of 3–8 m and may occupy footprints exceeding 50 000 m^2. The bottom of the lagoons is generally maintained at 1–2 m above the local water table, and is lined with impermeable materials such as clay, to prevent groundwater seepage. The slope on internal and external banks varies according to geotechnical properties of the earth, generally a stepper batter is desired to limit the footprint of the lagoon. Anaerobic lagoons may be open, where methane is released into the atmosphere and fugitive GHG emissions are high, or designed to be totally or partially covered with a geo-membrane cover that enables biogas collection and use. Anaerobic lagoons are typically operated with very limited external control (e.g., temperature, pH, mixing or aeration) and are therefore largely impacted by the local climate and flow conditions. Because of the large volumes, process correction under failure can be extremely expensive or impractical. Treatment lagoons are most suitable for warmer tropical climates, but can still operate effectively in temperate and colder climates if sized appropriately.

The treatment process in ALs relies on the removal of suspended solids via sedimentation and the subsequent degradation of settled organic material (see Figure 5.5). Traditionally, design of ALs has been heavily based on the hydraulic retention time (HRT), organic loading rate (OLR) and operating temperature (Mara & Pearson, 1998; Peña *et al.*, 2000). Designing ALs with long HRT is common in some applications, such as when treating highly concentrated waste streams, or when treating slowly

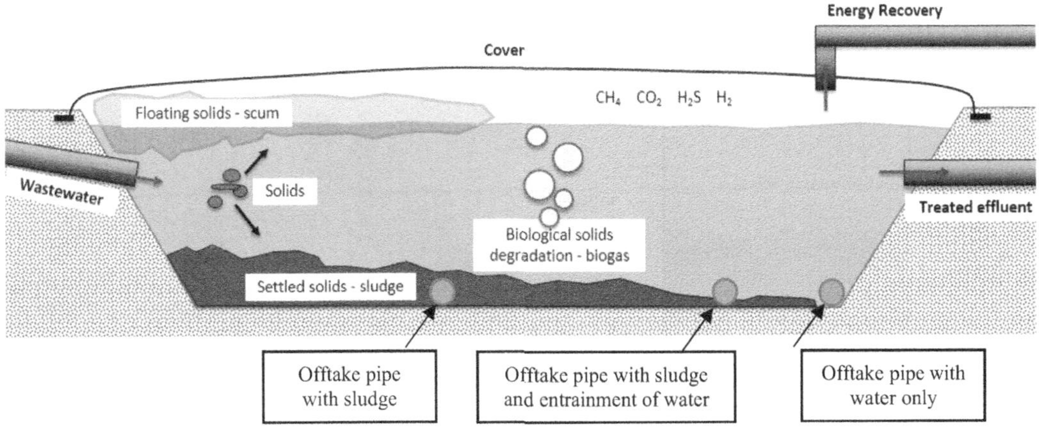

Figure 5.5 Schematic representation of a covered anaerobic lagoon. The boxes describe possible status of offtake pipes for anaerobic lagoons.

degrading manures or constructing lagoons in cooler climates. However, sedimentation processes and the resulting separation of HRT and SRT provides some flexibility in process design. High rate anaerobic lagoons can operate at HRT of 1–2 days but still achieve an extended SRT of years. High rate lagoons are applied in warmer climates, and for the treatment of less concentrated wastes, such as domestic sewage.

Sedimentation processes are essential for successful lagoon operation, particularly at shorter HRTs, however sedimentation can reduce the contact between fresh waste and active biomass and also result in progressive sludge accumulation and the need for manual sludge removal operations called desludging. The frequency of desludging requirements ranges from years to decades depending on the OLR of the lagoon and the portion of non-degradable solids in the feed stream. The cost and disruption of desludging operations has led to the development of innovative lagoon and cover designs such as the installation of piping systems, generally near the bottom of the lagoon, that enable continuous or periodic withdrawal and/or recirculation of sludge. Recirculation is not critical for the operation of an AL, but can improve the sludge-biomass contact and can therefore improve process performance.

5.3.3.2 High rate granular technologies

High-rate granular technologies are an intensified form of AD technology based on engineered bioreactors that occupy a much smaller footprint and offer much better controllability compared to lagoon-based processes. High-rate granular technologies operate with short HRTs and extended SRTs. This is achieved by integrating solids retention within the main digester in the form of anaerobic granules. Anaerobic granules are self-supporting particulate biofilms formed by auto-immobilisation of anaerobic bacteria in the absence of a structural carrier or support material. The granules consist of the symbiotic anaerobic microorganisms that work together for the methanogenic degradation of organic matter.

The most common example of high-rate anaerobic granular technology is the upflow anaerobic sludge blanket (UASB) reactor, which relies on a naturally forming blanket of granular sludge (particle size >1 mm). A typical UASB reactor contains three primary zones. The bottom zone is the reaction zone. Wastewater enters the UASB through the bottom of the reactor and into the reaction zone. Here the wastewater percolates through the sludge blanket in an upward flow path (Figure 5.6a). Liquid upflow velocities of 0.5–2 $m.h^{-1}$ are applied to promote granule formation and therefore

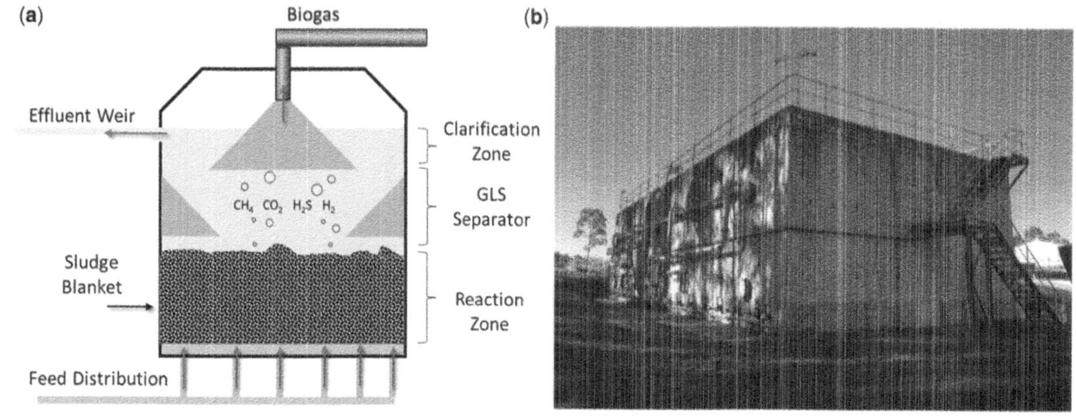

Figure 5.6 (a) schematic demonstrating the internal structure of a UASB; (b) a UASB installation for the treatment of brewery wastewater in Queensland, Australia.

biomass retention. However, the upflow velocity must be managed to prevent washout of biomass granules. The depth of the sludge blanket typically varies between 4 and 8 m. The middle zone of a UASB reactor is the gas-liquid-solid (GLS) separation system. This zone is generally 1 2 m deep and is designed with the dual purpose of collecting biogas produced in the bottom of the reactor, while settling and retaining any gas entrained solids. Above the GLS separation zone is a clarification zone. The clarification zone contains little or no biomass. The purpose of the clarification zone is to settle fine particles and improve the effluent quality leaving the reactor.

The solid–liquid separation in UASB reactors allows for very long solids retention times (i.e., often in the order of months), while maintaining short HRT ranging between 6 and 48 hours results in compact reactor designs. Design variations include the addition of packing/carriers in hybrid reactors or extended super-high rate/low footprint systems such as expanded granular sludge bed (EGSB) and internal circulation (IC) reactors. To give an indication of the sizing and capacity of a UASB, Figure 5.6b shows a full-scale UASB designed to treat 3000 m^3 of wastewater per day.

When designing high-rate anaerobic granular reactors, the volume of the reactor is based on the organic loading rate (kg COD/m^3.d). The design organic loading rate is a function of the wastewater characteristics, temperature and the solids concentration within the reactor. Importantly, the OLR will have a greater impact on process stability and performance than the HRT. The geometry of the reactor is determined in order to manage the liquid upflow velocity within the reactor and the superficial gas velocity. Examples of common design ranges for high rate granular anaerobic processes are shown in Table 5.2.

Table 5.2 Common design and operating ranges for high rate anaerobic bioreactors.

Reactor Type	Complex Solids and FOGs	COD Loading Rate (kg COD/m^3.d)	HRT (h)	COD Removal Efficiency (% COD Removal)
UASB	Limited (no FOGs)	2–15	6–24	65–90
Internal circulation	No	15–25	1–24	65–90
Fluidized bed	No	2–60	1–24	70–90
Anaerobic Membrane reactor	Yes	2–15	6–24	70–95

5.3.3.3 Anaerobic membrane technologies

Anaerobic membrane bioreactors (AnMBRs) are an alternative high-rate anaerobic digestion technology where biomass and suspended solids are retained in the process using diffusive membranes rather than settling processes (Figure 5.7). The separate retention of solids decouples the SRT and the HRT allowing for long sludge age within a compact reactor design. The use of membranes for solids separation improves solids retention compared to granular processes. The membrane allows AnMBRs to be applied to wastewater streams with high suspended solids content and/or poor settling characteristics (such as high fat content wastewater) and still produce very high-quality effluent within the compact reactor footprint. Higher levels of solids retention achieved using the membranes can make AnMBRs more robust to load variations and can result in higher methane conversion rates.

Membrane selection, design and management is a major aspect of AnMBR design and operation. AnMBRs may be configured using membranes submerged in the reactor, or membranes configured as part of an external recirculation loop (Judd, 2011). Membrane area requirements are determined based on the required permeate flow and the membrane flux.

During operation, membrane fouling is a common challenge with all membrane bioreactors. As wastewater is drawn through the membrane, solids can accumulate on the membrane surface in a fouling layer; this form of fouling is similar to the formation of a filter cake and increases the flow resistance which results in higher transmembrane pressures and either increased energy demand for filtration, reduced flux, or both. Fouling propensity varies based on a number of process characteristics including wastewater composition, rheology, membrane style and membrane flux. In terms of wastewater composition, fouling is more severe in applications with a high protein to polysaccharide ratio in the waster [20, 21].

In conventional membrane bioreactors (MBRs), gas scouring is a commonly used fouling control strategy and is achieved using aeration and coarse bubble diffusers. In an AnMBR that contains submerged membranes, gas scouring is achieved by re-circulating biogas across the membrane. However, in AnMBRs where the membranes are configured within a side-stream or recirculation loop, the side-stream units can use liquid shear directly in a cross-flow configuration. The ability to operate at a manageable trans membrane pressure (TMP) and an effective critical flux (the flux below which fouling is maintained at steady state) is the primary factor influencing economic feasibility of membrane processes (Lee *et al.*, 2001). Typical operating ranges for AnMBR processes are shown in Table 5.3.

Fouling management has a critical impact on the operating costs of the process. Operating the AnMBR at higher solids concentrations and/or at higher membrane flux will increase fouling and

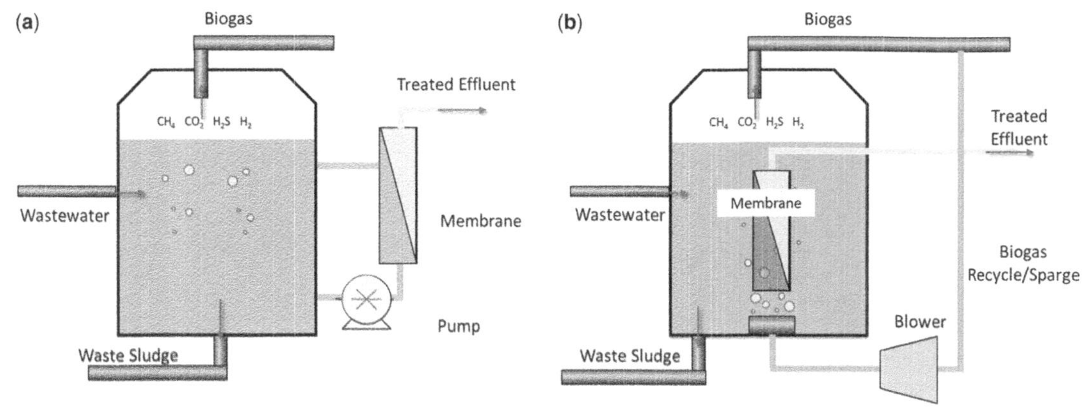

Figure 5.7 AnMBR configurations: (a) sidestream membrane bioreactor (sAnMBR); (b) immersed membrane bioreactor (iAnMBR).

Table 5.3 Membrane types and basic operating conditions (Judd, 2011; Schrotter & Bozkaya-Schrotter, 2010).

Membrane Type	TMP[a,b] (bar)	Flux[b] ($l.m^{-2}.h^{-1}$)	Permeate Recovery (%)[c]
Nanofiltration	5–50	10–20	40–80
Microfiltration and ultrafiltration	0.05–4	15–100	90–100

[a]TMP – Trans membrane pressure.
[b]Low-flux low pressure vs high flux high pressure.
[c]Water recovered as permeate vs water lost as a waste stream (i.e., in the sludge waste).

therefore fouling management costs. However, the higher solids concentrations also increase the concentration of active biomass within the reactor and this improves both the uptake/removal of soluble compounds from the wastewater and the robustness of the process to variations in organic load. Therefore, management of the AnMBR is a balance between operating costs and process performance.

AnMBR is an attractive technology due to excellent effluent quality, high tolerance to load variations, ability to produce a solids free effluent for the purposes of reuse, and high methane conversion. Currently, high membrane prices, membrane fouling challenges and high-cost, high-energy membrane cleaning requirements are the main drawbacks for the AnMBR processes. However, there is continuous development and improvement of membrane materials, membrane cleaning strategies and membrane anti-fouling technologies which will improve the economics and attractiveness of AnMBRs.

5.3.3.4 Mixed liquor reactor technologies

Mixed liquor anaerobic reactors are one of the most common and most flexible anaerobic digestion configurations (Figure 5.8). These technologies are most often applied to sewage sludge, energy crops, food waste or manure digestion. The reactors are typically cylindrical (normally with recirculated gas and/or liquid mixing) or egg-shaped (normally with mechanical mixing) and constructed of concrete, steel or fiberglass panels. In principle, mixed liquor digesters operate as a fully mixed reactor where

Figure 5.8 Examples of mixed liquor anaerobic digesters installed at municipal wastewater treatment plants in Australia.

the HRT is equal to the SRT. Advantages of a well-mixed reactor configuration include the significant reduction of sedimentation and stratification (i.e., the formation of settled sludge or floating scum layers) which promotes better contact between active microorganisms and waste, provides better control of scum at the top of the digesters and provides better control of temperature, pH and retention times. However, because there is no separation between SRT and HRT, the reactors must be designed to accommodate slow growing microorganisms. Minimum retention times under mesophilic conditions are 4–6 days to prevent washout of methanogens, however longer retention times are desired to maximize energy recovery and may exceed 60 days for slowly biodegradable material (i.e., cattle manure).

Process retention times are fixed based on the reaction times required to digest the organic solids, however organic loading rates to the reactor ($gCOD.L^{-1}.d^{-1}$) can vary depending on solids concentrations in the reactor feed. Pre-conditioning steps such as sedimentation, dissolved air flotation or gravity belt thickening are commonly applied to up-concentrate the solids content in the reactor feed in order to maximize the reactor loading rate. In a mixed liquor reactor, the maximum in-reactor solids concentration is approximately 4% TS (feed concentration of approximately 8% TS). This is due to mixing and materials handling considerations and is related to sludge viscosity. Higher feed concentrations can be achieved for highly degradable feedstocks, such as food waste; or where pre-treatment is applied and the viscosity of the feed is altered, such as thermal hydrolysis of sewage sludge. Mixed liquor reactors are not economically attractive when applied to dilute wastewater streams due to the long treatment times and large reactor volumes required and the low organic loading rates achieved from a dilute feed.

5.3.3.5 Solid phase anaerobic technologies

A wide range of municipal, industrial, and agricultural wastes have relatively high solids concentration (above 15–20%). Solid wastes, especially lignocellulosic biomass, solid food wastes (FW), the organic fraction of municipal solid waste (OFMSW) and some animal manures may not be suitable for treatment in mixed liquor reactor technologies without significant dilution or pre-treatment. Solid-phase anaerobic technologies are gaining popularity as a cost-effective technology for these wastes. These solid-phase technologies can be applied to dewatered sewage sludge, but are not commonly applied at domestic wastewater treatment plants.

Plug flow digesters are used to treat a semi-solid slurry (i.e., 10–30% dry solids content) in a long tube or trench shaped reactor (Figure 5.9). Material is loaded at the front of the digester, and is moved through the reactor in a plug flow using agitators, rakes or screws. Solid-phase plug flow reactors

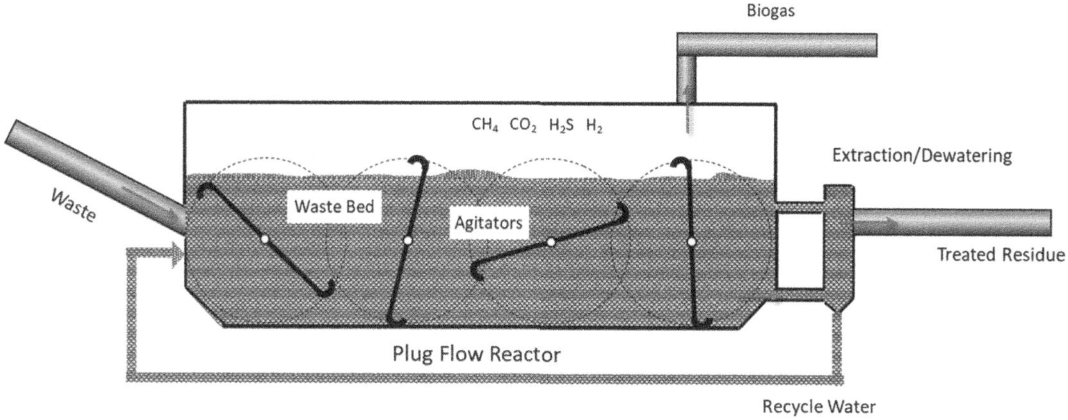

Figure 5.9 Schematic of a dry plug flow reactor (top).

have high rates of digestion, but may require inoculation or digestate recycling to ensure sufficient anaerobic microorganisms are present in the feed to reduce local acidification and/or souring. Mechanical mixing is not required but can be engineered to enhance contact between the waste and anaerobic biomass and therefore aid the digestion process. Plug flow digesters are most often applied to treatment of agricultural solids. The SRTs and organic loading rates vary depending on feedstock, with SRTs as short as 15 days achievable for some wastes (i.e., WAS activated sludge) and SRT over 90 days required for others (i.e., cattle manure, woody biomass).

In a leach-bed process, a solid bed of material is first loaded into a reactor, tumbler, or basket. The reactor is sealed, and a liquid stream termed leachate is circulated through the reactor and over the waste bed. Leachate circulation may be constant or periodic. The leachate percolates through the solid matrix supplying microorganisms to facilitate degradation and simultaneously wash away organic acids to prevent the inhibitory effect of local souring. The organic acids are subsequently degraded to produce methane. In general, hydrolysis of the waste occurs on the solid bed, while methanogenesis can occur on both the solid bed and within the leachate. The leachate may pool within the primary reactor, or be transported through an external reactor tank designed to promote methanogenesis, as depicted in Figure 5.10. Where an external methanogenic reactor is used, leachate recirculation rates can be varied to control organic acid loading into the external reactor tank and maintain optimal conditions for methane production.

Batch leach-bed reactors are generally low capital, but manual loading and unloading operations can be labour intensive and a significant cost. It is common to use multiple reactors operating at different points in the loading, reaction and unloading cycles to handle continuous waste production. Continuous leach bed reactors are an alternative configuration in which the solid waste bed moves along the length of the reactor in a plug flow pattern using similar equipment to the solid-phase plug flow reactor. Continuous leach-bed reactors generally require much higher capital investment due to the more complex solids handling requirements. Both batch and continuous leach-bed reactors are currently being successfully applied at full-scale.

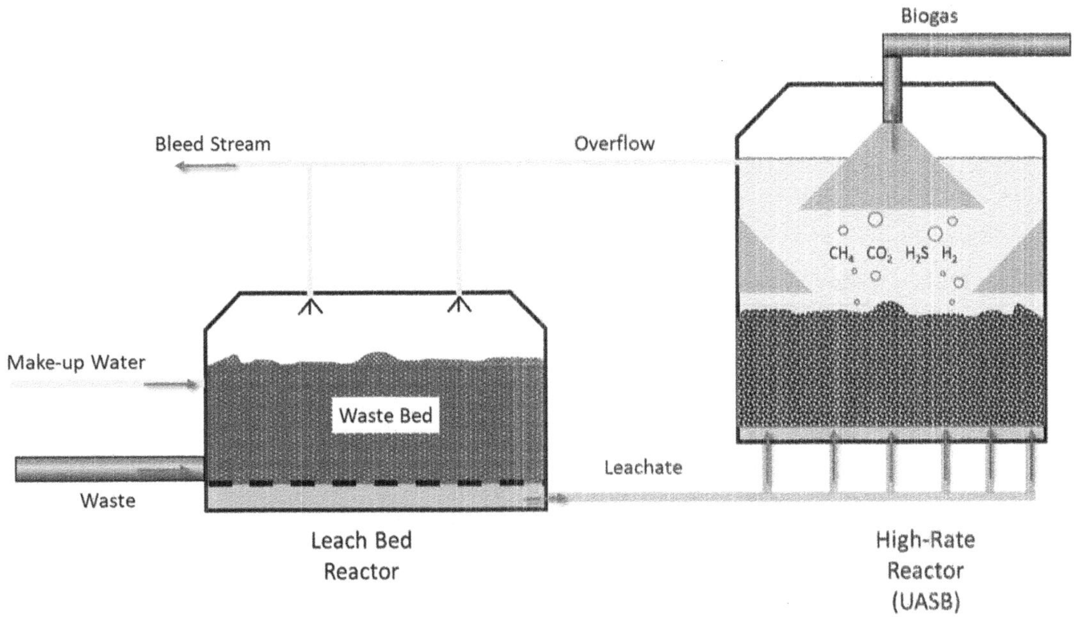

Figure 5.10 Combined leach-bed and high-rate system.

5.3.3.6 *Summary of AD technologies*

A summary of commonly used anaerobic digestion technologies and their main areas of application is shown in Table 5.4.

5.3.4 Design and operation of anaerobic digestion technologies

Anaerobic digestion can be implemented using a broad range of reactor configurations and ancillary equipment designed for optimized process performance in each specific application. Similarly, AD can be implemented to process a wide variety different wastes or bioenergy feedstocks at virtually any scale. Each AD application has unique requirements related to the feedstock, the plant location, footprint and seasonal climate and social and economic factors that must be assessed and considered as part of the process design and operation. This section discusses: (i) selection and sizing of appropriate reactor technology; (ii) ancillary equipment (requirements) for materials handling and downstream management of AD products; and (iii) monitoring and maintenance of process operations.

5.3.4.1 *Technology selection*

Selection of AD reactor technology is based on a number of considerations related to both the characteristics of the feedstock and constraints at the plant location. Important considerations related to the feedstock include the fractionation between particulate and soluble organic material, the presence of fat, oil and grease (FOG), the moisture content and the expected time required for degradation. Considerations related to the plant location include the available footprint, local climate, presence of supporting infrastructure such as upstream/downstream processes and disposal/re-use options for the digestion residues. Figure 5.11 is an example of a selection diagram used as part of the initial technology assessment. Note that the information provided in the figure does not consider all selection criteria (e.g., such as space availability) and should be considered as a guide only. For example, lagoon-based technologies may seem advantageous for low strength wastewater, however such technology would not be appropriate in colder climates (<10°C) or where there is only limited space available.

5.3.4.2 *Process sizing (loading rates)*

There are three main factors that determine the sizing of AD processes: (i) managing organic load to provide beneficial conditions for methanogenesis, including maintaining pH above 7.0 and below 8.0; (ii) maintaining sufficient retention time for growth and maintenance of AD microbiology; and (iii) maintaining sufficient retention time for hydrolysis of particulate substrates. The limiting factor is determined by the feedstock characteristics and the specific process constraints for each application. In general, feedstocks are classified as solid waste or wastewater, based on the distribution of particulate and soluble COD.

Hydrolysis is considered the process-limiting step for AD of particulate substrates, and reactor sizing is based on the apparent first order hydrolysis rate, the reactor configuration (i.e., mixed vs plug flow) and the extent of degradation required. Equation (5.5) shows the relationship between methane production and retention time in these processes for a batch or plug flow reactor. Equation (5.6) shows the relationship for a continuous mixed liquor reactor:

$$CH_4 = B_0\left(1 - e^{-k_{\text{hyd}}.t}\right) \tag{5.5}$$

$$CH_4 = B_0\left(1 - \frac{1}{1 + k_{\text{hyd}}.t}\right) \tag{5.6}$$

where CH_4 represents methane production (per kg or t added), B_0 represents the ultimate methane potential (determined from BMP testing (see section 5.3.2)), k_{hyd} represents the apparent hydrolysis rate (determined from batch test) and t represents the solids retention time of the digester (days).

Table 5.4 Overview of commonly applied anaerobic digestion technologies, note some forms are AD are not commonly applied in domestic waste treatment.

Technology	Typical Reactor Types	Advantages	Disadvantages	Loading Rate (kgCOD. $m^{-3}d^{-1}$)	Example Application
Lagoon-based anaerobic technologies	• Covered anaerobic lagoon (CAL)	• Low capital cost • Low operating costs • Flexible sizes and designs • Effective at settling and degrading solids	• Very high footprint • Limited controllability • Poor/variable methane capture • Can produce odours • Must be desludged • Generally limited to dilute feeds	• 0.1–0.5	• Domestic wastewater • Animal manure • Animal processing
High rate granular technologies	• Upflow anaerobic sludge blanket (UASB), • Expanded granular sludge bed (EGSB), internal circulation (IC) reactor	• Low footprint • Low capital cost • Operating requirements well understood • High quality effluent	• Poor tolerance to FOGs • Non-degradable solids can reduce performance	• 10 (UASB) • 20 (EGSB/IC)	• Agri-food industry • Distillery wastewater • Paper and textile, and so on.
Anaerobic membrane technologies	• Anaerobic membrane bioreactor (AnMBR)	• Very stable for biomass retention • Good gas yield for energy recovery • Can treat WW at extreme conditions • Highest effluent quality	• Expensive membrane and membrane fouling	• 1–15	• Domestic wastewater • Animal processing • Distillery wastewater
Mixed liquor reactor technologies		• Established technology • Flexible configurations (i.e. multi-stage) • Easy to control • Continuous gas production.	• Poor volumetric loading rate • Much larger tanks – expensive capital • Solids content of feed is limited • Liquid (not solid) residue	• 1–5	• Sewage sludge • Food waste • Animal manure • Crop residues and energy crops
Solid phase anaerobic technologies – continuous	• Continuous dry solid phase (plug-flow) reactor	• Continuous gas and residue production • Suitable for high solids and dry feedstocks • High loading rates	• High capital costs • Complex mechanical system • Potential solids handling issues	• 10	• Animal manures • Food waste

(Continued)

Table 5.4 Overview of commonly applied anaerobic digestion technologies, note some forms are AD are not commonly applied in domestic waste treatment. *(Continued)*

Technology	Typical Reactor Types	Advantages	Disadvantages	Loading Rate (kgCOD. $m^{-3}d^{-1}$)	Example Application
Solid phase anaerobic technologies – batch	• Batch solid phase (leach bed) reactor	• Low capital and operating cost • High loading rates • Good gas conversion due to retention of active biomass • Easy to control via leachate • No milling required	• Non-continuous system (gas flow changes in quality and flow over time) • Can be difficult to seal (gas seals) • Needs loading and unloading	• 6–10	• Agricultural wastes, • Energy crop wastes, • Poultry waste

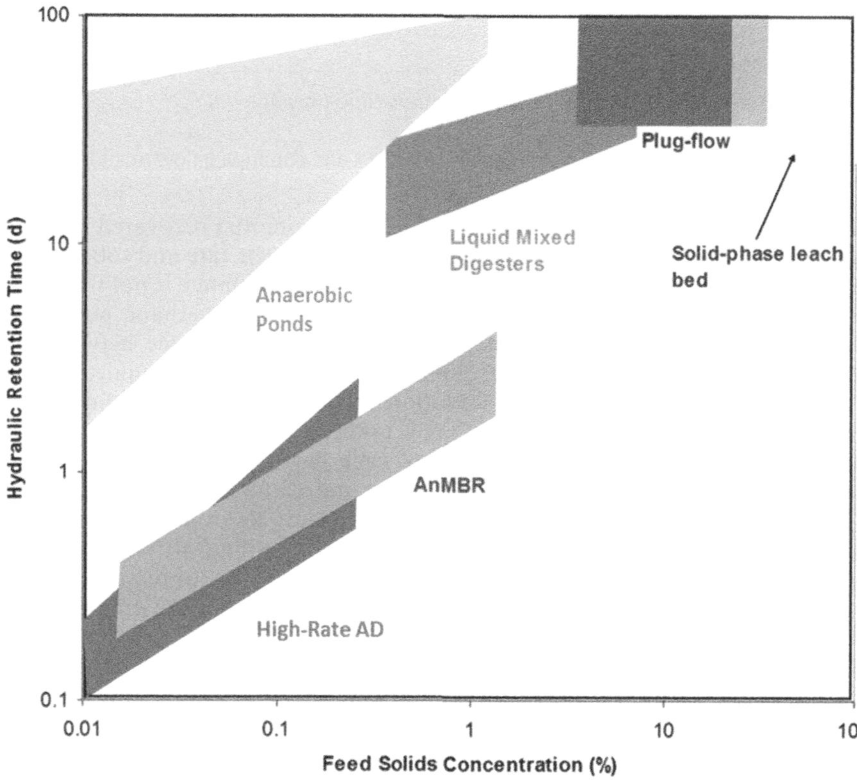

Figure 5.11 Anaerobic treatment technologies ranked by hydraulic retention time (vertical axis), and solids concentration (horizontal axis) (adapted and extended from Batstone and Jensen, 2011)).

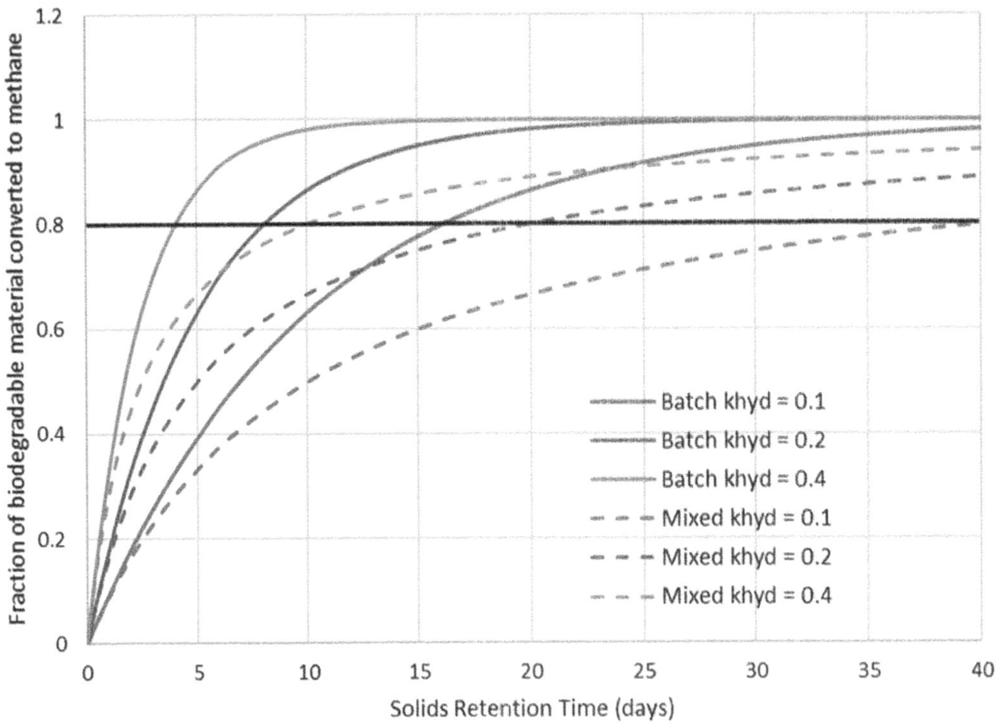

Figure 5.12 Model based prediction of methane recovery for CSTR and batch/plug flow reactors.

Figure 5.12 provides examples for the fraction of methane potential recovered in batch digesters and continuous mixed digesters as a function of apparent hydrolysis rate and solids retention time. Hydrolysis is a first order process and therefore degradation performance is not linear with reactor retention time. Digesters are not designed to recover 100% of the methane potential available. Actual recovery is typically in the range of 80–90%, which is a balance between degradation performance and capital investment. Note that continuous mixed processes require longer treatment times to achieve similar performance due to a dilution effect from constant withdrawal of effluent, which may be partially degraded. Batch and plug flow processes typically require shorter treatment times; however, there is a requirement to inoculate each batch, which increases the volume of material added to the reactor. Importantly, the volume required for a batch of plug flow process is still typically smaller than the volume for a continuous reactor receiving the same feedstock. To illustrate the difference in required SRT between these reactor configurations, a feedstock with an apparent hydrolysis rate of 0.1 day^{-1} would require an SRT of 40 days to recover 80% of methane potential in a continuous mixed digester, but would only require an SRT of approximately 17 days in a batch reactor.

For AD of soluble substrates, methanogenesis not hydrolysis is considered the process-limiting step. This can be attributed to the slower microbial growth and slower substrate utilization of this step. In all AD processes, minimum SRTs are required to ensure sufficient time for microbial growth and cell maintenance. Processes digesting soluble substrates often incorporate biomass retentions systems that separate the SRT and the HRT (see section 5.3.3). In these reactors the solids retention times are long to enable biomass growth and retention, while hydraulic retention times are much shorter to enable high throughput and smaller reactor volumes. In these systems, the minimum solids retention

Table 5.5 Minimum solids retention times to support microbial growth and prevent washout during anaerobic digestion as a function of temperature (Tchobanoglous *et al.*, 2003).

Operating Temperature (°C)	Minimum SRT (Days)	Desired SRT (Days)
18	11	28
24	8	20
30	6	14
35	4	10
40	4	10

times for biomass growth often have less impact on overall vessel sizes and the reactor size is based on the organic loading rate, with nominal design parameters as presented in section 5.3.3. During AD in mixed liquor reactors the SRT is equal to the HRT and therefore minimum retention times must be considered during vessel sizing. Minimum retention times are a function of temperature, as shown in Table 5.5.

5.3.4.3 Materials handling and transport

Material handling and transport is a critical component of any anaerobic digestion system. Materials handling operations include feedstock receival, pre-processing and pre-thickening or wastes, loading to the digesters and removal of digestate for downstream processing, such as dewatering. At centralized treatment facilities materials handling operations may also include waste receival facilities. Materials handling technologies are selected based on feedstock properties, digester configurations and specific site requirements. Common materials handling equipment in municipal wastewater handling applications are: (i) pumping for wastewater and slurry streams with <5% solids content; and (ii) screw or belt conveying for streams with higher solid contents, that is solid concentration higher than 5–10%. Industrial sites may have wastes from different processing areas with very different characteristics and may therefore require separate materials handling and transport operations. For example, animal processing facilities may generate large volumes of wastewater and multiple solid waste streams rich in either fat, oil and grease or high in lignocellulosic fibres. When combined, these wastes may form a pumpable mixture, however the solid components increase the risk for blockages, particularly if the transport processes operate intermittently. Therefore, separate transport processes for feeding different feedstocks to a digester are often desired. This may extend to separate treatment processes, or the various waste streams may be combined in the digester.

5.3.4.4 Reactor mixing

Reactor mixing is essential in common AD configurations such as mixed liquor reactors and AnMBRs. Two primary reasons for mixing are to: (i) maintain the integrity of the decomposition process by improving circulation, consistency and mass transfer in the reactor (both physical and thermal); and (ii) minimize the formation of scum and sediment layers through dispersion, thereby maximizing the effective volume of the reactor. For example, mixing prevents local accumulation of intermediates (i.e., organic acids) and toxic substances by diluting them in the reactor volume. Mixing mechanisms are summarized in Table 5.6 and can generally be divided into three main types:

(1) *Mechanical mixing*: Rotating propellers are used to mix the digester contents. The propeller designs are generally low speed-large diameter or high speed-smaller diameter. Low speed-large diameter mixers generate less shear and are more effective at bulk fluid movement. High-speed-smaller diameter mixers generate high shear and are more effective at rapid mixing processes, such as the distribution of chemicals dosed into a process. Propellers can be located

Table 5.6 A comparison of digester mixing systems.

Mixing Mechanism	Principle	Capex	Typical Values
Mechanical mixing	High or low speed submerged turbine or mixers suitable for fixed or floating covers	Low	0.005–0.008 kW per m^3 of reactor operating volume
Gas injection mixing	Unconfined: collect gas at the top of the digesters, compress the gas, and then discharge the gas through a pattern of bottom diffusers or through a series of radially placed top-mounted lances; suitable for fixed, floating, or gas holder covers	Medium to high	4.5–5 $L.min^{-1}.m^{-3}$ of reactor operating volume
	Confined: collect gas at the top of the digesters, compress gas and then discharge through confined tubes; suitable for fixed, floating, or gas holder covers		5–7 $L.min^{-1}.m^{-3}$ of reactor operating volume
Liquid circulation mixing	Agitation in which some of the liquid tank contents are withdrawn from the reactor by external mixing pumps and reinjected back into the reactor through discharge nozzles at the bottom	High	30–50 $L.min^{-1}.m^{-3}$ of reactor operating volume (20–30 mins turnover time)

within the digester to achieve bulk fluid movement or may be mounted in draft tubes that create a jet mixing effect.

(2) *Gas injection mixing*: A gas compressor is used for either unconfined or confined gas injection. The gas is commonly injected through sequentially discharge lances, floor-mounted diffusers, draft tubes, and bubble guns.

(3) *Liquid circulation mixing*: External pumps may be used to circulate the digester content to achieve agitation. Liquid circulation may be achieved using external recirculation loops and small diameter nozzles that generate high liquid velocities. Draft tubes or baffles may be used to assist in the direction of liquid flow.

Each of the mixing methods have advantages and disadvantages. The selection and design of mixing equipment will depend mainly on the volume of the reactor, the geometry of the reactor, the viscosity of the digester contents, and the throughput of the mixing system.

5.3.4.5 Temperature management

AD processes operate at a variety of temperatures which are categorized into three ranges: psychrophilic (up to 15°C); mesophilic (25–40°C); and thermophilic (50–70°C) (Batstone *et al.*, 2002). Temperature has a strong impact on the biological, chemical and physical processes that occur within AD processes, therefore operating temperatures are often controlled. The most common operating temperatures for AD are 37°C for mesophilic processes and 55°C for thermophilic processes. However, temperature control is not practical in all AD applications. For example, temperature control is not practical in many large lagoons or in high-rate reactors with very short HRT. In some cases, a level of temperature control can be attempted by adjusting the temperature of the feed wastewater prior to the digester, however the actual reactor temperature will be lower due to ambient heat loss. In other cases the temperature of the feed wastewater will not be controlled or adjusted and will vary due to seasonal changes in the ambient climate. In both cases, seasonal variations in ambient temperature can have a significant impact on process performance.

Anaerobic processes are not auto-heating and heat energy must be supplied when operating at temperatures above ambient conditions. The most common method of digester heating is via external pumping and circulation operations, where digestate is pumped through external heat exchangers. The heat exchangers are commonly designed with a single pass ΔT of 2–4°C and recirculation rates are then calculated in order to transfer sufficient heat back to the process. Higher single pass ΔT may be used, however the process must be carefully monitored to ensure the sludge temperature cycling is not destabilizing the microbial communities.

The heating requirements of an anaerobic digestion process consist of the heat required to raise the feedstock from ambient to the digester operating temperature and the heat required to compensate for losses through the reactor surface and process pipelines. Heat requirements to raise the temperature of the feedstock and the requirements to compensate for losses in the system are given by Equations (5.7) and (5.8), respectively.

$$Q_{\text{feed}} = \dot{m} C_P \left(T_{\text{dig}} - T_{\text{amb}}\right) \quad (5.7)$$

$$Q_{\text{losses}} = UA\left(T_{\text{dig}} - T_{\text{amb}}\right) \quad (5.8)$$

where Q is a heating requirement (kJ.s^{-1} or kW); C_p is specific heat capacity (kJ.kg^{-1}.°C^{-1}) where 4.2 kJ. kg^{-1}.°C^{-1} can be used to approximate specific heat of slurries; T_{dig} is the operating temperature of the digester (°C); and T_{amb} is the ambient temperature (°C). U is overall heat transfer coefficient (kW. m^{-2}.°C^{-1}) and varies according to the reactor design and construction materials (typical values shown in Table 5.7).

5.3.4.6 Dewatering, drying and conservation

The effluent from anaerobic digestion processes, also known as digestate, may contain a significant flow of non-degraded residues. The mass of solid residues will depend on the properties of the feedstock (f_d) (see section 5.3.2) and the applied SRT (see section 5.3.4). Digestate from mixed liquor AD processes typically contains 2–6% solids and significant volumes of free water. The high moisture content of digestate allows for simpler materials handling (i.e., easy to pump). However, in most applications, further downstream processing is required to reduce external transport costs prior to disposal, agricultural re-use or incineration (Figure 5.13). Digestate dewatering is

Table 5.7 Typical heat transfer coefficients used to calculate external heat losses from anaerobic digesters (adapted from Tchobanoglous *et al.* 2003).

Wall Material	Typical U Values (W.m^{-2}.K^{-1})
Concrete walls (above ground)	
300 mm thick – No insulation	4.7–5.1
300 mm thick – Air gap and brick facing	1.8–2.4
300 mm thick – insulation	0.6–0.8
Concrete walls (below ground)	
300 mm thick – moist earth	2.85
300 mm thick – dry earth	1.7
Floating covers	
No insulation	1.8–2.0
Insulation	0.9–1.0
Fixed concrete covers	
225 mm thick – No insulation	3.0–3.6
100 mm thick – No insulation	4.0–5.0
100 mm thick – Insulation	1.2–1.6
Fixed steel covers (6 mm)	4.0–5.4

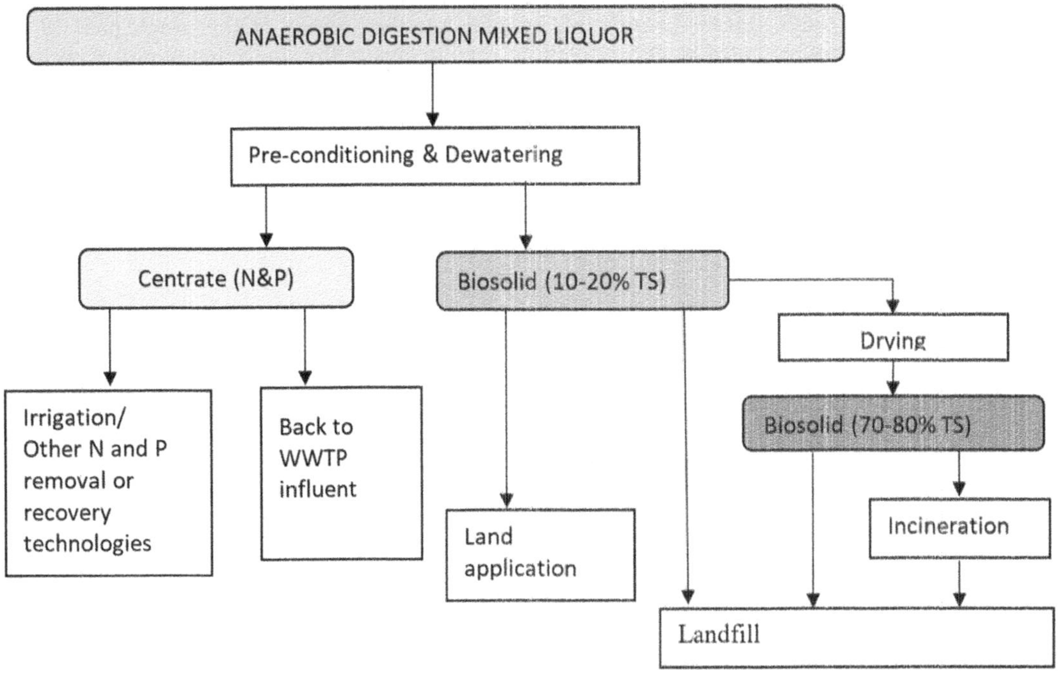

Figure 5.13 Schematic representation of downstream digestate handling operations associated with anaerobic digestion.

a form of solid/liquid separation applied to AD effluent. Dewatering involves pre-conditioning by polymer addition (e.g., polyacrylamide) followed by a physical separation technique such as a screw press, belt press or centrifuge. When applied to digestate from sewage sludge and/or manure treatment, conventional dewatering equipment achieves a dewatered digestate cake at 15–30% solids. Higher cake solids are achieved for some agricultural residues but this is very application specific.

A variety of drying technologies including thermal dryers, glasshouse solar dryers, drying pans or stockpiles can be applied after the dewatering process in order to increase the cake solids and therefore significantly reduce the volumes of treated digestate for transport and disposal. Drying technologies are capable of increasing the solids content of dewatered sludge from 10–30 to >80%, however drying times can vary greatly depending on the energy applied. For example, drying times in thermal dryers are much shorter than drying times in drying pans and stockpiles. Drying is desired when digestate is disposed to landfill or incinerated or when transported large distances. However, where solid digestate is reused via local land application such as forestation, land reclamation or agricultural production, additional digestate drying is less common and in some cases is not desired.

The liquid stream from digestate dewatering, also known as supernatant or centrate, can be high in ammonium nitrogen (N) and ortho-phosphorus (P) and requires further treatment before discharge to the environment. Established and emerging concepts to recover phosphate and ammonium from digestate (and other nutrient rich water streams) are discussed in Chapters 9 and 10.

5.3.4.7 Biogas treatment and utilization

The biogas produced during AD primarily comprises methane (CH_4) and carbon dioxide (CO_2) with minor amounts of hydrogen (H_2), hydrogen sulphide (H_2S) and ammonia (NH_3). Additionally, when leaving the digester, biogas is saturated with water vapour and may contain dust particles and volatile organic compounds (e.g., siloxanes). The methane content of biogas has a high calorific value of approximately 55 MJ/kg CH_4 and this makes biogas an important energy source to produce heat and/or electricity for wastewater treatment plants. While onsite energy production is the most common biogas use applied today, there are a range of alternate and emerging biogas uses discussed in detail in Chapter 6, examples include:

- onsite heat and/or steam generation, that is steam boiler;
- electricity and heat production (CHP);
- upgrade to higher methane content for use as vehicle fuel (LNG) or as a feed stream to the local natural gas grid (Persson *et al.*, 2006);
- production of chemicals, proteins and/or biopolymers such as PHA and PHB (Born, 2005; Heubeck *et al.*, 2007).

For some applications, the biogas may require treatment before use. Treatment may be relatively simple, including desulphurization and water reduction or be more complex, such as CO_2 removal and/or upgrading the methane content.

Biogas treatment requirements are generally determined based on the biogas end use. Examples of commercially applied biogas treatment/upgrading and use pathways are shown in Figure 5.14. Figure 5.14 also shows emerging biogas utilization processes. Further discussion on biogas utilization and upgrading is presented in Chapter 6.

5.3.5 Process monitoring and troubleshooting

Given the diversity of the AD microbial community, and the breadth of biochemical reactions required to degrade organic wastes and produce methane, close monitoring of the AD process is paramount. This section describes the most important parameters that require adequate monitoring in order to determine the performance and stability of the process, as well as the methods to troubleshoot a failing AD process.

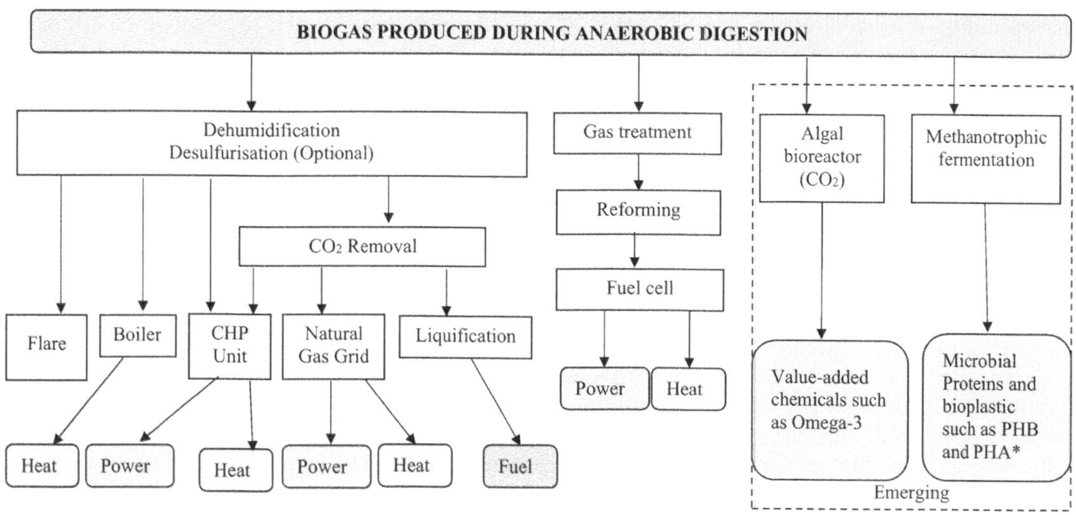

Figure 5.14 Schematic representation of general biogas treatment/upgrading pathways and two novel product pathways.

5.3.5.1 Process performance

Chemical oxygen demand (COD), defined as the amount of oxygen required to oxidize all the organic carbon in a sample, is a common parameter used to estimate the amount of organic matter in a sample (Henze *et al.*, 1987; Noguerol-Arias *et al.*, 2012). Analyzing the form of COD in a feedstock (i.e., total COD, particulate COD and soluble COD) is a useful parameter in assessing the most suitable form of anaerobic digester technology and appropriate digester sizing (see sections 5.3.4.1 and 5.3.4.2). Additionally, tCOD (and VS) is used to assess the removal of organic matter in anaerobic digesters, while sCOD can be used to monitor process stability through the concentration of intermediate compounds in the system (see section 5.3.5.2). COD is an effective process assessment tool as COD is conserved through anaerobic processes, that is, COD cannot not be formed nor destroyed, but can change forms between solid, liquid and gas. Therefore, in steady state systems the amount of tCOD entering the systems should be equal to the amount of COD leaving the system as effluent or biogas (Batstone *et al.*, 2002).

In many AD applications the economic viability of the process is more sensitive to solids destruction and the level of solids stabilization than biogas production. Therefore, total solids (TS) and volatile solids (VS) are common parameters to characterize the feedstock prior to anaerobic digestion and to characterize the performance of the anaerobic digestion process. Specifically, VS destruction is a primary measure of process performance. There are three common ways to calculate VS destruction: (i) mass-balance VS destruction; (ii) Van Kleeck VS destruction; and (iii) apparent VS destruction on gas flow, which relies on the principle that organic matter destroyed must be converted to biogas. Calculation procedures for monitoring VS destruction in AD processes are shown in Table 5.8. All measures assume the system is at steady state, or that values are averaged over a period of operation. If the system is not at steady state, mass balance VS destruction, and gas flow VS destruction need to be adjusted for flow, and Van Kleeck VS destruction cannot be used (Ge *et al.*, 2010). In some cases, TS and VS values must be corrected for losses of volatile compounds during the measurement procedure (Kreuger *et al.*, 2011; Peces *et al.*, 2014).

5.3.5.2 Process stability

As discussed in section 5.3.1, AD is a series of complex biochemical processes including hydrolysis, fermentation and methanogenesis that only occur in environments that are suitable for the corresponding microbial communities, that is at suitable temperature, pH, in absence of inhibiting substances, in presence of abundant substrates for microbial growth and at an appropriate treatment time. Therefore, a range of physical, chemical and biological parameters require monitoring in order to maintain reliable process performance in terms of organic degradation or to troubleshoot process perturbations/failures (Table 5.9).

Volatile fatty acids (VFA) including acetate, propionate, butyrate and valerate, are intermediate products in anaerobic digestion (see Figure 5.1). When digesters are functioning effectively, acid forming processes (hydrolysis and fermentation) exist in balance with acid consuming processes (methanogenesis); however, when the process is not balanced VFA may accumulate. Therefore, concentration of VFAs in the bulk reactor or effluent can be used as an indicator of process performance and stability. VFA accumulation trends and changes are more useful than instantaneous concentrations when monitoring the performance, health and stability of AD processes. This is partly because microbial communities will respond differently to VFA depending on the microbial community profile and the operating conditions (Angelidaki *et al.*, 1993) and partly due to an emerging trend to design AD reactors to produce VFA as an end product rather than methane. VFA analysis by gas chromatography (GC) or liquid chromatography (HPLC) provides information about the concentration and speciation of VFA; however, bulk VFA concentrations can also be estimated at much lower cost using simpler titration methods, as described in the following section.

Alkalinity (also known as buffer capacity): In the context of AD, alkalinity represents the capacity of the digestate to buffer pH decreases from the productions of organic acids and CO_2. Alkalinity is

Table 5.8 Approaches for monitoring volatile solids destruction in anaerobic digesters.

Calculation Method	Equation	Key Advantages and Common Sources of Error
Mass-balance VS destruction	$VS_{destroyed} = \frac{(VS_{conc,in} - VS_{conc,out})}{VS_{conc,in}} \quad (5.9)$ VS_{conc} is concentration of organics as measured by the volatile solids method (g L^{-1}), and subscript in and out indicate concentrations in the inlet and outlet streams	• Sensitive to systematic sampling issues. For example, it is common to have differential settling around sample points, such that the solids concentration is not representative of the bulk reactor, or outlet concentration
Van Kleeck VS destruction	$VS_{destroyed} = \frac{VS_{frac,in} - VS_{frac,out}}{VS_{frac,in} - VS_{frac,in} VS_{frac,out}} \quad (5.10)$ VS_{frac} is fraction of total solids that is volatile (VS_{conc}/ TS_{conc})	• Sensitive to accumulation of mineral solids and/or grit in the reactor (which will read as a false low destruction), or precipitation of mineral solids (which will read as a false high) • Not as susceptible to systematic sampling issues, as dilution or accumulation of mineral and organic solids will normally be consistent • It is not dependent on flow measurement
Gas flow VS destruction	$VS_{destroyed} = \frac{COD_{gas}(kgCOD.d^{-1})}{COD_{in}(kgCOD.d^{-1})} \quad (5.11)$ COD_{in} is incoming COD load, based on the feed volume and feed COD concentrationCOD_{gas} is the calculated gas flow COD in kgCOD.d^{-1}, calculated using Equation (5.10): $COD_{gas} = 2.9 Q_{gas} p_{CH4} \quad (5.12)$ Q_{gas} is volumetric gas flow converted to standard temperature and pressure (0°C, 1 bar) ($Nm^3.d^{-1}$), p_{CH4} is the methane composition of the biogas (%), and 2.9 is a conversion factor based on the density of methane and the oxidation chemistry (kgCOD Nm^{-3})	• Least reliable of the three methods • Dependent on correct flow measurement in both liquid and gas streams, as well as correct VS inlet measurement • Sensitive to the measured COD:VS ratio, and this can vary significantly • Methods assumes that the COD:VS ratio is the same across the digester and does not change during digestion, which is not strictly correct • While it should not be used as a primary measure of VS destruction, it is a useful comparison to other terms • It is not sensitive at all on outlet flow measurement or outlet liquid digester analysis

measured using acid titrations and the results presented as $mgCaCO_3/L$ equivalent. Partial alkalinity is determined using pH 5.7 as the titration end point and represents the concentration of bicarbonate in the process (APHA *et al.*, 2005). Total alkalinity is determined using pH 4.3 as the titration end point and represents both bicarbonate and VFA in the process (APHA *et al.*, 2005). VFA concentrations can therefore be approximated from the difference between total alkalinity and partial alkalinity, also known as the intermediate alkalinity (Ferrer *et al.*, 2010; Fonoll *et al.*, 2015). The ratio of intermediate alkalinity to partial alkalinity is a further measure of process stability, with ratios of <0.3 desired in healthy and stable processes (Martín-González *et al.*, 2013).

Total Kjeldahl nitrogen (TKN) and total ammonical nitrogen (TAN): Free ammonia (NH_3) inhibition is one of the most common inhibition processes observed in AD processes (Rajagopal *et al.*, 2013; Yenigün & Demirel, 2013). Ammonia nitrogen is present in many AD feedstocks and is

Table 5.9 Approaches for monitoring the stability and performance of anaerobic digestion processes.

Process Parameter	Measurement Technique	Process Health and Stability Indication
COD	On-line: UV-Vis spectroscopy Off-line: colorimetric method	Total COD is used to measure the removal of organic matter in anaerobic digesters, while soluble COD can be used to monitor the concentration of intermediate compounds and therefore process stability
TS/ VS	On-line: Limited use of turbidity sensors for TS measurement Off-line: Oven based drying and combustion methods	VS destruction is a primary measure of process performance. There are no universal VS destruction standards, therefore focus on changes in VS destruction performance. Use multiple calculation methods (Table 5.8) to assess data quality
Biogas composition	On-line: Infrared sensor (IR) Off-line: Gas chromatography (GC)	Changes in biogas composition may occur due to changes in the feedstock. However, changes in biogas composition may also indicate a change in process balance. In particular, H_2 may be present if acid producing processes exceed methanogenic capacity
pH	On-line: *in-situ* pH sensor Off-line: Portable pH probe	Maintain a neutral pH (7–8) to limit direct inhibition on microbial communities and to reduce the impact of ionic substances
Temperature	On-line: Thermocouple or RTD sensor	Temperature impacts biological, chemical and physical processes that occur within AD processes. Minimize sudden changes in temperature (e.g. ΔT within heating loops should be <4–6°C)
Alkalinity	On-line and off-line: Titration with HCl	Not a direct indicator of process performance, but indicates capacity of the system to resist a decline in pH. Alkalinity >2 $gCaCO3.L^{-1}$ desired and alkalinity ratio of >0.3 desired
VFA	On-line: Rare, estimated through titration Off-line: Gas Chromatographic (GC) or high-performance liquid chromatography (HPLC)	Changes in VFA accumulation or speciation represent a change in balance between acid producing processes and acid consuming processes. There are no universal inhibition thresholds
Other inhibitory compounds	Off-line: • Ions: Ionic chromatography or conductivity probe • NH_3: Ion selective probe or ionic conductivity • Phenolic/aromatic compounds by HPLC • Heavy metals: Inductively coupled plasma (ICP)	

produced during AD caused by the hydrolysis of organic nitrogen compounds such as protein. Total Kjeldahl nitrogen (TKN) is a measurement of the combined concentration of organic nitrogen and ammonia, therefore TKN represents the potential ammonia concentration in a digester if complete degradation of the feedstock occurs. Therefore, both TKN and total ammonical nitrogen (TAN) measurements are commonly included in AD monitoring programs to analyze the risk of ammonia inhibition and/or to troubleshoot failing AD processes. However, as for VFA concentration, the TAN inhibition threshold depends on multiple factors, including process conditions (pH, temperature) and microbial tolerance (Calli *et al.*, 2005; Lu *et al.*, 2018; Tian *et al.*, 2018). For this reason, trend data and specifically changes in TAN concentration or pH over time are often more valuable in process troubleshooting than looking at instantaneous concentrations.

pH level: Anaerobic digestion processes, particularly the methanogenesis steps, are sensitive to pH. At low pH values, that is <6, biological activity of methanogens can be inhibited, through the disruption of homeostasis or the denaturation of enzymes. Furthermore, a lowered pH shifts the acid-base equilibrium of ionic substances. In the case of acids, such as hydrogen sulphide or VFAs, low pH increases the concentration of uncharged, or 'free' forms, which increases their inhibitory effect. In contrast, bases such as ammonia are present in a free form at elevated pH, that is >7.5, and lower pH can reduce (but not remove) the inhibitory effects. Maintaining a neutral pH can reduce the impact of both hydrogen sulphide and free ammonia. It is also important to avoid rapid changes in pH. Continuous (online) measurement of pH and measurement of alkalinity are useful parameters to ensure the process pH is within desired limits and that there is sufficient buffering capacity to resist rapid pH shifts. As a guideline, an alkalinity of 2–3 $gCaCO_3.L^{-1}$ is desirable to ensure stable process operation in terms of pH levels. The specific alkalinity requirements will depend on the temperature and the CO_2 concentration of the biogas with lower alkalinity requirements at higher temperatures and lower CO_2 concentrations.

5.3.5.3 *Process troubleshooting*

AD is sensitive to environmental changes and the presence of inhibitory or toxic compounds in the feedstock. It is important to minimize destabilized or unbalanced process conditions to prevent significant and/or long-term damage to the microbial community, particularly the methanogenic population due to their relatively slow growth rate, low cell yields and thus prolonged recovery from a process failure. Early warning signs for an unstable AD process are: (i) reductions in methane production; (ii) increased concentration of H_2 in biogas; and (iii) accumulation of VFAs. Signs of long(er) term instability of AD processes include: (i) shifts in pH; and (ii) a decrease in VS destruction rate. Figure 5.15 summarizes a troubleshooting procedure designed to identify and correct common operational issues in AD process.

5.4 CASE STUDIES AND IMPLEMENTATION

5.4.1 Case study 1: anaerobic digestion of agricultural wastewater

Figure 5.16 shows a covered anaerobic lagoon installed as part of the waste treatment process at a pig production facility. The pig production facility is located in a rural area and does not have an option to discharge water to a sewer. Therefore, the wastewater must be treated to achieve standards required for irrigation. The facility produces an average of 500 000 m^3 of wastewater per day over a 12-month cycle, with higher water use in summer months and lower water use in winter months. Average wastewater compositions after screening are 8200 mg L^{-1} VS, 15 600 mg L^{-1} tCOD, and 5200 mg L^{-1} sCOD. However, both flow and composition are subject to large daily and seasonal variations. Removal of the organics was identified as a core treatment requirement, with residual nutrients being used for irrigation. The waste treatment train used at this facility includes static screens, followed by covered anaerobic lagoons, a partially aerated basin and large winter storage lagoons. Covered anaerobic lagoons were selected to minimize the cost and operating requirement of the treatment process. Lagoon treatment was particularly suitable due to the large footprints available and the large degree of internal equalization to handle variable flows and compositions of the wastewater. However, temperature control is difficult in lagoon treatment and the seasonal variation in ambient temperature has been a key operational challenge. At this site, monthly temperature averages vary from 8 to 24°C. Biological processes are significantly slower during the cooler months and this was addressed using a larger lagoon with longer retention time to reduce loading stress during the winter months. The covered lagoon has an operating volume of 18 000 m^3. The dimensions are approximately 130 m long and 40 m wide with a maximum depth of 7.5 m. The effluent is further treated using a

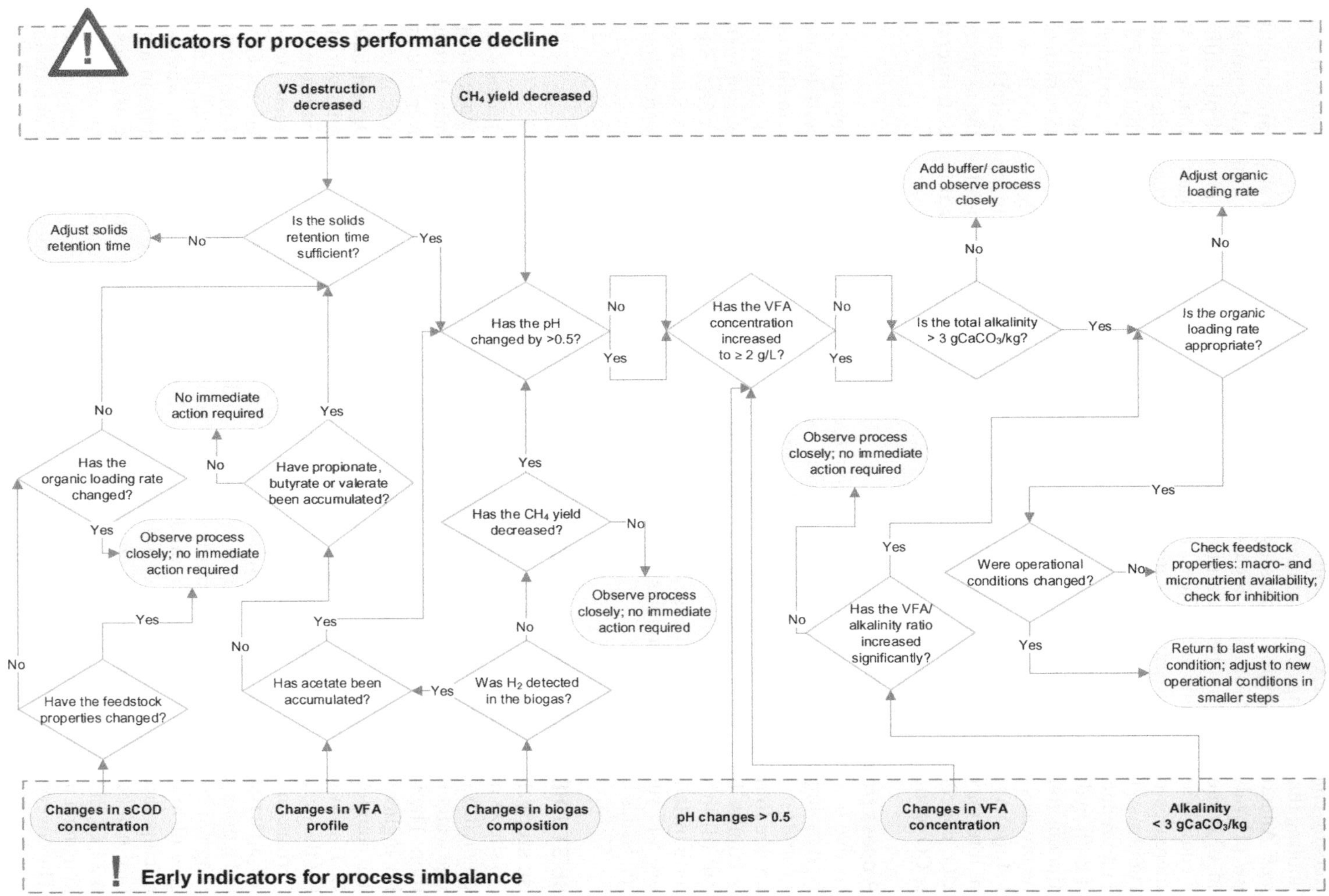

Figure 5.15 Decision tree for troubleshooting and correcting anaerobic digestion processes.

Figure 5.16 Example of a covered anaerobic lagoon installed at a rural pig production facility (Gopalan, 2013).

9000 m^3 partially aerated basin before being sent to winter storage lagoons (120 000 m^3). The average HRT in the CAL is 36 days, with much longer SRT (i.e., years). The annual average organic loading rate for this lagoon is 0.4 $g.L^{-1}.d^{-1}$ of COD. Average COD and VS removal in the CAL are 70 and 64% respectively, but all performance parameters show seasonal variations. For example, average biogas production in summer months is 3600 $m^3.d^{-1}$ compared to 2500 $m^3.d^{-1}$ in winter months.

The use of a CAL in this application is possible due to the temperate climate and the large parcels of land available for both the treatment process (CAL + storage) and irrigation of the treated wastewater.

5.4.2 Case study 2: high-rate anaerobic digestion in the food and beverage industry

Figure 5.17 is an example of an internal circulation reactor and associated infrastructure installed at a brewery. The brewery is located in a densely populated urban industrial area and discharges water to the sewer network. The wastewater discharge is subjected to trade waste fees for both the volume of water and the load of organic, solid and nutrient contaminants. It is important to note that these trade waste fees are relatively high and therefore a major incentive for the brewery to treat the wastewater on-site prior to sewer discharge. The plant generates 2000 m^3 of wastewater per day containing on average approximately 5000 $mg.L^{-1}$ tCOD, 50 $mg.L^{-1}$ N and 30 $mg.L^{-1}$ P. However, the COD content can reach a maximum concentration of >30 000 $mg.L^{-1}$ during the daily operating cycle. The organic component of the waste is carbohydrate and alcohol with no fat, oil or grease present. Biological removal of the organics was identified as a core requirement of reducing and managing the commercial waste charges due to the very high COD to N ratio. Energy consumption at the brewery is high due to high steam production requirements. This provided a clear opportunity to utilize biogas energy onsite within the boiler system. The potential for energy recovery through AD rather than the high additional energy consumption that would be required to oxidize the COD through aeration was a significant cost factor that was considered during the selection of AD. Specifically, a high rate granular reactor was selected due to the very tight space restrictions on site. The anaerobic treatment train includes a 600 m^3 equalization and pre-fermentation tank prior to the IC (internal recirculation reactor), with a second 400 m^3 diversion tank to handle excessive flow or composition variations. Biogas is produced in a 500 000 m^3 IC reactor, and hydraulic retention is managed using a splitter to recirculate a portion

Figure 5.17 Example of an internal circulation (IC) reactor and associated infrastructure installed at a brewery.

of effluent back through the IC. The IC effluent is sent to an aerobic membrane bioreactor (MBR) for polishing prior to discharge. The MBR is not designed for nitrification–denitrification, but to remove residual COD, N and P through sludge growth and collection. The treatment train removes over 95% of COD from the wastewater with approximately 90% removed in the IC. Average biogas production from the plant is 250 $m^3.h^{-1}$, which offsets approximately 5–10% of the site boiler fuel requirements.

5.4.3 Case study 3: anaerobic digestion of municipal wastewater sludge

Figure 5.18 shows an example of mesophilic anaerobic sludge digester installed during the upgrade of a municipal sewage treatment plant (300 000 PE (population equivalent)). The sewage treatment plant is required to meet tight nutrient discharge limits of 3 $mg.L^{-1}$ TN (total nitrogen) and 1 $mg.L^{-1}$ TP (total phosphorus). This is achieved without using primary sedimentation. Instead, raw wastewater is supplied directly to an activated sludge system to drive the biological removal of nutrients. The activated sludge system is operated at an extended sludge age of 15 days, thereby producing a waste activated sludge comprised primarily of microbial biomass which biodegrades slowly. Anaerobic digestion was installed at the WWTP as part of a plant upgrade in 2009. The upgrade had three core objectives:

(1) Provide an additional 16 ML/d (~60 000 EP) of treatment capacity;
(2) Match or exceed performance of the existing plant with effluent TN of <2.5 mg/L;
(3) Construct a sludge management train to produce 'Grade B' treated biosolids.

Anaerobic digestion was installed to meet the requirement of producing Grade B treated biosolids and also because of renewable energy and sustainability drivers. During the design phase, BMP testing (see section 5.3.2) was conducted to demonstrate that WAS-only anaerobic digestion was biologically and economically feasible and, moreover, to determine the treatment times required to meet sludge stability requirements. High rate granular processes were not considered for this application due to the high fraction of particulate organic matter and the low degradability of the substrate ($f_d \approx 33\%$,

Figure 5.18 Example of two continuous mixed liquor digesters installed at a sewage treatment plant to stabilize waste activated sludge (Gloag *et al.*, 2014).

determined by the BMP testing). Mixed liquor digesters were selected due to the urban location and limited footprint. A series of energy efficiency strategies were added to reduce energy demand by over 60% and compensate for the relatively poor sludge degradability. Measures included pre-thickening the WAS to reduce digester volume and heating requirements, insulating the digesters to reduce heat loss and installing high efficiency boilers to drive the CHP units.

The digestion plant consists of two 6000 m^3 mixed liquor digesters operating in parallel. The design SRT is 20–26 days. Digester mixing is achieved using gas lances that allow biogas to be pumped into the bottom of the reactor. The gas lances are distributed throughout the reactor and operate on a sequential cycle. Digester heating is achieved via sludge recirculation through external heat exchangers. The digesters achieve over 30% VS destruction, which corresponds to more than 90% of the biodegradable material (measured through BMP testing). Biogas production exceeds the average 3000 m^3/d predicted from batch testing, which is consistent with previous observations that a well operated reactor can exceed the performance measured in BMP testing. The two anaerobic digesters produce more than enough energy to meet mixing and heating requirements and therefore produces excess energy that is used to offset aeration energy in the mainline wastewater treatment plant.

5.5 CHALLENGES, OPPORTUNITIES AND RESEARCH NEEDS

Anaerobic digestion is a mature technology for waste treatment and renewable energy production. The growing importance of anaerobic digestion in the green/circular economy is readily visible by: (i) the exponential growth of scientific publications in peer reviewed journals, which increased from 230 in 2001 to 3100 in 2018 (Elsevier Scopus, 10th March 2019); and (ii) steady growth in patents from the same period (Albers *et al.*, 2016). This increasing interest in anaerobic digestion technology is a consequence of several key facts, including: (i) higher knowledge of the AD process and constraints; (ii) successful experience in constructing and operating AD plants; (iii) the potential of AD for waste treatment and energy production; and (iv) environmental driver to divert organic waste from landfill towards re-use. Although AD is considered a mature technology, further development is needed to improve the efficiency of AD and support its transition from waste treatment technology to a key technology in the circular economy.

The success of AD as a technology has promoted the diversification of its application from traditional substrates (e.g., sewage sludge, animal manure, food waste, food processing residues) to a wider and complex range of substrates such as slaughterhouse wastewater, poultry manure, energy crops, microalgae and microalgae residues, petrochemical wastewater and landfill leachate, among many others. This flexibility is a particular strength of AD. However, there are challenges, but also opportunities, when applying AD.

Economic challenges: When AD is applied to substrates with poor biodegradability or very slow digestion rates, the process economics are poor due to a combination of large reactor volume requirements (high CAPEX), poor biogas yields (low revenue) and high residual solids for disposal (incurring large OPEX). These challenges exist in both potential new AD applications and within existing AD plants where changes in environmental regulations, such as lower TN discharge, are changing upstream processes in such a way that feedstock degradability is reduced. Opportunities to process more complex and challenging substrates have driven extensive research into pre-treatments that aim to make the substrate more biodegradable (e.g., WAS thermal hydrolysis, crop residues silage) and/or remove inhibitory compounds (e.g., phenols in petrochemical wastewater, ammonical nitrogen in poultry manure). Carrère *et al.* (2010) published a comprehensive review on pre-treatments prior to AD, which also discusses the relative suitability of each pre-treatment type for different substrate types. However, pre-treatment development remains relatively immature. Commercial applications of AD pre-treatment are increasing, as pre-treatment technologies continue to improve, and optimization strategies develop such that the benefits seen during AD exceed the pre-treatment cost.

Presence of process inhibiting compounds: Many potential feedstocks for AD contain potential inhibitors of AD microbiology, including sodium and long chain fatty acids in fish processing waste, potassium in molasses, ammonia in some manures, heavy metals in industrial wastes (leather manufacture) and more. There is extensive ongoing research in AD inhibition and resilience and on how inhibitors impact the microbial community of the anaerobic digestion process. Inhibition reviews have been published by Chen *et al.* (2008, 2014), Rajagopal *et al.* (2013) and Yenigün and Demirel (2013). Results generally show that there are not universally applicable inhibition threshold concentrations for any inhibitory compound, rather that inhibition resilience is influenced by factors including: (i) digester substrate type; (ii) digester configuration; (iii) experimental techniques; (iv) microbial community; and microbial acclimation. There remain knowledge gaps regarding the relevance of each factor behind the large variability on inhibition threshold concentrations. The relationship between process performance, inhibition resilience and microbial community is also a topic that warrants further research.

Micronutrient requirements: There remains a poor understanding of the requirements and function of some micronutrients in AD processes. Micronutrients such as Fe, Ni, Mo, Co, and Se are cofactors in numerous enzymatic reactions and must be present for the process to occur (Romero-Güiza *et al.*, 2016), however these compounds also result in process inhibition at high levels. There is a lack of understanding regarding the requirements of micronutrients, and the complexing or bioavailability of micronutrients in poorly performing processes.

Anaerobic co-digestion (AcoD): The simultaneous digestion of two or more substrates can be an opportunity to overcome the drawbacks of mono-digestion (single substrate anaerobic digestion) and improve the economic feasibility of existing AD infrastructure. The feasibility of AD plants is improved through the increased biogas production and the treatment of several wastes in a single facility to 'share/reduce' treatment costs. AcoD ideally focuses on mixing substrates that promoted positive interactions between substrates (e.g., macronutrient and micronutrient equilibrium, moisture balance and dilution of challenging compounds (i.e., grease trap wastes). However, due to the industrial outlook of AD plants and because the improvement of methane production is primarily linked to the increased organic loading rate, rather than positive interactions, all kinds of mixtures have been considered and applied at full-scale (Mata-Alvarez *et al.*, 2014). The fact that each co-digestion scenario is different means that AcoD implementation will always require a minimum level of decision-making, risk management and research. There is a need to develop comprehensive tools to support co-substrate selection. Ideally, this tool would consider the plant-wide implications of AcoD, include positive impacts from enhanced biogas revenue and negative impacts from increased nutrient load, sludge production and residual solids handling (Arnell *et al.*, 2016). Anaerobic co-digestion including research and full-scale case studies has been extensively reviewed in the recent years and we would to refer the reader to the following documents: Mata-Alvarez *et al.* (2014), Nghiem *et al.* (2017) and Xie *et al.* (2018).

5.6 SUMMARY

Anaerobic digestion (AD) is a complex biochemical process where organic matter decomposes in the absence of oxygen through four sequential metabolic steps, namely: (i) hydrolysis; (ii) acidogenesis (primary fermentation); (iii) acetogenesis (secondary fermentation); and (iv) methanogenesis. The AD process is flexible enough to be applied to both wet and dry organic wastes, although the moisture content will impact reactor design, mass transfer phenomena and ultimate process rates. The principle products of anaerobic digestion are biogas (CH_4 and CO_2) and non-digested residues termed digestate. When biogas is captured and burned to generate heat and/or electricity, AD processes are generally considered as net-energy producing and this is a core advantage of the technology. Nutrients are preserved during AD and this provides the opportunity to use digestate as a fertilizer or a precursor of other value-add technologies.

Feedstock composition is a key parameter that impacts both the metabolic pathways required for anaerobic degradation and the degradation end-products including methane yields. Theoretical methane yields are related to the chemical structure and the COD/VS ratio of the substrate. Theoretical yields for completely degradable model substrates range from 415 L CH_4 kg^{-1} VS to 496 L CH_4 kg^{-1} VS and 1007 L CH_4 kg^{-1} VS (gas volumes expressed at 0°C, 1 atm) for carbohydrates, proteins and lipids respectively.

In practice, biogas production is lower than the theoretical yields due to the complex structure of feedstocks and the presence of non-anaerobically degradable compounds such as lignin, inert organics or inhibitory compounds. Practical estimates of methane potential are based on experimental anaerobic biodegradability assays, such as the BMP test which determines: (i) the ultimate methane potential (B_0) – the maximum volume of methane that can be produced from a specific sample; (ii) degradable fraction f_d – the fraction of COD that is biodegradable and can be converted to methane; and (iii) apparent first order hydrolysis rate coefficient (k_{hyd}) – the speed of degradation.

A broad range of AD technologies are commonly used including lagoon-based technology, high rate granular processes, high rate membrane processes, continuous mixed liquor digesters, solid phase reactors and multi-stage or hybrid configurations. Selection of AD reactor technology is based on several considerations related to both the characteristics of the feedstock and constraints at the plant location. Considerations related to the feedstock include the fractionation between particulate and soluble organic material, the presence of fat, oil and grease, the moisture content and the expected time required for degradation. Considerations related to the plant location include the available footprint, local climate and presence of supporting infrastructure.

There are three core factors that determine the sizing of AD processes: (i) managing organic load to provide beneficial conditions for methanogenesis (including maintaining pH above 7.0); (ii) maintaining sufficient retention time for the growth and maintenance of the required AD microbiology; and (iii) maintaining sufficient retention time to allow for hydrolysis of particulate substrates. The limiting factor is determined by the feedstock characteristics and the specific process constraints for each application. In general, feedstocks are classed as solid waste or wastewater, based on the distribution of particulate and soluble COD.

For AD of particulate substrates, hydrolysis is considered the process-limiting step and reactor sizing is based on the apparent first order hydrolysis rate, the reactor configuration (i.e., mixed vs plug flow) and the extent of degradation desired. Batch, continuous mixed and continuous plug-flow reactor configurations may be utilized. For AD of soluble substrates, methanogenesis is considered the process-limiting step due to the slower microbial growth and slower substrate utilization of this step. Processes treating soluble wastewater often incorporate biomass retentions systems, in order to decouple the SRT from the HRT. Reactors treating soluble wastewater are typically sized based on the organic loading rate, with nominal design parameters for each technology type.

While AD processes are a source of biogas and renewable energy, in many AD applications the process economics and therefore the performance goals are more strongly related to the level of solids destruction and stabilization. COD and VS are the most common metrics used to assess the performance of anaerobic digesters. These metrics are used for balancing and calculation of removal rates. A summary of physical, chemical and biological parameters that may be monitored in order to maintain reliable process performance, achieve pollutant removal standards or troubleshoot process failures includes:

- *Chemical oxygen demand* (*COD*): Total COD is used to measure the removal of organic matter in anaerobic digesters, while soluble COD can be used to monitor the concentration of intermediate compounds and therefore process stability.
- *Volatile solids* (*VS*): VS destruction is a primary measure of process performance and is measured using three common calculation methods: mass-balance VS destruction, Van Kleeck VS destruction and apparent VS destruction on gas flow.
- *Temperature*: Temperature has a strong impact on the biological, chemical and physical processes that occur within AD processes. Rapid changes in temperature can cause significant process instability, however progressive changes in temperature also impact process capacity, with slower reaction rates at lower temperatures.
- *pH*: Biological activity can be directly inhibited at low pH (<6). Acid-base equilibria of many common inhibitors is directly impacted by pH, with a corresponding impact on the severity of inhibition. Maintaining a neutral pH can reduce the impact of both acid and base inhibitors, it is also important to avoid rapid changes in pH.
- *Volatile fatty acids (VFAs)*: VFA accumulation occurs when there is a change in balance between acid forming processes (hydrolysis and fermentation) and acid consuming processes (methanogenesis). VFA accumulation can directly inhibit methanogenesis and can drive changes in pH which increase inhibition.
- *Total ammonical nitrogen*: Ammonia inhibition is one of the most common inhibition processes observed in AD processes. TKN and TAN measurements are commonly included in AD monitoring programs to analyze the risk of ammonia inhibition and/or to troubleshoot failing AD processes.

5.7 EXERCISES

Exercise 5.1: A wastewater stream contains 5 g.L^{-1} carbohydrate, 2 g.L^{-1} protein and 1 g.L^{-1} fat, oil and grease. If the waste is completely degradable, calculate:

(a) The maximum theoretical methane that could be produced from this waste stream.
(b) The total volume and composition of the biogas stream (HINT: consider CH_4 and CO_2 only).
(c) The potential heat that could be generated from the biogas (HINT 55 MJ.kg^{-1} CH_4).
(d) The potential electricity that could be generated from the biogas (HINT: electrical efficiency 0.35).

Exercise 5.2: A solid phase plug flow digester is loaded with a waste stream that contains tCOD at 120 000 g.kg^{-1}. The waste stream has a degradable fraction of 0.65 and an apparent hydrolysis rate coefficient of 0.08 day^{-1}. If the waste is well inoculated and the SRT in the reactor is 30 days, determine the expected methane yield (per gCOD added) and the fraction of COD removed.

Exercise 5.3: Typical wastewater production from an animal processing plant is shown below:

Flow: 2500 m^3.d^{-1}
COD: 6000 mg.L^{-1}
TSS: 1000 mg.L^{-1}
FOG: 2500 mg.L^{-1}
TKN: 200 mg.L^{-1}
TP: 65 mg.L^{-1}
pH: 7–9
Temp: 36°C

(a) Using organic loading rate as a critical design parameter, estimate the size of an AnMBR required to treat this wastewater.
(b) The AnMBR is designed with an external submerged membrane module containing microfiltration membranes. If the membranes operate at a flux of 15 LMH, estimate the required membrane surface area.
(c) If the external membrane module contains approximately 100 m^2 per cubic meter of tank volume, estimate the additional tank volume required for the membranes.
(e) Estimate the methane production when COD conversion in the reactor is 80% of the organic load.

Exercise 5.4: Experimental results from a series of bio-chemical methane potential tests are shown in the table below.

Time (Days)	Pig Manure	Grease Trap Waste	Organic Fraction of MSW
	CH_4 (L.kgVS^{-1})	CH_4 (L.kgVS^{-1})	CH_4 (L.kgVS^{-1})
0	0	0	0
2	112	419	36
4	187	607	66
6	238	691	90
8	271	729	110
10	294	746	126
15	323	758	155
20	334	760	173
25	338	760	184
30	339	760	190
40	340	760	196

(a) For each substrate, determine the ultimate methane potential (B_0), and the apparent hydrolysis rate coefficient (k_{hyd}) (HINT: you can use graphical methods to estimate the parameters).
(b) Do any of the tests show signs of microbial inhibition?

Exercise 5.5: For each of the wastewater streams in the table below, suggest a suitable anaerobic digestion technology. Justify your selection:

Industry	Municipal Primary Sludge	Paper Mill Wastewater	Mixed Horticultural Waste
Volumetric flow ($m^3 \cdot d^{-1}$)	80	1200	150 (seasonal)
tCOD ($mg.L^{-1}$)	70 000	10 000	265 000
sCOD ($mg.L^{-1}$)	22 000	6500	30 000
TSS ($mg.L^{-1}$)	44 000	150	215 000
VSS ($mg.L^{-1}$)	34 000	140	180 000
FOG ($mg.L^{-1}$)	3000	<1	1500
TKN ($mg.L^{-1}$)	1500	30	3500
TP ($mg.L^{-1}$)	260	<1	520

Exercise 5.6: A food production plant discharges wastewater to the sewer after primary treatment. As is typical in food processing plants, the WWTP sees large variations in wastewater quality and flow, with the following characteristics after primary treatment:

Flow: 900–1500 $m^3.d^{-1}$
COD: 600–1000 $mg.L^{-1}$
TSS: 250 $mg.L^{-1}$
pH: 7–9
Temp: 32°C

The food processing plant is in a rural location with a warm temperate climate. The plant is considering either a covered anaerobic lagoon or a high rate granular process:

(a) If a UASB reactor is used to treat this wastewater, recommend an appropriate reactor size.
(b) If a covered anaerobic lagoon is used to treat this wastewater, recommend an appropriate lagoon size.
(c) Present advantages and disadvantages of each treatment option and provide a recommendation.
(d) Discuss design strategies or ancillary equipment that could be added to the process to address the disadvantages mentioned in part (c).

Exercise 5.7: Consider the mixed liquor anaerobic digestion process summarized in the flowsheet below:

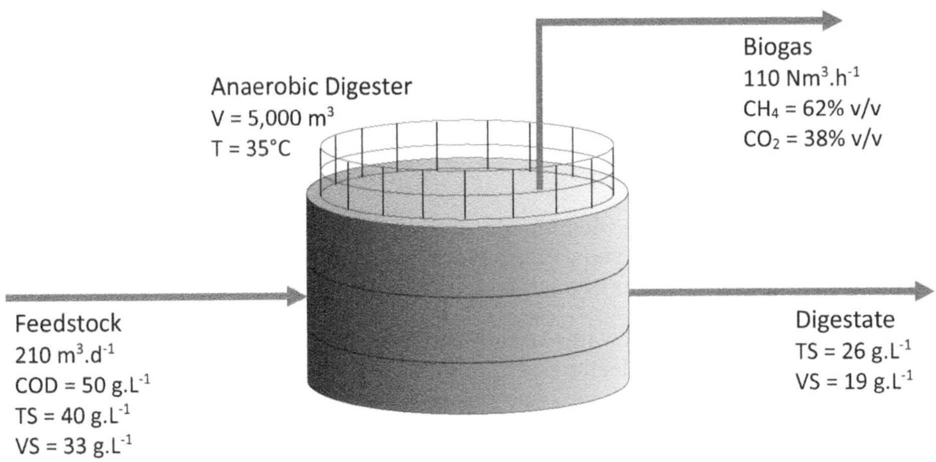

(a) Determine the solids retention time in the process.
(b) Calculate the organic loading rate for the process.
(c) Calculate the volatile solids destruction using: the mass balance equation, the Van Kleeck Equation, gas flow.
(d) Estimate the fraction of COD removed from the waste stream and the expected COD concentration of the digestate.

Exercise 5.8: You have been asked to evaluate a novel two-stage anaerobic digestion process shown in the flowsheet below. The digesters have been constructed above the ground using concrete without insulation. The average ambient temperature is 25°C.

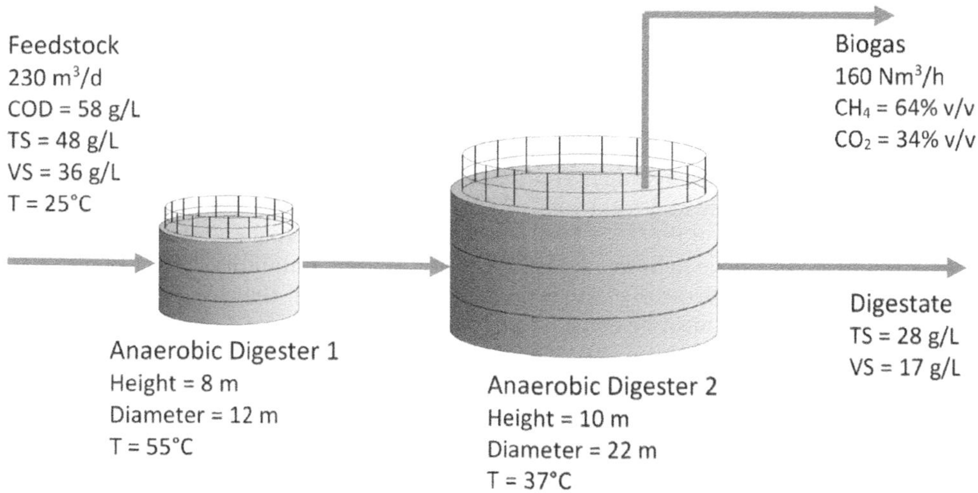

(a) Calculate the solids retention time for each digester and the combined retention time for the process. Is the retention time sufficient?
(b) Calculate the heating requirement for anaerobic digester 1
(c) Calculate the heating requirement for anaerobic digester 2
(d) Calculate the heating energy from combustion of the biogas.
(e) Is this process net energy-generating or net energy-consuming?

Exercise 5.9: You have been asked to evaluate the feasibility of single-stage anaerobic digestion applied to dairy manure. The raw dairy manure has a composition of 210 g.L^{-1} TS, 185 g.L^{-1} VS, 295 g.L^{-1} COD, 4.4 g.L^{-1} TKN and an average daily production rate of 60 metric ton/day on a fresh weight basis. Your laboratory has conducted BMP testing and reported that the B_0, f_d, and k_{hyd} values for the waste at 35°C are 340 L.kgVS^{-1}, 0.55, and 0.14 day^{-1} respectively. Consider the two options below:

(a) A continuous mixed liquor reactor with a maximum feed concentration of 5% TS.
(b) A continuous solid phase plug flow reactor with a maximum feed concentration of 15% TS.

Determine the size requirement of each option. Discuss how this might impact the footprint and heating requirement (no calculations required). Which option would you recommend, explain your answer?

5.8 DISCUSSION QUESTIONS

Question 5.1: A brewery is considering a number of different technology options for a new wastewater treatment plant. There are three contemporary aerobic options proposed, along with one anaerobic technology.

What issues would you consider in evaluating aerobic versus anaerobic technologies?
What data would you request in order to assess and evaluate the technology options?
What specific differences would you expect when evaluating technology options?

Question 5.2: You work for a large water corporation. The General Manager recently made a commitment to the public that your wastewater treatment processes will achieve 'energy neutral' operations within ten years.

- How would you define 'energy neutral operations'?
- What impacts do you think this may have on the treatment process or treatment goals?
- What are the major energy-consuming and energy-generating processes that need to be considered?
- How do you think you can achieve 'energy neutral operation' and what changes or technologies would be required?

Question 5.3: A domestic sewage treatment plant (65 $ML.d^{-1}$) comprises a long sludge age BNR activated sludge system. There is no primary sedimentation. Sludge production at the site is approximately 1000 metric tons of sludge per day at a total solids content of 2%. There is a landfill approximately 15 km from the treatment plant, but there is a levy for each ton of waste that enters the landfill. Land application is allowed if the sludge achieves stability, pathogen and contaminant requirements. The cost of distributing the material across the land is relatively low, however the closest land available is 150 km away and transport costs must be considered.

- What are some of the major factors to consider when evaluating the treatment and disposal options for this sludge?
- When considering AD of this material, there are likely going to be some specific challenges that impact the design, performance and economics of the AD process. What are these challenges?
- What techniques could be utilised to improve AD potential?

Question 5.4: Biosolids is a term used to describe waste organic sludges, typically sewage sludges that have been treated and stabilised. In some countries, a high fraction of biosolids are re-used on agricultural land, while in other countries land application is banned and biosolids are sent to landfill or to incineration. Discuss the relative merits and potential risks of applying biosolids to farm land. How are these risks addressed through conventional treatment processes such as anaerobic digestion? Are there alternative technologies that could better manage these risks?

Question 5.5: Many sewage treatment plants are built to handle population growth within the catchment area within the plant operating life. As a result, anaerobic digesters at STPs can have large amounts of spare treatment capacity. Discuss options to utilize this spare treatment capacity. Discuss the impacts of your suggestion on the STP and downstream processes. What are some pros and cons?

Question 5.6: In the energy crop industry, crops are grown specifically for the production fuel and energy, either via alcohol fermentation, anaerobic digestion or briquetting and combustion. However, energy crop industries utilise farm land that could otherwise be used for food production. Discuss the relative merits of food and energy production and the competition for farm land. How is this situation likely to change in the future?

Question 5.7: In a linear economy, resources are obtained, refined into products, the products are consumed and the wastes disposed. A circular economy seeks to extend the lifecycle of resources through more efficient production and use, but also through the recycling of waste streams into new products, thus extending or renewing the cycle. Does anaerobic digestion fit within a circular economy? Discuss and justify your answer.

Question 5.8: Anaerobic digestion is one of several waste-to-energy platforms that are currently applied at a commercial scale. Other existing and emerging waste-to-energy technologies include combustion, incineration, gasification, pyrolysis and hydrothermal liquefaction. However, energy is not the only valuable product that can be produced from the waste. The carboxylate platform is a concept where mixed cultures of microorganisms convert organic waste to carboxylates that can subsequently be converted to useful bioproducts and commodity chemicals. Considering these technologies and others, discuss the emerging competition for the carbon in waste and criteria that could assist in selecting the best use for carbon recycling from waste.

FURTHER READING MATERIALS

Agler M. T., Wrenn B. A. and Zinder S. H. (2011). Waste to bioproduct conversion with undefined mixed cultures: the carboxylate platform. *Trends in Biotechnology*, **29**(2), 70–78.

Angelidaki I. and Treu L. (2018). Biogas upgrading and utilization: current status and perspectives. *Biotechnology Advances*, **36**(2), 452–466.

Batstone D. J. and Jensen P. D. (2011). Anaerobic processes. In: Treatise on Water Science, P. Wilderer (ed.), Elsevier, Oxford, pp. 615–639.

Carrère H., Dumas C., Battimelli A., Batstone D. J., Delgenès J. P., Steyer J. P. and Ferrer I. (2010). Pretreatment methods to improve sludge anaerobic degradability: a review. *Journal of Hazardous Materials*, **183**(1–3), 1–15.

Chen Y., Cheng J. J. and Creamer K. S. (2008). Inhibition of anaerobic digestion process: a review. *Bioresource Technology*, **99**(10), 4044–4064.

Ho D. P., Ngo H. H. B. and Guo W. B. (2014). A mini review on renewable sources for biofuel. *Bioresource Technology*, **169**, 742–749.

Holliger C., Alves M., Andrade D., Angelidaki I., Astals S., Baier U., Bougrier C., Buffière P., Carballa M., de Wilde V., Ebertseder F., Fernández B., Ficara E., Fotidis I., Frigon J.-C., de Laclos H. F., Ghasimi D. S. M., Hack G., Hartel M., Heerenklage J., Horvath I. S., Jenicek P., Koch K., Krautwald J., Lizasoain J., Liu J., Mosberger L., Nistor M., Oechsner H., Oliveira J. V., Paterson M., Pauss A., Pommier S., Porqueddu I., Raposo F., Ribeiro T., Rüsch Pfund F., Strömberg S., Torrijos M., van Eekert M., van Lier J., Wedwitschka H. and Wierinck I. (2016). Towards a standardization of biomethane potential tests. *Water Science and Technology*, **74**(11), 2515–2522.

Mata-Alvarez J., Dosta J., Romero-Güiza M. S., Fonoll X., Peces M. and Astals S. (2014). A critical review on anaerobic co-digestion achievements between 2010 and 2013. *Renewable and Sustainable Energy Reviews*, **36**, 412–427.

Tchobanoglous G. and Burton F. L. (2003). Wastewater Engineering: Treatment and Reuse, 4th edn. McGraw-Hill, Boston.

Van Lier J. B. and Mahmoud, N. (2008). Anaerobic wastewater treatment. In: Biological Wastewater Treatment: Principles, Modelling and Design, M. Henze, M. C. M. van Loosdrecht, G. A. Ekama and D. Brdjanovic (eds), IWA Publishing, London, UK, pp. 415–456.

Vanwonterghem I., Jensen P. D., Ho D. P., Batstone D. J. and Tyson G. W. (2014). Linking microbial community structure, interactions and function in anaerobic digesters using new molecular techniques. *Current Opinion in Biotechnology*, **27**, 55–64.

Verbeeck K. and Buelens L. C. (2018). Upgrading the value of anaerobic digestion via chemical production from grid injected biomethane. *Energy and Environmental Science*, **7**.

REFERENCES

Albers S. C., Berklund A. M. and Graff G. D. (2016). The rise and fall of innovation in biofuels. *Nature Biotechnology*, **34**(8), 814–821.

Angelidaki I., Ellegaard L. and Ahring B. K. (1993). A mathematical model for dynamic simulation of anaerobic digestion of complex substrates: focusing on ammonia inhibition. *Biotechnology and Bioengineering*, **42**(2), 159–166.

Angelidaki I., Alves M., Bolzonella D., Borzacconi L., Campos J. L., Guwy A. J., Kalyuzhnyi S., Jenicek P. and van Lier J. B. (2009). Defining the biomethane potential (BMP) of solid organic wastes and energy crops: a proposed protocol for batch assays. *Water Science and Technology*, **59**(5), 927–934.

APHA (American Public Health Association), AWWA (American Water Works Association) & WEF (Water Environment Federation). (2005). Standard Methods for the Examination of Water & Wastewater. American Public Health Association, Washington, DC.

Arnell M., Astals S., Åmand L., Batstone D. J., Jensen P. D. and Jeppsson U. (2016). Modelling anaerobic co-digestion in benchmark simulation model No. 2: parameter estimation, substrate characterisation and plant-wide integration. *Water Research*, **98**, 138–146.

Batstone D. J. and Jensen P. D. (2011). Anaerobic processes. In: Treatise on Water Science, P. Wilderer (ed.), Elsevier, Oxford, pp. 615–639.

Batstone D. J., Keller J., Angelidaki I., Kalyuzhnyi S. V., Pavlostathis S. G., Rozzi A., Sanders W. T., Siegrist H. and Vavilin V. A. (2002). The IWA anaerobic digestion model no 1 (ADM1). *Water Science and Technology*, **45**(10), 65–73.

Born J. (2005). From sugar factories to biorefineries. In: Baltic Biorefinery Symposium, M. Madsen, C. J. Lomborg, J. B. Holm-Nielsen, K. Esbensen and E. Karlsen (eds), Esbjerg, Denmark, pp. 23–32.

Calli B., Mertoglu B., Inanc B. and Yenigun O. (2005). Effects of high free ammonia concentrations on the performances of anaerobic bioreactors. *Process Biochemistry*, **40**(3–4), 1285–1292.

Campanaro S., Treu L., Kougias P. G., Luo G. and Angelidaki I. (2018). Metagenomic binning reveals the functional roles of core abundant microorganisms in twelve full-scale biogas plants. *Water Research*, **140**, 123–134.

Carrère H., Dumas C., Battimelli A., Batstone D. J., Delgenès J. P., Steyer J. P. and Ferrer I. (2010). Pretreatment methods to improve sludge anaerobic degradability: a review. *Journal of Hazardous Materials*, **183**(1–3), 1–15.

Chen Y., Cheng J. J. and Creamer K. S. (2008). Inhibition of anaerobic digestion process: a review. *Bioresource Technology*, **99**(10), 4044–4064.

Chen J. L., Ortiz R., Steele T. W. J. and Stuckey D. C. (2014). Toxicants inhibiting anaerobic digestion: a review. *Biotechnology Advances*, **32**(8), 1523–1534.

Da Silva C., Astals S., Peces M., Campos J. L. and Guerrero L. (2018). Biochemical methane potential (BMP) tests: reducing test time by early parameter estimation. *Waste Management*, **71**, 19–24.

Ferrer I., Vázquez F. and Font X. (2010). Long term operation of a thermophilic anaerobic reactor: process stability and efficiency at decreasing sludge retention time. *Bioresource Technology*, **101**(9), 2972–2980.

Fonoll X., Astals S., Dosta J. and Mata-Alvarez J. (2015). Anaerobic co-digestion of sewage sludge and fruit wastes: evaluation of the transitory states when the co-substrate is changed. *Chemical Engineering Journal*, **262**, 1268–1274.

Ge H., Jensen P. D. and Batstone D. J. (2010). Pre-treatment mechanisms during thermophilic-mesophilic temperature phased anaerobic digestion of primary sludge. *Water Research*, **44**(1), 123–130.

Gloag G., Batstone D., Simonis J., Robertson D., Jensen P., O'Halloran K. and Longley V. (2014). WAS Only Mesophilic Anaerobic Digestion at Coonbabah WWTP, Brisbane Australia. Ozwater'14 – Australia's International Water Conference and Exhibition. Australian Water Association, Brisbane, Australia.

Gopalan P. (2013). Low Cost Applications of Anaerobic Digestion to Livestock Wastes. The University of Queensland, Brisbane.

Henze M., Grady C., Jr, Gujer W., Marais G. and Matsuo T. (1987). Activated Sludge Model no. 1: Iawprc Scientific and Technical Report no. 1. IAWPRC, London.

Heubeck S., Craggs R. and Shilton A. (2007). Influence of CO_2 scrubbing from biogas on the treatment performance of a high rate algal pond. *Water Science and Technology*, **55**(11), 193–200.

Holliger C., Alves M., Andrade D., Angelidaki I., Astals S., Baier U., Bougrier C., Buffière P., Carballa M., de Wilde V., Ebertseder F., Fernández B., Ficara E., Fotidis I., Frigon J.-C., de Laclos H. F., Ghasimi D. S. M., Hack G., Hartel M., Heerenklage J., Horvath I. S., Jenicek P., Koch K., Krautwald J., Lizasoain J., Liu J., Mosberger L., Nistor M., Oechsner H., Oliveira J. V., Paterson M., Pauss A., Pommier S., Porqueddu I., Raposo F., Ribeiro T., Rüsch Pfund F., Strömberg S., Torrijos M., van Eekert M., van Lier J., Wedwitschka H. and Wierinck I. (2016). Towards a standardization of biomethane potential tests. *Water Science and Technology*, **74**(11), 2515–2522.

Judd S. (2011). The MBR Book: Principles and Applications of Membrane Bioreactors for Water and Wastewater Treatment. Butterworth-Heinemann, Burlington.

Kirkegaard R. H., McIlroy S. J., Kristensen J. M., Nierychlo M., Karst S. M., Dueholm M. S., Albertsen M. and Nielsen P. H. (2017). The impact of immigration on microbial community composition in full-scale anaerobic digesters. *Scientific Reports*, **7**(1), 1–11.

Kreuger E., Nges I. A. and Björnsson L. (2011). Ensiling of crops for biogas production: effects on methane yield and total solids determination. *Biotechnology for Biofuels*, **4**, 44–44.
Lee S.-M., Jung J.-Y. and Chung Y.-C. (2001). Novel method for enhancing permeate flux of submerged membrane system in two-phase anaerobic reactor. *Water Research*, **35**(2), 471–477.
Lesteur M., Bellon-Maurel V., Gonzalez C., Latrille E., Roger J., Junqua G. and Steyer J. (2010). Alternative methods for determining anaerobic biodegradability: a review. *Process Biochemistry*, **45**(4), 431–440.
Lu Y., Liaquat R., Astals S., Jensen P. D., Batstone D. J. and Tait S. (2018). Relationship between microbial community, operational factors and ammonia inhibition resilience in anaerobic digesters at low and moderate ammonia background concentrations. *New Biotechnology*, **44**, 23–30.
Mara D. D. and Pearson H. W. (1998). Design Manual for Waste Stabilization Ponds in Mediterranean Countries. Lagoon Technology Interational Ltd., Leeds, UK.
Martín-González L., Font X. and Vicent T. (2013). Alkalinity ratios to identify process imbalances in anaerobic digesters treating source-sorted organic fraction of municipal wastes. *Biochemical Engineering Journal*, **76**, 1–5.
Mata-Alvarez J., Dosta J., Romero-Güiza M. S., Fonoll X., Peces M. and Astals S. (2014). A critical review on anaerobic co-digestion achievements between 2010 and 2013. *Renewable and Sustainable Energy Reviews*, **36**, 412–427.
Nghiem L. D., Koch K., Bolzonella D. and Drewes J. E. (2017). Full scale co-digestion of wastewater sludge and food waste: bottlenecks and possibilities. *Renewable and Sustainable Energy Reviews*, **72**, 354–362.
Noguerol-Arias J., Rodríguez-Abalde A., Romero-Merino E. and Flotats X. (2012). Determination of chemical oxygen demand in heterogeneous solid or semisolid samples using a novel method combining solid dilutions as a preparation step followed by optimized closed reflux and colorimetric measurement. *Analytical Chemistry*, **84**(13), 5548–5555.
Nopens I., Batstone D. J., Copp J. B., Jeppsson U., Volcke E., Alex J., Vanrolleghem P. A.. (2009). An ASM/ADM model interface for dynamic plant-wide simulation. *Water Research*, **43**(7), 1913–1923.
Peces M., Astals S. and Mata-Alvarez J. 2014 Assessing total and volatile solids in municipal solid waste samples. *Environmental Technology (United Kingdom)*, **35**(24), 3041–3046.
Peña M. R., Mara D. D. and Sanchez A. (2000). Dispersion studies in anaerobic ponds: implications for design and operation. *Water Science and Technology*, **42**, 273–282.
Persson M., Jonsson O. and Wellinger A. (2006). Biogas Upgrading to Vehicle Fuel Standards and Grid Injection. IEA Bioenergy Task 37 – Energy From Biogas and Landfill Gas. IEA Bioenergy, Aadorf, Switzerland.
Rajagopal R., Massé D. I. and Singh G. (2013). A critical review on inhibition of anaerobic digestion process by excess ammonia. *Bioresource Technology*, **143**, 632–641.
Romero-Güiza M. S., Vila J., Mata-Alvarez J., Chimenos J. M. and Astals S. 2016 The role of additives on anaerobic digestion: a review. *Renewable and Sustainable Energy Reviews*, **58**, 1486–1499.
Schrotter J. C. and Bozkaya-Schrotter B. (2010). Current and emerging membrane processes for water treatment. In: Membrane Technology: Membranes for Water Treatment, K.-V. Peinemann and S. P. Nunes (eds), Wiley-VCH, Hoboken, pp. 53–92.
Tchobanoglous G., Burton F. L. and Stensel H. D. (2003). Wastewater Engineering: Treatment and Reuse, 4th edn. McGraw-Hill, Boston.
Tian H., Fotidis I. A., Kissas K. and Angelidaki I. (2018). Effect of different ammonia sources on aceticlastic and hydrogenotrophic methanogens. *Bioresource Technology*, **250**, 390–397.
van Lier J. B., Mahmoud N. and Zeeman G. (2008). Anaerobic wastewater treatment. In: Biological Wastewater Treatment, Principles, Modelling and Design, M. Henze, M. C. M. Van Loosdrecht, G. Ekama and D. Brdjanovic (eds), IWA Publishing, London, UK, pp. 415–456.
Vanwonterghem I., Jensen P. D., Rabaey K. and Tyson G. W. (2016). Genome-centric resolution of microbial diversity, metabolism and interactions in anaerobic digestion. *Environmental Microbiology*, **18**(9), 3144–3158.
Xie S., Higgins M. J., Bustamante H., Galway B. and Nghiem L. D. (2018). Current status and perspectives on anaerobic co-digestion and associated downstream processes. *Environmental Science: Water Research and Technology*, **4**(11), 1759–1770.
Yenigün O. and Demirel B. (2013). Ammonia inhibition in anaerobic digestion: a review. *Process Biochemistry*, **48**(5–6), 901–911.

doi: 10.2166/9781780409566

Chapter 6

Upgrading anaerobic digestion within the energy economy – the methane platform

Largus T. Angenent[1], Joseph G. Usack[1], Tianran Sun[1], Christian Fink[1], Bastian Molitor[1], Rodrigo Labatut[2], Roy Posmanik[3], Manuel Hörl[4] and Doris Hafenbradl[4]

[1]*Center for Applied GeoSciences, University of Tübingen, Schnarrenbergstraße 94-96, 72076 Tübingen, Germany*
[2]*Pontifical Catholic University of Chile, Department of Hydraulic and Environmental Engineering, Vicuna Mackena 4860, Santiago 7820436, Chile*
[3]*Institute of Soil, Water and Environmental Science, Agricultural Research Organization (ARO) – Volcani Institute, Newe Ya'ar Research Center, Ramat Yishai 30095, Israel*
[4]*Electrochaea GmbH, Semmelweisstraße 3, 82152 Planegg, Germany*

6.1 INTRODUCTION

Anaerobic digestion is a mature biotechnology production platform that is mainly used in a decentralized manner at many wastewater treatment plants, industrial sites, and farms all over the world (Angenent *et al.*, 2004; Lettinga, 1995; also see Chapter 5 of this book). The produced biogas contains approximately 35–50% carbon dioxide besides methane. The carbon dioxide at such high concentrations reduces the energetic density of biogas and renders it to be utilized on-site in a low-efficiency way. However, this carbon dioxide can be removed or even converted into methane. The resulting renewable natural gas containing primarily methane (~97–98%; only little carbon dioxide and other minor gaseous impurities), which is also referred to as biomethane, can be injected into the existing natural gas grid.

The conversion of carbon dioxide from biogas into more methane within a power-to-gas (PtG) system could almost double the methane yields from anaerobic digesters. Pumping this renewable natural gas into the existing natural gas grid would connect all decentralized anaerobic digesters and use the total amount of produced methane as an important energy carrier in the energy economy. For example, this methane can be used as a preferred low-carbon fuel for trucks, buses, and other heavy vehicles. Alternatively, the carbon in biomethane from a centralized system can also be used as a source for chemical production or for alternative energy carriers, such as carbon monoxide or hydrogen, after post-treatment (Verbeeck *et al.*, 2018).

Even though injecting biomethane from biogas into the natural gas grid has been technically feasible for quite some time, this has yet to become an economically attractive proposition. Here, we will discuss and explain the recent technological advances that can make such a proposition more attractive in the future. Of course, regulatory constraints, the absence of political will, and cheap fossil fuels have also been very important deterrents in addition to the lack of technological advances, but we will not discuss these here. We would like to refer the reader to Chapter 13, where the regulatory aspects are discussed in more detail.

We will start this chapter by explaining why carbon dioxide is produced in anaerobic digestion in the first place. Then, we will focus on how to possibly increase the amount of methane and carbon dioxide (biogas) from biomass conversion with anaerobic digestion. This increase can be achieved by biological means, such as by increasing the temperature to improve biomass breakdown, or non-biological means, such as by integrating thermochemical process units upstream or downstream of the anaerobic digester. The resulting increase in the amount of methane (the energy carrier) will make it more attractive in the future to invest in technologies to remove or convert carbon dioxide from the biogas, and is, therefore, important here.

The integration of abiotic processes with anaerobic digestion to improve its economic feasibility was also recently described in a review paper (Angenent *et al.*, 2018). We continue the chapter by describing three mature abiotic examples (water scrubbing, amine scrubbing, and pressure-swing adsorption) to remove carbon dioxide and other impurities from biogas. The removed carbon dioxide is either: (1) released into the atmosphere; or (2) converted into more methane in a downstream carbon dioxide conversion system with the opportunity to almost double the methane yields.

One of the systems to convert carbon dioxide into more methane is with ex-situ biomethanation (in a separate bioreactor from the anaerobic digester) with thermophilic (65°C) methanogenic archaea (Martin *et al.*, 2013). Archaea are single-cellular microbes from a different domain of life than bacteria and eukarya and are therefore not closely related to bacteria (Pace, 1997). Engineers have performed carbon dioxide conversion by introducing hydrogen gas (from an abiotic electrolyzer by water splitting) together with the carbon dioxide (from biogas) into a bioreactor. Because the hydrogen gas would be produced from renewable electric power (i.e., windmills, photo-voltaic), the biomethanation system, which is also referred to as power-to-gas (PtG), becomes a storage-enabling technology for renewable electric power. The existing storage capacity for this renewable natural gas in natural gas grids exceeds several months during peak demand. We will discuss and explain the PtG technology by introducing the electrolyzer, the bioreactor, and the biocatalyst. Finally, we end this chapter with a perspective on some future considerations.

6.2 LEARNING OBJECTIVES

At the completion of this chapter you should be able to:

- Discuss the technological advances that are needed to connect anaerobic digesters to biogas upgrading and thermochemical processes, and to make it a major player in the energy economy.
- Explain why anaerobic digesters have been characterized by a high methane yield and have been referred to as extreme fermentation.
- Explain the fundamental principles of scrubbing, membrane separation, and pressure-swing adsorption for carbon dioxide removal from biogas.
- Name and explain the fundamental principle of two hydrothermal technologies that can increase the methane yield from biomass conversion and place them in their optimum location before or after the anaerobic digester.
- Discuss why it is advantageous to perform PtG with a thermophilic bioreactor compared to a mesophilic bioreactor.
- Synthesize all the knowledge in this chapter and describe how you would make the connected anaerobic digestion system produce the most methane.
- Characterize and explain specific challenges for the connected anaerobic digesters and biomethanation systems in a society that runs on renewable electric power and renewable natural gas.

6.3 WHAT DRIVES METHANE AND CARBON DIOXIDE YIELDS IN ANAEROBIC DIGESTION

6.3.1 Thermodynamic reason for high methane yields – extreme fermentation

Methane fermentation, commonly known as anaerobic digestion (AD), is a biochemical process whereby organic compounds are transformed by a series of biochemical reactions that ultimately result in: (1) methane and carbon dioxide in biogas; and (2) biomass due to cell growth. The open culture (open, because new microbes are constantly introduced with the substrate) of microbial consortia (the microbiota), which is also referred to as a reactor microbiome, consists of an anaerobic food web with many different bacteria and archaea that are working together.

Heterotrophic bacteria oxidize complex organic compounds, such as sugars and fatty acids, without oxygen into fermentation products (including, hydrogen, carbon dioxide, acetate, propionate, acetate, butyrate, and ethanol) to obtain energy (i.e., ATP) that is required for synthesis of new cells and cell maintenance. In addition, methanogenic archaea (methanogens) produce ATP by using either:

(1) acetate (acetoclastic methanogens) to cleave the acetate into methane and carbon dioxide (Equation (6.1)):

$$CH_3COOH \rightarrow CH_4 + CO_2 \tag{6.1}$$

(2) hydrogen and carbon dioxide (hydrogenotrophic methanogens) to produce methane and water (Equation (6.2)):

$$CO_2 + 4H_2 \rightarrow CH_4 + 2H_2O \tag{6.2}$$

Similar to all biochemical reactions, energy is gained by sets of redox reactions for which an electron donor (for humans this is food) and an electron accepter (for humans this is oxygen) are required.

Under anaerobic conditions, no external electron acceptor, such as oxygen, is available, and the only solution to gain energy is to use the organic substrate in waste as both the electron donor and electron acceptor (i.e., fermentation). In doing so, the methanogenic microbiome produces the most reduced carbon possible, which is the carbon in methane (oxidation state of −4), *and* the most oxidized carbon possible, which is the carbon in carbon dioxide (oxidation state of +4) (Figure 6.1) (van Haandel & van der Lubbe, 2012). The oxidation state of carbon is defined as the degree that electrons are being lost due to oxidation. This way, the microbiome essentially harvests all the energy that possibly can be retrieved from the organic substrate without the presence of a common electron acceptor, such as oxygen (Hanselmann, 1991).

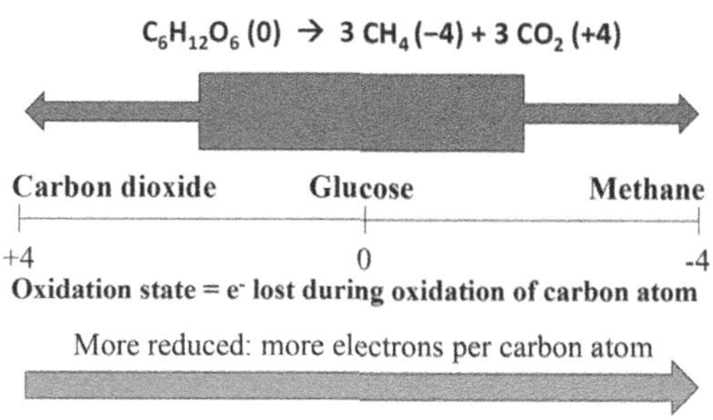

Figure 6.1 Schematic to explain that an electron balance should be maintained for the closed off anaerobic digester, resulting in both methane and carbon dioxide being produced with an organic substrate such as glucose. For carbon dioxide the carbon atom has an oxidation state of four, which means that four electrons were used up during oxidation, while the opposite is true for methane (oxidation state of minus four: four electrons were gained during oxidation). With equal molar numbers for the products, the electron balance for carbon is maintained (0 = −4 + 4).

This implies that, upon completion, any other type of fermentation (e.g., to lactate, butyrate, caproate) will produce less free energy for the microbes compared to AD, and thus a lower yield of the carbon end product. Furthermore, since both methane and carbon dioxide bubble out from the bioreactor broth when they are produced, there is no product accumulation (no additional thermodynamic limitations). When the conditions are favourable, such as a neutral pH in the fermentation broth, a very long residence time of the substrate, and in the presence of methanogens, the desired carbon product will be primarily methane, and not, for example, lactate, butyrate, or caproate. As a result, the methane yield (the methane formed compared to the degraded organic substrate) will be high, and the carbon product yield will be higher than for any other fermentation. This also implies that AD should never be underestimated as a biotechnology production platform, because it is a way of steering the conversion of a complex substrate into one carbon product with the highest yield possible.

6.3.2 Why carbon dioxide is produced besides methane – electron balance

Biogas is composed of a mixture of gases for which methane and carbon dioxide are the main constituents. Although biogas from complex organic substrates is typically composed of ~60% methane and ~40% carbon dioxide (v/v), the exact proportion between the two gases is entirely dependent on the mixture of organic compounds within the AD influent (substrate). For example, carbohydrates, such as the sugar glucose (with carbon at an oxidation state of zero), will produce equal molar parts of methane (50%) and carbon dioxide (Equation (6.3)):

$$C_6H_{12}O_6[0] \rightarrow 3CH_4[-4] + 3CO_2[+4] \tag{6.3}$$

Proteins and lipids will produce higher proportions of methane (69 and 70% methane [v/v], respectively).

The explanation why there is always carbon dioxide being formed from organic carbon compounds during AD lies again in the understanding that life is based on redox reactions within a completely anaerobic and closed-off vessel. In the case of methane fermentation, this involves both the reduction of carbon in the organic compound to form methane (gaining electrons); and the oxidation of carbon in the organic compound to form carbon dioxide (using up electrons). Since besides a carbon balance we also need an electron balance, carbon dioxide has to be formed besides methane, otherwise we would end up with more electrons in the product compared to the organic substrate, and that is not possible. Only by adding electrons, for example, in the form of hydrogen gas and by utilizing hydrogenotrophic methanogens (Equation (6.2)), can we reduce the amount of carbon dioxide in the biogas.

6.3.3 How to improve methane yields with biology – thermophilic AD

As temperature increases, microbial rates, and therefore methane production rates of AD, also increase. This, however, does not occur in a temperature continuum – there are three main temperature ranges for optimal microbial growth: psychrophilic (<20°C), mesophilic (30–40°C), and thermophilic (55–65°C). With each incremental temperature range, both the thermodynamics and the kinetics of methane production improve. Better thermodynamics, however, has little impact on methane yields. Why? Because thermodynamics only dictates if a reaction will proceed, and thermodynamics for the two main biochemical pathways of methane production in AD are already well favoured, even at psychrophilic temperatures. Conversely, improved kinetics has a significant impact on the methane production rates. Essentially, a digester operated at thermophilic temperatures can achieve higher volumetric methane production rates (in m^3 methane per m^3 reactor volume per day) than the same digester operated at mesophilic temperatures when all other conditions are held equal.

Still, operators of anaerobic digesters can look forward to higher methane yields due to the faster kinetic rates of certain biochemical reactions (Labatut *et al.*, 2011). This is because hydrolysis (solubilization) of non-soluble organic waste biomass is the rate-limiting step in AD, resulting in a non-complete degradation. Because of a considerable faster reaction rate for thermophilic bacteria that perform this hydrolysis compared to their mesophilic counterparts in similar bioreactor volumes (the residence times are similar), more of the non-soluble substrate is solubilized, resulting in higher methane yields. Despite this advantage, most anaerobic digesters are operated under mesophilic conditions because of the higher energy input required and poorer process stability observed in thermophilic AD (Labatut *et al.*, 2014). Therefore, a chance exists at most AD locations to improve the methane yields by increasing the operating temperature of the digester.

6.4 HOW TO IMPROVE METHANE YIELDS OR PRODUCE OTHER ENERGY-CARRIERS WITH HYDROTHERMAL SYSTEMS

6.4.1 What are hydrothermal systems?

One of the emerging technologies that can be coupled with AD to improve the overall energy return is hydrothermal processing – an attractive option for the production of energy and bio-based chemicals from organic substrates. The main advantage of hydrothermal processes is the use of water as the reaction media, in contrast to dry thermochemical processes (i.e., pyrolysis or gasification) which require the removal of water *prior* to the process. Thus, hydrothermal processes offer opportunities for an energy efficient valorization of wet biomass, such as animal manure and food wastes, before or after AD (Posmanik *et al.*, 2018). The hydrothermal reaction is based on substantial changes in the properties of water, occurring between the boiling temperature and the critical point ($T_c = 374°C$; $P_c = 22$ MPa) (Figure 6.2). Water above its saturation pressure (dark line in Figure 6.2) remains in a liquid state at elevated temperatures (usually referred to as subcritical water or subcritical liquid). Under such conditions, several properties of water rapidly change and turn the water into an

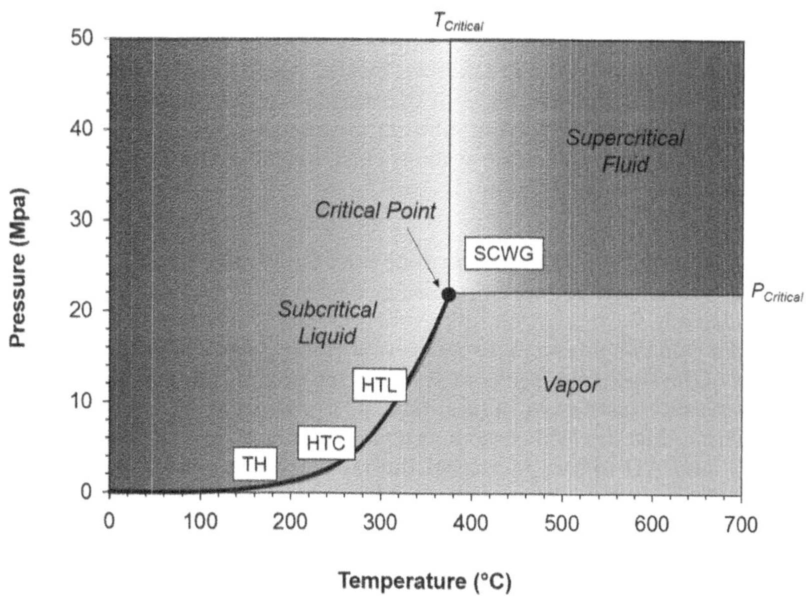

Figure 6.2 Phase diagram of water with the environmental conditions pressure and temperature, showing where specific hydrothermal processes occur in subcritical and supercritical regions. From low to high temperature and pressure, the hydrothermal processes are: thermal hydrolysis (TH), hydrothermal carbonization (HTC), hydrothermal liquefaction (HTL), and supercritical water gasification (SCWG).

attractive media for chemical conversion processes, including fast hydrolysis, phase fractionation, and re-polymerization reactions (Peterson *et al.*, 2008; Tekin *et al.*, 2014).

Hydrothermal processes can be classified depending on their processing temperature and pressure, yielding different product distributions. A reaction at relatively low temperatures (up to 170°C) is referred to as thermal hydrolysis (TH) and generally results in hydrolysis of carbohydrates. Hydrothermal carbonization (HTC) occurs at mild temperatures (170–250°C) and pressures around 2.5 MPa to generate a solid product (char) that can be used as an energy product or for agricultural purposes. Between 250 and 370°C and operating pressures between 5 and 22 MPa, hydrothermal liquefaction (HTL) is favoured, generating an energy-dense bio-crude oil, which can be used as a liquid fuel. Beyond the critical point of water, supercritical water gasification (SCWG) typically takes place to convert the feedstock into a combustible gas: methane-rich in the near-critical region and H_2-rich at temperatures above 600°C (Figure 6.2).

6.4.2 Thermal hydrolysis (TH) as a pre-treatment to AD

Thermal hydrolysis (TH) of sludge *prior* to AD is an established mature technology which has been commercially available for more than 20 years. TH combines sterilization of the biomass (e.g., waste activated sludge at a domestic wastewater treatment plant) with improving its biodegradability to favour AD performance, resulting in higher methane yields (Barber, 2016). The main advantages of TH are: (1) increasing the organic loading rates (i.e., minimizing the size of the digester) due to a better degradability of the biomass; (2) increasing the yield of biogas from the biomass; (3) improving the dewaterability of the sludge (i.e., reducing the downstream processing costs); and (4) sterilization of the sludge, possibly allowing downstream land application.

The optimal operating conditions involve temperatures between 160 and 180°C for 20–40 minutes (Bougrier *et al.*, 2008; Stuckey & McCarty, 1984). TH is generally favoured to treat carbohydrate- or protein-rich sludge and has little influence on the lipids. Therefore, it is more suitable to treat waste activated sludge, rather than primary sludge (Wilson & Novak, 2009). Because TH increases the solubility of proteins and improves the breakdown by the digester microbiome, an increase in free ammonia and alkalinity is a major challenge of the technology. Currently, ammonia toxicity is the rate-limiting design consideration, requiring dilution of the hydrolyzed sludge *prior* to feeding it to the anaerobic digester. Regardless, full-scale systems exist, including the TH system at the Blue Plains wastewater treatment facility in Washington, DC, USA.

6.4.3 Hydrothermal liquefaction (HTL) as a pre- or post-treatment for AD

HTL is an emerging technology, yet very promising, due to the valuable product (bio-crude oil) it generates within a relatively short reaction time (<60 minutes). Therefore, HTL can be coupled to an AD to yield additional valuable products, to recover nutrients, to stabilize the waste, and to expand the overall efficiency of the waste management system. One process integration option is by using the HTL aqueous product as a substrate for a high-rate AD system such as an upflow anaerobic sludge blanket (UASB) reactor – thus the HTL system as a pre-treatment for AD. Researchers demonstrated that combining HTL and AD indeed increased the recovery of the total energy from food waste (Posmanik *et al.*, 2017). However, the total energy balance of such process integration provides a trade-off between bio-oil (gained from HTL) and biogas (gained from AD).

Another process integration option is by using AD effluent as the feedstock for HTL – thus as a post-treatment for AD. Looking into that scenario, researchers showed that changing the hydraulic retention times (the length of time that the liquid waste remains in the vessel) in the AD directly affected the yield of the bio-oil that was produced in the HTL, with the highest yield at a residence time of 38 days and the lowest yield at a residence time of 60 days (Eboibi *et al.*, 2015). Of course, a longer residence time in the AD resulted in less substrate for HTL.

6.5 REMOVING CARBON DIOXIDE FROM BIOGAS

6.5.1 The basic principle of carbon dioxide separation

To introduce renewable natural gas into the natural gas, grid each jurisdiction has different regulations and standards, but as a general rule the gas should consist of ~97–98% methane. Carbon dioxide can be separated and removed from biogas using a variety of techniques. The techniques described in this section will include: (1) physical absorption; (2) chemical absorption; and (3) adsorption. Although the specific mechanisms involved with each technique are different, they all rely on the same basic principle, which is the preferential retention or exclusion of one or more components in a fluid, based on differences in its physical and chemical properties. Specifically, we want to produce a concentrated (or enriched) stream of methane gas by selectively removing carbon dioxide from biogas.

6.5.2 Physical and chemical absorption

Absorption is a process involving the entrainment or incorporation of a fluid component (liquid or gas) into the bulk phase of a liquid or solid. The bulk phase material responsible for entraining the fluid component is called the sorbent. Absorption is achieved via two main mechanisms: (1) physical absorption; and (2) chemical absorption. In physical absorption, the fluid component is physically trapped inside the matrix of the bulk phase material without the aid of chemical bond formation. Physical absorption could be the entrapment of H_2 gas molecules between the interstitial spaces of liquid water, for example, or the diffusion of oil into the pores of a clay mineral. Chemical absorption, on the other hand, involves the formation of chemical bonds between the fluid component and the bulk phase of the material. For example, hydrogen sulphide (H_2S) in a gas stream can be chemically absorbed and precipitated by exposing it to a bed of steel shavings, which contain ~99% iron (II), in the solid-state reaction (Equation (6.4)):

$$Fe^{2+}_{(s)} + H_2S_{(g)} \rightarrow FeS_{(s)} + H_{2(g)} \tag{6.4}$$

6.5.3 Physical absorption of carbon dioxide using water scrubbing

The most common application of physical absorption for carbon dioxide removal from biogas is water scrubbing. In this process, carbon dioxide from the biogas is readily absorbed by water (sorbent) with minimal absorption of methane. This preferential absorption of carbon dioxide is due to the high solubility of carbon dioxide in water (1.45 g L^{-1}, at 25°C and 100 kPa [1 atm]) compared to methane (0.022 g L^{-1}, at 25°C and 100 kPa). The solubility of carbon dioxide and methane is also greatly affected by temperature and pressure. Solubility of these gases will decrease (S↓) at higher temperatures (T↑), and increase (S↑) at higher pressures (P↑). In other words, solubility is indirectly related with temperature (S↓ · T↑; S↑ · T↓), and directly related with pressure (S↑ · P↑; S↓ · P↓).

Another important parameter in gas absorption is the mass transfer rate of the gas into the bulk medium. Overall mass transfer is directly related to the concentration gradient, the interface contact area, and the contact time. For example, we know intuitively that mixing will increase the absorption rate of carbon dioxide into water. This is because mixing prevents accumulation of dissolved carbon dioxide at the gas/liquid interface, thereby maintaining a high concentration gradient. Another way to increase mass transfer is to create small gas bubbles at the bottom of a water column using diffusers. The small gas bubbles have a higher surface area per unit volume compared to larger gas bubbles, thereby increasing the interface contact area. Small gas bubbles also experience higher drag forces compared to large gas bubbles, which slows their ascent through the water column and increases the contact time between the gas bubble and the water. Now that we have explained these basis concepts, we can describe a typical water scrubbing system.

Most water scrubber systems are configured as tall, vertical columns (Figure 6.3). Water is sprayed into the top of the column, while the biogas is introduced at the bottom of the column. The biogas (at

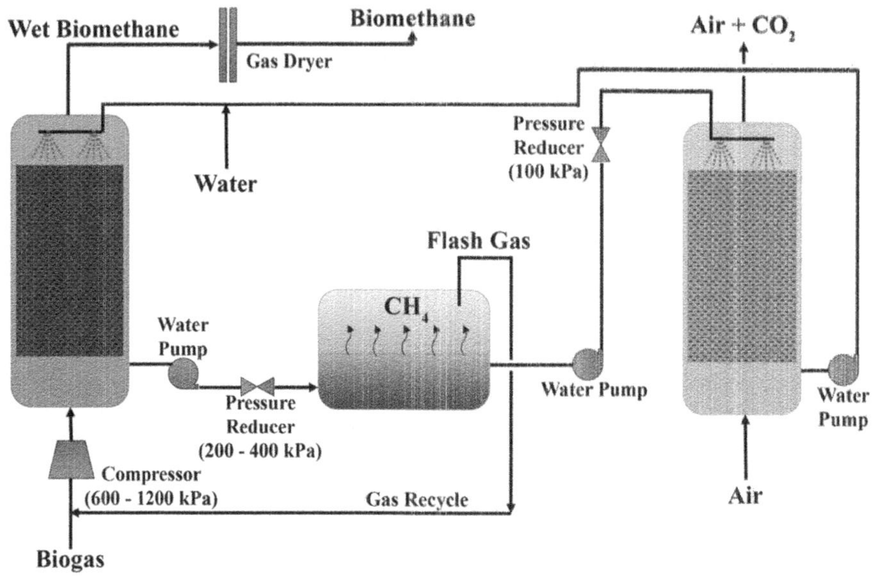

Figure 6.3 Diagram of a typical water scrubbing system for biogas upgrading. The system contains three main components: (1) a carbon dioxide absorption column (left block); (2) a flash tank (middle block); and (3) a carbon dioxide stripping column (right block). Compressed biogas is injected into the carbon dioxide absorption column while air is directed through the carbon dioxide stripping column. Water is recirculated between the three components and introduced to the top of each column via sprayers. Wet biomethane leaving the carbon dioxide absorption column is passed through a gas dryer (depicted here as a plate heat exchanger).

approximately 35°C) is usually pressurized (i.e., 600–1200 kPa) to promote carbon dioxide absorption as it rises through the column. Packing material is often added to the column as well to increase turbulence and better mix the biogas and water streams as they pass each other in a counter-current flow. As they interact, proportionally more carbon dioxide is absorbed compared to methane. By the time the gas stream leaves the top of the column it is almost completely depleted of carbon dioxide and enriched with methane (e.g., >97%).

The gas leaving the column, which can now be called biomethane, will then be transferred to a drying process (e.g., cooled in a heat exchanger) to remove the moisture picked up during the scrubbing process (Figure 6.3). On the other end of the column, the water leaving the bottom is now rich in dissolved carbon dioxide, with only low levels of dissolved methane. This carbon dioxide-laden water is sent to a low-pressure flash tank (i.e., 200–400 kPa), which causes a quick dissolution of the dissolved methane while at the same time retaining most of the carbon dioxide. The off-gas from the flash tank is then recycled back to the water scrubber and combined with fresh biogas. Finally, the carbon dioxide-laden water is transferred to the top of a stripping column at low pressure (i.e., 100 kPa), where the carbon dioxide is *de*sorbed or driven-off from the water using an upward-flowing air stream. The regenerated water is then returned to the water-scrubbing column to complete the cycle (Figure 6.3).

6.5.4 Chemical absorption of carbon dioxide using amines

Amines are compounds similar to ammonia (NH_3). In both cases, nitrogen serves as the central atom and allows the formation of three chemical bonds. However, unlike ammonia, which has three hydrogen bonds, amines replace one or more of the hydrogen bonds with an organic functional group (e.g. methylamine is basically an ammonia molecule with one H-bond replaced with a methyl group). Amino acids, which are the building blocks of proteins, are classified as amines. In the context of carbon dioxide scrubbing, amines, such as diethanolamine (DEA) and methyldiethanolamine (MDEA), are typically used due to their high selectivity for carbon dioxide compared to methane.

Because of this, they can treat gas streams containing very low concentrations of carbon dioxide, and thereby achieve higher biomethane purity levels (e.g., >99.9%). An amine scrubber system is very analogous to a water scrubbing system, consisting of an absorption column and stripping column, but rather than using water in the absorption column and air in the desorption column, they use an amine solution and heat, respectively. The heat is required to break the chemical bonds formed between the carbon dioxide and amine molecules, with the goal to release the carbon dioxide at the end of the process. Finally, unlike water scrubbing systems, amine scrubber systems do not require a flash tank to recover methane because the amine solution absorbs very little methane.

6.5.5 Adsorption

Unlike absorption processes, which entrain the fluid inside a liquid or solid bulk phase material, adsorption processes involve the retention of a fluid on the surface of a solid substratum. The solid materials are typically selected to have high surface area and porosity to help separate the components in a fluid mixture based on molecular size and surface-related interactions such as Van der Waals forces and polar bonds. Typical examples of solid materials that are used for adsorption include silica gels, activated carbon, carbon molecular sieves, and zeolites. Selective retention of fluid components can be achieved using two different approaches. The first approach, known as equilibrium-based adsorption, relies on differences in surface interactions between the fluid components and the adsorbent material to selectively retain a component (or components) based on the strength of its interaction. For example, fluid components with large molecular size can be separated from small molecular components because the Van der Waals forces holding the large molecules to the solid surface are stronger than the forces holding the small molecules. At equilibrium, a higher concentration of large molecules will be adsorbed to the surface compared to small molecules.

The second approach, known as kinetic adsorption, relies on differential rates of adsorption between fluid components onto the solid surface. For example, solid materials can be selected based on the pore size to control the diffusion rate of fluid components into the solid matrix. In the case of methane and carbon dioxide, which have a kinetic diameter of 3.8 and 3.4 Angstroms (Å), a solid substratum (e.g., carbon molecular sieve) with an intermediate pore size could be used to promote faster diffusion of carbon dioxide (3.4 Å) into the pore matrix compared to methane (3.8 Å), which effectively concentrates the carbon dioxide. Finally, like absorption, the process of adsorption depends strongly on the applied temperature, pressure, and contact time. These parameters can be manipulated to improve the selectivity of the adsorption process, taking into account the specific physical/chemical properties of the fluid components.

6.5.6 Biogas upgrading using pressure-swing adsorption

Within the field of adsorption technology, pressure-swing adsorption (PSA) is very common and is often used for biogas upgrading. As indicated in the name, differential adsorption is achieved by alternating (swinging) between pressure levels. It works on the principle that at higher pressures relatively more adsorption occurs, and when the pressure is reduced, the fluid component will detach or *de*sorb. A typical biogas PSA system is carried out in four steps: (1) pressurization; (2) feeding; (3) blow-down; and (4) purging. In the first step, the biogas is pressurized to approximately 400–1000 kPa and delivered to the bottom of an adsorption column where carbon dioxide is selectively adsorbed. Meanwhile, methane passes through the column with minimal retention. During the feeding step, the column is fed with pressurized biogas until the adsorption material reaches its carbon dioxide saturation point. At this stage, the blow-down step begins, which involves shunting the inflow of biogas to allow the pressure within the column to dissipate. As the pressure decreases, carbon dioxide desorbs from the adsorption material, and regenerates the column. Once the minimum pressure is reached, the remaining carbon dioxide is driven off by purging the column with some of the processed biomethane to further regenerate the column and complete the cycle. The PSA process can be configured to run continuously using two or more adsorption columns, where each column is at a different stage of adsorption or regeneration.

6.6 EX-SITU BIOMETHANATION

6.6.1 PtG concept

The basic concept of PtG is to utilize renewable electric power to produce gaseous energy carriers (Schiebahn *et al.*, 2015). Currently, two PtG approaches exist (i.e., power-to-hydrogen and power-to-methane). Both use electric power in a first step to split water into hydrogen and oxygen with an electrolyzer. In the case of a power-to-hydrogen system, hydrogen is the final energy carrier. In contrast, the hydrogen is further converted into methane with a power-to-methane system in a second step by introducing it together with carbon dioxide (from biogas or other CO_2 sources) into a reactor. The individual reaction steps are:

(1) electrolysis (Equation (6.5)):

$$4\,H_2O \rightarrow 4\,H_2 + 2\,O_2 \tag{6.5}$$

(2) *plus* methanation (Equation (6.2)), resulting in a net reaction (Equation (6.6)):

$$CO_2 + 2\,H_2O \rightarrow CH_4 + 2\,O_2 \tag{6.6}$$

The methanation can be performed with methanogenic archaea (biomethanation) or chemical catalysis (Sabatier process). Biomethanation can use biogas directly without methane separation and purification because the methanogens can withstand impurities, or even profit from, for example, sulphur contaminants (hydrogen sulphide) as a growth nutrient. This is different for the abiotic Sabatier process (thermo-chemical methanation). There, a solid metal catalyst (usually nickel) is present inside the reactor and catalyzes the methanation reaction. However, the catalyst can be easily poisoned by impurities. This process is operated at high temperatures to activate the catalyst (300–550°C) and also at a high pressure (up to 10 000 kPa) to shift the equilibrium of the methanation reaction towards the products. The Sabatier process, therefore, requires highly purified carbon dioxide. It seems likely that for systems that are connected to decentralized anaerobic digesters for which stringent biogas cleaning is difficult, the biological route of PtG will be chosen (Figure 6.4) (Bailera *et al.*, 2017).

PtG is a storage technology for renewable energy, where electric power that is generated by wind mills or photo-voltaic systems is used to produce gaseous energy carriers (Götz *et al.*, 2016). In the case of methane, the gas can be injected and stored as renewable natural gas in the existing natural gas grid, which offers a storage capacity for methane that exceeds the energy consumption needs for several months. Although hydrogen could also be stored in the natural gas grid after certain changes are made, the amount of pure hydrogen that can be injected is strictly limited by regulations, which restricts the theoretical storage capacity of the power-to-hydrogen technology. Of course, the natural gas grid could be updated and changed to handle higher concentrations of hydrogen gas, which is a highly diffusive gas due to its relatively small molecule size. An additional advantage of the power-to-methane technology is that it can circumvent the release of carbon dioxide at the decentralized anaerobic digester sites. Carbon dioxide, which would have otherwise been released into the atmosphere, is converted into methane and used at central locations.

6.6.2 Hydrogen supply via electrolysis

How to transform liquid water to gas? Vaporization is obviously one answer. However, we can also employ another physicochemical process to split liquid water into more useful and valuable hydrogen and

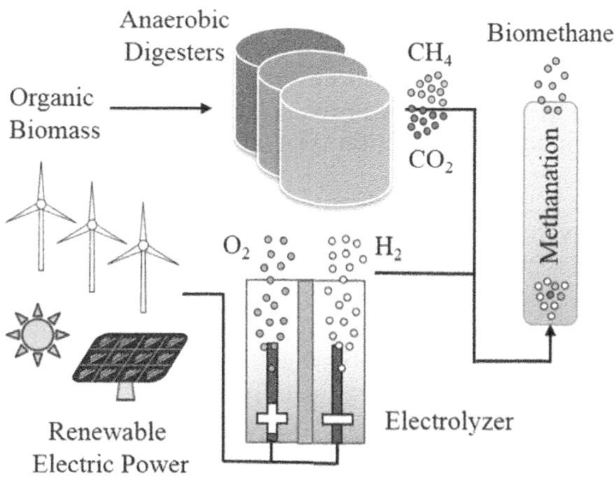

Figure 6.4 Schematic of a PtG system to upgrade biogas into biomethane. The process includes an electrolyzer and a biomethanation reactor for untreated biogas from an anaerobic digester. When the biogas is first cleaned and without methane, an abiotic methanation reactor is also possible based on the Sabatier process.

oxygen gases. This process is called electrolysis for which electric current drives chemical reactions to complete the separation of its elements. An electrical circuit, which consists of working electrodes (anode and cathode) and supporting electrolyte, achieves electrolysis by passing of an electric current (Figure 6.5). Electrons are current carriers in electrodes, whereas charged ionic species carry current through the electrolyte solution. Chemical reactions are induced by the passage of current but occur only at the electrode/electrolyte interfaces and not in the bulk of either electrodes or electrolyte. The products of electrolysis vary depending on the type of electrolyte and the utilized materials for the electrodes.

In alkaline solutions that are electrolyzed by inert carbon or metal electrodes, hydroxide ions (OH^-) will migrate as major current carriers in the solution, and oxidize at the anode to produce oxygen.

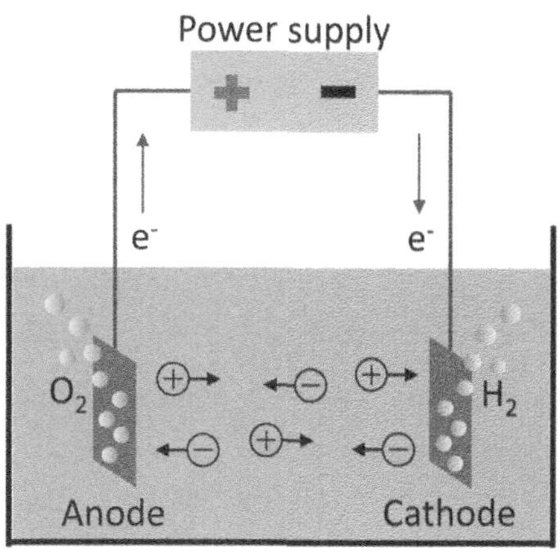

Figure 6.5 Schematic of an electrolysis cell. A power supply generates an electric current that sustains chemical reactions at the electrode/electrolyte interfaces. At the anode, oxygen gas is produced which could be used at the wastewater treatment plant to boost aerobic bioprocesses. At the cathode, hydrogen gas is produced at pressure that can be introduced into the methanation process. Finally, heat is produced, which can be utilized to heat the digester or for other applications.

At the cathode, water reduction occurs, which produces hydrogen. The overall required energy that drives electrolysis must equal the free energy changes of the chemical reactions *plus* the over-potential that sustains the reactions to proceed in a certain rate *plus* the heating loss of electrical resistance such as the electromigration of ionic species in the solution. With the depletion of electrolyte, more and more heat will be generated in the solution. Due to the production of both gases, pressure will build up during electrolysis.

6.6.3 Bioreactor technology

Methanogenic archaea are present in an aqueous matrix within the ex-situ biomethanation reactor, and water is being produced as a side product of the biological reaction to methane (Equation (6.2)). The nomenclature of ex-situ is used here to indicate that the hydrogenotrophic methanogenic reaction for PtG is taking place outside of the main anaerobic digester. When hydrogen gas is introduced into the main anaerobic digester to sustain this biological reaction (Equation (6.2)), the nomenclature is in-situ biomethanation. You can read why in-situ biomethanation is not deemed feasible for a commercial system in another review of our group (Angenent *et al.*, 2018). Once hydrogen and carbon dioxide have been introduced into the ex-situ biomethanation reactor, for example, via bubble formation, the gases first need to dissolve before they can be taken up and converted into methane by the archaea. Since the solubility of hydrogen is low compared to carbon dioxide, the rate of hydrogen transfer into the liquid is the rate-limiting step of the biomethanation process (Schill *et al.*, 1996). Increasing the hydrogen partial pressure can enhance the solubility of hydrogen gas. Therefore, most ex-situ biomethanation reactors are operated at overpressure.

In addition, the gas-to-liquid transfer rate can be improved by increasing the gas/liquid interfacial area. Therefore, special reactor types that provide large gas/liquid interfaces are being developed, resulting in volumetric methane production rates that can be larger than the 50 m^3 per m^3 volume per day that was achieved with real biogas in bench-scale bioreactors (Martin *et al.*, 2013). For example, the demonstration plant in Denmark (stirred bubble-column bioreactor) is operating at 250 m^3 per m^3 volume per day (Figure 6.6). That is much higher than for anaerobic digestion for biosolids treatment at a wastewater treatment plant (0.25 m^3 per m^3 volume per day) (Uman *et al.*, 2018), resulting in about 1000 times smaller volumes for biomethanation reactors compared to these anaerobic digesters. Stirred bubble-column bioreactors are applied to create very small gas bubbles through bubbling and mixing, which increases the specific surface area of the gas/liquid interface (Figure 6.6).

Figure 6.6 Picture of a 1-MW power-to-methane system at a wastewater treatment plant. The stirred bioreactor with bubble formation in a column is on the left. This picture was taken at Electrochaea's patented BioCat plant in Avedøre, Denmark, which is a demonstration-scale system to store biogas as renewable natural gas in the Danish natural gas grid.

An alternative concept is to create high specific liquid surface areas by creating thin liquid films in a trickle-bed reactor. Trickle-bed reactors use a solid matrix on which microbes grow as a biofilm. A liquid solution is then recycled from the bottom to the top of the bioreactor, while it trickles down over the biofilm without immersing the matrix into standing liquid. Basically, the gas within the anaerobic bioreactor surrounds the liquid on the biofilm to allow a proper gas/liquid transfer. Currently, the demonstration- and full-scale plants are built as stirred bubble-column bioreactors or trickle-bed reactors for ex-situ biomethanation (Bailera *et al.*, 2017).

6.6.4 Why operate at 65°C rather than 37°C

One very important parameter for ex-situ biomethanation is the temperature at which the bioreactor is operated. Often it is assumed that a bioreactor has to be heated to maintain a given temperature such as 37°C. Following this logic, it seems obvious that a process that has to be heated to 65°C consumes more energy compared to a process that only has to be heated to 35°C. While this is true in many cases for traditional bioprocesses including anaerobic digestion, in the case of biomethanation we need to look more closely at the underlying reaction for hydrogenotrophic methanogenesis again (Equation (6.2)). The energy gain from this reaction is relatively low, given by the overall Gibb's free energy change of –131 kJ mol^{-1} under biological standard conditions (pH = 7; temperature = 25°C; pressure is 100 kPa). Therefore, the cell has to maintain a high rate of methane formation to produce enough cellular energy in the form of ATP to sustain growth (Schill *et al.*, 1999). The resulting high ratio of methane (product) to biomass is an advantage of anaerobic bioprocesses.

What we also see from Equation (6.2) is that the methanogens produce one mole of gas and two moles of water from five moles of gas. This results in a considerable decrease of the system's entropy (negative ΔS), which means, in easy terms, that the system reaches a higher state of order. To maintain the driving force for energy conservation (negative ΔG; $\Delta G = \Delta H - T\ \Delta S$; and with – – = + for ΔS), the reaction must be accompanied with a decrease in the enthalpy (negative ΔH) by energy dissipation in the form of heat–exothermic conditions (Schill *et al.*, 1999). Taken together, it follows that due to the high methane production rate and heat generation during the metabolism, the reactor has to be cooled rather than heated. Therefore, an operating temperature of 65°C is beneficial compared to an operating temperature of 37°C. Furthermore, the higher operating temperature results in reduced chances for contamination and more efficient stripping of product gas (methane).

6.6.5 Thermophilic methanogen – *Methanothermobacter thermautotrophicus*

For the biomethanation step in PtG systems, *Methanothermobacter thermautotrophicus* (*M. thermautotrophicus*) is being used as the methanogen. It is a rod-shaped microbe with a length of 3–7 μm and a diameter around 0.5 μm (Figure 6.7). The optimal growing temperature is 65°C, which makes it a thermophilic (warm-loving) microbe (Zeikus & Wolfe, 1972). All methanogens belong to the domain of archaea, which is the third domain of life besides bacteria and eukarya (Pace, 1997). Typically, methanogens must be kept under strictly anaerobic conditions (oxygen-free) since they lack mechanisms for oxygen detoxification. *M. thermautotrophicus* is a hydrogenotrophic methanogen using anaerobic respiration (Equation (6.2)).

This anaerobic respiration process occurs via the Wolfe cycle, which is a pathway with several consecutive steps: (1) carbon dioxide is reduced to formate and connected to a one-carbon-carrier cofactor; (2) the resulting complex molecule is further reduced in several steps to a methyl-group; and (3) methane is formed. One of the reasons why *M. thermautotrophicus* is utilized in biomethanation is the fast growth rate, which is 8–10 times faster compared to most other methanogens. This makes it easier to maintain a well-grown biocatalyst in the bioreactor. Because the energy gain per molecule methane produced is relatively low, the methane production rate of *M. thermautotrophicus* has to be relatively high, which is beneficial for the biomethanation process (Martin *et al.*, 2013; Schill *et al.*, 1999). Another reason why this microbe is used for biomethanation is because it is not as sensitive to oxygen compared to other methanogens, which makes the biomethanation system easier to operate.

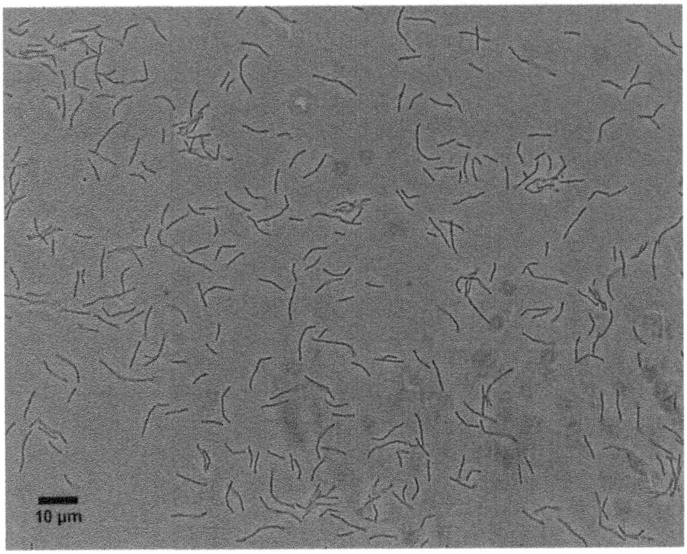

Figure 6.7 Light microscope view of living cells of *M. thermautotrophicus* in the exponential growth phase. Magnification 1000×. The reference is 10 μm.

6.7 PERSPECTIVES AND FUTURE NEEDS

Decentralized anaerobic digesters have been built all over the world but are mostly seen as a biomass and wastewater treatment technology, rather than an important energy provider for societies. With localized subsidies to boost the income from electric power production with combined heat and power systems, the digesters have become somewhat connected via the electric grid. However, these subsidies are often temporary. In addition, the conversion efficiency from methane to electric power is relatively low, resulting in a high heat production compared to heat demand in, for example, rural areas. By placing abiotic or biological systems close to the anaerobic digester, carbon dioxide can be removed or converted to produce more biomethane, which can be stored and transported into the existing natural gas grid. After transportation of the product (biomethane), a centralized chemical plant can utilize the energy or even the carbon from biomethane at a large scale. Distributing the methane directly to decentralized users can also promote the use of methane as a low-carbon fuel for vehicles. Both possibilities would make AD a much more important energy-production partner and boost AD implementation for carbon recovery.

By almost doubling the methane yield of digester through PtG, methane as a low-carbon fuel can have a real impact in the future energy economy. An increase in the total volume of anaerobic digesters is also required to increase this impact. When finally most organic waste will be treated with anaerobic digesters, this will likely occur through co-digestion of several different wastes together, such as food waste and animal wastes, possibly in large centralized digesters. However, a massive increase in digester capacity requires the smart reuse and recovery of nutrients (in addition to the carbon) from anaerobic digesters because otherwise the resulting biosolids from digesters would leak these nutrients into the environment after land application, resulting in uncontrolled nutrient runoff and eutrophication of surface waters (lakes, oceans). Thus, a systems-oriented approach is necessary that ties agriculture, food-waste management, land management, nutrient recovery, energy production, and water. This has also been referred to as the food energy water (FEW) nexus. Research is now necessary on how to combine all these aspects together without causing further environmental harm.

6.8 CHAPTER SUMMARY

In this chapter, we have introduced and explained several systems to improve the methane yield in AD, remove carbon dioxide from biogas, and increase the overall biomethane production. In addition, we have also introduced a possibly compatible thermochemical system to produce bio-oil and biomethane. Some of the systems are commercially available, while others are being developed as an operating-unit next to the decentralized anaerobic digester. We cannot predict yet which of these technologies will actually be used. However, it is clear that for AD to play an important role during the transition from a fossil fuel-based economy towards a renewable energy-based economy, some of these technologies will be necessary to connect all digesters with the existing natural gas grid.

6.9 EXERCISES

Exercise 6.1: If you have an anaerobic digester with an infinite volume, resulting in an infinite residence time for the substrate, would you produce more methane under mesophilic or thermophilic temperatures? Discuss the reason

Exercise 6.2: Why is there a thermodynamic advantage of anaerobic digestion as compared to other types of fermentation?

Exercise 6.3: A large dairy farm is using anaerobic digestion to produce biogas from cattle manure with a capacity of 6 000 000 $m^3 \cdot yr^{-1}$. The farm is considering adding a PSA system to convert the biogas into biomethane, which qualifies for renewable energy credits. However, their biogas contains 450 ppm(v) of hydrogen sulphide (H_2S), which is detrimental to the adsorption column (e.g. it forms irreversible bonds with the sorbent), and needs to be removed prior to PSA. A local industry produces iron-oxide impregnated wood chips, which could be used as an H_2S scrubbing material. The absorptive capacity of the wood chips is related to its iron oxide content, which is 190 kg $Fe_2O_3 \cdot m^{-3}$. To complete their economic analysis of the proposed system, they need to calculate the amount of woods chips required for H_2S scrubbing. Calculate the annual consumption of wood chips in cubic meters. Assume the wood chips have a 90% capacity factor relative to the theoretical maximum absorption. Note: the molecular weight of $Fe_2O_3 = 159.687$ g mol^{-1}, and $H_2S = 34.1$ g mol^{-1}.

Exercise 6.4: A dairy farm with 3000 cows has a manure waste flow of 500 m^3/d (10% solids). The total organic carbon (TOC) concentration in manure is 60 g/L. The hydraulic retention time (HRT) in AD is 20 days. The TOC removal efficiency in AD is 80% and for each kg of TOC removed, the digester generates 0.25 m^3 of CH_4. The residual TOC is then subjected to hydrothermal liquefaction (HTL), which converts 60% of the carbon into bio-crude oil carbon. Assume manure has the density of water.

(a) What should be the size of the AD?
(b) What is the CH_4 production per unit of digester?
(c) What is the effluent carbon flow from the AD to HTL?
(d) How much carbon (as CH_4 and bio-crude oil) can be recovered per cow per day.
(e) Describe the advantages and disadvantages for each one of the following integrated process sequences: (1) TH-AD; and (2) AD-HTL.

Exercise 6.5: Calculate the composition of the product gas leaving the PtG bioreactor if pure carbon dioxide (no biogas) is introduced with hydrogen at a ratio of 1:4 and assuming that (i) 25%, (ii) 50%, (iii) 75%, or (iv) 100% of the carbon dioxide are consumed by biomethanation. All the water generated by methanation can be assumed to remain as liquid.

Exercise 6.6: (A) Compare power-to-methane and power-to-hydrogen and name two advantages of each technology; and (B) Can you think of reasons, why the hydrogen content in the natural gas grid is restricted?

Exercise 6.7: Performing electrolysis in an electrolyte solution (50 mL) that contains 1 M NaOH by passing 1 mmol e^-:

(a) Discuss which species will be oxidized at the anode based on the standard reduction potential, OH^- ions or water?
(b) Calculate how much oxygen will be produced at the anode?
(c) Discuss which species will be reduced at the cathode based on the standard reduction potential, Na^+ ions or water (assuming H^+ reduction is negligible)?
(d) Calculate how much hydrogen will be produced at the cathode?

Exercise 6.8: Considering the information you have from this chapter and given that $\Delta G = \Delta H - T\Delta S$, conclude whether the change in Gibb's free energy is more negative at 37°C or at 65°C for hydrogenotrophic methanogenesis? Evaluate what this implies for the maximum biomass yield per methane of thermophilic hydrogenotrophic methanogens compared to mesophilic species?

6.10 DISCUSSION QUESTIONS

Question 6.1: Is it possible to avoid the formation of carbon dioxide by biological means? How would you do it?

Question 6.2: In each of the technologies described in this chapter, carbon dioxide was considered a valueless by-product. However, given that many of the off-streams have been enriched in CO_2, should these off-streams really be considered valueless? Can you think of a particular application where this CO_2 maybe be useful? If so, which biogas upgrading technology would be the most advantageous for the proposed application?

Question 6.3: As the lead engineer of a WWTP, you are in charge to integrate a PtG system to treat the biogas stream (50% carbon dioxide, 50% methane) of the WWTP's anaerobic digester. You are asked to give a presentation to the board of directors in which you provide a concept on how the process will be integrated with all its mass and energy streams into the WWTP's infrastructure. Can you identify synergistic effects?

Question 6.4: After reading about the thermophilic methanogens,

(a) Do you think the thermophilic character of *M. thermautotrophicus* is an advantage or disadvantage for biomethanation applications?
(b) Which are the only substrates that *M. thermautotrophicus* can metabolize?
(c) Do you need to add a carbon source to the growth medium of *M. thermautotrophicus*? If yes, which one. If no, why not?

Question 6.5: With the biochemical reaction of carbon dioxide and hydrogen gas into methane and water at a high production rate, would you need to heat or cool the reactor if you want to operate at a temperature of 65°C, and why?

Question 6.6: Can you think of another way to convert a liquid into a gas besides vaporization?

FURTHER READING MATERIALS

Blanco H. and Faaij A. (2018). A review at the role of storage in energy systems with a focus on power to gas and long-term storage. *Renewable and Sustainable Energy Reviews*, **81**, 1049–1086.

Ghaib K. and Ben-Fares F.-Z. (2018). Power-to-methane: a state-of-the-art review. *Renewable and Sustainable Energy Reviews*, **81**, 433–446.

The German Solar Energy Society and Ecofys. (2004). *Chapter 3: Anaerobic Digestion. Planning and Installing Bioenergy Systems: A Guide for Installers, Architects and Engineers*. James & James/Earthscan, London, UK.

Weiland P. (2006). Biomass digestion in agriculture: a successful pathway for the energy production and waste treatment in Germany. *Engineering in Life Sciences*, **6**, 302–309.

REFERENCES

Angenent L. T., Karim K., Al-Dahhan M. H., Wrenn B. A. and Domínguez-Espinosa R. (2004). Production of bioenergy and biochemicals from industrial and agricultural wastewater. *Trends in Biotechnology*, **22**, 477–485.

Angenent L. T., Usack J. G., Xu J., Hafenbradl D., Posmanik R. and Tester J. W. (2018). Integrating electrochemical, biological, physical, and thermochemical process units to expand the applicability of anaerobic digestion. *Bioresource Technology*, **247**(Supplement C), 1085–1094.

Bailera M., Lisbona P., Romeo L. M. and Espatolero S. (2017). Power to gas projects review: lab, pilot and demo plants for storing renewable energy and CO_2. *Renewable and Sustainable Energy Reviews*, **69**, 292–312.

Barber W. P. F. (2016). Thermal hydrolysis for sewage treatment: a critical review. *Water Research*, **104**, 53–71.

Bougrier C., Delgenès J. P. and Carrère H. (2008). Effects of thermal treatments on five different waste activated sludge samples solubilisation, physical properties and anaerobic digestion. *Chemical Engineering Journal*, **139**, 236–244.

Eboibi B. E., Lewis D. M., Ashman P. J. and Chinnasamy S. (2015). Integrating anaerobic digestion and hydrothermal liquefaction for renewable energy production: an experimental investigation. *Environmental Progress & Sustainable Energy*, **34**, 1662–1673.

Götz M., Lefebvre J., Mörs F., McDaniel Koch A., Graf F., Bajohr S., Reimert R. and Kolb T. (2016). Renewable power-to-gas: a technological and economic review. *Renewable Energy*, **85**, 1371–1390.

Hanselmann K. W. (1991). Microbial energetics applied to waste repositories. *Cellular and Molecular Life Sciences*, **47**, 645–687.

Labatut R. A., Angenent L. T. and Scott N. R. (2011). Biochemical methane potential and biodegradability of complex organic substrates. *Bioresource Technology*, **102**, 2255–2264.

Labatut R. A., Angenent L. T. and Scott N. R. (2014). Conventional mesophilic vs. Thermophilic anaerobic digestion: a trade-off between performance and stability? *Water Research*, **53**, 249–258.

Lettinga G. (1995). Anaerobic digestion and wastewater treatment systems. *Antonie Van Leeuwenhoek*, **67**, 3–28.

Martin M. R., Fornero J. J., Stark R., Mets L. and Angenent L. T. (2013). A single-culture bioprocess of *methanothermobacter thermautotrophicus* to upgrade digester biogas by CO_2-to-CH_4 conversion with H_2. *Archaea* **2013**, 1–11 (Article ID 157529).

Pace N. R. (1997). A molecular view of microbial diversity and the biosphere. *Science*, **276**, 734–740.

Peterson A. A., Vogel F., Lachance R. P., Fröling M., Antal M. J., Jr and Tester J. W. (2008). Thermochemical biofuel production in hydrothermal media: a review of sub- and supercritical water technologies. *Energy & Environmental Science*, **1**, 32–65.

Posmanik R., Labatut R. A., Kim A. H., Usack J. G., Tester J. W. and Angenent L. T. (2017). Coupling hydrothermal liquefaction and anaerobic digestion for energy valorization from model biomass feedstocks. *Bioresource Technology*, **233**, 134–143.

Posmanik R., Martinez C. M., Cantero-Tubilla B., Cantero D. A., Sills D. L., Cocero M. J. and Tester J. W. (2018). Acid and alkali catalyzed hydrothermal liquefaction of dairy manure digestate and food waste. *ACS Sustainable Chemistry & Engineering*, **6**, 2724–2732.

Schiebahn S., Grube T., Robinius M., Tietze V., Kumar B. and Stolten D. (2015). Power to gas: technological overview, systems analysis and economic assessment for a case study in Germany. *International Journal of Hydrogen Energy*, **40**, 4285–4294.

Schill N., van Gulik W. M., Voisard D. and von Stockar U. (1996). Continuous cultures limited by a gaseous substrate: development of a simple, unstructured mathematical model and experimental verification with *Methanobacterium thermoautotrophicum*. *Biotechnology and Bioengineering*, **51**, 645–658.

Schill N. A., Liu J. S. and Stockar U. (1999). Thermodynamic analysis of growth of *methanobacterium thermoautotrophicum*. *Biotechnology and Bioengineering*, **64**(1), 74–81.

Stuckey D. C. and McCarty P. L. (1984). The effect of thermal pretreatment on the anaerobic biodegradability and toxicity of waste activated sludge. *Water Research*, **18**(11), 1343–1353.

Tekin K., Karagöz S. and Bektaş S. (2014). A review of hydrothermal biomass processing. *Renewable and Sustainable Energy Reviews*, **40**, 673–687.

Uman A. E., Usack J. G., Lozano J. L. and Angenent L. T. (2018). Controlled experiment contradicts the apparent benefits of the Fenton reaction during anaerobic digestion at a municipal wastewater treatment plant. *Water Science and Technology*, **78**(9), 1861–1870.

van Haandel A. and van der Lubbe J. (2012). *Handbook of Biological Wastewater Treatment*, 2nd edn. IWA Publishing, London, UK.

Verbeeck K., Buelens L. C., Galvita V. V., Marin G. B., Van Geem K. M. and Rabaey K. (2018). Upgrading the value of anaerobic digestion via chemical production from grid injected biomethane. *Energy & Environmental Science*, **11**, 1788–1802.

Wilson C. A. and Novak J. T. (2009). Hydrolysis of macromolecular components of primary and secondary wastewater sludge by thermal hydrolytic pretreatment. *Water Research*, **43**, 4489–4498.

Zeikus J. G. and Wolfe R. S. (1972). *Methanobacterium thermoautotrophicus* sp. n., an anaerobic, autotrophic, extreme thermophile. *Journal of Bacteriology*, **109**, 707–713.

doi: 10.2166/9781780409566

Chapter 7

Anaerobic fermentation technologies for the production of chemical building blocks and bio-based products from wastewater

Pieter Candry[1,2], *José Maria Carvajal-Arroyo*[1,2], *Steven Pratt*[3], *João Sousa*[4], *Çağrı Akyol*[5,6], *Francesco Fatone*[5] *and Ramon Ganigué*[1,2]

[1]*Centre for Microbial Ecology and Technology (CMET), Ghent University, Coupure Links 653, 9000 Ghent, Belgium*
[2]*Centre for Advanced Process Technology for Urban REsource Recovery, Frieda Saeysstraat 1, 9052 Gent, Belgium*
[3]*School of Chemical Engineering, The University of Queensland, St Lucia, QLD 4072, Australia*
[4]*Paques Technology B.V., T. de Boerstraat 24, 8561 EL, Balk, The Netherlands*
[5]*Department of Science and Engineering of Materials, Environment and Urban Planning-SIMAU, Marche Polytechnic University, Via Brecce Bianche 12, 60100 Ancona, Italy*
[6]*Department of Green Chemistry & Technology, Ghent University, Coupure Links 653, 9000 Ghent, Belgium*

7.1 INTRODUCTION

Anaerobic digestion (AD) is currently one of the most widely applied technologies for the treatment and valorization of organic waste streams as electrical and/or thermal energy (as illustrated in Chapter 5). Yet, the economic value of the produced biogas is rather low (Table 7.1), which has led to an increasing interest in developing anaerobic based technologies that can produce high(er)-value products – beyond biogas – to increase the economic profits from organic waste streams. Within this context, the production of short-chain carboxylic acids (SCCA) and their derivative products have gained significant interest in recent years.

As discussed in detail in Chapter 5, AD is a multi-step biochemical process, breaking down the complex organic matter into, ultimately, CH_4 and CO_2. In the acidogenic step, organic monomers are fermented to SCCA, organic acids with one carboxyl group and a saturated chain of 2–5 carbon atoms. SCCA hold a higher economic value than CH_4 (Table 7.1) and can be further upgraded to a wide variety of consumer products through chemical and/or biological processes. Some of these applications include, but are not limited to, the production of polyhydroxyalkanoates (PHAs, a type of bioplastics), esters (fragrances and flavours), and solvents (Albuquerque *et al.*, 2011; Zacharof & Lovitt, 2013). SCCA can also be converted to medium chain carboxylic acids (MCCA) microbially. These can be used as such as anticorrosion agent or antimicrobial agent in, for example, animal rearing, but can also be processed into biofuels, bioplastics and other consumer goods (Angenent *et al.*, 2016). Halting anaerobic digestion at the level of SCCA formation would allow for the redirection of the AD process from a low-value energy recovery to a higher-value chemical production platform. In this way, the economics of the process could be substantially improved.

Table 7.1 Economic values for various products that can be obtained through anaerobic treatment of waste streams with mixed cultures.

Product	Market Price (€·tonne^{-1})	Market Price (€·tonne C^{-1})	Market Price (€·tonne COD^{-1})
CH_4	90–200	120–267	20–50
Acetic acid	330–670	825–1675	310–630
Propionic acid	1250–1380	2570–2840	830–910
Butyric acid	1670–2090	3060–3830	920–1150
Lactic acid	840–1510	2100–3775	790–1410
Caproic acid	1880–2090	3030–3370	850–950

Prices were obtained from Moscoviz *et al.* (2018).

This chapter will first introduce the microbiology and biochemistry of carboxylic acid production (both SCCA and MCCA), knowledge not only relevant to the chapter at hand, but also to preceding chapters that discuss anaerobic digestion (Chapters 5 and 6). Gaining a fundamental understanding of the microbiology and biochemistry is essential prior to describing the more engineering orientated design considerations and applications of this platform. A product is only valuable if it can be recovered and separated. As such, the general principles of the main downstream approaches to recover carboxylates from fermentation broths and/or upgrade them to value-added products are subsequently discussed. Last, the chapter will touch upon some of the research needs as well as the key challenges and opportunities for SCCA and MCCA production to become viable technologies that can be implemented by the water industry at full-scale.

7.2 LEARNING OBJECTIVES

At the completion of this chapter you should be able to:

- Understand what the carboxylate platform is and why it is an interesting emerging approach to produce chemical building blocks from wastewater (and other waste streams).
- Gain a fundamental understanding of the different metabolic steps in the production of carboxylic acids, their microbiology and biochemistry and understand the implications for practice.
- Outline the main physico-chemical routes to recover carboxylates/carboxylic acids from fermentation broths and understand their basic engineering principles.
- Explain how the intermediates of the acidogenic fermentation can be biologically converted to alternative higher-value end-products.
- Describe the current status of SCCA and MCCA production from wastewater (and derivatives) and outline the key challenges and opportunities.
- Discuss the economic rationale driving the development of carboxylic acid-producing technologies, their inherent advantages – and disadvantages – compared to AD, and how are these connected to the technical challenges ahead of it.
- Synthesize all the knowledge in this chapter and describe how you would turn anaerobic digesters into reactors for the production of carboxylic acids.

7.3 MICROBIOLOGY AND BIOCHEMISTRY OF CARBOXYLIC ACID PRODUCTION

Organic waste streams – such as waste activated sludge – are a complex mixture of carbohydrates, proteins and fats. As explained in Chapter 5, AD can break down this complex organic matter in a four-step microbial process: (i) hydrolysis of the complex organic matter to release the biochemical building blocks, that is monosaccharides, amino acids and fatty acids; (ii) acidogenesis, in which these building blocks are fermented to a mixture of SCCA; (iii) acetogenesis, that is a secondary fermentation converting the mixture of SCCA to acetic acid, H_2 and CO_2; and lastly, (iv) methanogenesis, converting the acetic acid, H_2 and CO_2 to CH_4. This metabolic network changes when the target is shifted from CH_4 to carboxylic acids; complex organic matter is still hydrolyzed and fermented to a mixture of SCCA, but acetogenesis and methanogenesis are suppressed. Potentially, under certain conditions the end-products of acidogenesis can be further converted to, for example, other SCCA or MCCA in secondary microbial processes. The following sections will introduce the main processes involved in the transformation of complex organic material (e.g., sludge) into mixtures of SCCA, lactic acid and ethanol (i.e., hydrolysis and acidogenic fermentation), and their potential further conversion to MCCA.

7.3.1 Hydrolysis

Hydrolysis is the process of breaking up polymers, for example polysaccharides, proteins and lipids, into their monomeric building blocks – respectively monosaccharides, amino acids and fatty acids. During this process, a molecule of water is used to break the bonds between the monomer units (Figure 7.1).

Although these reactions are catalyzed enzymatically, organisms do not receive energy from this process. It is, however, necessary to gain access to the monomer units, which can be then used for energy generation. With most substrates – also waste activated sludge – hydrolysis is the rate limiting step in AD. It is usually modelled by a first order degradation kinetic, that is the hydrolysis rate depends on the concentration of degradable particulate organic matter (Equation (7.1)) (Batstone *et al.*, 2002; Pavlostathis & Giraldo-Gomez, 2009):

$$\frac{dF}{dt} = -k_h \cdot F \tag{7.1}$$

where F is the concentration of degradable particulate organic matter ($g{\cdot}L^{-1}$) and k_{hyd} is the hydrolysis constant (d^{-1}).

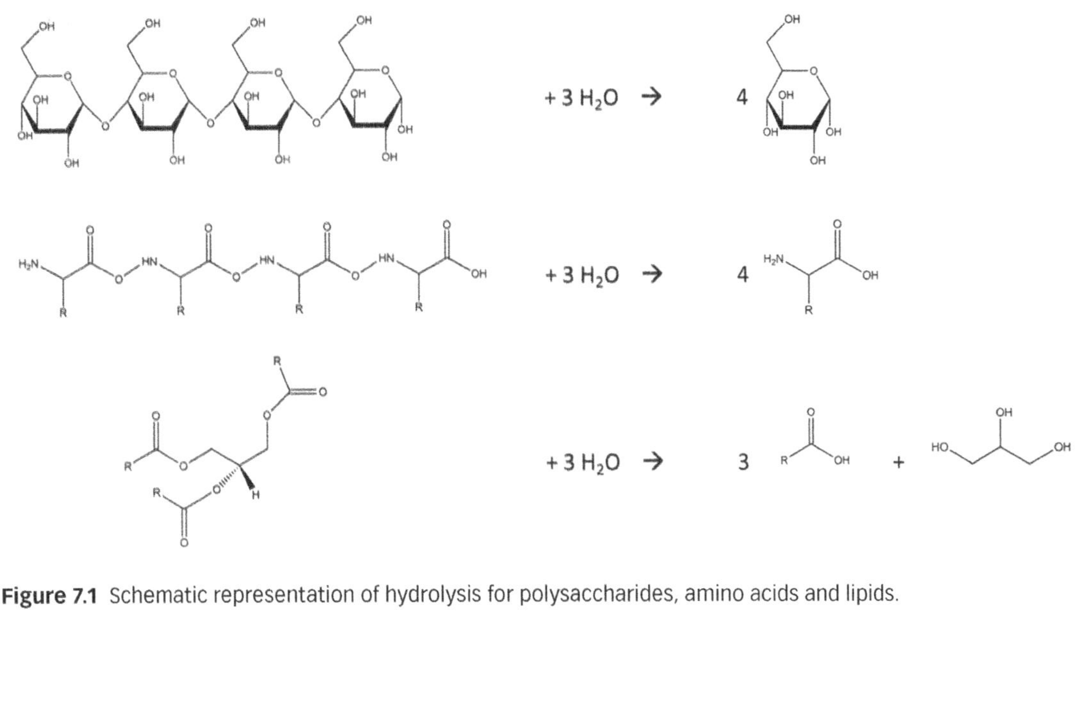

Figure 7.1 Schematic representation of hydrolysis for polysaccharides, amino acids and lipids.

Hydrolysis constants are mostly dependent on: (i) the nature of the material to be degraded (i.e., carbohydrate, protein, lipid); (ii) the temperature; (iii) pH; and (iv) the particle size and its surface area (Vavilin *et al.*, 2008). As such a broad range of constants can be found in literature. For instance, k_{hyd} in the order of 0.005–0.01, 0.015–0.075 and 0.025–0.2 d^{-1} have been reported for the hydrolysis of lipids, proteins and carbohydrates in solid organic waste at 55°C. The anaerobic mesophilic hydrolysis constants of protein-rich waste activated sludge from municipal WWTPs are usually in the order of 0.1–1 d^{-1} (Christ *et al.*, 2000; Sanders, 2001; Tomei *et al.*, 2009; Vavilin *et al.*, 2008). It should be mentioned that feedstocks can be highly varied in their composition (i.e., proportions of carbohydrates, protein, lipids), and by extension also in the rate of hydrolysis and the rate of the process.

7.3.1.1 Hydrolysis of polysaccharides

Carbohydrates may be an important constituent in industrial wastewater but are usually not present at high concentrations in waste activated sludge, due to the aeration steps removing most of these compounds. However, co-digestion of waste activated sludge with more degradable carbohydrate-rich feedstocks has been suggested as a means to improve digestibility (Sosnowski *et al.*, 2003). Carbohydrates in such organic waste streams are often present in the form of polysaccharides such as starches and (hemi)-celluloses. These compounds are essentially chains of saccharides, often high in D-glucose, but other saccharides can be present as well. As a first step, these chains need to be broken up into mono- or oligosaccharides before they can be fermented. In the case of starches, this can be done by a group of enzymes called amylases, ubiquitous throughout all domains of nature (Tester *et al.*, 2006). Even though cellulose chains are also predominantly made up of D-glucose molecules, like starches, the cellulase enzymes necessary to break up these chains are less widespread throughout nature. Fungi are well-known cellulose degraders, and cellulolytic activity is widespread throughout the family. However, only few genera in Bacteria contain cellulolytic organisms, and generally only a few species within one genus are active cellulose degraders (Lynd *et al.*, 2002). This is partly due to the different configuration of the cellulose chain – glucose units are linked with β(1-4)-bond unlike the α(1-4) bond in starches – as well as the insoluble nature of celluloses, which require cellulolytic enzymes to be functional outside the cell. Some examples of anaerobic cellulose degraders are *Clostridium thermocellum*, *Clostridium cellulolyticum* and *Butyrivibrio fibrosolvens* (Koeck *et al.*, 2014). Despite the apparent phylogenetic sparseness of cellulose-hydrolyzing organisms, there is usually no need to add hydrolyzing organisms or enzymes to waste activated sludge: the native community always contains some hydrolytic organism that can take up this role to liberate the solid organic matter for further conversion.

7.3.1.2 Hydrolysis of proteins

Proteins are essentially a chain of different amino acids, which then assemble into a 3D structure. During the hydrolysis, protease enzymes break the peptide bonds in the protein, releasing the individual amino acids, or short chains of amino acids, called peptides. Commercial production of proteases has been demonstrated with *Bacillus* and *Clostridium* genera (Rao *et al.*, 1998; Siebert & Toerien, 1969), yet, again, there is usually no need to add specific hydrolytic organisms or enzymes to waste feedstocks, as protease-producing organism can be readily found in nearly all environments.

7.3.1.3 Hydrolysis of fats

Wastewater and microbial biomass contain fats, which are made up of triglycerides, composed of a glycerol backbone linked to three fatty acid molecules via ester bonds. Fats can be hydrolyzed by lipases, which cut the ester bonds, resulting in the production of one molecule of glycerol and three of fatty acid, the length of which is dependent on the type of fat (Madigan *et al.*, 2012). These lipase enzymes are ubiquitous throughout all domains of life, and a wide range of bacteria have been employed to produce these enzymes on an industrial scale for enzymatically catalyzed processes (Javed *et al.*, 2018), but for (anaerobic) degradation of waste sludge, lipase producing organisms are commonly present in mixed communities.

7.3.2 Primary fermentations

Monomers generated during the hydrolysis step can subsequently be utilized by microbes. Under anaerobic conditions the lack of external electron acceptors forces microorganisms to metabolize these monomers via fermentation. Fermentation is an anaerobic redox process, in which the oxidation of the substrate is coupled to the reduction of another substrate or an intermediate derived from the oxidation, with the difference in redox potential of the substrate and the end product providing energy for ATP synthesis (Müller, 2008). Some special cases exist where the oxidation of one substrate is linked to the reduction of another (organic) substrate, for instance Stickland reactions involved in protein degradation.

7.3.2.1 Primary fermentation pathways for saccharides

Monosaccharides can be utilized in a wide range of anaerobic fermentative metabolisms. Monosaccharides contain either five carbon atoms (C5 monosaccharides, e.g., xylose, arabinose, etc.) or six carbon atoms (C6 monosaccharides, e.g., glucose, fructose, etc.). Different metabolisms are involved in the fermentation of C5 and C6 monosaccharides, affecting product and energy yields for fermentative bacteria. During fermentation of C6 monosaccharides, glycolysis first converts glucose through glyceraldehyde-3-phosphate to pyruvic acid, as depicted in Figure 7.2. This can in turn be: (i) reduced to succinic acid and/or propionic acid; (ii) reduced to lactic acid; or (iii) decarboxylated to acetyl-CoA resulting in the release of either a formic acid molecule or CO_2 and H_{2-} molecules, depending on the pH. The formed acetyl-CoA can subsequently be converted to various products such as ethanol, acetic acid, or reduced to butyryl-CoA, followed by conversion to butyric acid (Temudo *et al.*, 2007). Which of these reactions will occur depends on the fermentative pathway(s) present in the organism in the system.

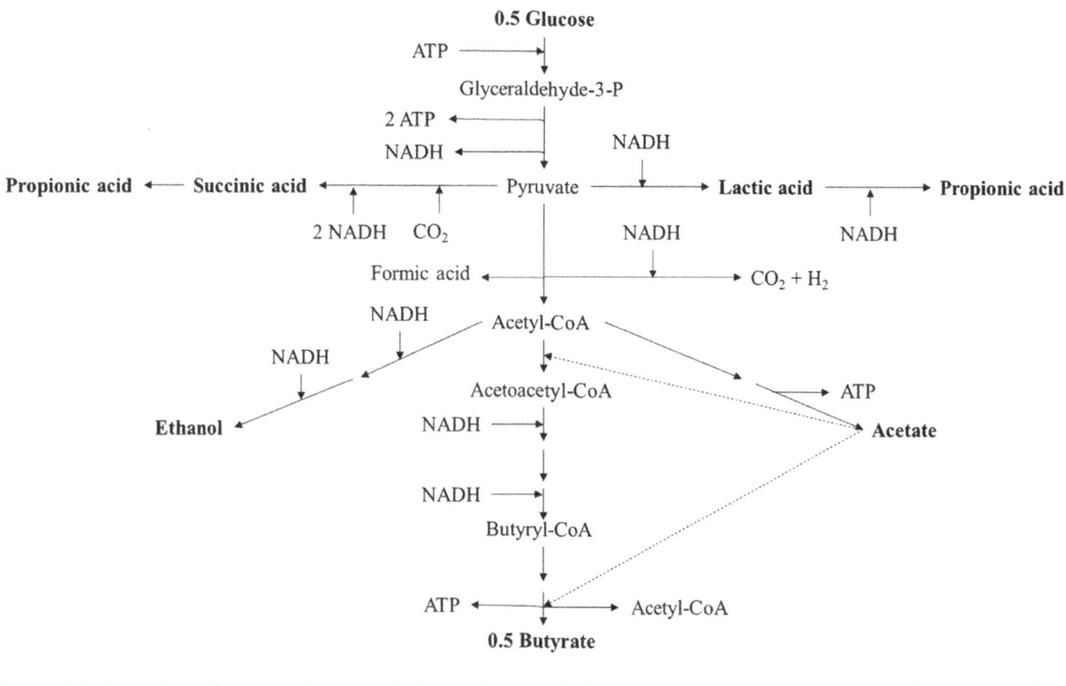

Figure 7.2 Overview of potential fermentation pathways during mixed culture fermentation of C6 saccharides, using glucose as a model compound. Adapted from Temudo *et al.* (2007).

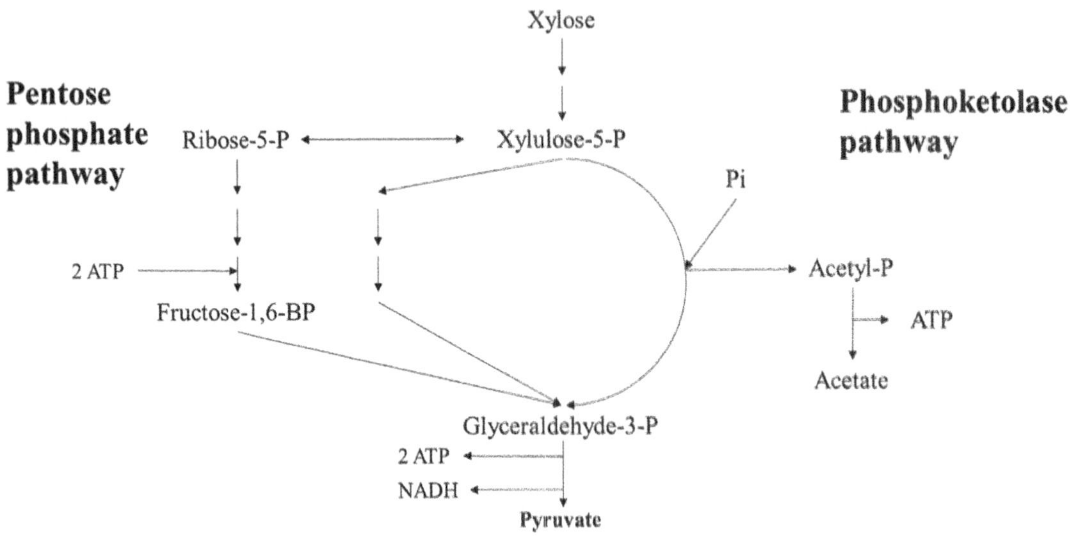

Figure 7.3 Initial conversion steps of C5 saccharides to pyruvate, with xylose as a model compound. After pyruvate, the same pathways as for C6 saccharides are possible. Figure adapted from Temudo *et al.* (2009).

Fermentation of C5 monosaccharides differs from that of C6 mostly in the initial steps of the fermentation process. While they are also converted to glyceraldehyde-3-P and subsequently pyruvate, this conversion goes through specific pathways for C5 monosaccharides, as depicted in Figure 7.3 for xylose. In short, the pentose phosphate pathway (PPP) converts part of the xylose to ribose-5-P and part to xylulose-5-P. These are then used to form fructose-1,6-BP, followed by conversion into glyceraldehyde-3-P. Ultimately, for every three molecules of xylose converted, five molecules of pyruvate are produced, with production of five ATP and five NADH. An alternative pathway, the phosphoketolase pathway (PKP), also goes through xylulose-5-P. However, conversion of xylulose-5-P to glyceraldehyde-3-P in the PKP also generates acetyl-P, which will in turn be used to produce acetic acid and ATP. This means that every molecule of C5 monosaccharides converted through the PKP will always yield one molecule of acetic acid on top of the product obtained from pyruvate. Since both the PPP and PKP yield pyruvate, the same products can be obtained as from C6 monosaccharides (Temudo *et al.*, 2009).

An intermediate version of these two metabolisms is heterolactic fermentation, as performed by *Leuconostoc* bacteria. In this fermentation process, glucose is converted to xylulose-5-P by liberating one molecule of CO_2. The xylulose-5-P is then further converted over the phosphoketolase-pathway as mentioned above. The pyruvate obtained is converted to lactate, while the acetyl-CoA can be converted to ethanol. This yields equimolar amounts of lactic acid and ethanol from glucose (Demoss *et al.*, 1951; Dols *et al.*, 1997).

Ultimately, the metabolic potential of an open, undefined microbial community will enable conversion of the saccharide building blocks to a mixture of the products mentioned above. The product profile will be determined by: (i) the microbial community present; (ii) operational parameters, for example pH, temperature, substrate concentration; and (iii) type of saccharides present after hydrolysis.

7.3.2.2 Primary fermentation pathways for amino acids

The amino acids released during hydrolysis of proteins can be further fermented towards a range of products, yielding one molecule of carboxylic acid, NH_3 and CO_2 per amino acid fermented.

The carboxylic acid produced is amino acid-specific, although it is in most cases acetic acid. The most common mechanism for amino acid fermentation is through Stickland reactions (coupled fermentations). In this pathway, two amino acids are concomitantly degraded, where one of the acids is oxidized and acts as electron donor and the other is reduced by accepting these electrons. The electron-donating amino acid is deaminated (i.e., release of NH_3) and decarboxylated (i.e., release of CO_2), and subsequently oxidized to an SCCA with a carbon chain shortened by one carbon atom. During this process, electrons are released, which are used to reduce the deaminated electron accepting amino acid to a SCCA with a carbon chain of the same length as the original amino acid. Besides SCCA, also branched carboxylic acids (e.g., iso-valeric acid or iso-caproic acid), alcohols and aromatic compounds (e.g., phenol, cresol) can be released during fermentation of amino acids, depending on the initial structure of the amino acid (Madigan *et al.*, 2012). Alternatively, amino acids can be fermented individually, although not all amino acids can be converted this way. Again, the products obtained depend on the amino acid substrate (Elsden & Hilton, 1978). In a wastewater context, the range of amino acids present in waste sludge is usually very broad, and steering protein fermentations is highly challenging. A last pathway for conversion of amino acids is the Ehrlich pathway, found in yeasts (e.g., *Saccharomyces cerevisiae*). Here, proteins are first converted to an aldehyde, which is then used to produce an acid or ester, and an (amyl) alcohol (Hazelwood *et al.*, 2008). It needs to be noted that these yeasts are usually not present in wastewater settings, and the contribution of the Ehrlich pathway to protein degradation from waste sludge will be negligible.

In general, amino acid fermentations are mostly carried out by gram-positive bacteria, principally from the genus Clostridia. These organisms are key members of nearly all anaerobic, fermenting communities, so amino acids in waste streams can be readily degraded in most applications. The fermentation stoichiometry of each specific amino acid and the microbiology of amino acid fermentation are extensively reviewed elsewhere and will not be further discussed in this chapter (Mead, 1971; Ramsay & Pullammanappallil, 2001).

7.3.2.3 Primary fermentation pathways for long-chain fatty acids (LCFA)

Breakdown of LCFA happens through a cyclical pathway called beta-oxidation. In this pathway, a coenzyme A-group is attached to the carboxylic group, forming an acyl-CoA. This acyl-CoA is then degraded in a cyclical process, releasing one molecule of acetyl-CoA and one NADH molecule per turn of the cycle, resulting in an acyl-chain shortened by two carbon units (Madigan *et al.*, 2012). The intracellular pool of $NADH/NAD^+$ is limited, which means the generated NADH needs to be quickly re-oxidized back to NAD^+ to balance the $NADH/NAD^+$-pool. Under aerobic conditions, NADH oxidation can be combined with the reduction of O_2 to H_2O. Due to the lack of oxygen under anaerobic conditions, this can only be achieved through reduction of H^+ to H_2. This process becomes thermodynamically unfeasible at H_2 partial pressures over approximately 100 Pa (equal to 0.01% H_2 in the headspace of the system), implying H_2 has to be removed immediately for this reaction to continue. In anaerobic digestion, it has been shown that a syntrophy between an anaerobic fatty acid oxidizer and hydrogenotrophic methanogens provides the low H_2 partial pressures required for this process (Conrad *et al.*, 1986). However, in fermentations aimed at carboxylic acid production no H_2-scavenger can sufficiently lower H_2 partial pressures to enable continued LCFA-degradation. This means lipids-rich wastes are challenging feedstocks for the production of high-value products. So far, no reports have targeted such wastes for bioproduction processes beyond lower-value applications such as biogas production through anaerobic digestion.

7.3.2.4 Practical implications

In the previous sections, we offered an overview of the metabolic pathways involved in the primary fermentations from different substrates commonly found in organic waste feedstocks such as waste sludge. While many products can be formed, the heterogeneity and complexity of the feedstocks relevant in wastewater settings means that steering these processes is highly challenging. The large volumes and complexity of the substrate excludes the application of pure cultures for bioproduction, and technologies rely on mixed consortia to provide us with high-value products. Yet, obtaining a pure product profile made up of only one product from real wastes has not yet been demonstrated. Even finding general trends in operational parameters (e.g., pH, organic loading rate, organics concentration, etc.) to steer the fermentation process has proven highly challenging (Arslan *et al.*, 2016). Consequently, most applications developed so far address this by upgrading the mixed product profile, an approach that will be discussed in section 7.4.

7.3.3 Secondary anaerobic conversions

The products of a primary fermentation can be further transformed anaerobically into other compounds. An example found in AD is acetogenesis, where the mixture of products (e.g., SCCA, lactic acid, ethanol) from acidogenesis are converted to acetic acid through anaerobic oxidation. On the other hand, in the context of resource recovery and bioproduction from waste streams, other anaerobic biotransformations aim to increase the value of the primary fermentation products. Depending on the target product, it can be necessary to add carbon or electron sources, such as ethanol or lactic acid, to stimulate these secondary processes and/or physically separate primary and secondary biotransformations.

7.3.3.1 Secondary fermentations to SCCA from lactic acid

Lactic acid is a common product formed in fermentations of carbohydrates, as discussed in section 7.3.2.1. Lactic acid bacteria are renowned for their relatively high growth rates and tolerance to low pH values, giving them a competitive advantage over other fermentative bacteria (Castillo Martinez *et al.*, 2013). Lactic acid can in turn be further fermented to other SCCA. Lactic acid can either be: (i) converted to acetyl-CoA, after which it used to produce acetic acid, or butyric acid (cf. Figure 7.2); or (ii) converted to propionic acid with either acrylyl-CoA – via the acrylate pathway – or succinate – Wood–Werkman cycle – as key intermediates (Parizzi *et al.*, 2012; Prabhu *et al.*, 2012).

7.3.3.2 Reverse beta-oxidation with ethanol as electron donor

As mentioned earlier in this chapter, the production of MCCA has gained interest as a way to recover high-value chemicals from waste streams. The microbial production of MCCA in itself is not new and has been known since the early 1940s (Barker, 1941), when a mesophilic anaerobic bacterium capable of using ethanol and acetic acid to produce caproic acid (C6) (*Clostridium kluyveri*) was isolated from canal mud. For a long time, *C. kluyveri* was of special interest to microbiologists due to its unique metabolism, named reverse beta-oxidation, by which it adds two carbon units to the acyl chain of a monocarboxylic acid per turn of the cycle. This metabolism has been elucidated over the years, its reaction stoichiometry is shown in Table 7.2, and is schematically represented in Figure 7.4. The process starts by the oxidation of six molecules of ethanol to acetyl-CoA, generating 12 molecules of NADH. One of the six molecules of acetyl-CoA is further converted into acetic acid, generating one ATP from substrate-level phosphorylation. The other five molecules of acetyl-CoA go into the reverse beta-oxidation cycle, combining with five molecules of acetyl-CoA to form acetoacetyl-CoA, which are further transformed into butyryl-CoA with the consumption of 15 NADH molecules. In a final step, these five molecules of butyryl-CoA then combine with five molecules of acetic acid, releasing five molecules of butyric acid, and five molecules of acetyl-CoA that then go back to the start of the cycle. This process can also be repeated by combining acetyl-CoA from ethanol oxidation with butyryl-CoA from the first cycle, which will ultimately yield hexanoyl-CoA and subsequently caproic acid (Angenent *et al.*, 2016).

During this process, an imbalance is created in the NADH/NAD^+ pool, because 15 NADH molecules are consumed while only 12 are produced. Because of the limited size of the NADH/NAD^+ pool, the process needs to be balanced. This is achieved by reducing three NAD^+ to NADH with

Table 7.2 Overview of common metabolic reactions in fermentation from mixed organic feedstocks, for example industrial wastewater, waste sludge or co-digestion feedstocks, and examples of organisms responsible for these pathways.

Substrate	Product	Stoichiometry	Example Organisms
Glucose	Ethanol	$C_6H_{12}O_6 \rightarrow 2\ C_2H_5OH + 2\ CO_2 + 2\ H_2$	*Sacharromyces cerevisiae, Zymomonas mobilis*
	Lactic acid	$C_6H_{12}O_6 \rightarrow 2\ C_3H_6O_3$ $C_6H_{12}O_6 \rightarrow C_3H_6O + C_2H_5OH + CO_2 + H_2$	*Lactobacillus* *Leuconostoc*
	Acetic acid	$C_6H_{12}O_6 \rightarrow 2\ CH_3COOH + 2\ CO_2 + 2\ H_2$	*Clostridium, Acetobacterium*
	Butyric acid	$C_6H_{12}O_6 \rightarrow C_3H_7COOH + 2\ CO_2 + H_2$	*Clostridium, Faecalibacterium*
	Succinic acid	$C_6H_{12}O_6 + CO_2 \rightarrow 2\ C_4H_6O_4$	*Actinobacillus succinogenes*
Ethanol	Acetic acid	$C_2H_5OH + H_2O \rightarrow CH_3COOH + 2\ H_2$	*Acetobacter*
Ethanol, acetic acid	Butyric acid	$6\ C_2H_5OH + 4\ CH_3COOH \rightarrow 5\ C_3H_7COOH + 2\ H_2 + 5\ H_2O$	*Clostridium kluyveri*
Ethanol, butyric acid	Caproic acid	$6\ C_2H_5OH + 5\ C_3H_7COOH \rightarrow 5\ C_5H_{11}COOH + 2\ H_2 + 5\ H_2O$	*Clostridium kluyveri*
Ethanol, acetic acid		$12\ C_2H_5OH + 3\ CH_3COOH \rightarrow 5\ C_5H_{11}COOH + 4\ H_2 + 10\ H_2O$	*Clostridium kluyveri*
Lactic acid	Propionic acid	$C_3H_6O_3 \rightarrow C_2H_5COOH$	*Propionibacterium*
	Acetic acid	$C_3H_6O_3 + H_2O \rightarrow CH_3COOH + CO_2 + 2\ H_2$	*Acetobacter*
	Butyric acid	$2\ C_3H_6O_3 \rightarrow C_3H_7COOH + 2\ CO_2 + 2\ H_2$	*Megasphaera elsdenii*
	Caproic acid	$3\ C_3H_6O_3 \rightarrow C_5H_{11}COOH + 3\ CO_2 + 2\ H_2$	*Ruminococcaceae* CPB6

ferredoxin – originating from the penultimate step of the reverse beta-oxidation cycle. This is done at the membrane-bound RnF-enzyme complex. During this process, six protons or Na^+ molecules are pumped outside the cell. This creates a proton, or sodium, motive force over the membrane that can be used to generate an additional 1.5 ATP by a membrane-bound ATPase complex. In this way, balancing the NADH/NAD^+ pool is a crucial part of the energy generation in the metabolism of *C. kluyveri* (Angenent *et al.*, 2016).

7.3.3.3 Chain elongation using alternative electron donors

Due to the diverse nature of wastewater and organic waste streams, converting them into a mixture of ethanol and acetic acid suitable for subsequent chain elongation can be challenging. Alternatively, ethanol could be supplemented after primary fermentation to convert the short-chain products to MCCA, however, this supplementation reduces the potential profits, and increases the environmental impacts of the process (Chen *et al.*, 2017). Other substrates have been shown to allow production of MCCA as well, often through metabolisms similar to the reverse beta-oxidation described earlier, such as saccharides, lactic acid, proteins, and pyruvate (Angenent *et al.*, 2016).

Saccharides in waste streams can originate from the hydrolysis of starches and celluloses and are often present in a mix of mono- and oligosaccharides. Several chain elongating bacteria have been isolated capable of using saccharides, for example *Megasphaera hexanoica* using fructose, and *Caproiciproducens galactitolivorans* using D-galactitol, a sugar alcohol. Both isolates were capable of producing over 7 g caproic acid·L^{-1} (Jeon *et al.*, 2017; Kim *et al.*, 2015).

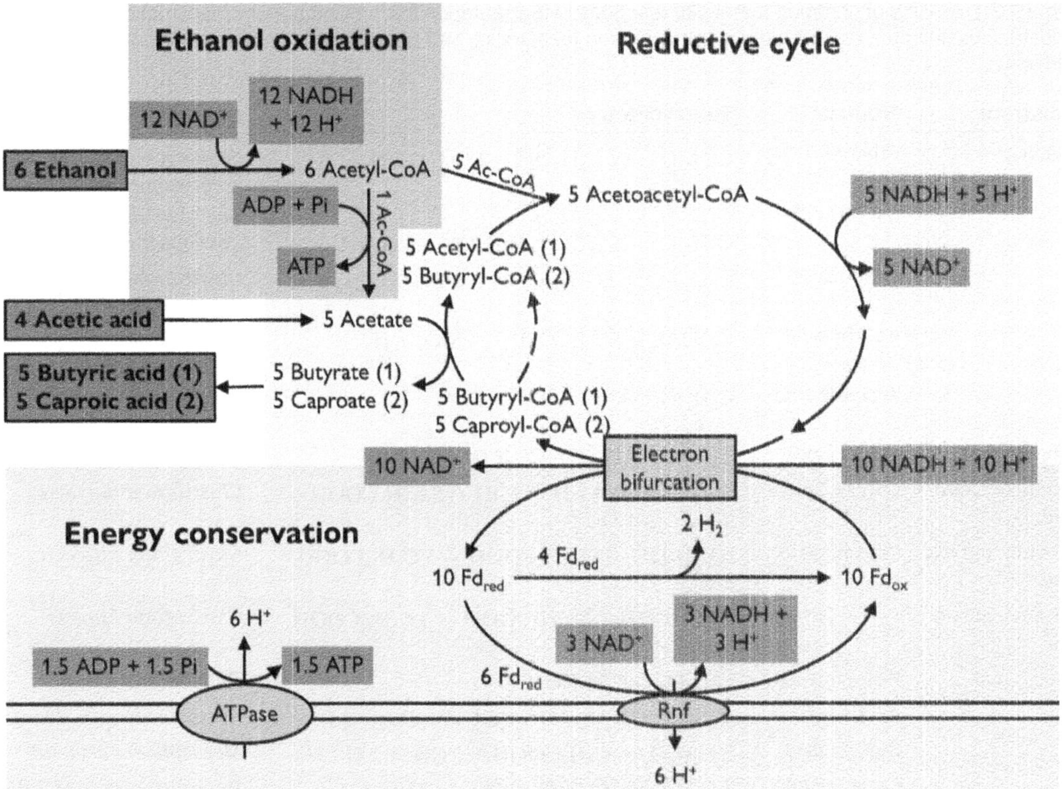

Figure 7.4 Simplified representation of the reverse beta-oxidation metabolism (adapted from Angenent *et al.*, 2016).

Lactic acid, which can easily be obtained from saccharides by lactic acid bacteria such as *Lactobacillus*, can also be used as electron donor for MCCA production. This metabolism was first reported in *Megasphaera elsdenii*, although MCCA production from lactic acid is negligible (Elsden *et al.*, 1956; Weimer & Moen, 2013). In a series of studies carried out with mixed cultures fed with thin stillage – a side-stream from the bio-EtOH production process – it was hypothesized that saccharides in the stillage could first be converted to lactic acid, which was then used for MCCA production (Andersen *et al.*, 2015, 2017). The first conclusive observation of lactic acid-based chain elongation came from a study investigating caproic acid production in Chinese liquor pit fermentations. Of the potential electron donors present in these fermentations – ethanol, lactic acid and glucose – lactic acid was the only substrate that enabled production of caproic acid (Zhu *et al.*, 2015). In a follow-up study, the responsible bacterium (*Ruminococcaceae* sp. CPB6) was isolated and characterized (Zhu *et al.*, 2017).

Peptides are the product of protein hydrolysis and can in turn be further fermented. While fermentations to SCCA from peptides were described in section 7.3.2.2, other metabolisms are also possible. For instance, one isolate (*Eubacterium pyruvativorans*) has been shown to convert peptides into caproic acid, likely through a pathway similar to reverse beta-oxidation, while using the nitrogen to produce ammonia (Wallace *et al.*, 2003).

Towards practice, the above substrates are rarely found at sufficiently high concentrations in waste activated sludge – or other wastewater-derived feedstocks – to enable direct MCCA-production. To achieve this, co-digestion of waste with sugar- or lactic acid-rich feedstocks could be a potential approach for the valorization of waste activated sludge. Industrial wastewater from the dairy, food and beverage industry may allow for MCCA production via lactic acid or saccharides.

7.3.4 The reason behind it all: energy maximization and redox balancing

Over the last section, a range of anaerobic metabolic pathways have been discussed, some of which can be performed by a large group of bacteria, such as carbohydrate fermentation, while other metabolisms belong to a phylogenetically narrow group, or even to unique species (e.g., ethanol chain elongation by *Clostridium kluyveri*). Some metabolisms are enzymatically very simple (for instance, lactic acid fermentation), while others require many enzymes (for instance, reverse beta-oxidation). In any system, organisms will aim to maximize the net energy generation rate (Großkopf & Soyer, 2016). This net energy includes both energy yield, that is ATP, but also energy consumption, for example enzyme synthesis and cell maintenance (Kleerebezem & Van Loosdrecht, 2010). Considering the energy yield of glucose as substrate, Table 7.3 shows the ATP yields for different metabolic end products. When producing lactic acid, ethanol, or a mixture of the two (i.e., heterolactic fermentation), this will yield two ATP per molecule of glucose, originating from substrate-level phosphorylation (SLP) during the conversion of glucose to pyruvate over glyceraldehyde 3-phosphate. Fermentation to acetic acid yields an additional two ATP per molecule of glucose, by SLP when acetyl-CoA is converted to acetate over acetyl-phosphate. However, fermentation to lactic acid is a shorter pathway than acetic acid fermentation, requiring lower energy investment in enzymes, and enabling higher rates (Kreft *et al.*, 2020).

In parallel, every metabolic pathway needs to achieve a redox balance. This means that every electron going into the system must go out, for instance glucose as electron donor coming into the cell and being excreted as an extracellular carboxylic acid. The first step in glucose fermentation is the oxidation to pyruvate. Any oxidation reaction needs a reduction to balance the electrons in the system, which is why cells use intracellular electron carriers to store these electrons. Examples of such carriers are NADH, NADPH or ferredoxin. For instance, in the oxidation of glucose to pyruvate, two molecules of NAD^+ are reduced to NADH (Temudo *et al.*, 2007). However, a cell cannot accumulate electrons infinitely, due to the limited $NAD(P)H/NAD(P)^+$ pool, and the NAD(P)H generated must be converted back to $NAD(P)^+$. In fermentations, lacking any other electron acceptors, two options exist for this regeneration. A first option is the use of NADH to reduce an organic electron acceptor (Müller, 2008). This is what happens during lactic acid fermentation, reducing pyruvate to lactic acid. Lacking this, NADH can also transfer its electrons to H^+ as an electron acceptor, producing H_2. The latter happens during fermentation of glucose to acetic acid, or for instance, during anaerobic

Table 7.3 Overview of product, ATP and NADH yields for fermentation of glucose and xylose as model compounds for C6 and C5 saccharides.

Product	Glucose			Xylose (PPP)			Xylose (PKP)		
	Y_{prod}	Y_{ATP}	Y_{NADH}	Y_{prod}	Y_{ATP}	Y_{NADH}	Y_{prod}	Y_{ATP}	Y_{NADH}
Succinic acid	2	2	−2	1.67	1.67	−1.67	1 + 1 Ac	2	−1
Lactic acid	2	2	0	1.67	1.67	0.00	1 + 1 Ac	2	0
Ethanol	2	2	−2	1.67	1.67	−1.67	1 + 1 Ac	2	−1
Acetic acid	2	4	2	1.67	3.33	1.67	2	3	1
Propionic acid	2	2	−2	1.67	1.67	−1.67	1 + 1 Ac	2	−1
Butyric acid	1	3	0	0.83	3.33	0.00	0.5 + 1 Ac	3	0

Xylose can be fermented through the pentose phosphate pathway (PPP) and phosphoketolase pathway (PKP). The latter yields one acetic acid (Ac) and one ATP during conversion of xylose to pyruvate (see Figure 7.3).

oxidation of carboxylic acids. In processes were NAD^+ is in excess, for example reverse beta-oxidation or ethanol fermentation, ferredoxin is used as a high energy electron carrier. At the membrane bound Rnf complex, this ferredoxin is oxidized, reducing NADH to NAD^+, as well as pumping protons or sodium ions out of the cell to generate a chemical energy gradient. This energy gradient can in turn be used to generate additional ATP through an ATPase complex (Biegel *et al.*, 2011).

There is, however, a last consideration to be made: the thermodynamics of the chemical reactions. Ethanol can be converted by two pathways in the absence of electron acceptors: (i) anaerobic ethanol oxidation; or (ii) reverse beta-oxidation. The former yields 1 mol ATP per mol ethanol (cf. Figure 7.2; ethanol to acetyl-CoA to acetate), while the latter yields only 0.42 mol ATP per mol ethanol (Seedorf *et al.*, 2008). On top of that, the reverse beta-oxidation is a complex pathway, requiring synthesis of a large number of enzymes. Why then, would any organism go through the effort of setting up such a complex pathway? The answer to this can be found in thermodynamics and energy generation. Converting ethanol to acetate under anaerobic conditions yields H_2. When the partial pressure of H_2 increases, the amount of energy generated per molecule of ethanol decreases. Eventually, the reaction becomes thermodynamically infeasible at partial pressures of approx. 0.1 bar (Cavalcante *et al.*, 2017). However, this does not mean that all available substrate has been depleted. Under these specific conditions (high organic load, high H_2 partial pressures) reverse beta-oxidation provides a way to generate more energy from the available substrates, even if it requires a larger number of enzymes. This knowledge can help in the design of systems targeting specific product profiles, by engineering operational conditions to stimulate the targeted metabolisms. In the case of chain elongation, this means it is crucial to maintain sufficiently high partial pressures of H_2 to prevent anaerobic oxidation of ethanol (Steinbusch *et al.*, 2011). In the end, it becomes clear that organisms have developed an array of ways to maintain the redox balance under the niche conditions they thrive in in nature. Understanding these conditions is crucial to harness the metabolic potential of these organisms and stimulate these metabolic pathways in any bioproduction system.

7.4 CHEMICAL AND BIOLOGICAL DOWNSTREAM/UPGRADING ROUTES FOR THE RECOVERY OF CARBOXYLIC ACIDS

7.4.1 Solid–liquid separation before product recovery

The organic acids generated through fermentation can be further upgraded via additional bioprocessing, referred to as biological upgrading, or via physicochemical processes, as schematized in Figure 7.5. For both of these trajectories, cell separation is the first step after fermentation. Cell separation is needed to protect downstream processes, for example preventing fouling, clogging in physicochemical unit operations, or microbial contamination of bioprocesses. Different methods, based on different principles, can be applied for separation of cell biomass with different effluent qualities. Additionally, the separation technique used will depend on the objective, generating a clear effluent, or dehydrating the solid biomass fraction. Here we will shortly discuss two of the most commonly used technologies that aim at generating clear streams for further downstream processing, namely centrifugation and filtration.

Centrifugation is widely applied in food processing, bioprocessing and wastewater treatment and is based on the differential density between solid particles and liquids (Svarovsky, 2001a, 2001b). Depending on scale and application, centrifugation can be done in batch or continuous mode. On the other hand, centrifugation often leads to only partial solid separation producing a liquid fraction that still contains suspended solids.

Filtration can be applied to retain biomass in the fermenter, or to clarify a fermentation effluent, as a stand-alone unit, or downstream of a centrifugation step. Given the size of bacterial cells, pore sizes between 0.1 and 0.5 μm are often enough to ensure complete biomass retention. Microfiltration (0.1–10 μm) and ultrafiltration (0.001–0.1 μm) are applied in processes that require a clear stream devoid of

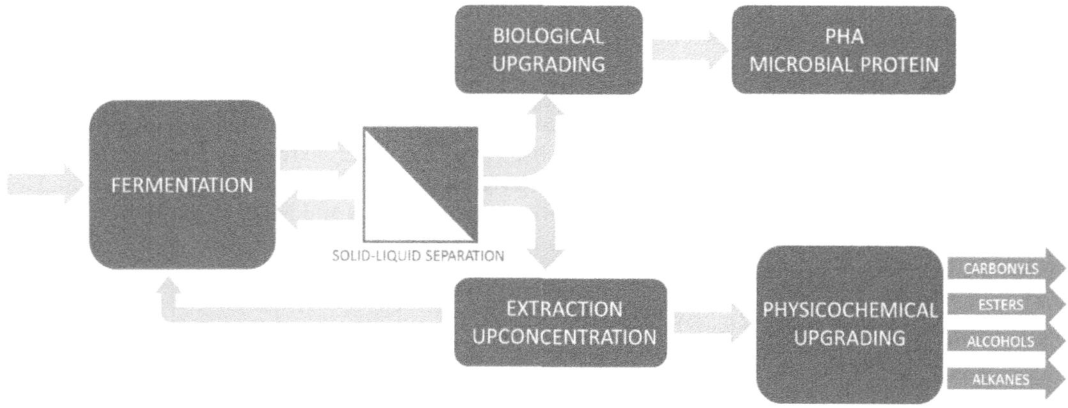

Figure 7.5 Combination of fermentation with downstream processing options.

suspended solids. The filtration can be carried in different configurations, that is internally, by using membranes submerged in the reactor broth, or externally, by recirculating the broth tangentially to the membrane, also known as cross flow filtration or tangential flow filtration. Membranes are built in different shapes and materials. Polymeric membranes (PVDF, PP, PE, etc.) are built as flat sheet, tubular or hollow fibre membranes. Ceramic membranes are built in flat sheets or in tubular shapes (Lin *et al.*, 2013). The type of membrane, material and configuration is determined by the process conditions, suspended solid concentration and composition of the broth. Especially when broths contain medium chain fatty acids at pH <6, ceramic membranes are preferred over polymeric membranes, due to the degradation of the polymers by the organic acids. One key issue to be considered for filtration processes in waste-contexts is fouling, where solids in the stream build up on the membrane, leading to increased pressure drops, and increased energy requirements for filtration. Membrane cleaning, back-washing and other fouling-mitigating strategies have been developed and are routine practice for these systems, as described elsewhere (Judd & Judd, 2011).

7.4.2 Physicochemical product upgrading

This sub-section briefly summarizes some of the most promising physicochemical approaches for the recovery and upgrading of carboxylic acids from fermentation broths. These are, in most cases, based on existing technologies applied in other settings that can be bought off-the-shelf. Therefore, the focus of this sub-section will lay on introducing the general principals of each technology rather than discussing how to engineer them, which has been extensively described elsewhere (Green & Perry, 2008). The latter will be dependent on many factors such as the carboxylic acid composition, concentration, final target product, and so on.

7.4.2.1 Product extraction and up-concentration

Traditional anaerobic digestion generates methane, a gaseous product that spontaneously separates from the broth. Open culture acidogenic fermentations produce organic acids with variable, but generally, high water solubility. In this context, acetic, propionic, butyric and valeric acids are water miscible, and only acids with carbon length of five carbon atoms or longer form an oil phase when the concentration of their undissociated form reaches the solubility limit (e.g., 49.7 g L^{-1} for valeric acid, 10.8 g L^{-1} for caproic acid, etc. (Saboe *et al.*, 2018)). Due to the high solubility of organic acids, and their low concentration achieved in mixed culture fermentations, extraction, concentration and purification methods are often needed. For instance, the addition of $CaCO_3$ or NH_4HCO_3 to counteract acidification may lead to the production of calcium or ammonium carboxylate salts that will precipitate when their concentration in the fermentation broth is high enough (e.g., >320 g L^{-1} for Ca acetate, at 25°C, >61 g L^{-1} for Ca lactate, at 25°C, etc.) (Lóреz-Garzón & Straathof, 2014). On the other hand, when fermentations are operated at low pH, the concentration of organic acids in the broth are seldom higher than 10–20 g L^{-1}, too low to enable their direct recovery as salts, requiring other approaches. In general, the choice of downstream route (i.e., extraction, concentration and purification) depend on the physicochemical characteristics of the acid and the final application. Here we will describe methods for extracting and up-concentrating SCCA and MCCA. Extraction methods include gas stripping, adsorption, pressure-driven membrane processes, liquid-liquid extraction, and electrochemical membrane processes.

7.4.2.1.1 Gas stripping combined with absorption

Air and gas stripping applications aim to remove volatile compounds from a solution by contacting clean air with the aqueous solution (i.e., fermentation broth) across a high surface area. This process is governed by Henry's law (Equation (7.2)), which describes the partial pressure of a volatile component (i.e., SCCA) in a gas phase (P_{SCCA}) in equilibrium with a dilute solution of that component at a concentration $C_{SCCA,aq}$ where H_{cp} is Henry's law constant (mol·m^{-3}·Pa^{-1}):

$$H_{cp} = \frac{C_{SCCA,aq}}{P_{SCCA}} \tag{7.2}$$

The maximum transfer rate (TR) of a specific SCCA that can be achieved at a stripping gas volumetric flow rate, Q, is given by Equation (7.3):

$$TR = QC_{SCCA(g)} = Qx\frac{C_{SCCA(aq)}}{H_{cp}RT} \tag{7.3}$$

Since only the undissociated species of the acids are volatile, the process is pH dependent. The carboxylic acids in open culture fermentations have similar pKa values (4.87–4.89) but differ in their Henry's law constants that range between 23 and 70 mol·m^{-3}·Pa^{-1} (Sander, 2015).

A scrubber can be used to simultaneously regenerate the stripping gas and quantitatively recover the SCCA. The absorption of the SCCA can be done in a $CaCO_3$ solution that enables recovering the products as carboxylate calcium salts. The stripping gas remains water saturated at constant temperature (Li *et al.*, 2015).

Pervaporation is a modified version of gas stripping, where clean gas is not stripped through the liquid, but instead contacts the liquid via a membrane. The SCCA can diffuse through the membrane and transfer to the gas phase, after which the SCCA can be recovered as calcium salts, as described above.

7.4.2.1.2 Adsorption

Carboxylic acids in gas or liquid phase can be sorbed onto a solid surface. This process is called adsorption and depends on the interaction between the solid surface moieties and the carboxyl group and hydrophobic tail of the carboxylic acids. Adsorption is typically carried out in a column packed with a sorbent, which is often a resin with positively charged amine functional groups on the surface (López-Garzón & Straathof, 2014). The ion exchange mechanism of an acid can be described with the following equilibria (Equations (7.4)–(7.6)), in which HA is the carboxylic acid, A^- is the carboxylate anion, R_xNH_{3-x} is the non-protonated amine (for a primary amine x = 1, for a secondary amine x = 2, and for tertiary amine x = 3), $R_xNH_{4-x}^+$ is the protonated amine and $RxNH_{4x}^+A$ is the acid-resin salt:

$$HA \leftrightarrow A^- + H^+ \tag{7.4}$$

$$R_xNH_{3-x} + H_2O \leftrightarrow R_xNH_{4-x} + OH^- \quad (7.5)$$

$$R_xNH_{4-x} + A^- \leftrightarrow R_xNH_{4-x}A \quad (7.6)$$

Additionally, van der Waals forces occur between the hydrophobic tail of the acids and the matrix of the resin (Rebecchi *et al.*, 2016). The adsorption efficiency depends on the adsorption capacity of the resin, that is mass of adsorbate per kilogram of adsorbent (Rebecchi *et al.*, 2016). Once the resin is saturated, the carboxylic acids can be recovered by using a suitable solvent. The overall efficiency of the process depends on the type of resin used, the adsorption capacity of the resin, operating pH and temperature, type of desorption and the desorbing chemical used, presence of other competing anions and the adsorption/contact time (Reyhanitash *et al.*, 2017).

7.4.2.1.3 Pressure-driven membrane processes

Reverse osmosis and nanofiltration are used to separate solutes from solvents by semi-permeable membranes. These can be used to recover SCCA from streams after solids-removal as described in section 7.4.1. In a second filtration step for product recovery, a membrane is used to concentrate the SCCA in solution, by allowing the transfer of water, but rejecting the SCCA (Zacharof & Lovitt, 2013). Several factors affect the recovery rate, including solution pH, temperature, product concentration and hydraulic pressure. While pH and pressure are positively correlated with the organic acid recovery, the increase in temperature decreases its concentration. This approach is established for industrial fermentations and there is a growing literature on these methods for SCCA recovery from waste streams (Atasoy *et al.*, 2018). As a final note, it is important to highlight that issues such as biofouling or inorganic scaling can occur, resulting in increased energy input and the need for regular membrane cleaning.

7.4.2.1.4 Liquid–liquid extraction

In liquid–liquid extraction systems, an aqueous phase containing the organic acids is put in contact with an organic solvent. The organic acids in their undissociated form are hydrophobic and partition into the organic phase. The longer the chain, the higher is the partition coefficient (Table 7.4). Since only the undissociated form dissolves in the organic solvent, it is evident from the equations listed below (Equations (7.7)–(7.9)) that the efficiency of the extraction step is pH dependent:

$$K_{isw} = \frac{C_{is}}{C_{Hiw}} \quad (7.7)$$

$$C_{Hiw} = \frac{C_{Tiw}}{1 + 10^{(pH - pK_{ia})}} \quad (7.8)$$

Table 7.4 Standard octanol – water partition coefficients for different organic acids (Hansch *et al.*, 1995).

Carboxylic Acid	Log K_{ow}
Acetic acid	−0.17
Propionic acid	0.33
Butyric acid	0.79
Valeric acid	1.39
Caproic acid	1.92
Caprylic acid	3.05

$$C_{is} = K_{isw} \cdot \frac{C_{Tiw}}{1 + 10^{(pH - pK_{ia})}} \tag{7.9}$$

where K_{isw} is the partition coefficient of the compound i between the solvent s and the aqueous phase at equilibrium, C_{is} is the concentration of i in the organic phase, C_{Hiw} is the concentration of the protonated species of the acid in the aqueous phase, C_{Tiw} is the total concentration of the carboxylic acid in the aqueous phase, and pK_{ia} is the pH at which the concentration of the protonated and dissociated species of a monoprotic acid are equal. Table 7.4 lists the log of the partition coefficient of different carboxylic acids between octanol and water under standard conditions.

Alkylphosphine oxides, trialkylamines, apolar solvents and combinations thereof are the most investigated solvents for the extraction of carboxylic acids from fermentation broths. Kerosene and paraffin oil have been successfully used as organic solvents in combination with trioctylphosphine oxide (TOPO) in concentrations in the range of 3–20% (Agler *et al.*, 2012; Levy *et al.*, 1981). Higher concentrations of TOPO in the solvent enhance the extraction of shorter chain carboxylic acids. The main advantage of the extraction with solvents is that it is more selective than other extraction techniques and enables selective extraction of medium chain over short chain carboxylic acids. Since the separation between the organic and liquid phases is slow, these can be indirectly contacted across a hydrophobic membrane, which enables the diffusion of the undissociated acids to the organic solvent (Agler *et al.*, 2012, 2014). This allows for a very significant reduction in equipment size, but membranes are easily fouled and wet, requiring periodical cleaning and regeneration of the hydrophobic surface.

7.4.2.1.5 Membrane electrochemical processes

Technologies like electrodialysis (Qian-Zhu *et al.*, 2016) or membrane electrolysis (Andersen *et al.*, 2014) allow extracting and concentrating the dissociated carboxylates to an aqueous extract by using the potential difference between two electrodes. In electrodialysis, cations are transported across a cation exchange membrane to the cathode compartment while anions are transported across an anion exchange membrane to the anode compartment. Carboxylic acids can be recovered in their undissociated form in the acidic anode compartment, while reduction of water at the cathode generates a caustic stream that can be used to control the pH in the fermenter. In membrane electrolysis, the fermentation broth is fed to the cathode compartment and the carboxylates extracted to the anode compartment where they become protonated. Additionally, the broth in the cathode compartment becomes alkaline due to the OH^- generation from water reduction and the acidification of the fermentation can be counteracted without the addition of chemicals. The main disadvantage of these membrane electrochemical technologies is the low selectivity of the membranes for organic acids which cause low power efficiencies due to co-extraction of smaller inorganic ions. Smaller molecules have higher electrophoretic mobility and therefore electrochemical techniques are more suitable for extraction of SCCA.

7.4.2.1.6 Hybrid downstream trains

Combinations of the techniques described above can be applied. For example, if MCCA are to be produced in acid form, liquid–liquid extraction can be combined with membrane electrolysis (Xu *et al.*, 2015). The organic acids in their protonated form are extracted from the fermentation broth to an organic solvent. Once in the organic phase they can be recovered in a second extraction step with an alkaline aqueous solution. The difference in pH between the broth and the alkaline side generates a gradient in the concentration of undissociated organic acids that drives the extraction. In order to sustain the driving force, the pH of the aqueous extract needs to be maintained alkaline. This can be done by circulating it through the cathode compartment of a membrane electrolysis cell. The OH^- generated in the cathode maintains the alkaline conditions. Additionally, if the electrolysis cell is fitted with an anion exchange membrane, the dissociated organic acids will be transported to the anode compartment, where they become concentrated and acidified by the H^+ generated by anodic water oxidation. Once the organic acids accumulate over their solubility limit, they form an organic phase composed of MCCA. This extraction pipeline requires the fermentation to be run at pH values in the range of 5–5.5, since only undissociated MCCA are extracted in the solvent extraction part of the system.

Liquid–liquid extraction can be combined with distillation, a well-established technology in the chemical industry. A first step of extraction with a solvent can be followed by distillation, which enables separating the organic acids from the solvent. For instance, it has been calculated that acetic acid can be recovered by using a pipeline including liquid–liquid extraction followed by a heat-integrated two-stage distillation, at 2.6 MJ Kg^{-1} acetic acid (Saboe *et al.*, 2018).

7.4.2.2 Chemical conversions

Organic acids derived from fermentations can be used as substrates for the production of multiple chemical derivatives. Organic acids can be esterified with alcohols under acidic conditions to generate alkyl esters. Alternatively, after pre-concentration, acids can be converted into ketones by thermochemical reduction. Ketones can be further hydrogenated to alcohols (Holtzapple *et al.*, 1999). Additionally, organic acids can be used to produce alkanes by an electrochemical hydrolytic process known as Kolbe electrolysis (Agler *et al.*, 2011; Levy *et al.*, 1981; Urban *et al.*, 2017).

The list of possible transformation products of organic acids is larger than their current applications and the progressive substitution of petroleum based feedstocks is likely to promote the generation of novel pipelines and applications.

7.4.3 Biological product upgrading

This sub-section will introduce two potential biological routes for the upgrading of carboxylic acid and their recovery as a marketable product: (i) polyhydroxyalkanoates as bioplastics; and (ii) microbial protein as feed. Both routes are further discussed in upcoming chapters, which also cover the more engineering-related aspects of such processes.

7.4.3.1 Polyhydroxyalkanoates for bioplastics

Polyhydroxyalkanoates (PHA) have been proposed as bio-based and bio-degradable alternative polymers to current petrochemical thermoplastics (Reddy *et al.*, 2003). Their mechanical and physical properties, for example insoluble in water, resistant to hydrolytic degradation, biocompatible and easier to process than traditional polymers (Bugnicourt *et al.*, 2014), are of particular interest because they can be tailored. They are also truly biodegradable in ambient conditions in soil and marine environments, which is not the case for some alternatives (like poly-lactic acid) that require specific conditions for biodegradation, such as higher temperatures (Tansengco & Tokiwa, 1998). PHA consist of esterified chains of hydroxyalkanoic acids (HA), that is carboxylic acids with an additional alcohol group, with most biologically produced PHA being made up of either polyhydroxybutyrate (PHB), polyhydroxyvalerate (PHV), or a copolymer of the two (PHBV). However, many variations of PHA have been demonstrated, and the monomers present in the PHA will determine their physical properties as a plastic.

Microbial synthesis of PHA has been long known (Lemoigne, 1927; Weibull, 1953), with PHA functioning as an energy storage molecule during aerobic growth of organisms under nutrient limited conditions (Doudoroff & Stanier, 1959). SCCA are often used as substrates for PHA production. These polymers are stored intracellularly, presenting themselves as intracellular granules. PHA production can be achieved in pure culture systems where over 90% of the final dry biomass weight is PHA. However, these processes require sterile culturing conditions and are limited to the use of defined substrates, mostly crop based. Alternatively, mixed cultures can be enriched in PHA-accumulating bacteria by applying selective strategies such as feast-famine and/or nitrogen limitation. These strategies select for organisms that can take up the carbon quickly, and store it internally for later use, for exmaple when nitrogen becomes available in the environment. With such a process, up to 90%

of the final dry biomass weight is PHA when using a defined feedstock (acetic acid), while a product with up to 80% PHA can be obtained when using waste-based SCCA (Johnson *et al.*, 2009; Korkakaki *et al.*, 2016). A case study on the PHA production from wastewater is presented in Chapter 8.

7.4.3.2 Microbial protein for feed and food

Microbial protein, that is microbial biomass with high protein content, has been suggested to create feed and food with high nutritional value from low-value or waste streams (see Chapter 9) (Matassa *et al.*, 2016). The strength of this concept is the direct recycling of carbon, nitrogen and energy from wastes to food. This can happen indirectly, by using it as animal feed, or directly using it for human consumption. Both methods have been shown to significantly reduce the pressure on the environment associated with food production. In a recent study, it was estimated that by 2050, microbial protein could make up 10–19% (175–307 Mton) of protein fed to livestock. This substitution could result in up to a 13% reduction in land use for crops, up to 8% reduction in nitrogen losses from croplands and up to 9% reduction in greenhouse gas emissions when comparing to a full crop-based livestock diet (Pikaar *et al.*, 2018). While the core of the microbial protein platform is the recovery of nitrogen into a valuable product, microorganisms also need a carbon and electron source to grow. Several pipelines have already been explored to supply both nitrogen and carbon to processes producing microbial protein (see Chapter 10). One example is the conversion of low-value industrial side or process streams to microbial protein, for instance, wastewater from a potato processing facility. Another route that has been suggested is the use of carboxylic acids produced through fermentation processes as a sustainable carbon source to grow biomass. Two types of bacteria have been proposed for the production of microbial protein with carboxylic acids as substrates: aerobic heterotrophic bacteria and purple non-sulphur bacteria (PNSB). The former group uses the carboxylic acids as a conventional aerobe, using them as energy source as well as carbon source. PNSB on the other hand are phototrophic and use light as a source for energy. This means that carboxylic acids are solely directed to biomass growth, resulting in a higher carbon yield. This coupling of fermentation and microbial protein production can provide a sustainable source of feed and food. However, this technology is still in development, and several technological and policy barriers need to be tackled before it can move forward (Alloul *et al.*, 2018).

7.5 CONCEPTUAL OVERVIEW OF THE PRODUCTION OF SHORT-CHAIN CARBOXYLIC ACIDS (SCCA) FROM WASTEWATER

7.5.1 Technological principles

SCCA are considered high value-added products, which can be converted into biofuels, bioplastics, and so on. or used as chemicals in various industries, including as a supplementary electron donor in wastewater treatment. Currently, most SCCA are petrochemically produced, but there is a growing research and industrial interest in the development of alternative biological routes that are potentially more sustainable (Reyhanitash *et al.*, 2017). Organic-rich industrial wastewater or sewage sludge can be hydrolyzed and fermented to SCCA in acidogenic fermentation. In the case of waste activated sludge, it has been estimated that if organic matter in sewage is first captured as sludge using, for example, chemically enhanced primary treatment, and subsequently fermented, about 20–30 g SCCA-COD·IE^{-1}·d^{-1} can be produced (Alloul *et al.*, 2018). The separation of the produced SCCA is a crucial process to obtain value-added products that can be sold (see section 7.4 for a summary of various separation technologies). It is important to highlight that the recovery of SCCA may only be economically viable at medium- and large-scale WWTPs due to the large capital costs required. An alternative to separation and commercialization of these SCCA as chemicals, is their use either directly in a WWTP as electron donor or as feedstock to produce other valuable compounds, such as polyhydroxyalkanoate (PHA) bioplastics. Therefore, the specific processes that are necessary to implement will vary according to the intended application(s). The benefit of direct use on-site is that the WWTP becomes its 'own client' and, as such, further purification, transformation or commercialization efforts are not needed.

7.5.2 Fundamental principles

Over the past 20 years much research has been devoted to the production of SCCA from complex organic wastes by mixed cultures, instead of methane. The main fundamentals of open mixed culture fermentation and product recovery have been introduced in sections 7.3 and 7.4. The following further builds on this knowledge by summarizing the key process parameters and conditions to achieve SCCA production.

First and foremost, SCCA accumulation in an open mixed culture requires the inhibition/suppression of organisms consuming them as substrate, for instance methanogenic archaea. The fermentation pH is one of the most important parameters to steer the process, and SCCA accumulation is usually achieved by inhibiting/suppressing methanogenesis at pH above 8.0 or decreasing the fermentation pH below 6.0 (Atasoy *et al.*, 2018). The fermentation pH also has a direct effect on the hydrolysis and acidogenesis steps, and pH deviating from the neutrality tend to negatively affect the rates of hydrolysis and/or acidogenesis. Temperature has also a great impact on enzymatic activities, growth of microorganisms, and hydrolysis rate. Within the mesophilic range, reaction rates increase with temperature. Shifting from mesophilic to thermophilic conditions may also result in higher SCCA concentrations and changes in the SCCA product spectrum, but at the expense of increasing heating cost (Hao & Wang, 2014; Zhang *et al.* 2009). Short biomass retention (<5 days) times can also contribute to SCCA accumulation by resulting in a partial/complete wash-out of methanogens, which have doubling times often in the order of 4–5 days.

The quantity of SCCA produced is dependent on the degree of acidification of the input substrates, which is reflected in the percentage of initial COD converted to organic acids and other fermentation products (Atasoy *et al.*, 2018). More reduced substrates such as cheese whey or energy crops will result in higher SCCA yields than, for instance, waste activated sludge. A membrane filtration unit (Longo *et al.*, 2015) or centrifuge (Morgan-Sagastume *et al.*, 2014) can be installed after the fermentation tank to separate the SCCA-rich fermentation liquid from the sludge, as discussed in section 7.4. The SCCA recovery rate is highly dependent on both the recovery approach and fermentation conditions (as the latter determine the fraction of SCCA that are in their (un)dissociated form and can be extracted by a certain method).

7.5.3 Applications

SCCA have various potential applications. Organic acids are used to generate very diverse chemical commodities. Acetic acid is used in the production of vinyl acetate, ketene and acetic anhydride (Sneeden, 1982). Although organic acids generally have an unpleasant smell, their ester products are often used in fragrances. Esters like methyl-, ethyl- or isobutyl-acetate, are commonly used as solvents. Esters of valeric acid are primarily used as food additives due to their fruity smell. Acetic, propionic and butyric acids are used in the preparation of cellulose (-propionate, -butyrate) acetates, which are thermoplastics used in the production of films and fibres (Jiang *et al.*, 2018). Typical market prices, applications, and production methods associated with common SCCA are shown in Table 7.5.

SCCA can also be used as feedstocks in the production of renewable plastics and bio textiles. The most paradigmatic example is their use as carbon and energy source in PHA production (see Chapter 9 for further information). The required types of SCCA are different depending on the polymers to be produced. Whereas acetic acid and butyric acid are necessary for polyhydroxybutyrate (PHB) production, propionic acid is favoured for polyhydroxy valerate (PHV) (Peces *et al.*, 2016).

SCCA-containing fermentation liquids can also be used on-site for the optimization of biological nutrient removal (BNR) in WTTPS, as they contain easily biodegradable carbon sources (Kehrein

Table 7.5 SCCA features (adapted from Atasoy *et al.*, 2018).

SCCA	Market Price (€/ton)	Application	Production Methods	Reference
Acetic acid	400–800	Vinyl acetate monomer (polymers, adhesives, dyes), ester production, chemicals, food additive, vinegar, solvent	Chemical synthesis and microbial fermentation	Bhatia and Yang (2017)
Butyric acid	1500–1650	Animal and human food additive, chemical intermediate, solvent, flavouring agent	Chemical synthesis, microbial fermentation, extraction from butter	Zigová and Šturdík (2000)
Propionic acid	2000–2500	Esters used for food industry as aroma additive, food additive, flavouring, pharmaceuticals, animal feed supplement, fishing bait additive	Chemical synthesis and microbial fermentation, by-product of acetic acid production	Cheryan (2009)

et al., 2020). In many cases, wastewater does not include a sufficient amount of biodegradable soluble organic carbon for denitrification and hence external electron donors such as methanol, glucose or acetic acid, are supplied externally for efficient BNR (Hu *et al.*, 2018). In this case, SCCA recovered from wastewater can act as a source of electrons in biological nutrient removal units, replacing methanol or other external substrates. Acetic acid is the most preferred SCCA for denitrification since it is an easily degradable compound by many bacteria, followed by butyric and propionic acid. Propionic acid has been found to enhance denitrifying biological phosphorus removal via nitrite due to the increase in the amount of phosphorus-accumulating organisms (Frison *et al.*, 2016).

7.5.4 Case studies

Despite the research interest in SCCA production from wastewater or its derivatives, to the best of the authors' knowledge there are no pilot-scale installations or case studies demonstrating the production of SCCA from wastewater and their recovery as organic acids. However, several case studies have explored the production of SCCA and its use in secondary biological processes.

First, the PHARIO project (www.phario.eu) demonstrated the production of PHA from SCCA derived from waste activated sludge. One of the focuses of the project was to demonstrate the production of SCCA from waste streams (i.e., carbohydrate-rich process effluent delivered from a local candy factory and waste activated sludge). Experiments were conducted in a pilot-scale (1200 L) well-stirred batch anaerobic fermentation vessel temperature-controlled at 37°C. The process water from the candy factory was fermented for in batch for 7 days with a pH controlled between 5.5 and 6.0. The SRT of the biomass was between 7 and 10 days. The fermentation product contained 16 g readily biodegradable COD·L^{-1} and the SCCA composition of the different batches was composed of mixtures of C2–C7 carboxylic acids. The fermentation experiments with the primary sludge from the Waterschap de Dommel were conducted in batches with continuous mechanical stirring for 6 days at 37°C, with pH monitoring but with no pH control. The sludge matrix was self-buffering and generally the pH was inherently maintained between 4.8 and 5.5. As opposed to the candy factory process water, no biomass was retained between batches since the WAS contained also anaerobic fermenters (STOWA, 2017). The PHARIO project has achieved SCCA yields around 0.25 g SCCA·g^{-1} VSS and SCCA concentrations in the fermentate in the order of 8–10 g COD·L^{-1}, mostly mixtures of acetic, propionic, butyric and valeric acids (STOWA, 2014).

Another example of this is the production of SCCA and their in-situ use in a WWTP for enhanced nutrient removal is the Short-Cut Enhanced Nutrients Abatement (SCENA). This process aims at biological nitrogen removal and P-bioaccumulation via nitrite during the treatment of anaerobic supernatant (Figure 7.6). Sludge liquor from dewatering of digested sludge is heavily loaded with

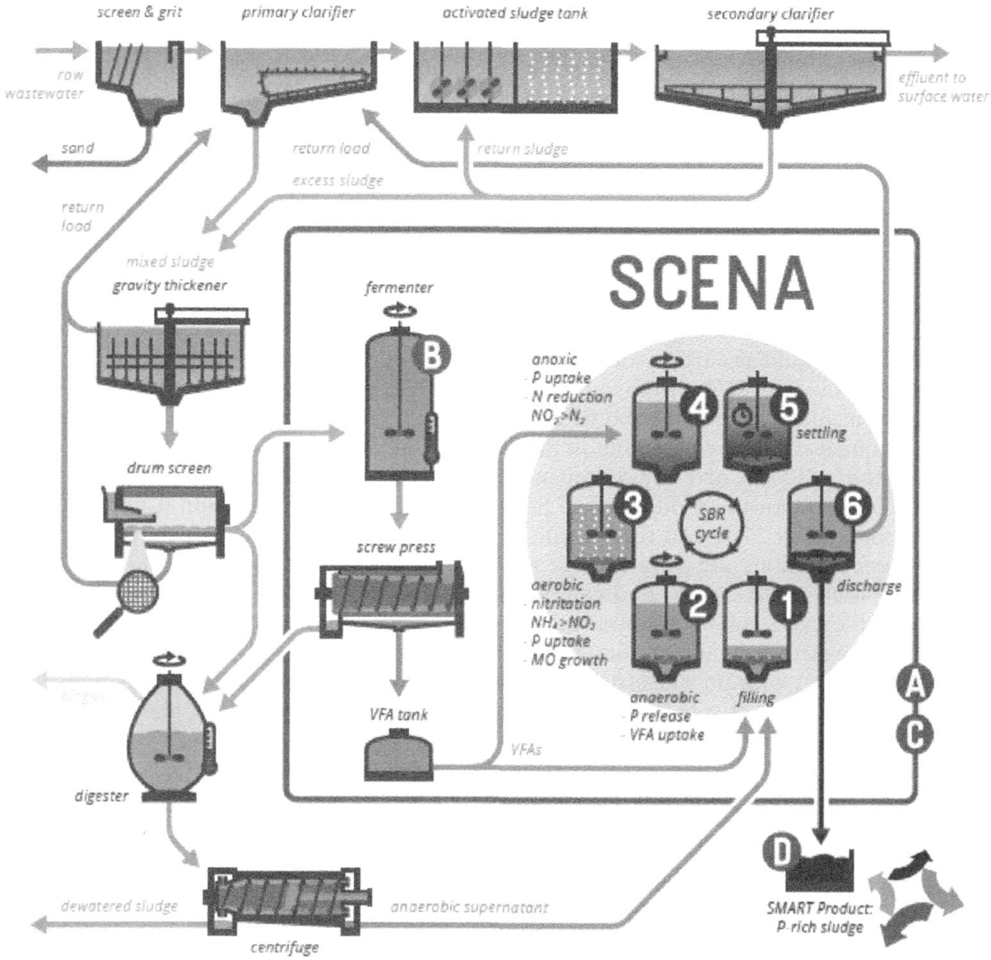

Unique Selling Points

A Low-energy nutrient removal from sludge liquor

B Biological N and P elimination without chemicals or external carbon source

C Stable and robust operation compared to other biological processes

D P-rich sludge can be valorized as organic fertilizer

	Nitrification Denitrification	Deammonification	SCENA
External C Source	Yes	No	Bio based VFAs from sewage sludge
Type of inocolum	conventional activated sludge	Deammonification inocolum	conventional activated sludge
Cost and Energy	High	Medium	Low

Figure 7.6 Schematic diagram of the SCENA process. The SCENA process was demonstrated at full-scale in the WWTP of Carbonera (Italy) within the H2020 SMART-Plant project.

nitrogen and phosphorus, especially when the sludge is pre-treated by thermal pressure hydrolysis (THP) before digestion, putting an additional load on the mainline and limiting the overall treatment capacity of the plant. Sludge liquor after THP also contains a low fraction of biodegradable organic carbon, so that conventional biological processes for nutrient removal in sludge liquor have to be operated with the costly addition of an external carbon source. The SCENA process can remove nitrogen and phosphorus from sludge liquors with low energy demand using an internal carbon source such as supernatant from primary sludge thickening or SCCA from fermentation of sludge. The SCENA system integrates the following processes: (i) optional upstream concentration of cellulosic sludge; (ii) fermentation of dynamic thickened sewage sludge to produce SCCA as carbon source; and (iii) via nitrite nitrogen and phosphorus removal (by P-bioaccumulation) from sludge reject water using an SBR. In this configuration, nitrogen is removed through the bioprocesses of nitritation/denitritation and enhanced biological phosphorus removal (EBPR) via nitrite using the SCCA from sludge fermentation liquid as carbon source.

The targeted recovery of SCCA from municipal wastewater using the SCENA process was achieved by controlled fermentation in the Carbonera WWTP at pilot-scale (Longo *et al.*, 2015) and was the basis of a subsequent scale-up carried out in 2017 within the Horizon2020 SMART-Plant innovation action (Longo *et al.*, 2017). Currently, a sequencing batch fermentation reactor (SBFR) (total volume of 3 m^3) is in operation for the production of SCCA by acidogenic fermentation (Conca *et al.*, 2020). The SBFR is operated at 37°C and the mixed liquor is kept under continuous agitation by a mixer. High SCCA concentrations are obtained for the main species as acetic acid (2000–3000 mg L^{-1}), propionic acid (4000–5000 mg L^{-1}) and valeric acid (700–800 mg L^{-1}). The environmental life cycle assessment and life cycle costing analysis revealed that the implementation of the SCENA process in the WWTP side-stream is an economic and environmentally friendly solution compared to the traditional plant layout with no side-stream treatment, due to the reduction of energy and chemicals for N and P removal, respectively, allowing the production of BioP (biological phosphorus removal) biosolids in the sludge line. Based on the results of the pilot operation, it was estimated that the installation of the SCENA process on an existing plant would have an associated additional CAPEX investment of €6/PE and provide an OPEX saving of €1.3/PE/y. The payoff time of such installation would be approximately five years. The application of SCENA affects the overall WWTP because it reduces the nitrogen load to the mainstream biological reactor. Implementation of a SCENA system for side-stream nutrient removal does not have a positive energy balance when operated with an external carbon source.

7.6 CONCEPTUAL OVERVIEW OF THE PRODUCTION OF MEDIUM-CHAIN CARBOXYLIC ACIDS (MCCA) FROM WASTEWATER

MCCA have been suggested as interesting bioproducts due to their market value and interesting chemical properties. Their production requires coupling SCCA production to a secondary anaerobic biotransformation (i.e., chain elongation). In this section, we will explore the fundamentals, applications and state of technology development for production of MCCA from waste streams.

7.6.1 Technological principles

Since MCCA production requires, in most cases, two distinct metabolic processes, an MCCA production process can be conceived as a one- or two-stage system depending on the wastewater properties and chosen route. In a two-stage process, the wastewater (or sludge) would be first fermented to SCCA in a first acidification step, and subsequently converted to MCCA in a chain elongation step. The latter could proceed via ethanol chain elongation if ethanol is added externally, or via lactic acid if this is in-situ generated in the first step. Alternatively, these processes could be engineered in a one-stage system, decreasing the capital costs (only one reactor is needed instead of two), but making process optimization more challenging as it is not possible to optimize process conditions (e.g., organic loading rate, retention time, pH, temperature, etc.) for each metabolic step individually.

A second important consideration from a process design perspective is whether or not MCCA are extracted in-line or not, and using what approach. Accumulation of MCCA and their associated toxicity is a strong driver to recover the product directly from the reactor broth (i.e., in-line product recovery). The choice of product recovery strategy will highly depend on the process pH. When a neutrophilic chain elongation process is chosen, recovery has been mostly undertaken via adsorption processes. When targeting in-line product recovery, mildly acidic chain elongation coupled to liquid-liquid extraction (i.e., pertraction, see section 7.4.2.1.4) is, at present, the best recovery approach because: (i) it is selective to more hydrophobic compounds such as MCCA over SCCA; and (ii) does not extract anions and cations.

7.6.2 Fundamental principles

Microbial MCCA production entails SCCA production by an acidogenic fermentation coupled to a biological upgrading step (i.e., chain elongation). Exceptions to that are the production of iso-caproic acid via leucine fermentation and the direct production of MCCA from carbohydrates by *Megasphaera hexanoica* or *Caproiciproducens galactitolivorans* (Jeon *et al.*, 2010, 2016, 2017; Kim *et al.*, 2015). In general, however, MCCA production in mixed microbial communities requires the availability of SCCA and either ethanol or lactic acid as electron donors.

MCCA production via ethanol chain elongation is, to date, the most studied route (although not necessarily the most relevant for their production from wastewater since ethanol is very seldom produced at relevant concentrations in mixed culture fermentations (Chen *et al.*, 2017). As such, this approach requires the addition of external ethanol as substrate. Using this ethanol as electron donor, ethanol chain elongators can convert the SCCA produced in an acidogenic fermentation from carbohydrates or proteins into MCCA (see section 7.3.3.2 for further details).

In contrast, lactic acid can be produced from carbohydrate-rich wastewater by lactic acid bacteria, even in open cultures (Kim *et al.*, 2016). This lactic acid can then be used as electron donor to elongate SCCA present in the broth (or derived from lactic acid oxidation) in processes analogous to ethanol chain elongation from a biochemical standpoint (Zhu *et al.*, 2017). Other compounds often present in wastewater such as peptides or carbohydrates (e.g., fructose or galactitol) have also been shown to be potential substrates for MCCA production (see section 7.3.3.3 for further details), although their relevance in the wastewater context is, at this point, unclear.

Usually, molar ratios of electron donor (either ethanol or lactic acid) to SCCA of four are needed to elongate, that is acetic acid to caproic acid (Barker *et al.*, 1945). In practice, most reports use ratios $\geq$4 to enhance the production of MCCA (mostly targeting higher fractions of C8) (Spirito *et al.*, 2018). However, this approach comes at the expense of excess electron donor oxidation, decreasing the overall process selectivity (Candry *et al.*, 2020; Roghair *et al.*, 2018a, 2018b).

7.6.3 Applications

Caproic (C6), enanthic (C7) and caprylic (C8) acid(s) are some of the most target MCCA for bioproduction processes. These compounds are currently sourced from natural oils (e.g., palm oil, coconut oil), have small markets (<10 000 tonnes/year) and high market values (around €2000/tonne) (Moscoviz *et al.*, 2018). This market opportunity has triggered the recent interest in the microbial production of MCCA from waste streams. MCCA are recovered as an oil, very rarely pure, but rather containing variable fraction of SCCA (usually below 30% w/w (Agler *et al.*, 2012)). Depending on the application, for example feed additive, antibiotic substitute and corrosion inhibitor as well as feedstock for the production of fragrances, flavours, fuels and plasticizers (Angenent *et al.*, 2016), it may be possible to directly use the bio-oils as raw material, although in most cases they will need to be fractionated via, that is, distillation.

7.6.4 Case studies

First and foremost, it is important to start this sub-section by emphasizing that microbial MCCA production is still an emerging concept and at the moment this chapter is being written, no full-scale installations are yet operational. There are however a couple of full- or demo-scale installations currently being constructed that are worth mentioning to highlight the industrial relevance of the topic at hand. It should be emphasized that these cases do not start from wastewater or wastewater-derived feedstocks. The first one to be considered is a semi-commercial demonstrator at the port of Amsterdam built by Chaincraft, a Dutch company that targets the production of MCCA from food waste by acidification and chain elongation with external ethanol addition (ChainCraft, 2019). As discussed later in this sub-section, several of the reports on the production of MCCA from waste activated sludge follow a similar approach, combining waste acidification with chain elongation using external ethanol as electron donor. Secondly, Afyren is a French company focused on the production of carboxylic acids (among which is caproic acid) from co-products from the sugar industry. Afyren is currently constructing a first plant with a capacity of 16 000 tons mixed carboxylic acids per annum (Sofinnova, 2020).

Beyond these two large-scale endeavours, a last commercial effort worth mentioning is that of Capro-X, an agritech spin-off from Cornell University. Capro-X aims to produce MCCA from acid whey, a dairy industry wastewater that requires treatment before discharge (Capro-X, 2020; Carvalho *et al.*, 2013; Duber *et al.*, 2018). This process is currently being tested at a 5600 L demonstration-scale plant at a dairy plant. Although specific information on the Capro-X process is currently not available, recent publications from the company founders suggest a temperature-phased two-stage process via lactic acid, coupling thermophilic acid whey fermentation to lactic acid, and subsequent mesophilic conversion of the lactic acid to MCCA. In that particular study both steps were operated at a pH of 5, and MCCAs were extracted using pertraction. MCCA production rates of 1.7 g L^{-1} d^{-1} were reported at an MCCA selectivity of 66% on a carbon basis (Xu *et al.*, 2018). Similar production rates and selectivity (2.6 g L^{-1} d^{-1}, 58–83%) were obtained in another lab-scale study by Duber and co-workers converting acid whey to MCCA in a one-stage system operated at 30°C and pH 5.5 without in-line product recovery (Duber *et al.*, 2018). Following a similar approach, the production of MCCA from thin stillage (an organic-rich watery stream generated as a side-stream of the bioethanol industry) has been demonstrated in a one-stage 60 L semi-pilot operated at pH 5.0–5.5 with product extraction (both pertraction coupled to membrane electrolysis and direct membrane electrolysis). The system, fed at an organic loading rate of 5 kg COD m^{-3} d^{-1} achieved a production of 1.7 ± 0.6 g MCCA L^{-1} d^{-1}. Using pertraction, the n-caproic acid was extracted at an efficiency of 73% (Carvajal-Arroyo *et al.*, 2020). Besides, successful MCCA production (7 g MCCA acid L^{-1} – mostly caproic acid, 81% MCCA selectivity) has been reported from Chinese liquor-making wastewater, which contains both ethanol and lactic acid (Wu *et al.*, 2018) and wine lees (Kucek *et al.*, 2016), and so on. proving that the production of MCCA from carbohydrate-rich industrial wastewater is a promising concept to generate value of an organics-rich wastewater.

Comparatively, the production of MCCA from domestic wastewater (mostly from waste activated sludge) is in a more preliminary stage and low technology readiness level. Low concentrations of caproic acid (representing less than 5% of the SCCA chemical oxygen demand) have been reported in fermentation broths derived from the fermentation of thermally hydrolyzed waste activated sludge (WAS) (Morgan-Sagastume *et al.*, 2011). Recently, several researchers have attempted to boost MCCA production in WAS fermentations by adding external ethanol, in a similar approach as that of the company Chaincraft. Wu and co-workers fermented WAS in alkaline conditions and with the external addition of ethanol, produced about 2–2.5 g L^{-1} caproic acid (Wu *et al.*, 2020). Similarly, Wang and co-workers fermented WAS (8% TS) and further converted the SCCA produced to MCCA in a second step using a chain elongation inoculum and with external ethanol addition (ratio 1:3). In that case, they reached final caproic acid concentrations of 5 g L^{-1} (Wang *et al.*, 2020). It is important to mention that these two studies were small-scale laboratory proofs of concept, far from what one would consider realistic case studies. Finally, a last report worth mentioning is the co-fermentation of waste activated sludge with organic waste consisting of alcohol and soda beverage, food, dairy, fruit, fat and oil wastes. By operating a 15-L bio-reactor at pH 5.0 semi-continuously, using as feed a mixture of primary sewage sludge to organic waste at ratio of 1:1, it was possible to achieve an MCCA concentration of 4 g L^{-1} at a selectivity of around 55% (Owusu-Agyeman *et al.*, 2020). Despite not considering product recovery (the focus of the study was the production of SCCA to aid denitrification in the WRRF), this latter study represents a more realistic take on the production of MCCA from sludge as it avoided the need for external addition of expensive ethanol and was conducted in a bio-reactor. However, while promising, this MCCA production avenue is still in its infancy.

7.7 CHALLENGES, OPPORTUNITIES AND RESEARCH NEEDS

Bioproduction of carboxylic acids from wastewater and industrial side-streams is, as mentioned before, still in its early stages. There are several key challenges that need to be overcome to further develop and mature these technologies. In what follows, we give a brief overview into these challenges, where research can help overcome these challenges, and the opportunities these systems offer for the future.

7.7.1 Bioprocess engineering

An efficient bioproduction process should aim to convert the feedstock at high production rates, with high product selectivity and substrate consumption efficiency. Optimizing these targets simultaneously may not be possible, and the best bioprocess may be obtained by balancing these optimization goals.

Volumetric production rates (g product·L^{-1} reactor·h^{-1}) are the result of the biomass concentration (g biomass·L^{-1}) and biomass-specific production rates (g product·g^{-1} biomass·h^{-1}). High production rates could be achieved by engineering ways to keep these two simultaneously high. For instance, biomass concentrations can be strongly increased by uncoupling hydraulic retention time (HRT) and sludge retention time (SRT), that is by retaining biomass in the reactor. This can be achieved through membrane retention (Pan *et al.*, 2020), anaerobic filter reactors (Grootscholten *et al.*, 2013) or granular reactors (Carvajal-Arroyo *et al.*, 2019; Roghair *et al.*, 2016). Granules are self-aggregating biofilms with high settling velocities that enable their retention in a reactor system, an approach already used in full-scale anaerobic digestions, for example UASB, EGSB or IC reactor configurations (see Chapter 5 for more details). Reduced biological activity can be the result of low substrate concentration – which should be avoided by amending feedstocks with more concentrated streams or engineering the feeding strategy – or product toxicity. Carboxylic acids can be toxic through different mechanisms, such as acidification of cell cytoplasm, disruption of cell membranes or interactions with intracellular enzymes (Desbois & Smith, 2010; Palmqvist & Hahn-Hägerdal, 2000; Royce *et al.*, 2013). One of the main approaches to alleviate product toxicity in carboxylic acid producing bioprocesses has been in-situ product recovery (see sections 7.4.2 and 7.6.2). Future research should aid in developing strategies to overcome both the low biomass concentration and low biomass activity challenges and pave the way to high-rate production processes.

Another key challenge to take into account is product selectivity, and associated with this, substrate conversion efficiency. Mixed culture fermentations rarely yield a single product, but usually result in a mixture of several SCCA and/or MCCA. Steering fermentations towards a target product is important to achieve an economically feasible process, as the substrate utilization and its conversion to the product of interest should be maximized. How to steer this depends heavily on what your target product is, and whether the goal is a single product or a mixture of products. For instance, for the production of polyhydroxyalkanoate (PHA) bioplastics, a mix of SCCA is actually desirable. Choosing an appropriate strategy to steer fermentation towards a product requires information about the substrate(s) used and good control over the process. Future research could aid in defining process-specific approaches to optimize product selectivity. A point that is associated with this, and the maximization of substrate conversion efficiency, is the need to inhibit methanogenesis in carboxylic acid-producing systems. Generation of methane implies a loss of substrate to a low-value and, in these processes, undesirable product. Several feasible strategies to suppress methanogenesis have been developed, and usually a combination of these is applied to prevent incursion of methanogens in the system. For instance, (non)-specific chemical inhibitors can be used, such as 2-bromoethanesulfonate (BES) or iodoform (Holtzapple *et al.*, 1999; Zinder *et al.*, 1984). Methanogens can also be removed from the process inoculum by heat treatment, retaining only spore-forming organisms. Both of these approaches have the downside that they do not prevent methanogens from re-entering the system through the feedstock, and, they may be too costly to apply on a frequent basis. Alternatively, operation at low or high pH has been applied to keep methanogens out (Agler *et al.*, 2012; Zhang *et al.*, 2009) and using a low SRT can lead to the wash out of slower growing methanogens (Appels *et al.*, 2008), yet both strategies may have negative impacts on the activity and/or composition of the community. Realistically, a 'silver bullet' to address the presence of methanogens without affecting process performance has yet to be developed.

A last challenge that is worth mentioning is the development of realistic co-fermentation approaches for valorization of waste sludge as MCCA. A first research hurdle to overcome is identifying substrates that could be used to amend waste sludge with. Likely targets would be carbohydrate-rich wastes such as food waste, vinasse from sugar refineries, or side-streams from bio-ethanol production. However, a factor that is at least as important as 'which streams are suitable?' is 'which streams are available?'. Developing industrial symbioses for chemicals recovery from wastes is a context-dependent challenge and restricting suitable co-fermentation wastes may result in foregoing high-potential, but only locally suitable, approaches.

7.7.2 SCCA/MCCA product recovery

In many bioprocesses, downstream processing represents a key component of the process economics and environmental footprint, often constraining their full-scale application (Woodley *et al.*, 2008). The product recovery impacts the economics of the overall processes in several ways. An ideal downstream process has high extraction capacity, high selectivity and requires a low energy and chemical input, although these are often opposed concepts.

A high extraction capacity minimizes the size of the downstream processing units, but often results in poor selectivity. The selectivity impacts the overall production train in different ways: it ensures that only the final target products are extracted, enabling fermentation intermediates and unwanted compounds to remain in the fermentation broth and minimizes the cost of product purification. For instance, in a bioprocess targeting caprylic acid, the extraction of shorter elongation intermediates, that is acetic acid, butyric acid and caproic acid, would impact the selectivity of the bioprocess, since those extracted intermediates cannot be converted to the final target product. Additionally, the presence of unwanted by-products in the extract means that further fractionation is necessary. When the extract is a mixture of salt carboxylates, purification of the salts would require recrystallization (Kim *et al.*, 2018), and when the extract is an oily mixture of free acids, distillation would be needed to fractionate and purify the final products, at an additional energetic cost (Saboe *et al.*, 2018).

When using liquid–liquid extraction, the selection of the solvent is crucial, for example Saboe *et al.* (2018) showed that Cyanex 923, a solvent composed mostly of trialkyphosphine oxides, could extract four times more caproic acid than trioctylamine, but the latter was two times more selective for caproic acid with respect to valeric acid. Similar results were obtained when comparing mineral oil with ketone-based solvents. Mixtures of phosphine based extractants with an organic apolar solvent provided better selectivity towards longer carboxylic acids than the phosphine-based solvents alone, but butyric and valeric acids are co-extracted (Naert, 2020).

Membrane processes enable extracting carboxylic acids from fermentation broths while minimizing or dispensing the need for chemical dosage, although they suffer from low selectivity and are generally energy intensive. Further development of membrane processes should improve their selectivity and their fouling potential, which dramatically reduces their life.

Even though there has been a vast effort in developing microbial processes for the production of carboxylic acids as resource recovery products, there is a need for the development of economically sound downstream processing pipelines. Interdisciplinary research crossing the boundary between microbiology, chemistry and process engineering will be crucial for the future of these technologies.

7.7.3 From lab to real life

For several decades, biogas production from organic waste streams and wastewater has been considered one of the most efficient technologies for resource recovery, together with land application as organic fertilizer. However, as stated in this chapter, technologies to produce higher value products from organic-rich wastes such as SCCA and MCCA are breaking through. Although several pilot and demonstration scale projects are underway, most technologies for the production of carboxylic acids from waste and industrial streams are currently at laboratory scale.

Bridging the gap from lab to real life remains challenging and requires, beyond the technological and scaling up challenges, careful consideration of market development so that supply and demand match. Using caproic acid as an example: its current global market is 25 000 tonnes per year (Moscoviz *et al.*, 2018). Even if caproic acid can be supplied at a more attractive price and necessary quality, its mid-users need time to build-up capacity and/or develop new processes and product applications that will result in an increased demand. As a result, production will need to be adjusted to market development to ensure price stability. Another important aspect is whether or not the mid- and final-user will accept a product derived from wastes, which has an impact not only on the acceptance, but also on the selling price of the wastewater-derived carboxylic acids (and hence on the overall economics and feasibility of the process).

Securing partners to cover all links of the supply chain (i.e., feedstock, production, recovery, mid-user, distributor of final product and consumer of final product) is also critical. All these components of the chain need to be secured by agreements that will ensure that once the SCCA/MCCA production plant starts, its product will be taken up along the chain. Only with such agreements in place will it be possible to guarantee a return on investment for the feedstock/producer and attract the necessary investment. However, getting product off-take agreements can prove itself a challenge. One needs to consider that mid-users would only commit themselves after successful preliminary tests or prototypes. To conduct preliminary tests at such industrial scale, high volumes of product may be required (up to a ton of SCCA/MCCA), which can only be achieved through demonstration-scale projects. The investment needed for such large-scale projects that do not result in short-term profits is a key challenge of the technology development, a stage commonly known as 'the innovation valley of death' in which promising research cannot reach the market due to a lack of investment. It is thus essential to find partners willing to invest in this step and take the risk associated with it. For industry, this can only be achieved if the company has a long-term vision connected to their investment. A demonstration project for SCCA/MCCA production is the final step towards commercialization, being the gains from the investment expected at this later stage. However, it is challenging for companies to accept that it may take five years or more until profits can be obtained. Hence, finding the right supply chain partners to bridge the gap from lab to real life will be crucial for the development of full-scale SCCA/MCCA bioproduction applications.

7.8 CHAPTER SUMMARY

In this chapter, some emerging concepts for the production of biobased products from organics present in wastewater and/or derivatives (i.e., waste activated sludge) have been discussed, with SCCA as central intermediate. The chapter focuses first on the microbial fundamentals of the process (hydrolysis and acidogenesis), describing how organic compounds can be converted to various SCCA as a function of the type of substrate, microbiology and operational conditions. Subsequently, various physicochemical and biological recovery and/or upgrading routes of these SCCA are briefly illustrated, including their fundamental principles and basic engineering concepts. Based on this knowledge, this chapter then describes how SCCA and MCCA could be produced from wastewater and the status of development of such technology. Last but not least, the key challenges and shortcomings with regards to bioprocess engineering, product recovery and market development are briefly touched upon.

7.9 EXERCISES

Exercise 7.1: Hydrolysis is the rate-limiting step in anaerobic conversions, and is usually described using a first order kinetic (see Equation (7.1)). Assuming a substrate with an initial solids concentration of 30 g $COD{\cdot}L^{-1}$ and an anaerobic hydrolysis constant of 0.1 d^{-1}, how long should the HRT of a reactor be to ensure at least 85% substrate conversion?

Exercise 7.2: Calculate the ATP yield of a lactic acid fermentation of 1 kg glucose [mol $ATP{\cdot}kg\ glucose^{-1}$], according to the information provided in Table 7.3. If a wastewater contains 30 g $glucose{\cdot}L^{-1}$, how much energy can be potentially harvested by microorganisms performing such fermentation [mol $ATP{\cdot}L^{-1}$]?

Exercise 7.3: A continuous stirred tank reactor (CSTR) of 50 m^3 of volume treating industrial wastewater is operated at a hydraulic retention time of 4 days. If the concentration of organic matter in the wastewater is 45 g $COD{\cdot}L^{-1}$, what is the organic loading rate of the system [g $COD{\cdot}L^{-1}{\cdot}d^{-1}$]? If the biomass concentration in your reactor is 5 g $VSS{\cdot}L^{-1}$, what is then your F/M ratio [kg $COD{\cdot}Kg\ VSS^{-1}{\cdot}d^{-1}$]?

Exercise 7.4: In a given fementation process, carbohydrate-rich wastewater is fermented to either: (i) lactic acid; or (ii) propionic acid. Calculate the NaOH requirements as meq $OH^{-}{\cdot}mol^{-1}$ acid produced and kg $NaOH{\cdot}kg^{-1}$ acid produced for each scenario at two operational pHs, 5.5 and 7. The *pka* of lactic acid and propionic acid are 3.85 and 4.87 respectively.

Exercise 7.5: A food and beverage industry produces a wastewater flow of 250 m^3 daily, containing 27 $g{\cdot}L^{-1}$ COD, of which 87% is biodegradable. Calculate what would be the maximum annual caproic acid production if all organic material can be selectively converted to it. Assume that 5% of the bCOD ends up in biomass.

Exercise 7.6: Caproic acid is produced at a rate of 0.7 $g{\cdot}L^{-1}{\cdot}h^{-1}$ in a CSTR operated at an HRT of 2 days. What should be the caproic acid extraction rate [$g{\cdot}L^{-1}{\cdot}h^{-1}$] to ensure a concentration at the effluent below 1 g caproic $acid{\cdot}L^{-1}$?

Exercise 7.7: A fermentation broth contains 10 g butyric $acid{\cdot}L^{-1}$ and 5 g caproic $acid{\cdot}L^{-1}$. Assuming pka of 4.82 and 4.88, respectively, calculate the concentration of their undissociated form at a pH of 5.5 and 7. By which factor should we concentrate this stream to allow their recovery by phase separation?

Exercise 7.8: A dairy industry has invested in a process to convert their wastewater streams (200 $m^3{\cdot}d^{-1}$) into MCCA, which consists of a continuous stirred bioreactor coupled to an in-line extraction system ($V = 100\ m^3$). Caproic acid is the dominant end product and it is produced at a rate of 0.5 $g{\cdot}L^{-1}{\cdot}h^{-1}$. The reactor pH is controlled at 5.5 and extraction is set in a way that the caproic acid concentration in solution is always equal or below 1 $g{\cdot}L^{-1}$. A process issue results in a failure in the extraction system but does not affect the bioreactor operation. As a result, caproic acid accumulates in the system. If we assume that the impact of caproic acid toxicity can be described using a non-competitive inhibition term and the K_I for this specific microbial community is 0.15 $g{\cdot}L^{-1}$, how would activity be reduced if extraction is stopped for 24 hours. For the sake of simplicity, assume that the production rate will remain unaffected during that period.

Exercise 7.9: A bio-production plant consist of fermentation and downstream processing. The fermentation step has a yield of 4 mmol propionic acid·g COD^{-1}. The propionic acid in the fermentation broth can be recovered with a 75% efficiency. If we want to produce 100 tonne propionic acid·d^{-1}, what should be the substrate loading [tonne COD·d^{-1}]?

Exercise 7.10: The size of an installation determines the investment needed, although often not in a linear manner. Larger installations are comparatively cheaper than smaller plants when one takes into consideration the cost per unit of volume. This is usually described using a scaling law of the formula: $Cost_2 = Cost_1 \cdot (Size_2/Size_1)^a$, where $Cost_1$ is the cost to build a certain installation of $Size_1$, and $Cost_2$ is the cost to build it at a different scale, $Size_2$. These costs are related by a parameter a, which reflects the linearity in scale-up cost. Imagine that constructing a 1 m^3 installation to produce and extract a mixture of SCCA (acetic, butyric and propionic acids) costs €20 000. Assuming an a of 0.7, a unit production of 1 tonne SCCA·m^{-3}·d^{-1}, and a selling price of €1700·$tonne^{-1}$, what should the plant size be to reach break-even in five years.

7.10 DISCUSSION QUESTIONS

Question 7.1 (*process design, choice of wastewater*): You are an R&D engineer working for a company that has developed a process to convert aqueous waste and side-streams into MCCA. You are currently exploring different possible applications for your technology, including wastewater from a potato factory rich in starch, secondary sludge from a local wastewater treatment facility and an aqueous waste stream from a food-processing company specialized in the production of oil and margarine. The chief technical officer of your company has asked you to select the most suitable of these streams based on your a-priori knowledge. Which one do you think is more suitable to your technology and why?

Question 7.2 (*process design, dimensioning, capital costs*): When constructing a new plant or facility, the initial investment needed for the purchase and equipment can represent a large part of the total costs. Discuss how key feedstock constraints and design and operational factors (e.g., organic loading rate, substrate degradability, product concentration, production rate, etc.) can influence the CAPEX. Take into consideration not only the bio-reactor, but also other equipment needed for, for example, solid–liquid separation or downstream processing.

Question 7.3 (*fermentation products, biochemistry*): According to Table 7.3, the fermentation of glucose to acetic acid results in the highest energy yields [ATP per mol glucose fermented]. Is the fermentation of glucose to purely acetic acid possible and if so can you explain why?

Question 7.4 (*fermentation products, biochemistry*): Does the fermentation of glucose to carboxylic acids always result in the production of CO_2? If so, why does this happen? Discuss whether the production of this CO_2 could be considered an issue in terms of climate change.

Question 7.5 (*process design, selectivity*): High product selectivity is important to ensure the inputs resources are efficiently transformed into the product of choice. What are the critical factors to take into consideration when designing a bio-process to ensure a selective production process. Take into consideration aspects such as inputs, microbiology, fermentation and downstream. Moreover, discuss the difficulties in order achieving selectivity with a special focus on the complexity and often fluctuating nature of wastewater streams in terms of varying flows, composition and concentrations.

Question 7.6 (*wastewater properties, product quality*): Product recovery is one of the critical steps of a technology to convert wastewater into raw materials. How does the input wastewater affect downstream processing, both from a technical performance, economic, product quality as well as from a consumer acceptance standpoint? Based on these points, compare the use of sludge from municipal wastewater treatment plants versus a carbohydrate-rich industrial wastewater.

Question 7.7 (*product quality, chemical industry and economy-of-scale*): SCCA, such as acetic acid and propionic acid, are produced through chemical process. Considering that chemical process can produce these chemicals at a large scale and of constant quality, discuss the potential impact of the latter in terms of market potential. In your answer, include the following aspects: (i) quality and quantity aspects; (ii) logistics (i.e. means of storage, transportation and distribution); and (iii) ultimate end-user.

Question 7.8 (*product properties, product recovery*): MCCA are seen as a more desirable product than SCCA. Discuss what are the key technical and economic drivers for their production over SCCA. Are there downsides to be considered? On what basis do you think that a company will make the end-product choice?

Question 7.9 (*technology uptake, new business models*): Traditionally, the first and foremost task of municipal wastewater treatment plants is to remove pollutants as a means to safeguard human health and protect the receiving aquatic environment. Likewise, industries in the food and beverage sector, dairy, and so on. focus on the manufacturing of certain goods, with wastewater being an unavoidable, yet undesired, side-product containing pollutants (but thus also resources) that require treatment prior to disposal similar to municipal WWTPs. The development of novel technologies for resource recovery from wastewater should never conflict with is 'protecting' task, yet it requires more than that, including the development of new business models and management approaches. Reflect on how the need for the latter can potentially (negatively) impact the development and acceptance of new processes and consider what type of business model would encourage their implementation in real life installations.

Question 7.10 (*environmental impact, externalities*): Beyond economics, are there other drivers for the development of novel routes to valorize organic materials in wastewater? Discuss how different stakeholders can positively/negatively affect this development effort. In your discussion include terms such as green image, sustainability, social license to operate, regulatory and political drivers.

Question 7.11 (*scaling-up technologies, technology development and technology readiness level*): 'The valley of death' is the stage of technology development between academic-based innovations and their commercial application in the marketplace. The production of SCCA and MCCA is (or will soon reach) this stage. Identify the key shortcomings of the different routes to convert wastewater into carboxylic acids and consider what sort of technical advances are required to convince potential investors. In your opinion, what role (if any) should policymakers play in the effort to bridge the gap of 'the valley of death'?

REFERENCES

Agler M. T., Wrenn B. A., Zinder S. H. and Angenent L. T. (2011). Waste to bioproduct conversion with undefined mixed cultures: the carboxylate platform. *Trends in Biotechnology*, **29**, 70–78. https://doi.org/10.1016/j.tibtech.2010.11.006

Agler M. T., Spirito C. M., Usack J. G., Werner J. J. and Angenent L. T. (2012). Chain elongation with reactor microbiomes: upgrading dilute ethanol to medium-chain carboxylates. *Energy and Environmental Science*, **5**, 8189–8192. https://doi.org/10.1039/c2ee22101b

Agler M. T., Spirito C. M., Usack J. G., Werner J. J. and Angenent L. T. (2014). Development of a highly specific and productive process for *n*- caproic acid production: applying lessons from methanogenic microbiomes. *Water Science and Technology*, **69**, 62–68. https://doi.org/10.2166/wst.2013.549

Albuquerque M. G. E., Martino V., Pollet E., Avérous L. and Reis M. A. M. (2011). Mixed culture polyhydroxyalkanoate (PHA) production from volatile fatty acid (VFA)-rich streams: effect of substrate composition and feeding regime on PHA productivity, composition and properties. *Journal of Biotechnology*, **151**, 66–76. https://doi.org/10.1016/j.jbiotec.2010.10.070

Alloul A., Ganigue R., Spiller M., Meerburg F. A., Cagnetta C., Rabaey K. and Vlaeminck S. E. (2018). Capture – ferment – upgrade: a three-step approach for the valorization of sewage organics as commodities. *Environmental Science and Technology*, **52**, 6729–6742. https://doi.org/10.1021/acs.est.7b05712

Andersen S. J., Hennebel T., Gildemyn S., Coma M., Desloover J., Berton J., Tsukamoto J., Stevens C. and Rabaey K. (2014). Electrolytic membrane extraction enables production of fine chemicals from biorefinery sidestreams. *Environmental Science and Technology*, **48**, 7135–7142. https://doi.org/10.1021/es500483w

Andersen S. J., Candry P., Basadre T., Khor W. C., Roume H., Hernandez-Sanabria E., Coma M. and Rabaey K. (2015). Electrolytic extraction drives volatile fatty acid chain elongation through lactic acid and replaces chemical pH control in thin stillage fermentation. *Biotechnology for Biofuels*, **8**, 1–14. https://doi.org/10.1186/s13068-015-0396-7

Andersen S. J., Groof V., De Khor W. C., Roume H., Props R., Coma M. and Rabaey K. (2017). A Clostridium group IV species dominates and suppresses a mixed culture fermentation by tolerance to medium chain fatty acids products. *Frontiers in Bioengineering and Biotechnology*, **5**, 1–10. https://doi.org/10.3389/fbioe.2017.00008

Angenent L. T., Richter H., Buckel W., Spirito C. M., Steinbusch K. J. J., Plugge C. M., Strik D. P. B. T. B., Grootscholten T. I. M., Buisman C. J. N. and Hamelers H. V. M. (2016). Chain elongation with reactor microbiomes: open-culture biotechnology to produce biochemicals. *Environmental Science and Technology*, **50**, 2796–2810. https://doi.org/10.1021/acs.est.5b04847

Appels L., Baeyens J., Degrève J. and Dewil R. (2008). Principles and potential of the anaerobic digestion of waste-activated sludge. *Progress in Energy and Combustion Science*, **34**, 755–781. https://doi.org/10.1016/j.pecs.2008.06.002

Arslan D., Steinbusch K. J. J., Diels L., Hamelers H. V. M., Strik D. P. B. T. B., Buisman C. J. N. and De Wever H. (2016). Selective short-chain carboxylates production: a review of control mechanisms to direct mixed culture fermentations. *Critical Reviews in Environmental Science and Technology*, **46**, 592–634. https://doi.org/10.1080/10643389.2016.1145959

Atasoy M., Owusu-Agyeman I., Plaza E. and Cetecioglu Z. (2018). Bio-based volatile fatty acid production and recovery from waste streams: current status and future challenges. *Bioresource Technology*, **268**, 773–786. https://doi.org/10.1016/j.biortech.2018.07.042

Barker H. A. and Taha S. M. (1941). Clostridium kluyveri, an organism concerned in the formation of caproic acid from alcohol. *J Bacteriol*, **43**, 347–363.

Barker H. A., Kamen M. D. and Bornstein B. T. (1945). The synthesis of butyric and caproic acids from ethanol and acetic acid by *Clostridium Kluyveri*. *Proceedings of the National Academy of Sciences of the United States of America*, **31**, 373–381.

Batstone D. J., Keller J., Angelidaki I., Kalyuzhnyi S. V., Pavlostathis S. G., Rozzi A., Sanders W. T., Siegrist H. and Vavilin V. A. (2002). The IWA anaerobic digestion model No 1 (ADM1). *Water Science and Technology*, **45**, 65–73. https://doi.org/10.2166/wst.2008.678

Bhatia S. K. and Yang Y.-H. (2017). Microbial production of volatile fatty acids: current status and future perspectives. *Reviews in Environmental Science and Biotechnology*, **16**, 327–345. https://doi.org/10.1007/s11157-017-9431-4

Biegel E., Schmidt S., González J. M. and Müller V. (2011). Biochemistry, evolution and physiological function of the Rnf complex, a novel ion-motive electron transport complex in prokaryotes. *Cellular and Molecular Life Sciences*, **68**, 613–634. https://doi.org/10.1007/s00018-010-0555-8

Bugnicourt E., Cinelli P., Lazzeri A. and Alvarez V. (2014). Polyhydroxyalkanoate (PHA): review of synthesis, characteristics, processing and potential applications in packaging. *Express Polymer Letters*, **8**, 791–808. https://doi.org/10.3144/expresspolymlett.2014.82

Candry P., Ulcar B., Petrognani C., Rabaey K. and Ganigué R. (2020). Ethanol:propionate ratio drives product selectivity in odd-chain elongation with *Clostridium kluyveri* and mixed communities. *Bioresource Technology*, **313**, 1–5. https://doi.org/10.1016/j.biortech.2020.123651

Capro-X. (2020). *Capro-X: The Sustainable Way for Whey*. Available at: http://capro-x.com/ (accessed 26 January 2022).

Carvajal-Arroyo J. M., Candry P., Andersen S. J., Props R., Seviour T., Ganigué R. and Rabaey K. (2019). Granular fermentation enables high rate caproic acid production from solid-free thin stillage. *Green Chemistry*, **21**, 1330–1339. https://doi.org/10.1039/C8GC03648A

Carvajal-Arroyo J. M., Andersen S. J., Ganigué R., Rozendal R. A., Angenent L. T. and Rabaey K. (2020). Production and extraction of medium chain carboxylic acids at a semi-pilot scale. *Chemical Engineering Journal*, **416**, 1–9. https://doi.org/10.1016/j.cej.2020.127886

Carvalho F., Prazeres A. R. and Rivas J. (2013). Cheese whey wastewater: characterization and treatment. *Science of the Total Environment*, **445–446**, 385–396. https://doi.org/10.1016/j.scitotenv.2012.12.038

Castillo Martinez F. A., Balciunas E. M., Salgado J. M., Domínguez González J. M., Converti A. and Oliveira R. P. de S. (2013). Lactic acid properties, applications and production: a review. *Trends in Food Science and Technology*, **30**, 70–83. https://doi.org/10.1016/j.tifs.2012.11.007

Cavalcante W. D. A., Leitão R. C., Gehring T. A., Angenent L. T. and Santaella S. T. (2017). Anaerobic fermentation for n-caproic acid production: a review. *Process Biochemistry*, **54**, 106–119. https://doi.org/10.1016/j.procbio.2016.12.024

ChainCraft B. V. (2019). *ChainCraft*. Available at: https://www.chaincraft.nl/ (accessed 26 Januaryr 2020).

Chen W. S., Strik D. P. B. T. B., Buisman C. J. N. and Kroeze C. (2017). Production of caproic acid from mixed organic waste: an environmental life cycle perspective. *Environmental Science and Technology*, **51**, 7159–7168. https://doi.org/10.1021/acs.est.6b06220

Cheryan M. (2009). Acetic acid production. In: Encyclopedia of Microbiology, 3rd edn, M. Schaechter (ed.), Academic Press, Cambridge, Massachusetts, USA, pp. 144–149. https://doi.org/10.1016/B978-012373944-5.00128-0

Christ O., Wilderer P. A., Angerhöfer R. and Faulstich M. (2000). Mathematical modeling of the hydrolysis of anaerobic processes. *Water Science and Technology*, **41**, 61–65. https://doi.org/10.2166/wst.2000.0056

Conca V., Ros C., Valentino F., Eusebi A. L., Frison N. and Fatone F. (2020). Long-term validation of polyhydroxyalkanoates production potential from the sidestream of municipal wastewater treatment plant at pilot scale. *Chemical Engineering Journal*, **390**, 1–13. https://doi.org/10.1016/j.cej.2020.124627

Conrad R., Schink B. and Phelps T. J. (1986). Thermodynamics of H2-consuming and H2-producing metabolic reactions in diverse methanogenic environments under in situ conditions. *FEMS Microbiology Ecology*, **38**, 353–360.

Demoss R. D., Bard R. C. and Gunsalus I. C. (1951). The mechanism of the heterolactic fermentation: a new rout of ethanol formation. *Journal of Bacteriology*, **62**, 499–511.

Desbois A. P. and Smith V. J. (2010). Antibacterial free fatty acids: activities, mechanisms of action and biotechnological potential. *Applied Microbiology and Biotechnology*, **85**, 1629–1642. https://doi.org/10.1007/s00253-009-2355-3

Dols M., Chraibi W., Remaud-Simeon M., Lindley N. D. and Monsan P. F. (1997). Growth and energetics of Leuconostoc mesenteroides NRRL B-1299 during metabolism of various sugars and their consequences for dextransucrase production. *Applied Environmental Microbiology*, **63**, 2159–2165.

Doudoroff M. and Stanier M. Y. (1959). Role of poly-beta-hydroxybutyric acid in the assimilation of organic carbon by bacteria. *Nature*, **183**, 1440–1442.

Duber A., Jaroszynski L., Zagrodnik R., Chwialkowska J., Juzwa W., Ciesielski S. and Oleskowicz-Popiel P. (2018). Exploiting the real wastewater potential for resource recovery – *n*-caproate production from acid whey. *Green Chemistry*, **20**, 3790–3803. https://doi.org/10.1039/c8gc01759j

Elsden S. R. and Hilton M. G. (1978). Volatile acid production from threonine, valine, leucine and isoleucine by Clostridia. *Archives of Microbiology*, **117**, 165–172.

Elsden S. R., Volcani B. E., Gilchrist F. M. C. and Lewis D. (1956). Properties of a fatty acid forming organism isolated from the rumen of sheep. *Journal of Bacteriology*, **72**, 681–689.

Frison N., Katsou E., Malamis S. and Fatone F. (2016). A novel scheme for denitrifying biological phosphorus removal via nitrite from nutrient-rich anaerobic effluents in a short-cut sequencing batch reactor. *Journal of Chemical Technology and Biotechnology*, **91**, 190–197. https://doi.org/10.1002/jctb.4561

Green D. W. and Perry R. H. (2008). Perry's Chemical Engineers' Handbook, 8th edn. McGraw-Hill Education, New York.

Grootscholten T. I. M., Steinbusch K. J. J., Hamelers H. V. M. and Buisman C. J. N. (2013). Chain elongation of acetate and ethanol in an upflow anaerobic filter for high rate MCFA production. *Bioresource Technology*, **135**, 440–445. https://doi.org/10.1016/j.biortech.2012.10.165

Großkopf T. and Soyer O. S. (2016). Microbial diversity arising from thermodynamic constraints. *ISME Journal*, **10**, 2725–2733. https://doi.org/10.1038/ismej.2016.49

Hansch C., Leo A. and Hoekman D. (1995). Exploring QSAR: Hydrophobic, Electronic, and Steric Constants. American Chemical Society, Washington, DC.

Hao J. and Wang H. (2015). Volatile fatty acids productions by mesophilic and thermophilic sludge fermentation: Biological responses to fermentation temperature *Bioresource Technology*, **175**, 367–373.

Hazelwood L. A., Daran J. M., Van Maris A. J. A., Pronk J. T. and Dickinson J. R. (2008). The ehrlich pathway for fusel alcohol production: a century of research on Saccharomyces cerevisiae metabolism. *Applied and Environmental Microbiology*, **74**(8), 2259–2266. https://doi.org/10.1128/AEM.00934-08

Holtzapple M. T., Davison R. R., Ross M. K., Aldrett-Lee S., Nagwani M., Lee C.-M., Lee C., Adelson S., Kaar W., Gaskin D., Shirage H., Chang N.-S., Chang V. S. and Loescher M. E. (1999). Biomass conversion to mixed alcohol fuels using the MixAlco process. *Applied Biochemistry and Biotechnology*, **79**, 609–631.

Hu X., Sobotka D., Czerwionka K., Zhou Q., Xie L. and Makinia J. (2018). Effects of different external carbon sources and electron acceptors on interactions between denitrification and phosphorus removal in biological nutrient removal processes. *Journal of Zhejiang Univsity Science B*, **19**, 305–316. https://doi.org/10.1631/jzus.B1700064

Javed S., Azeem F., Hussain S., Rasul I., Hussnain M., Riaz M., Afzal M., Kouser A. and Nadeem H. (2018). Bacterial lipases: a review on purification and characterization. *Progress in Biophysics and Molecular Biology*, **132**, 23–34. https://doi.org/10.1016/j.pbiomolbio.2017.07.014

Jeon B. S., Kim B. C., Um Y. and Sang B. I. (2010). Production of hexanoic acid from d-galactitol by a newly isolated Clostridium sp. BS-1. *Applied Microbiology and Biotechnology*, **88**, 1161–1167. https://doi.org/10.1007/s00253-010-2827-5

Jeon B. S., Choi O., Um Y. and Sang B. I. (2016). Production of medium-chain carboxylic acids by Megasphaera sp. MH with supplemental electron acceptors. *Biotechnology for Biofuels*, **9**, 1–9. https://doi.org/10.1186/s13068-016-0549-3

Jeon B. S., Kim S. and Sang B. I. (2017). Megasphaera hexanoica sp. Nov., a medium-chain carboxylic acid-producing bacterium isolated from a cow rumen. *International Journal of Systematic and Evolutionary Microbiology*, **67**, 2114–2120. https://doi.org/10.1099/ijsem.0.001888

Jiang L., Fu H., Yang H. K., Xu W., Wang J. and Yang S. T. (2018). Butyric acid: applications and recent advances in its bioproduction. *Biotechnology Advances*, **36**, 2101–2117. https://doi.org/10.1016/j.biotechadv.2018.09.005

Johnson K., Jiang Y., Kleerebezem R., Muyzer G. and Van Loosdrecht M. C. M. (2009). Enrichment of a mixed bacterial culture with a high polyhydroxyalkanoate storage capacity. *Biomacromolecules*, **10**, 670–676. https://doi.org/10.1021/bm8013796

Judd S. and Judd C. (2011). *The MBR Book*, 2nd edn. Butterworth-Heineman, Oxford, UK.

Kehrein P., van Loosdrecht M., Osseweijer P., Garfí M., Dewulf J. and Posada J. (2020). A critical review of resource recovery from municipal wastewater treatment plants – market supply potentials, technologies and bottlenecks. *Environmental Science: Water Research and Technology*, **6**, 877–910. https://doi.org/10.1039/c9ew00905a

Kim B. C., Jeon B. S., Kim S., Kim H., Um Y. and Sang B. I. (2015). Caproiciproducens galactitolivorans gen. Nov., sp. nov., a bacterium capable of producing caproic acid from galactitol, isolated from a wastewater treatment plant. *International Journal of Systematic and Evolutionary Microbiology*, **65**, 4902–4908. https://doi.org/10.1099/ijsem.0.000665

Kim D. H., Lee M. K., Hwang Y., Im W. T., Yun Y. M., Park C. and Kim M. S. (2016). Microbial granulation for lactic acid production. *Biotechnology and Bioengineering*, **113**, 101–111. https://doi.org/10.1002/bit.25540

Kim E. H., You S. S. and Kang J. W. (2018). Effective separation of succinic acid by combined crystallization. *Korean Journal of Chemical Engineering*, **35**, 204–209. https://doi.org/10.1007/s11814-017-0232-9

Kleerebezem R. and Van Loosdrecht M. C. M. (2010). A generalized method for thermodynamic state analysis of environmental systems. *Critical Reviews in Environmental Science and Technology*, **40**, 1–54. https://doi.org/10.1080/10643380802000974

Koeck D. E., Pechtl A., Zverlov V. V. and Schwarz W. H. (2014). Genomics of cellulolytic bacteria. *Current Opinion in Biotechnology*, **29**, 171–183. https://doi.org/10.1016/j.copbio.2014.07.002

Korkakaki E., Mulders M., Veeken A., Rozendal R., van Loosdrecht M. C. M. and Kleerebezem R. (2016). PHA Production from the organic fraction of municipal solid waste (OFMSW): overcoming the inhibitory matrix. *Water Research*, **96**, 74–83. https://doi.org/10.1016/j.watres.2016.03.033

Kreft J. U., Griffin B. M. and González-Cabaleiro R. (2020). Evolutionary causes and consequences of metabolic division of labour: why anaerobes do and aerobes don't. *Current Opinion in Biotechnology*, **62**, 80–87. https://doi.org/10.1016/j.copbio.2019.08.008

Kucek L. A., Xu J., Nguyen M. and Angenent L. T. (2016). Waste conversion into n-caprylate and n-caproate: resource recovery from wine lees using anaerobic reactor microbiomes and in-line extraction. *Frontiers in Microbiology*, **7**, 1–14. https://doi.org/10.3389/fmicb.2016.01892

Lemoigne M. (1927). Studies on microbial autolysis. Origin of B-oxybutyric acid formed by autolysis. *Annales de L'Insitut Pasteur*, **2**, 148–165.

Levy P. F., Sanderson J. E., Kispert R. G. and Wise D. L. (1981). Biorefining of biomass to liquid fuels and organic chemicals. *Enzyme and Microbial Technology*, **3**, 207–215.

Li X., Swan J. E., Nair G. R. and Langdon A. G. (2015). Preparation of volatile fatty acid (VFA) calcium salts by anaerobic digestion of glucose. *Biotechnology and Applied Biochemistry*, **62**, 476–482. https://doi.org/10.1002/bab.1301

Lin H., Peng W., Zhang M., Chen J., Hong H. and Zhang Y. (2013). A review on anaerobic membrane bioreactors: applications, membrane fouling and future perspectives. *Desalination*, **314**, 169–188. https://doi.org/10.1016/j.desal.2013.01.019

Longo S., Katsou E., Malamis S., Frison N., Renzi D. and Fatone F. (2015). Recovery of volatile fatty acids from fermentation of sewage sludge in municipal wastewater treatment plants. *Bioresource Technology*, **175**, 436–444. https://doi.org/10.1016/j.biortech.2014.09.107

Longo S., Frison N., Renzi D., Fatone F. and Hospido A. (2017). Is SCENA a good approach for side-stream integrated treatment from an environmental and economic point of view? *Water Research*, **125**, 478–489. https://doi.org/10.1016/j.watres.2017.09.006

López-Garzón C. S. and Straathof A. J. J. (2014). Recovery of carboxylic acids produced by fermentation. *Biotechnology Advances*, **32**, 873–904. https://doi.org/10.1016/j.biotechadv.2014.04.002

Lynd L. R., Weimer P. J., van Zyl W. H. and Pretorius I. S. (2002). Microbial cellulose utilization: fundamentals and biotechnology. *Microbiology and Molecular Biology Reviews*, **66**, 506–577. https://doi.org/10.1128/MMBR.66.3.506

Madigan M., Martinko J., Stahl D. and Clark D. (2012). *Biology of Microorganisms*, 13th edn. Pearson Education, San Francisco.

Matassa S., Boon N., Pikaar I. and Verstraete W. (2016). Microbial protein: future sustainable food supply route with low environmental footprint. *Microbial Biotechnology*, **9**, 568–575. https://doi.org/10.1111/1751-7915.12369

Mead G. C. (1971). The amino acid-fermenting Clostridia. *Journal of General Microbiology*, **67**, 47–56.

Morgan-Sagastume F., Pratt S., Karlsson A., Cirne D., Lant P. and Werker A. (2011). Production of volatile fatty acids by fermentation of waste activated sludge pre-treated in full-scale thermal hydrolysis plants. *Bioresource Technology*, **102**, 3089–3097. https://doi.org/10.1016/j.biortech.2010.10.054

Morgan-Sagastume F., Valentino F., Hjort M., Cirne D., Karabegovic L., Gerardin F., Johansson P., Karlsson A., Magnusson P., Alexandersson T., Bengtsson S., Majone M. and Werker A. (2014). Polyhydroxyalkanoate (PHA) production from sludge and municipal wastewater treatment. *Water Science Technology*, **69**, 177–184.

Moscoviz R., Trably E., Bernet N. and Carrère H. (2018). The environmental biorefinery: state-of-the-art on the production of hydrogen and value-added biomolecules in mixed-culture fermentation. *Green Chemistry*, **20**, 3159–3179. https://doi.org/10.1039/c8gc00572a

Müller V. (2008). *Bacterial Fermentation, Encyclopedia of Life Sciences (ELS)*. John Wiley & Sons, Ltd., Chichester. https://doi.org/10.1002/9780470015902.a0001415.pub2

Naert P. (2020). *Downstream Processing of Bio-Derived Short-Chain Carboxylic Acids via Esterification*. Ghent University, Belgium.

Owusu-Agyeman I., Plaza E. and Cetecioglu Z. (2020). Production of volatile fatty acids through co-digestion of sewage sludge and external organic waste: effect of substrate proportions and long-term operation. *Waste Management*, **112**, 30–39. https://doi.org/10.1016/j.wasman.2020.05.027

Palmqvist E. and Hahn-Hägerdal B. (2000). Fermentation of lignocellulosic hydrolysates. II: inhibitors and mechanisms of inhibition. *Bioresource Technology*, **74**(1), 25–33.

Pan X. R., Huang L., Fu X. Z., Yuan Y. R., Liu H. Q., Li W. W., Yu L., Zhao Q. B., Zuo J., Chen L. and Lam P. K. S. (2020). Long-term, selective production of caproate in an anaerobic membrane bioreactor. *Bioresource Technology*, **302**, 1–5. https://doi.org/10.1016/j.biortech.2020.122865

Parizzi L. P., Grassi M. C. B., Llerena L. A., Carazzolle M. F., Queiroz V. L., Lunardi I., Zeidler A. F. and Teixeira P. J. P. L., Mieczkowski P., Rincones J. and Pereira G. A. G. (2012). The genome sequence of Propionibacterium acidipropionici provides insights into its biotechnological and industrial potential. *BMC Genomics*, **13**, 1–20.

Pavlostathis S. G. and Giraldo-Gomez E. (2009). Kinetics of anaerobic treatment: a critical review. *Critical Reviews in Environmental Control*, **21**, 411–490. https://doi.org/10.1080/10643389109388424

Peces M., Astals S., Clarke W. P. and Jensen P. D. (2016). Semi-aerobic fermentation as a novel pre-treatment to obtain VFA and increase methane yield from primary sludge. *Bioresource Technology*, **200**, 631–638. https://doi.org/10.1016/j.biortech.2015.10.085

Pikaar I., Matassa S., Bodirsky B. L., Weindl I., Der F. H., Rabaey K., Boon N., Bruschi M., Yuan Z., Zanten H. Van, Herrero M., Verstraete W. and Popp A. (2018). Decoupling livestock from land use through industrial feed production pathways. *Environmental Science and Technology*, **52**, 7351–7359. https://doi.org/10.1021/acs.est.8b00216

Prabhu R., Altman E. and Eiteman M. A. (2012). Lactate and acrylate metabolism by *Megasphaera elsdenii* under batch and steady-state conditions. *Applied and Environmental Microbiology*, **78**, 8564–8570. https://doi.org/10.1128/AEM.02443-12

Qian-Zhu L., Xing-Lin J., Xin-Jun F., Ji-Ming W., Chao S., Hai-Bo Z., Mo X. and Hui-Zhou L. (2016). Recovery processes of organic acids from fermentation broths in the biomass-based industry. *Journal of Microbiology and Biotechnology*, **26**, 1–8. https://doi.org/10.4014/jmb.1505.05049

Ramsay I. R. and Pullammanappallil P. C. (2001). Protein degradation during anaerobic wastewater treatment: derivation of stoichiometry. *Biodegradation*, **12**, 247–256. https://doi.org/10.1023/A:1013116728817

Rao M. B., Tanksale A. M., Ghatge M. S. and Deshpande V. V. (1998). Molecular and biotechnological aspects of microbial proteases. *Microbiology and Molecular Biology Reviews*, **62**, 597–635.

Rebecchi S., Pinelli D., Bertin L., Zama F., Fava F. and Frascari D. (2016). Volatile fatty acids recovery from the effluent of an acidogenic digestion process fed with grape pomace by adsorption on ion exchange resins. *Chemical Engineering Journal*, **306**, 629–639. https://doi.org/10.1016/j.cej.2016.07.101

Reddy C. S. K., Ghai R. and Rashmi Kalia V. C. (2003). Polyhydroxyalkanoates: an overview. *Bioresource Technology*, **87**, 137–146. https://doi.org/10.1016/S0960-8524(02)00212-2

Reyhanitash E., Kersten S. R. A. and Schuur B. (2017). Recovery of volatile fatty acids from fermented wastewater by adsorption. *ACS Sustainable Chemical and Engineering*, **5**, 9176–9184. https://doi.org/10.1021/acssuschemeng.7b02095

Roghair M., Strik D. P. B. T. B., Steinbusch K. J. J., Weusthuis R. A., Bruins M. E. and Buisman C. J. N. (2016). Granular sludge formation and characterization in a chain elongation process. *Process Biochemistry*, **51**, 1594–1598. https://doi.org/10.1016/j.procbio.2016.06.012

Roghair M., Hoogstad T., Strik D. P. B. T. B., Plugge C. M., Timmers P. H. A., Weusthuis R. A., Bruins M. E. and Buisman C. J. N. (2018a). Controlling ethanol use in chain elongation by CO_2 loading rate. *Environmental Science and Technology*, **52**, 1496–1506. https://doi.org/10.1021/acs.est.7b04904

Roghair M., Liu Y., Strik D. P. B. T. B., Weusthuis R. A., Bruins M. E. and Buisman C. J. N. (2018b). Development of an effective chain elongation process from acidified food waste and ethanol into n-caproate. *Frontiers in Bioengineering and Biotechnology*, **6**, 1–11. https://doi.org/10.3389/fbioe.2018.00050

Royce L. A., Liu P., Stebbins M. J., Hanson B. C. and Jarboe L. R. (2013). The damaging effects of short chain fatty acids on Escherichia coli membranes. *Applied Microbiology and Biotechnology*, **97**, 8317–8327. https://doi.org/10.1007/s00253-013-5113-5

Saboe P. O., Manker L. P., Michener W. E., Peterson D. J., Brandner D. G., Deutch S. P., Kumar M., Cywar R. M., Beckham G. T. and Karp E. M. (2018). In situ recovery of bio-based carboxylic acids. *Green Chemistry*, **20**, 1791–1804. https://doi.org/10.1039/c7gc03747c

Sander R. (2015). Compilation of henry's law constants (version 4.0) for water as solvent. *Atmos. Chem. Phys.* **15**, 4399–4981. https://doi.org/10.5194/acp-15-4399-2015

Sanders W. (2001). Anaerobic Hydrolysis Digestion of Complex Substrates. Wageningen University, The Netherlands.

Seedorf H., Fricke W. F., Veith B., Brüggeman H., Liesegang H., Strittmatter A., Miethke M., Buckel W., Hinderberger J., Li F., Hagemeier C., Thauer R. K. and Gottschalk G. (2008). The genome of *Clostridium kluyveri*, a strict anaerobe. *Proceedings of the National Academy of Sciences of the United States of America*, **105**, 2128–2133.

Siebert M. L. and Toerien D. F. (1969). The proteolytic bacteria present in the anaerobic digestion of raw sewage sludge. *Water Research*, **3**, 241–250. https://doi.org/10.1016/0043-1354(69)90020-7

Sneeden R. P. A. (1982). Organic syntheses where carbon monoxide is the unique source of carbon. In: Comprehensive Organometallic Chemistry, G. Wilkinson, F. G. A. Stone and E. W. Abel (eds), Pergamon, Oxford, pp. 19–100.

Sofinnova. (2020). AFYREN and the SPI industrial projects fund, operated by Bpifrance, announce the creation of an industrial company, AFYREN NEOXY. Available at: https://renewable-carbon.eu/news/afyren-and-the-spi-industrial-projects-fund-operated-by-bpifrance-announce-the-creation-of-an-industrial-company-afyren-neoxy/ (accessed 26 January 2022).

Sosnowski P., Wieczorek A. and Ledakowicz S. (2003). Anaerobic co-digestion of sewage sludge and organic fraction of municipal solid wastes. *Advances in Environmental Research*, **7**, 609–616. https://doi.org/10.1016/S1093-0191(02)00049-7

Spirito C. M., Marzilli A. M. and Angenent L. T. (2018). Higher substrate ratios of ethanol to acetate steered chain elongation toward n-caprylate in a bioreactor with product extraction. *Environmental Science and Technology*, **52**, 13438–13447. https://doi.org/10.1021/acs.est.8b03856

Steinbusch K. J. J., Hamelers H. V. M., Plugge C. M. and Buisman C. J. N. (2011). Biological formation of caproate and caprylate from acetate: fuel and chemical production from low grade biomass. *Energy and Environmental Science*, **4**, 216–224. https://doi.org/10.1039/C0EE00282H

STOWA. (2014). Bioplastic From Sludge: Exploring PHA Production From Sewage Sludge. STOWA, The Amersfoot, The Netherlands.

STOWA. (2017). PHARIO: Stepping Stone To A Sustainable Value Chain for PHA Bioplastic Using Municipal Activated Sludge. STOWA, Amersfoot, The Netherlands.

Svarovsky L. (2001a). Introduction to solid-liquid separation. In: Solid-Liquid Separation, 4th edn., Ladislav Svarovsky (ed.), Butterworth-Heineman, Oxford, pp. 1–29. https://doi.org/10.1016/B978-075064568-3/50025-0

Svarovsky L. (2001b). Separation by centrifugal sedimentation. In: Solid–Liquid Separation, 4th edn., Ladislav Svarovsky (ed.), Butterworth-Heineman, Oxford, pp. 246–280. https://doi.org/10.1016/B978-075064568-3/50032-8

Tansengco M. L. and Tokiwa Y. (1998). Comparative population study of aliphatic polyesters-degrading microorganisms at 50°C. *Chemical Letters*, **27**, 1043–1044.

Temudo M. F., Kleerebezem R. and van Loosdrecht M. (2007). Influence of the pH on (open) mixed culture fermentation of glucose: a chemostat study. *Biotechnology and Bioengineering*, **98**, 69–79. https://doi.org/10.1002/bit

Temudo M. F., Mato T., Kleerebezem R. and Van Loosdrecht M. C. M. (2009). Xylose anaerobic conversion by open-mixed cultures. *Applied Microbiology and Biotechnology*, **82**, 231–239. https://doi.org/10.1007/s00253-008-1749-y

Tester R. F., Qi X. and Karkalas J. (2006). Hydrolysis of native starches with amylases. *Animal Feed Science and Technology*, **130**, 39–54. https://doi.org/10.1016/j.anifeedsci.2006.01.016

Tomei M. C., Braguglia C. M., Cento G. and Mininni G. (2009). Modeling of anaerobic digestion of sludge. *Critical Reviews in Environmental Science and Technology*, **39**, 1003–1051. https://doi.org/10.1080/10643380801977818

Urban C., Xu J., Sträuber H., Dos Santos Dantas T. R., Mühlenberg J., Härtig C., Angenent L. T. and Harnisch F. (2017). Production of drop-in fuels from biomass at high selectivity by combined microbial and electrochemical conversion. *Energy and Environmental Science*, **10**, 2231–2244. https://doi.org/10.1039/c7ee01303e

Vavilin V. A., Fernandez B., Palatsi J. and Flotats X. (2008). Hydrolysis kinetics in anaerobic degradation of particulate organic material: an overview. *Waste Management*, **28**, 939–951. https://doi.org/10.1016/j.wasman.2007.03.028

Wallace R. J., McKain N., McEwan N. R., Miyagawa E., Chaudhary L. C., King T. P., Walker N. D., Apajalahti J. H. A. and Newbold C. J. (2003). Eubacterium pyruvativorans sp. nov., a novel non-saccharolytic anaerobe from the rumen that ferments pyruvate and amino acids, forms caproate and utilizes acetate and propionate. *International Journal of Systematic and Evolutionary Microbiology*, **53**, 965–970. https://doi.org/10.1099/ijs.0.02110-0

Wang Q., Zhang P., Bao S., Liang J., Wu Y., Chen N., Wang S. and Cai Y. (2020). Chain elongation performances with anaerobic fermentation liquid from sewage sludge with high total solid as electron acceptor. *Bioresource Technology*, **306**, 1–8. https://doi.org/10.1016/j.biortech.2020.123188

Weibull C. (1953). Characterization of the protoplasmic constituents of *Bacillus megaterium*. *Journal of Bacteriology*, **66**, 696–702.

Weimer P. J. and Moen G. N. (2013). Quantitative analysis of growth and volatile fatty acid production by the anaerobic ruminal bacterium *Megasphaera elsdenii* T81. *Applied Microbiology and Biotechnology*, **97**, 4075–4081. https://doi.org/10.1007/s00253-012-4645-4

Woodley J. M., Bisschops M., Straathof A. J. J. and Ottens M. (2008). Future directions for in-situ product removal (ISPR). *Journal of Chemical Technology and Biotechnology*, **83**, 121–123. https://doi.org/10.1002/jctb.1790

Wu Q., Guo W., Bao X., Meng X., Yin R., Du J., Zheng H., Feng X., Luo H. and Ren N. (2018). Upgrading liquor-making wastewater into medium chain fatty acid: insights into co-electron donors, key microflora, and energy harvest. *Water Research*, **145**, 650–659. https://doi.org/10.1016/j.watres.2018.08.046

Wu S. L., Sun J., Chen X., Wei W., Song L., Dai X. and Ni B. J. (2020). Unveiling the mechanisms of medium-chain fatty acid production from waste activated sludge alkaline fermentation liquor through physiological, thermodynamic and metagenomic investigations. *Water Research*, **169**, 1–12. https://doi.org/10.1016/j.watres.2019.115218

Xu J., Guzman J. J. L., Andersen S. J., Rabaey K. and Angenent L. T. (2015). In-line and selective phase separation of medium-chain carboxylic acids using membrane electrolysis. *Chemical Communications*, **51**, 6847–6850. https://doi.org/10.1039/C5CC01897H

Xu J., Hao J., Guzman J. J. L., Spirito C. M., Harroff L. A. and Angenent L. T. (2018). Temperature-phased conversion of acid whey waste into medium-chain carboxylic acids via lactic acid: no external e-donor. *Joule* **2**, 280–295. https://doi.org/10.1016/j.joule.2017.11.008

Zacharof M.-P. and Lovitt R. W. (2013). Complex effluent streams as a potential source of volatile fatty acids. *Waste and Biomass Valorization*, **4**, 557–581. https://doi.org/10.1007/s12649-013-9202-6

Zhang P., Chen Y. and Zhou Q. (2009). Effects of pH on the waste activated sludge hydrolysis and acidification under mesophilic and thermophilic conditions. 2009 Int. Conf. Energy Environ. Technol. ICEET 2009, Vol. **3**, pp. 306–309. https://doi.org/10.1109/ICEET.2009.539

Zhu X., Tao Y., Liang C., Li X., Wei N., Zhang W., Zhou Y., Yang Y. and Bo T. (2015) The synthesis of n-caproate from lactate: a new efficient process for medium-chain carboxylates production. *Scientific Reports*, **5**, 1–9. https://doi.org/10.1038/srep14360

Zhu X., Zhou Y., Wang Y., Wu T., Li X., Li D. and Tao Y. (2017). Production of high-concentration n-caproic acid from lactate through fermentation using a newly isolated Ruminococcaceae bacterium CPB6. *Biotechnology for Biofuels*, **10**, 1–12. https://doi.org/10.1186/s13068-017-0788-y

Zigová J. and Šturdík E. (2000). Advances in biotechnological production of butyric acid. *Journal of Industrial Microbiology and Biotechnology*, **24**, 153–160. https://doi.org/10.1038/sj.jim.2900795

Zinder S. H., Anguish T. and Cardwell S. C. (1984). Selective inhibition by 2-bromoethanesulfonate of methanogenesis from acetate in a thermophilic anaerobic digestor. *Applied and Environmental Microbiology*, **47**, 1343–1345.

doi: 10.2166/9781780409566

Chapter 8

Upscaled and validated technologies for the production of bio-based materials from wastewater

Çağrı Akyol[1,2], E. Gozde Ozbayram[3], Anna Laura Eusebi[1], Alessia Foglia[1], Giulia Cipolletta[1], Nicola Frison[4], Vincenzo Conca[4], Cinzia Da Ros[4], Coos Wessels[5], Ramon Ganigué[6,7], Ilje Pikaar[8], Adrian Oehmen[9], Gilda Carvalho[10] and Francesco Fatone[1]

[1]*Department of Science and Engineering of Materials, Environment and Urban Planning-SIMAU, Marche Polytechnic University, Via Brecce Bianche 12, 60100 Ancona, Italy*
[2]*Department of Green Chemistry and Technology, Ghent University, Coupure Links 653, 9000 Ghent, Belgium*
[3]*Department of Marine and Freshwater Resources Management, Faculty of Aquatic Sciences, Istanbul University, Fatih, 34134 Istanbul, Turkey*
[4]*Department of Biotechnology, University of Verona, 37129 Verona, Italy*
[5]*CirTec BV, Nijverheidsweg 26, 1442 LD Purmerend, The Netherlands*
[6]*Center for Microbial Ecology and Technology (CMET), Campus Coupure – Ghent University, Coupure Links 653, 9000 Ghent, Belgium*
[7]*Centre for Advanced Process Technology for Urban REsource recovery, Frieda Saeysstraat 1, 9052 Gent, Belgium*
[8]*The School of Civil Engineering, The University of Queensland, St Lucia, QLD 4072, Australia*
[9]*The School of Chemical Engineering, The University of Queensland, St Lucia, QLD 4072, Australia*
[10]*Advanced Water Management Centre (AWMC), The University of Queensland, St Lucia, QLD 4072, Australia*

8.1 INTRODUCTION

The most commonly implemented and established method for the valorization of organics present in wastewater streams is anaerobic digestion (AD) as a means to recover chemical energy present in organic matter as electricity and/or heat via biogas (see Chapter 5). However, the current value of biogas is relatively low. As such, in recent years there has been an increasing interest in the development of technologies that can recover organics embedded in the wastewater matrix in the form of higher value products. Initially, R&D efforts were predominantly restricted to smaller scale testing with only few concepts tested at large and/or full scale. In order for these innovative concepts to become mature technologies that can be implemented by the water industry, long-term demonstration testing under real-life conditions at a practically releveant scale is essential. Within this context, significant efforts are currently underway through public–private collaborations and large scale EU funded projects. A plethora of organic compounds can be recovered from drinking water, municipal wastewater and various industrial wastewater streams. These include, but are not limited to, polysaccharides, cellulose, volatile fatty acids (VFA, also referred to as short-chain carboxylic acids), polyhydroxyalkanoates (PHA), amino acids, proteins, aliginate-like compounds, and so on. In this chapter, key achievements and progress made with respect to the recovery of organics from municipal wastewater are presented. In particular, this chapter focusses on two specific products, namely, (i) cellulose and (ii) polyhydroxyalkanoates (PHA). The fundamental principles, applications and design of approaches enabling the recovery of these compounds are discussed, including engineering considerations and implications for practice in a wastewater context. In addition to the technological aspects, the market potential of each of these compounds in terms of product value and requirements (e.g. quality, volumes and market demand) are discussed as well. Lastly, the research needs are highlighted, focused on challenges and opportunities for these concepts to become mature technologies that can be implemented by the water industry and replicated in different areas of the world.

8.2 LEARNING OBJECTIVES

At the completion of this chapter you should be able to:

- Understand the rationale, conceptual design and benefits of implementing approaches that enable the recovery and valorization of organics from municipal and industrial wastewater streams.
- Explain the fundamental principles and key design criteria of approaches for recovery of cellulose and PHA from wastewater streams.
- Explain key specific challenges of the development and implementation of each technology.
- Use waste characteristics and site-based constraints to select appropriate forms of organic recovery technology for a given situation.
- Set a design basis for organic recovery technologies, conduct preliminary reactor sizing and estimate their potential performance.

8.3 CONCEPTUAL OVERVIEW OF CELLULOSE RECOVERY FROM MUNICIPAL WASTEWATER

Cellulose is a polymer of glucose constituted by a linear chain of glucose, also defined as the β-1,4-polyacetal of cellobiose (4-O β-D-glucopyranosyl-D-glucose) with $(C_6H_{10}O_5)_n$ as its chemical formula (Huang, 2013). It is a renewable, biodegradable, and non-toxic material that constitutes many environmental-friendly and biocompatible products (Nechyporchuk *et al.*, 2016). Cellulose fibers are the major components (78.1% of virgin pulps and 67.4% of recycled pulps) of toilet paper (as well as hygiene tissues) together with smaller amounts of hemicellulosic fibers and lignin from wood pulps (Chen *et al.*, 2018). The average consumption of toilet paper per capita in Europe is estimated to be around 10–15 kg per year, which is three times more than the global average consumption (4.4 kg/capita·year) (Crutchik *et al.*, 2018). The toilet paper useage in the USA is even higher with an average use of about ≈23 kg/capita·year (Table 8.1). Moreover, although the annual consumption of toilet

Table 8.1 Per capita consumption of toilet paper in the United States, China, Japan, Western Europe, Latin America, and Africa (Li *et al.*, 2020).

Country/region	Estimated per capita consumption of toilet paper (kg) (2017 data)
United States	≈23
China	≈3
Japan	≈11
Western Europe	≈10–13
Latin America	≈4
Africa	≈0.4

paper per capita is considerably lower in China (i.e. 2.9 kg/capita/year), the total amount of toilet paper entering WWTPs is also significant because of its vast number of consumers (Li *et al.*, 2020).

Despite the large consumption of toilet paper, cellulose fibers originating from toilet paper usage are often overlooked as a contributor to total COD and total suspended solids (TSS). These fibers represent a significant amount of the total COD and TSS load in municipal wastewater. In fact, the influent cellulose content typically accounts for approximately 30 and 25–40% of the TSS and influent COD, respectively (Ahmed *et al.*, 2019). Due to the characteristics of cellulose fibers and the prevalent conditions in sewer networks, these cellulose fibers typically enter WWTPs in suspended solids form. Depending on the design of the wastewater treatment plant and process configuration, the cellulose is either aerobically oxidized to CO_2, used as a carbon source for denitrification, incorporated in heterotrophic biomass or partly converted to biogas during anaerobic digestion. While there is no consensus in the literature regarding the fate of cellulose during AD, it is generally accepted that long solids retention times (SRT) are needed to convert most of the cellulose into biogas (i.e. above 40 days may greatly improve the degradation to over 80%). Such long SRTs are typically not applied nor economically viable. Instead of burdening the biological processes during the secondary treatment in WWTPs, toilet paper fibers could be physically recovered as a potential resource during primary treatment (Li *et al.*, 2020). The latter opens up opportunities for recovery of these fibres through solid–liquid separation approaches. Within this context, the use of microsieves for cellulose recovery has gained significant interest and has shown great practical potential. Microsieves such as rotating belt filters (RBF) enable efficient separation of the cellulose fibers (as well as overall TSS) at high surface loading rates and are regarded as an interesting alternative to traditional primary sedimentation as a pre-treatment step prior to activated sludge processes (Ruiken *et al.*, 2013).

8.3.1 Fundamental principles

The working principle of RBFs is based on physical separation through size exclusion. As a pre-treatment step in municipal WWTP applications, commonly used mesh size ranges from 200 to 500 μm. It should be noted that this is not a strict design parameter and smaller or larger sizes can be used as well (Behera *et al.*, 2018; Palmieri *et al.*, 2019; Rusten *et al.*, 2017). In addition to the microsieve's mesh size, it is important to understand that the size exclusion principle imposed in RBFs also relies on the effect of cake filtration, which allows the removal of solid particles up to three times smaller than the nominal pore size of the filter mesh. The speed of the rotating belt can be adjusted to the applied hydraulic surface loading and solids loading as a means to manipulate the cake thickness, thereby changing the effective 'operational pore size' and thus the solids removal efficiency. Prior to entering the RBF, municipal wastewater goes through a grit removal step, in which the easily settable solids (larger solids) that may cause damage to the RBF as well as equipment and processess further downstream at the WWTP are removed (e.g. sand and coarse grit material). The wastewater in the RBF flows through the filter cloth and its particles create a 'pre-coat' on top of the filter cloth, thereby reducing the effective pore size of the microsieve. The sievings filtered out of the wastewater are typically pumped into a screw press where they are dewatered to a total solids (TS) content of typically 40–50%. After dewatering, the sievings are subsequently dried to achieve a dry solids content over 90% prior to being transported to a hammer mill for further processing and volarization. The quality of the produced sievings can be determined based on various parameters such as ash content, solids content, cellulose, and so on. In general, the sludge collected from an RBF treating municipal wastewater contains a fraction of cellulose fibers as high as ≈80–90% of the recovered organic fraction (Behera *et al.*, 2018; Ruiken *et al.*, 2013). A simplified schematic representation of the working principle of an RBF unit is provided in Figure 8.1.

8.3.2 Applications and design

RBFs are considered a high-rate primary treatment process for the treatment of municipal wastewater with typical surface loading rates of 100–200 $m^3/m^2/h$, offering a small-footprint alternative to primary

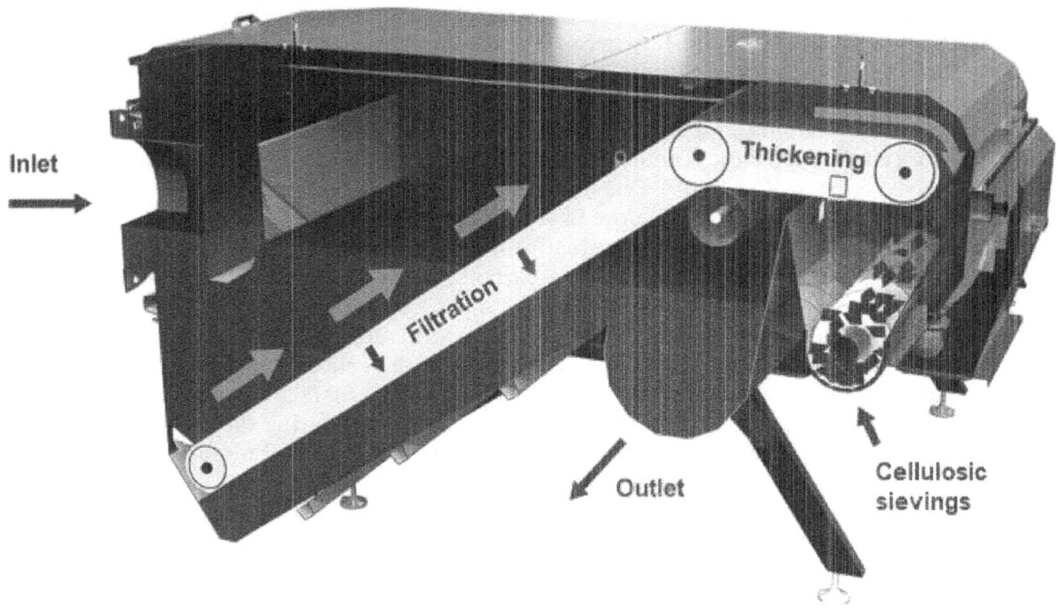

Figure 8.1 Simplified schematic representation of the working principle of microsieving, adapted from Salsnes (Norway) filter product sheet.

settling (with surface loading rates in the order of 1.5–2.5 $m^3/m^2/h$) (Franchi and Santoro, 2015). Interestingly, RBFs were initially developed in Norway as a mechanical low-footprint alternative to primary settling to achieve high TSS and BOD removal due to the lack of space rather then being developed with the aim of becoming a cellulose recovery approach. Importantly, in those days only COD and TSS removal was required without nitrogen or phosphorous discharge limits being enforced (Rusten and Odegaard, 2006). In addition, due to the low temperatures in Norway and its typical steep sewers with low hydraulic retention times, hydrolysis of solids (and toilet paper) was limited, enabling high TSS and COD removal efficiencies.

The selection of RBF technology over conventional primary clarification depends on several factors such as carbon and energy footprints, plant capacity, primary effluent quality, influent SS concentration, and selective particle size fractionation (Behera *et al.*, 2018). An important advantage of microsieves is their small footprint; for a given hydraulic load, they occupy less than 5% of the footprint of conventional clarifiers (Palmieri *et al.*, 2019). TSS removal/cellulose recovery and cellulose purity depend on multiple operating conditions such as loading rate, sieving rate, water flow rate, fine mesh size and influent wastewater constituents. RBFs are sometimes equipped with a cleaning system made up of an air knife and compressed hot water to remove the residual sludge cake from the belt (Da Ros *et al.*, 2020). Important occupational health and safety (OH&S) concerns to consider are the potential formation of dust and fire hazards during the drying step. Therefore, adequate ventilation and the avoidance of any contact between the heaters and the cellulose are critical design criteria. While important, these OH&S issues can be appropriately managed and should not be considered a limiting factor.

It is important to highlight that the performance of RBFs in terms of removal efficiencies can be highly site-specific and depends on the wastewater characteristics. For example, higher solids concentration may lead to the formation of a thicker 'filtering' layer on the surface of the belt, thereby reducing the effective pore size and thus enhancing the TSS removal efficiency. The presence of fibrous materials may lead to the formation of a thick and porous filtration layer that, in turn, also enhances TSS and organics removal. One can appreciate that the site-specific nature means that pilot scale testing is highly recommended at the target location prior to full-scale implementation in order to accurately determine the expected process performance. Obviously, the latter is not possible in greenfield applications and a performance assessment based on expected wastewater composition is necessary.

In addition to the benefits of recovering cellulose, by decreasing the TSS and organic loading to the subsequent activated sludge step for biological nutrient removal, the power consumption needed for aeration in the WWTP can be significantly decreased (Franchi and Santoro, 2015). Another advantage of cellulose removal prior to aerobic treatment is the increase in aerobic sludge age, thereby improving the nitrification process by 10–15% (Rusten *et al.*, 2016). The latter may be particular beneficial for colder climates. Lastly, the organic carbon recovered by an RBF can be fed to an anaerobic digester, contributing to a significant increase in biogas production (Behera *et al.*, 2018). Obviously, the latter is only possible in WWTPs that are equipped with an anerobic digester. It should also be noted that the use of cellulose for bioenergy production is positioned at the bottom of the so-called 'biomass value pyramid' (Gavrilescu, 2014). The biomass value pyramid is a concept specific to the bio-economy, in which different biomass applications and sectors (such as food, chemicals, biofuels) are ranked in terms of value added (Bout *et al.*, 2019). Therefore, biogas production from recovered cellulose should only be pursued if none of the aforementioned valorization alternatives is technically possible and/or economically feasible. For instance, recovered cellulose can be refined and used as a raw material in paper products (Kehrein *et al.*, 2020), as fibrous reinforcement material in bricks (Kim *et al.*, 2017), as an adhesion binder for asphalts (Makron, Finland) or in various biocomposites together with (bio) plastics such as PHA (Akyol *et al.*, 2020). Alternatively, cellulose can be utilized to produce valuable chemicals or biofuels, such as short-chain carboxylic acids (SCCAs, also referred to as volatile fatty acids (VFA)), polylactic acid, and bioethanol (Crutchik *et al.*, 2018; van der Hoek *et al.*, 2015).

8.3.3 Successful case studies of implementation

The cellulose recovered from toilet paper in municipal wastewater has found its way to the market through various technology providers of microsieves including Salsnes, Eco Mat RBF, Hydrotech Beltfilters and CirTec. For instance, CirTec has developed a dedicated process called Cellvation® that separates the cellulose from the incoming municipal wastewater, followed by a cleaning step that allows the recovery of the cellulose fibers as a clean product commercialized under the name Recell®, which can be used in various applications such as in asphalt, concrete, insulation material, and other building materials (Palmieri *et al.*, 2019). Within the EU Horizon 2020 SMART-Plant project, Recell® was successfully used to produce a variety of biocomposites such as façade, compression-resistant plates, and insulation flakes.

A simplified schematic representation of the overall Cellvation® process is presented in Figure 8.2. Raw wastewater enters the treatment plant and flows through a coarse screen and grit that removes large particles. After that, the Cellvation® process starts when the wastewater is pumped through a grit chamber to remove the easily settleable solids. The remaining wastewater goes through a cellulose washer that separates the cellulose fibres from larger particles and hairs. The influent wastewater is then fed to an RBF which significantly improves the water quality and wastewater discharge options. The filtrate is discharged and the solids are removed from the filter by means of a cleaning system that uses air pressure at the end of the filtration area. The cellulosic fibres coming from the RBF are pre-dewatered in a dewatering unit coupled to the filter and then are further dewatered by a CellPress. The dry solids leaving the CellPress are hygienized, dried and polished to either fluff or pellets.

The Cellvation® was installed at the WWTP of Geestmerambacht of the Water Authority Hollands Noorderkwartier located in the Netherlands (see Figure 8.3) with a capacity of up to 90 m^3/h. The medim-size plant treats a dry weather flow of 4160 m^3/h and maximum flow of 16 300 m^3/h and is

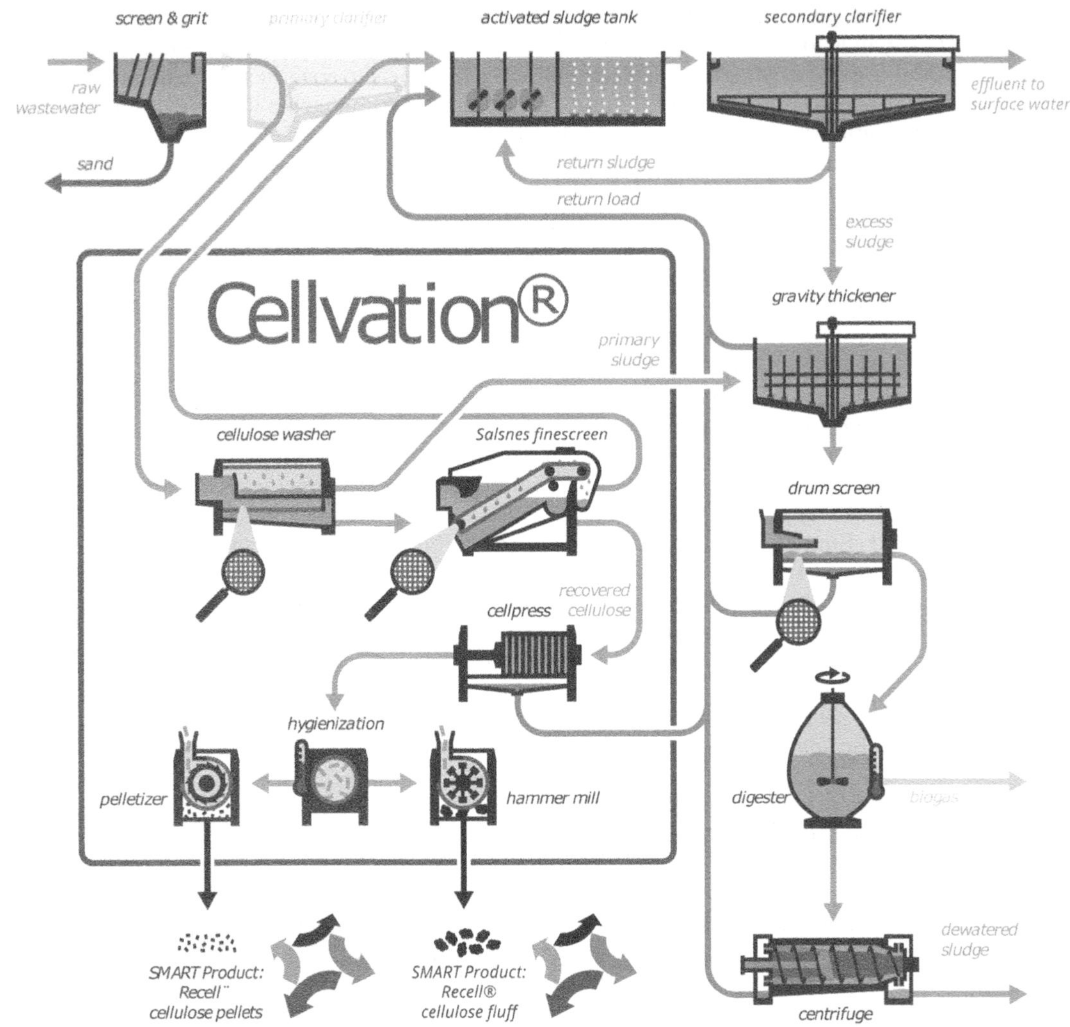

Figure 8.2 Schematic diagram of the Cellvation® process. The Cellvation® process is demonstrated in the WWTP Geestmerambacht within the H2020 SMART-Plant project.

a good example of cellulose recovery which harvests up to 400 kg cellulose per day. A Cellvation® installation, among others consisting of a fine screen, is installed in the feed, and treats the screened material. The screenings generally consist of fibres (80% of which are cellulose fibers), and remaining portions of protein, fats and ash. The amount of sludge produced on site is reduced by 35% when using the Cellvation® installation. The primary cellulose harvesting has a further impact on the overall treatment train, such as reduction in aeration energy of up to 20%, reduction in sludge volume up to 10%, reduction in TSS in treated influent up to 55%, reduction in COD in treated influent up to 30% as well as reduction of CO_2 footprint of up to 15% (http://www.smart-plant.eu/).

In what concerns process economics, the installation of a Cellvation® process at a small-to-medium sized new plant (50 000 PE) is estimated to have an associated additional capital expenditure (CAPEX) of €18/PE, albeit it decreases the operating expenditure in terms of aeration energy by

Figure 8.3 An RBF system is demonstrated at the sewage treatment plant of Geestmerambacht (The Netherlands).

about 20% and also reduces excess sludge production, resulting in a cost saving of about €0.5/PE/year. Besides the potential cost savings, the treatment capacity of an existing plant can be increased due to the reduction in organics loading to the activated sludge tank. In addition to the energy savings and increased treatment capacity, the reduction in the amount of sludge results not only in a lower polyelectrolyte requirement during sludge dewatering, but it also substantially reduces the amount of dewatered sludge to be disposed. The latter is estimated to result in substantial savings in the order of €2.7/PE/year (highly site-specific and depending on location and ultimate disposal route). That, in combination with an estimated revenue for high-quality recovered and cleaned cellulose (which has an EPA rating of class A) valued at €265/ton, would lead to an estimated payback time of approximately six years (SMART-Plant 2020a). It should be emphasized that the payback time highly depends on the local situation, wastewater characteristics, WWTP size and potential revenues of the recovered cellulose. As such, the value provided for the WWTP Geestmerambacht should only be seen as an indicative pay back time. Nevertheless, it provides some indication of the economic competitiveness of the use of microsieves for cellulose recovery. Moreover, many WWTPS are currently reaching their capacity in terms of solids loading due to expansion of the sewer catchment. Especially in these cases and where space is an issue, the retrofitting of existing plants with microsieves may provide an interesting alternative due to their low footprint.

In the WWTP of Falconara Marittima (Ancona, Italy), a pilot-scale RBF recovered cellulosic sludge from municipal wastewater at an influent flow rate of 15–78 m^3/h (Palmieri *et al.*, 2019). Different mesh sizes from 90 to 350 μm were tested in the pilot plant, and the recovered cellulosic material reached concentrations up to $\approx$27 g/m^3 (average 15.9 ± 10.1 g/m^3) with maximum solids removal of 74% at 90 μm. After the final washing, the cellulosic material composition was, on average: $9 \pm 1\%$ of lipids; $6 \pm 1\%$ of ashes; $9 \pm 1\%$ of hemicellulose; $38 \pm 6\%$ of pure cellulose; $40 \pm 3\%$ of lignin and humic compounds; 0% of protein and other compounds. The recovered cellulose was then valorized as reinforcing components in building materials. The flexural strength in mortars was improved with the addition of 20% of recovered cellulose fibers. Based on a feasibility study, it was estimated that a WWTP with a capacity of 150 000 PE could potentially produce 305–1069 sacks per day of pre-mixed mortar with 5–20% of recovered cellulosic fibers by volume, respectively. This amount corresponds to an average production of a small-medium cementitious products company.

Similarly, a pilot-scale RBF installed at the Carbonera municipal WWTP (Treviso, Italy) produced cellulosic sludge by the sieving of around 400 m^3 wastewater/day. The microsieving was accomplished through an RBF SF1000 (Salsnes Filter SA, Norway) with a flow capacity up to 15 L/s and a submerged sieve cloth area of 0.24 m^2 using two mesh sizes (350 and 210 μm). The RBF removed approx. 44% of TSS and 35% of COD. The cellulosic sludge had high content of cellulose (41% TS) and lignin (18% TS), while lipids and proteins accounted for 9 and 13% of TS, respectively. The cellulosic sludge was then utilized as a feedstock for the production of SCCA (see Chapter 7 for more detailed information on SCCA production). By adjusting the pH of the fermentor to 9 before the feeding and operating with a hydraulic retention time (HRT) of 6 days under mesophilic conditions, the system achieved bio-based SCCA productivity of 2.57 kg COD/m^3/day (Da Ros *et al.*, 2020). A first indicative economic assessment of that particular pilot trial highlighted that the production of bio-based SCCA from cellulosic primary sludge could be more profitable than its conversion to biogas.

A last case study to be considered is an RBF installed in the WWTP of Aarle-Rixtel, the Netherlands (Ahmed *et al.*, 2019). Without entering into the specific details of the case study, technical specifications, and so on, this specific case study is worth mentioning to illustrate the impact of site-specific conditions and their relation to the overall cellulose recovery potential and product quality that can be achieved. The cellulose content in the RBF sludge was found to be only 35% of the TSS (Ahmed *et al.*, 2019), which is (much) lower than the cellulose content achieved elsewhere (i.e. around 55–75%), significantly reducing product quality and potential for volarization of cellulose as a marketable commodity. This highlights the importance of, and need for, site-specific pilot studies to verify the practical feasiblity of RBF as a cellulose recovery and valorization approach.

8.4 CONCEPTUAL OVERVIEW OF POLYHYDROXYALKANOATES (PHA) PRODUCTION

PHA are microbial intracellular storage polymers that can serve as precursors of bioplastics. The development of technologies to produce PHA as biopolymer precursors has been ongoing for decades, but the high cost of PHA production using pure microbial cultures and refined feedstocks stands as a major obstacle to broader PHA applications. In recent years, this problem has been approached by using waste feedstock materials as carbon source, mixed microbial cultures (MMC) which do not require aseptic conditions, and aqueous two-phase systems for purification and recovery to make PHA production more economical (Pakalapati *et al.*, 2018). Rapidly increasing population, urbanization, and industrialization lead to the production of large quantities of sludge in WWTPs every year, making it a readily available and more economical feedstock for bioplastic production compared to the typical carbon sources used in commercial production (Bengtsson *et al.*, 2008). Hence, PHA production through MMC stands as a perfect example of resource recovery in WWTPs, combining conventional municipal wastewater treatment with innovative advanced technologies. Biological treatment of municipal wastewater and sludge management for recovering wastewater organic carbon as PHAs is a route to transform the 'treat and discharge' approach into a biorefinery approach.

8.4.1 Fundamental principles

MMC present in wastewater sludge hold great potential for PHA production (Reddy *et al.*, 2012). Microbial species in the sludge, such as bacteria, yeasts, and fungi, naturally synthesize biopolymers such as triacylglycerol (TAG), wax esters (WEs) or PHA in the presence of excess carbon source, particularly when either nitrogen or phosphorus (or both) are limiting in the growth media (Kumar *et al.*, 2017, 2018). Bacteria are the major contributers to PHA storage in MMC PHA processes, and many bacterial genera commonly found in activated sludge have been identified as PHA-storing microorganisms. This suggests that there is great potential for practical PHA production using activated sludge-derived MMC. However, the productivity of the process can be substantially improved with an enrichment step prior to accumulation. For example, PHA accumulation from acetate by activated sludge taken from municipal WWTPs was reported to be between 45 and 67 wt% with fed-batch feeding (Cavaillé *et al.*, 2013), while activated sludge subjected to enrichment can accumulate PHA up to 90 wt% from acetate (Johnson *et al.*, 2009). PHA storing oragnisms in MMC need to be fed with a feedstock rich in readily biodegradable organic matter, most preferably dominated by SCCAs, which are readily convertible to PHA, to minimize the production of other products such as glycogen.

A PHA-production process from wastewater is, thus, composed of three steps (Figure 8.4): a first anaerobic step, in which SCCAs are produced from wastewater or sludge (see Chapter 7 for more information on their production from wastewater), a second step where the microbial community is enriched in PHA storing organisms, and a final accumulation stage, using enriched biomass, where PHA storage is maximized. The second and third steps are usually aerobic and are supplied with the SCCAs produced in the first stage. Intracellular PHA needs to be then extracted and purified using physico-chemical methods.

Typically, feast and famine conditions are alternated to achieve a good enrichment of PHA accumulating organisms (Dionisi *et al.*, 2001). A relatively prolonged famine phase is necessary to impose a limitation of internal growth factors (RNA, enzymes, etc.) on the organisms that are not able to store PHA. In the subsequent feast phase, the uptake of carbon by these organisms will be delayed, as they re-establish the level of growth factors necessary for regular metabolic activity. These organisms will gradually be outcompeted by PHA-storers, which can remain active during the famine phase. This selection approach is known as the feast-famine strategy or aerobic dynamic feeding (Dionisi *et al.*, 2001).

Scaling-up PHA production from waste streams still requires further research to become economically feasible, which is currently the main challenge for the implementation of PHA production from waste or wastewater at the industrial scale (Rodriguez-Perez *et al.*, 2018). The selection of an appropriate waste stream is critical to ensure an adequate and constant carbon supply and obtain satisfactory recovery efficiencies. Recently, wastewater, cheese whey and organic wastes such as

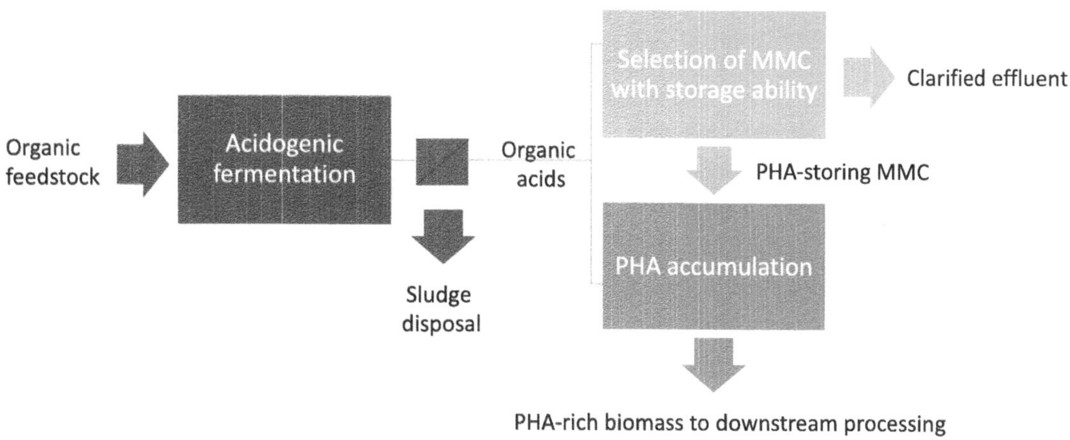

Figure 8.4 Three-step MMC process for PHA recovery: (1) acidogenic fermentation to obtain a SCCA-rich stream, (2) a dedicated biomass production yielding MMCs enriched with PHA-storing potential, and (3) a PHA accumulation step where (1) and (2) outputs are combined in a final biopolymer production bioprocess (modified from Valentino *et al.*, 2017).

fruit and vegetable wastes, have gained great attention as carbon sources. Hence, the process can be optimized in terms of the nutrient balance by combining two or more of these waste streams during the integration of PHA production into WWTPs (Rodriguez-Perez *et al.*, 2018).

8.4.2 Applications and design

PHA production from wastewater can be applied in different configurations, with the two most typical being illustrated in Figure 8.5. An enrichment-based process (Figure 8.5, left) relies on the common three-stage process detailed above. The configuration that relies on accumulation with sludge (Figure 8.5, right) skips step 2 associated with PHA biomass selection, using secondary WWTP sludge directly in a fed-batch accumulation reactor, with fermented wastewater as the SCCA source. While higher PHA contents (>80%) are potentially achievable in the enrichment-based process, the savings in capital costs associated with the elimination of the selection reactor can be advantageous in certain cases. In other situations, PHA selection can be combined with simultaneous side-stream treatment of nitrogen and phosphorus, lowering the nutrient loading rate recycled back to the mainstream WWTP following treatment of anaerobic digester supernatants (see section 8.4.3 for detailed case studies).

In the three-step process configuration, adequate PHA-storers selection is key for a productive system. The feast and famine approach usually used as a microbial selection pressure is normally established in sequencing batch reactors (SBRs), although it can also be imposed through a sequence of two continuous reactors, one with carbon supply and another without (Albuquerque *et al.*, 2010). An effective selection of PHA storing biomass is accomplished when the feast/famine ratio is lower than 0.2 min/min. Generally, the length of the feast phase increases at lower SRTs, as more substrate is supplied per unit of biomass per cycle and the cultures do not increase their SCCA uptake rate (Johnson *et al.* 2010). The SRT is also known to shape the selected microbial communities composition, which can affect the overall performance of the process. SRTs ranging from 1 to 6 days have been typically

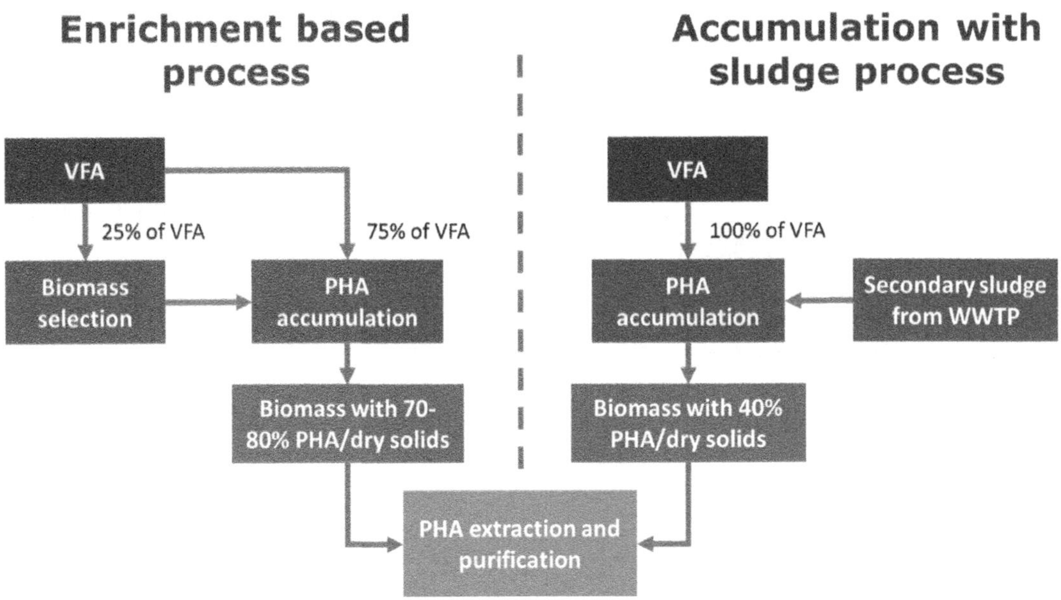

Figure 8.5 Main processes for producing PHA using mixed cultures fed with organic acid from fermented waste-based feedstocks. Left) PHA production based on biomass selection and enrichment; Right) PHA production from VFA (volatile fatty acids, which are SCCA) using secondary sludge as PHA accumulation biomass.

reported for PHA production from wastewater. Besides SRT, the organic loading rate (OLR) is a key parameter influencing the performance of a MMC PHA process. It is normally desired to maximize both the acidogenic and the selection reactor OLR. Higher OLR in the selection step normally leads to higher volumetric productivities due to increased productivity of PHA-storing biomass. However, above a certain OLR threshold, the process can either become inhibited or the feast phase be extended to a point that the selection pressure is inefficient (Campanari *et al.*, 2014; Carvalho *et al.*, 2014; Khatami *et al.*, 2021). Other parameters that can impact the success of the PHA accumulation process are pH, temperature, dissolved oxygen, cycle length and HRT (Kourmentza *et al.*, 2017; Oehmen *et al.*, 2014; Wang *et al.*, 2019). The decoupling of the nitrogen and the carbon sources is another means to drive the enrichment in PHA-storing organisms. If there is no nitrogen available during the feast phase, organisms can only take up and store carbon, which gives PHA-storers an additional competitive advantage when nitrogen is supplied in the famine phase (Oliveira *et al.*, 2017).

As outlined in Figure 8.4, the anaerobic fermentation of wastewater or sludge is of key importance for generating the SCCAs necessary for PHA production. The SCCA spectrum also governs the PHA monomeric composition, which in turn influences the properties of the PHA bioplastic product. For instance, the fraction of polyhydroxybutyrate (PHB) and polyhydroxyvalerate (PHV) are linearly dependent on the level of PHB (containing an even number of carbon atoms) and PHV (odd carbon atoms) precursors produced through anaerobic fermentation (Albuquerque *et al.*, 2007). For example, both acetic and butyric acids correspond to PHB precursors and propionic and valeric acids correspond to PHV precursors. Improved mechanical properties of the recovered polymer were noted in co-polymers of PHB and PHV, as opposed to PHB alone (Albuquerque *et al.*, 2011). These properties included decreased stiffness and brittleness, increased flexibility, increased tensile strength and toughness and a lower melting temperature that improved its processability. Thus, tuning the anaerobic fermentation step towards production of a SCCA profile leading to a desirable PHB/PHV fraction can have positive impacts on the PHA produced, where desired polymeric traits depend on the intended final application.

A very important step for use of these polymers is the final extraction and purification process. Several methods exist for extraction of PHA from the PHA-rich biomass produced. They are categorized mainly in physical separation processes, cell lysis-based processes and solvent-based processes (Pagliano *et al.*, 2021). The choice of method is highly dependent on the type of PHA-rich biomass produced. From waste-based feedstocks and mixed cultures, simple physical separation processes are usually not suitable as most of the impurities are still present, although physical separation is often applied in conjunction with cell lysis and solvent-based processes. Cell lysis-based techniques have the advantage of lower investment costs and simplicity of operation but often lead to the production of wastewater that requires additional treatment. Solvent-based methods are robust methods that can handle variations of the PHA-rich biomass but have higher investment costs and are more complex in operation. Therefore, the choice of method will depend on two main factors: type and variation of PHA-rich biomass (e.g. PHA content, type of contaminants, microorganisms present) and the quality requirements for the final PHA polymer (e.g. purity, molecular weight) (Fernández-Dacosta *et al.*, 2015; Samorì *et al.*, 2015). Cell lysis or solvent-based techniques are currently the most commonly applied in downstream processing for PHA extraction from mixed cultures (Pagliano *et al.*, 2021). Some commonly applied examples include chlorinated compounds (such as sodium hypochlorite (NaClO), chloroform ($CHCl_3$), dichloromethane (CH_2Cl_2), and so on), alkali or acid treatment, and surfactants (e.g. sodium dodecyl sulphate (SDS)). Typically this involves a series of steps based on mixing in a tank with the PHA-rich biomass, followed by separation through centrifugation, filtration or hydrocyclonation and a final drying step (Fernández-Dacosta *et al.*, 2015; Samorì *et al.*, 2015). A similar post-treatment stage to remove contaminants may also be necessary. Some key process factors requiring optimization and control include the temperature, as heating and cooling steps can also be required during downstream processing, retention time in the process tanks combining the biomass and key active ingredients, as well as process conditions specific to the separation and drying stages.

Table 8.2 Market possibilities and features of recovered PHA.

Features	Industrial use	Quality indicators	Price range	Reference
• Biocompatible • Biodegradable • Non-toxic • Appropriate mechanical, thermal, physical properties • Easy surface modification • Noncarcinogenic	• Material industry (packaging, paper coating etc.) • Fuel industry (biofuels additives) • Industrial microbiology	• Purity • Toxicity • Chemical constituents • Blending	2.2–5.0 €/kg	Anjum *et al.* (2016), Możejko-Ciesielska and Kiewisz (2016), Valentino *et al.* (2017)

The multiple stages that are often required to achieve high PHA recovery and purity represent a significant cost to the overall process.

The utilization of wastewater for the synthesis of PHAs can lead to production of a value-added compound; however, the final product quality and application will be dependent on the source of SCCA and the potential presence of contaminants. For instance it is unlikely that PHA obtained from wastewater could be used in medical applications where high purity products with a non-toxic nature are of utmost consideration (Raza *et al.*, 2018). Possible market alternatives and features for recovered PHA are summarized in Table 8.2.

8.4.3 Case studies

Two case studies will be discussed here, the first will focus on an enrichment-based process integrated with side-stream wastewater treatment, while the second deals with a system that uses sludge as source of microorganisms for accumulation.

First, the side-stream Short-Cut Enhanced Phosphorus and PHA Recovery (SCEPPHAR) process offers recovery of both PHA and struvite, efficient nitrogen removal via nitrite from the anaerobic supernatant and reduction of sludge volume, offsetting disposal costs (Figure 8.6). It consists of: (i) (alkaline) fermentation of sewage sludge (pH 6–10) and/or cellulosic sludge to enhance the production of selected SCCA (e.g. propionic acid first) and release nitrogen and phosphorus in soluble forms (ammonia and phosphate); (ii) solid and liquid separation of the fermentation products and recovery of struvite from the sewage sludge fermentation liquid (SFL) by the addition of $Mg(OH)_2$ to favor the precipitation in a crystallizer reactor; (iii) sequential nitritation of the sludge reject water; (iv) selection of PHA storing biomass in a SBR by the alternation of aerobic-feast and anoxic-famine conditions for denitritation driven by internally stored PHA as carbon source; (v) PHA production in a fed-batch reactor to accumulate the intracellular PHA content in the selected biomass using SFL.

In SCEPPHAR, the anaerobic supernatant of an anaerobic digester is fed into the nitritation SBR and then, after nitritation, its effluent (nitrified supernatant) is discharged into a storage tank. Nitrite is later used as electron acceptor by the PHA storing biomass in the selection SBR. PHA-accumulating biomass selection is essential to obtain a microbial community capable of synthesizing and hyper-accumulating PHAs and at the same time, removing nitrogen via nitrite. To achieve this result, biomass is subjected to feast and famine conditions under aerobic and anoxic conditions, respectively. During the feast phase, PHA are synthetized and accumulated by the biomass as internal carbon storage, while under famine conditions nitrite is used as electron acceptor for internal carbon (PHA) degradation under anoxic conditions, thus promoting the microbial growth of denitrifiers and PHA storing organisms without aeration input in the famine period. Aerobic conditions in such processes are known to be energy-intensive; it is estimated that approximately 39 MJ are needed to produce 1 kg of PHA via traditional aerobic accumulation processes (Frison *et al.*, 2015), thus presenting an advantage for the SCEPPHAR process. The accumulation SBR allows maximum PHA storage to be reached (PHA yields ranging between 0.58 and 0.61 g COD_{PHA}/g COD_{VFA}) (Conca *et al.*, 2020), and

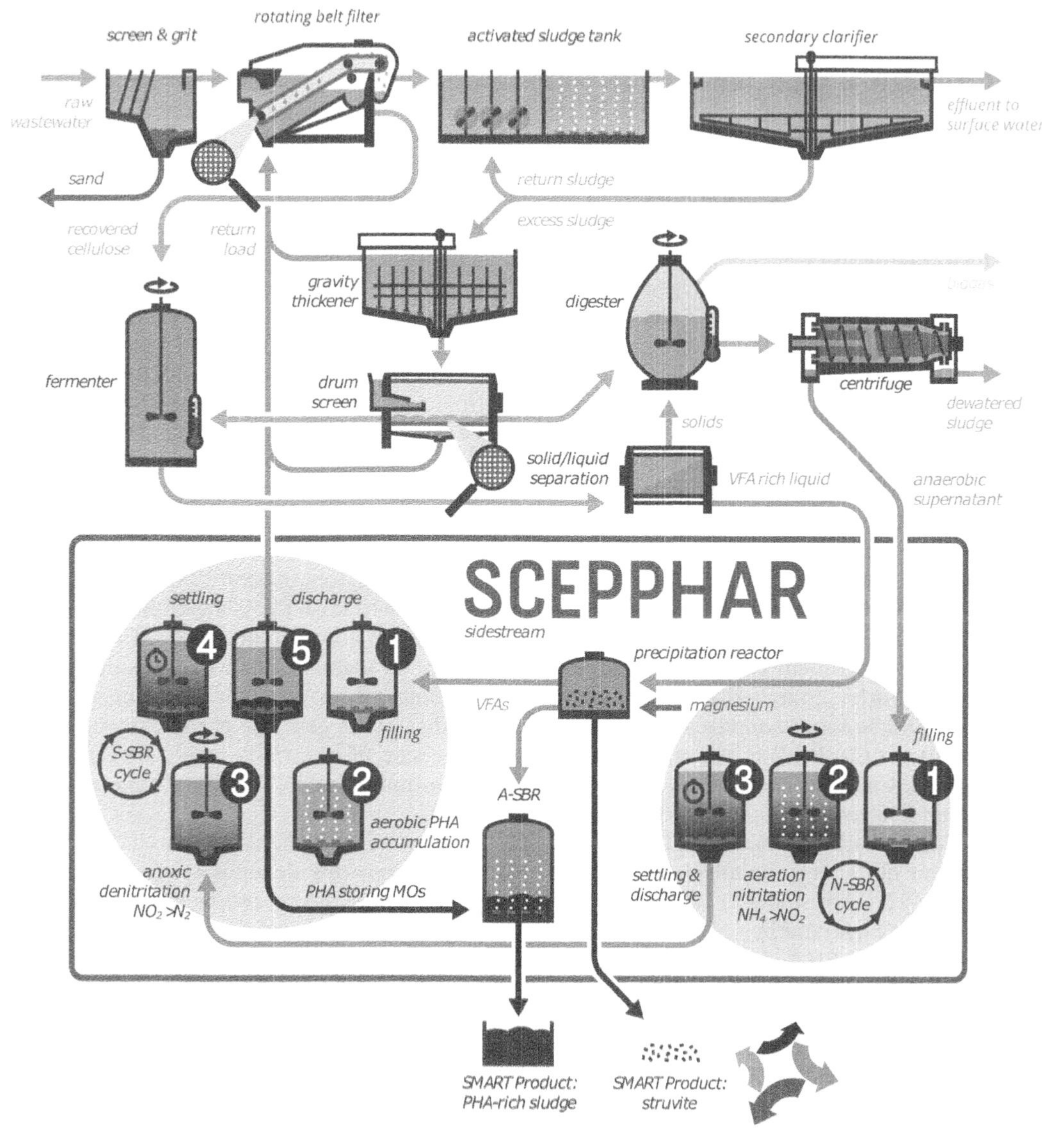

Figure 8.6 Schematic diagram of the side-stream SCEPPHAR process. The side-stream SCEPPHAR process was demonstrated in the WWTP of Carbonera (Italy) within the H2020 SMART-Plant project.

high rates under feast conditions, during which aeration is controlled based on the DO (e.g. 2 mgO_2/L) and pH (e.g. >7.5) profiles. The carbon source is intermittently supplied to the aerobic accumulation SBR for a total batch length of between 4 and 6 hours; at the end of the PHA accumulation step, biomass activity is interrupted through the dosage of a quencher (e.g. sulfuric acid) to cease biological activity and prevent PHA degradation. PHA is obtained as PHA rich sludge, where up to 50% of PHA (PHB + PHV) is accumulated in the bacterial cells. From an operational point of view, side-stream SCEPPHAR provides several advantages, such as: (i) a short start-up period; (ii) high stability of the process under transient conditions; (iii) self-management of the bioprocesses; (iv) simplified operations through on-line monitoring; (v) real time process control based on indirect parameters; and (vi) possibility to switch between PHA production and sole nitrogen removal from reject water via-nitrite.

The side-stream SCEPPHAR process was demonstrated in the Carbonera WWTP (as part of the SMART-Plant project), in a pilot installation that treated 4–5 m^3 of sludge liquor per day. The process achieved up to 85% N removal and enabled the recovery of phosphorus as struvite, in addition to producing PHA-rich sludge. The carbon source for optimized biopolymer production was produced on-site by fermentation of cellulosic sludge captured using an RBF (sieving of around 400 m^3/day of municipal wastewater, and an acidogenic fermentation unit with a working volume of 2.6 m^3 for the production of SCCA). SRT and HRT were maintained as 6–7 and around 2 days, respectively. Average PHA yields of 0.58–0.61 $gCOD_{PHA}/gCOD_{SCCA}$ were obtained, while the maximum PHA production was up to 1.2 kgPHA/PE/y when treating solely municipal wastewater (Conca *et al.*, 2020). In addition, this process decreased the energy costs for side-stream treatment by up to 20%. The recovered struvite can be used as P-based fertilizer while the PHA, which achieved a content in the biomass up to 50% on a dry weight basis, was studied for use in flame retardant formulations for building materials. On-site extraction of PHA was not deemed sustainable in the short term in terms of CAPEX & OPEX (around 1.5 €/kg), as the payoff time was too long. In a scenario where the side-stream SCEPPHAR is installed in an existing plant with a size of 50 000 PE and PHA is not extracted but bio-composites are produced directly from the PHA-rich sludge and sold for 1000€/ton, the additional CAPEX would be €6/PE (€ 300 000) and OPEX savings would range from €0 to €3.7/PE/y, depending on the cost of sludge disposal. The income from PHA sales would represent a maximum of €2/PE/y, while the savings of sludge treatment could be as high as €3.85/PE/y. All together this would result in a short payoff time, that is less than two years. In case PHA-rich biomass is not accepted by the market, PHA-rich sludge could be sent to an already existing anaerobic digester to generate electricity, further sold in the market for €0.10/kWh, where the income from the sale of electricity would be €0.12/PE/y, whilst the OPEX savings could be as high as €1.2/PE/y. In this scenario, the payoff time would still be lower than five years.

Implementing a side-stream SCEPPHAR system for wastewater treatment can also have a positive impact on the net energy balance of the entire plant, although this depends on the valorization route for the PHA-rich sludge. For instance, PHA valorization for biogas production on-site would reduce net energy demand by −15%, mainly by generating additional electricity from the biogas. The drying of dewatered PHA sludge and its direct use bio-composite production would reduce the net energy demand by −18%. The chemical extraction of PHA and production of a purified PHA powder is not beneficial for energy balance, increasing the net energy demand by +6%. For this route, the requirement of chemicals for PHA extraction, and especially the energy required for treatment of the liquid waste (basically dissolved excess sludge), would be substantial and off-set all credits for PHA use downstream. Additional electricity required for the process (+15–18%) is mostly compensated by savings in coagulation chemicals, which are not required with the bio-P removal of SCEPPHAR. However, on the downside, the SCEPPHAR process can increase net GHG emissions, mainly due to the predicted higher N_2O emissions (+33%) in the heterotrophic and autotrophic SBR compared to a conventional activated sludge process (SMART-Plant, 2020b).

As a side-note, similar enrichment-based approaches have been tested at larger scale. A pilot project at Brussels-North WWTP by Aquiris (a subsidiary of Veolia Water) demonstrated the technical feasibility of producing biomass with elevated PHA-accumulation potential from municipal wastewater treatment and fermented waste sludge as feedstock for PHA accumulation (Morgan-Sagastume *et al.*, 2016). The main features of the Brussels-North system were: Sludge fermentation at 42°C, pH = 5.5–6.5; $Y_{SCCA/VS} = 270 \pm 30$ g COD(SCCA)/g VS; 12 cycles per day; Feast/Famine 15/85%; OLR = 3 kg COD/m^3 d; SRT = 1–2 days; Final PHA content = 34% (gPHA/gVSS); Overall PHA accumulation rate = 0.07 g COD(PHA)/g $COD_{treated}$. (Morgan-Sagastume *et al.*, 2014). Following the same process concept, another pilot-scale prototype comprising biomass production from wastewater treatment and accumulation of PHA in the surplus biomass was installed at the demonstration site at Leeuwarden WWTP in Friesland (Bengtsson *et al.*, 2017). The treatment of wastewater and biomass production with PHA accumulation was conducted in one or two SBRs. Pre-denitrification and nitrification was carried out in the first SBR throughout the entire operational period under anoxic feast and aerobic famine conditions, then the first SBR was combined with the second SBR operated under anoxic conditions to achieve further N removal by post-denitrification. In the end, 83% COD and 80% N removal, as well as PHA accumulation up to 49% PHA of VSS with acetic acid or fermented organic residues as substrates, were achieved.

A second potential configuration for PHA production from sewage sludge was initially investigated by Dutch water utilites and other partners like Anox Kaldnes (Veola Water Technologies AB) within the PHARIO project. PHARIO was centred on processing surplus biomass from Dutch full-scale municipal WWTPs in the Netherlands to produce PHA biopolymers. The process steps involved in the PHARIO project consisted of the harvesting of secondary sludge as functional biomass (inoculum) able to produce PHA from SCCA-rich streams, obtained from either fermented organic waste, primary sludge or other sources (Figure 8.7). In the PHARIO concept PHA polymers are recovered via solvent extraction (using acetone) (Werker *et al.*, 2015).

The results of the PHARIO project revealed a PHA rich biomass with on average 0.41 g PHA/gVSS, with reproducible thermal properties and high thermal stability. As the main output of the project, Veolia Water Technologies developed basic engineering designs for a full-scale commercial plant with a production capacity of 2000 ton PHA per year. The results indicated that building the first full scale plant would require around 13 M€ of initial investment for the facilities and another 4 M€ to cover the operational costs for a period of three years. Based on the economic evaluation, PHA could be produced at a price of 4.5€/kg, thus generating a net profit of 1.1€/kg PHA, or 5.5 M€/year. The cost analysis concluded that the produced PHA had a competitive price compared to commercial products

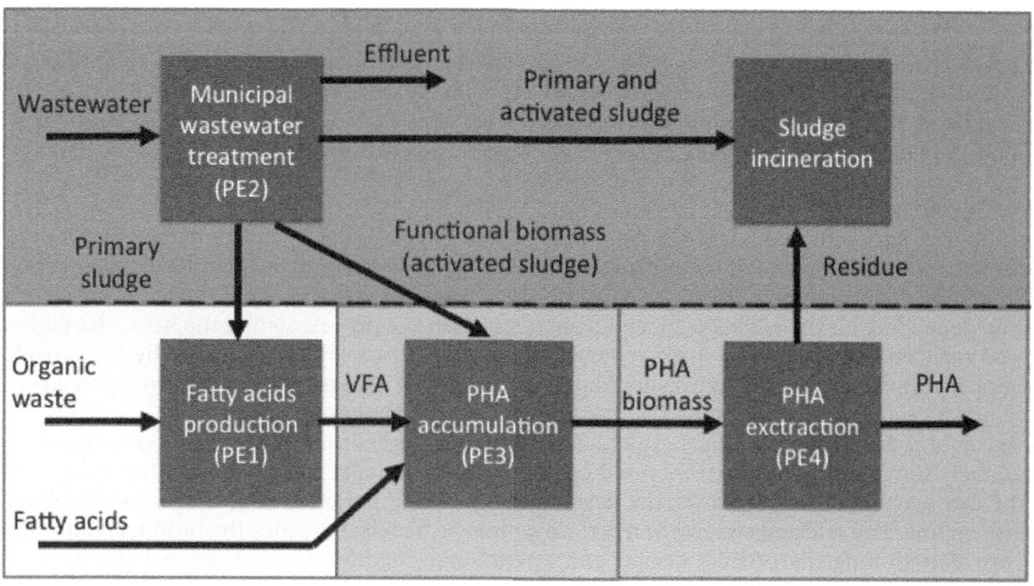

Figure 8.7 Process steps involved in the PHARIO project for PHA production from secondary sludge and organic waste as carbon source (STOWA, 2017).

and could be used as a green alternative in the manufacturing industry (Mannina *et al.*, 2020; Werker *et al.*, 2018). The construction of a demo-scale installation for the production of PHBV has just started in September 2020 in the Netherlands, and is expected to be operational by the end of 2021.

8.5 CHALLENGES, OPPORTUNITIES, AND RESEARCH NEEDS

There are a wide variety of possible alternatives for organic resource recovery from municipal wastewater, and the technical options are growing even in terms of demonstration in the operational environment. However, technical, social and legislation barriers remain and should be addressed by further research and innovative solutions to make and replicate business cases for resource recovery.

From a technical point of view, organic resource recovery technologies should be able to provide materials with high technical quality and stability that can be used in new or conventional industrial or consumer products. The novel technologies should be able to integrate existing infrastructure and allow for a smooth transition from WWTPs to resource recovery facilities. Water utility operators should have the technical skills to integrate and successfully run the resource recovery systems. From the economic and market point of view, cost should be calculated in a holistic manner, considering not only technical costs, but also social aspects, organizational strength, and institutional support. This approach can easily help to better understand the specific market potential of each recovered material.

Resource security should prioritize targeted recovery, and future research should support decisions about policy integration and regulatory change that must be addressed at international, national, and local scales. The major challenge to the overall resource recovery platform is public acceptance, since most of the recovered compounds are still labeled as wastes. The contamination risks associated with pathogens, chemicals, metabolites, and so on. concern society. Thus, new approaches should be developed to change general public opinion to support more sustainable management systems within circular economy frameworks. The communication problem can be overcome by a well-planned strategy focused on the most applicable technologies and geographical features, as well as cultural habits and sensitivities. Another important issue concerns improving markets for the recovered materials. This can be achieved by closing the gap between actual and potential recovery rates, which will directly affect the market price since improved recovery rates can lower the price.

In general, the bottlenecks of organic resource recovery in a WWTP are related to economics and value chain development, environment and health, and society and policy issues.

8.5.1 Cellulose

The implementation of a cellulose recovery process has the potential to reduce the net environmental impact of a conventional WWTP significantly, although to date there are no quantitative indications of the potential GHG emission reduction. Overall, the net energy demand of the system can decrease by 4–23% depending on the efficiency of cellulose removal in the process, while the net GHG emissions can be reduced by 2–19%. The further implementation of this technology (either by the retroffiting existing facilities or in new designs) should provide more real data to support the anticipated benefits of cellulose recovery, strengthening the business case for its implementation.

Beyond the benefits linked to energy and GHG emissions reduction, cellulose recovery can yield a product. The properties of the recovered cellulosic materials vary depending on the characteristics of the raw wastewater as well as to the design and operating parameters of the processes along the treatment line. The selected end-use of a certain cellulosic material dictates the product requirements, in turn determining the process design and operational conditions to be applied (Keijsers *et al.*, 2013). Public acceptance is not often easy for toilet-paper production from recovered cellulose, but valorization of cellulosic sludge via bio-drying and use as as feedstock for SCCA production or biofuel can generate additional credits in avoided energy demand and related GHG emissions.

8.5.2 PHA

While the first commercial full-scale process to produce PHA is currently being constructed with the PHARIO configuration, upscaling of the 3-step configuration is also taking place. In this case, the most critical upstream process factor that requires optimization is the overall process productivity, which depends on the PHA yield and on the PHA content in the biomass. A recent pilot study integrated multiple strategies that had been developed separately for different points of the 3-step process, attaining global productivities above 8 gPHA/L.d (Matos *et al.*, 2021). Based on this unprecedented PHA productivity from real feedstocks, the 3-stage configuration could become more competitive and be taken to the commercial scale in the near future.

In addition to productivity, the long-term stability of the process is another point where further research is needed for implementation of MMC PHA production from wastewater. The functional redundancy of multiple microbial populations within the MMC is one of the positive aspects of this technology with respect to pure cultures (Carvalho *et al.*, 2018). However, future research should validate the resiliency of enriched PHA-storing communities under the variable conditions of flow rate and influent composition that is typical of WWTPs. Moreover, stability relates not only to the productivity but also to the polymer characteristics (Werker *et al.*, 2018). Methods for achieving a stable PHA biopolymer product in terms of quantity and composition would aid downstream processing, and can be challenging with mixed cultures subject to dynamics in wastewater composition.

The economic viability of retrofitting existing WWTPs to produce PHA has been recently evaluated in a study comparing both large (>3 M PE) and small (50 k PE) sized plants. The final production cost of crude PHA (i.e. excluding downstream extraction) was found to be viable in both scenarios, in the range of 1.26–2.26 US$/kg PHA-crude for the large and small WWTP, respectively (Crutchik *et al.*, 2020). Despite the progress attained in the efficiency of the production step, current production costs are not yet competitive compared to conventional plastics, with the cost of recovery and purification currently being the critical bottleneck (Kehrein *et al.*, 2020). Indeed, the cost of downstream processing has been recognized as a major challenge for PHA recovery, and avoiding the use of harmful chlorinated solvents would also simplify the process for wastewater treatment plant operators, reducing the risk associated with hazardous materials. An alkali/surfactant based process has been shown to be more cost effective (1.40 €/kg PHB recovered) with reduced global warming potential (2.4 kg CO_2-eq/kg PHB) as compared to a solvent-based process (1.95 €/kg PHB, 4.30 kg CO_2-eq/kg PHB) (Fernández-Dacosta *et al.*, 2015), where the trade-off is typically reduced purity as compared to solvents (Pagliano *et al.*, 2021). Focussing on final applications necessitating low purity would be beneficial for PHA produced from wastewater, and would facilitate its integration into wastewater treatment plants. Simplification of downstream process flowsheets containing numerous stages also requires further study. The elimination of unit processes during downstream processing would further reduce the costs, increasing its economic viability.

8.6 CHAPTER SUMMARY

This chapter summarizes some of the emerging technologies for the recovery and valorization of organics in WWTPs. The main organic products that can be recovered/produced from municipal wastewater in WWTPs are cellulose and PHA. Cellulosic primary sludge, mostly generated from the flushing of toilet papers, can be efficiently utilized to recover cellulose or follow other pathways for the valorization of cellulose and separated organics. Recovered cellulose can be further processed and used as a starting material for SCCA or bioplastic production. Meanwhile, PHA recovery from sewage is another important technology since PHAs have comparable properties to petrochemical plastics. Producing PHA from sewage could provide a cost-competitive route for their production, although further research is needed to strengthen the economic business case of this organics valorization route.

8.7 EXERCISES

Exercise 8.1: A centralized municipal wastewater treatment plant services a city of 750 000 inhabitant equivalents. In order to improve the sustainability of urban wastewater management of the city, the city council has raised the recovery of cellulose from municipal wastewater through micro-sieves as one of the priority resources. Based on a toilet paper usage of 10 kg toilet paper/person/year (i.e. typical consumption in Western European countries):

(a) Calculate the amount of cellulose that can be recovered at the WWTP on a daily and annual basis at a practical relevant cellulose recovery efficiency of a micro-sieve unit of 80%.
(b) Calculate the theoretical amount of methane (in m^3) that can be generated from the amount of cellulose embedded in the wastewater matrix when not being recovered as cellulose but being fed to an anaerobic digestion. Assume that the COD content of cellulose is equal to 1.1 g COD/gram cellulose, that the cellulose can be fully degraded during mesophilic anaerobic digestion, and that 0.5 m^3 methane are produced per kg of COD converted.
(c) Assuming that 1 m^3 of biogas is equivalent to 6.27 kWh, that the conversion of methane to electricity in a CHP unit is 40% and the electricity consumption of a household of four people is 20 kWh/day, how many households can theoretically be powered from the produced electricity from the anaerobic digestion of cellulose.

Exercise 8.2: In various regions in the world existing WWTPs in urban areas are reaching their maximum capacity in terms of solids and COD load due to the increase in population growth within the catchment area of the WWTP during the last decade(s). Some of these WWTPs have serious constraints in terms of available land for extension of the WWTP. In this context, beyond the context of the importance of resource recovery, the implementation of micro-sieves can be of particular interest.

(a) Based on typical surface loading rates for primary settling tanks and micro-sieves of 1.5–2.5 and 100 m^3/m^2/hour, respectively, calculate the footprint of primary settling and micro-sieve for a WWTP with a capacity of 500 000 PE. Assume a wastewater production of 120 L/PE/d.
(b) Based on your answer in (a), what is the percentage in land space saving that can be achieved?
(c) Compare this with the overall size of a typical WWTP (tip: use google earth for an aerial shot of a WWTP near to where your life).

Exercise 8.3: The Canadian city of Halifax is building an activated sludge system at its main wastewater treatment plant that services 170 000 PE. Two resource recovery strategies are considered for the sludge produced from this treatment facility. The first consists of one-stage anaerobic digestion for methane recovery from biogas and the other is polyhydroxyalkanoate (PHA) recovery through a 3-stage process. The 3-stage PHA process consists of: (1) anaerobic sludge fermentation to SCCA; (2) aerobic culture selection through a feast/famine process fed with the effluent of (1); and (3) PHA accumulation from the selected culture in (2) using the effluent of (1). Assume standard (i.e. Table 8.1) wastewater flows and characteristics and that 80 gVSS of sludge are produced per m^3 of wastewater treated by the facility. Considering that the value of PHA is \$3.5/kg and that the value of methane is \$0.4/m^3, where 1 m^3 of biogas (with a methane content of 65%) is produced per kg of VSS, while 0.1 kg of PHA are produced per kg of VSS in the 3-stage process:

(a) Determine the relative value of the sludge stream using either the PHA recovery or biogas recovery strategy.
(b) Discuss the key factors impacting the capital and operational expenditures associated with biogas or PHA production. Which process is likely to incur higher production costs? Why? What would you consider to be the key points impacting your decision on the process to be implemented?

Exercise 8.4: Contrary to popular belief, a large fraction of the COD load entering WWTPs is in the form of toilet paper, depending on the region in the world. For example, the amount in South

American countries is typically much lower as it is often prohibited to flush toilet paper in the toilets as the sewer pipes in the building are not designed for this and as such need to be collected separately in a bin. Let us consider USA, with a consumption of toilet paper of 12.7 kg/person/year:

(a) Calculate the concentration of cellulose based COD assuming a daily water consumption of 200 L/per person per day.
(b) Determine the fraction of $COD_{cellulose}/COD_{total}$ assuming a typical municipal wastewater composition.

Exercise 8.5: In various places in the world, the excess waste activated sludge after dewatering is being transported to landfills for ultimate disposal. Moreover, not all WWTPs have implemented primary settling as a pre-treatment step, as such, in a typical biological nitrogen removal (or biological P removal) configuration all incoming cellulose enters the activated sludge tanks. According to the Foundation for Applied Water Research (STOWA), depending on the WWTP configuration and local conditions (e.g. wastewater composition, local climate, industrial activities in sewer catchment), about 30–70% of the cellulose is being aerobically converted with an accompanying biomass yield of 0.3 kg sludge/kg COD removed. Assume an annual toilet paper usage of 10 kg/PE, 70% aerobic degradation of cellulose, with a COD of 1.1 kg COD/kg cellulose, consider that 0.44 kg O_2 are needed for degradation of 1 kg of COD, a power consumption for aeration of 2.5 kg O_2/kWh for bubble aeration. Based on the above:

(a) Calculate the energy savings that can be achieved for a WWTP with a capacity of 100 000 PE by implementing microsieves as a pre-treatment step.
(b) Calculate the fraction of the incoming toilet paper that ultimately ends up in landfill (in case this is the ultimate disposal route) in case there is no primary treatment.

Exercise 8.6: A pilot-scale WWTP is designed to a fully integrated process to valorize the harvested screenings from municipal wastewater and to produce pure marketable cellulose. In this regard, the plant is equipped with an RBF for enhanced TSS separation and further cellulose recovery. Together with solids, cellulose fibers are separated with RBF and thus recovered. The RBF system works at a flowrate of 30 m^3/d. Calculate the maximum cellulose recovery yield considering that:

(1) The sludge production after 5 hours is 24 kg.
(2) The % of TS is 20%.
(3) The percentage of pure cellulose in the sludge, after a post-treatment of sludge washing (to concentrate the amount of fibers), is 35% of the cellulosic material.
(4) The specific recovery yield of the cellulosic material in terms of g VS after washing to g TS initially recovered in the sludge is 0.88 gVS/gTS.

Exercise 8.7: A given wastewater contains 50 g soluble COD/L, of which 30% are carbohydrates. Assuming that all fermentable fractions of carbohydrates are converted to acetic acid and butyric acid, to be used as substrate for PHA production, what would be their final concentrations in solution, both in g/L and in g COD/L. Assume that all carbohydrates are glucose, and the stochiometries are as given in Chapter 7.

Exercise 8.8: The PHA polymeric properties are strongly influenced by the relative PHB vs PHV fractions. Typically, acetate and butyrate are precursors for PHB production, while propionate and valerate are precursors of PHV production. Assuming that the PHB and PHV content are linearly related to the quantity of their SCCA precursors (on a mass basis) produced during acidogenic fermentation, calculate the relative fraction of PHB and PHV produced for the case of an acidogenic feedstock containing 4.1 g/L acetate, 1.6 g/L propionate, 1.1 g/L butyrate and 0.6 g/L valerate. Assume a constant yield coefficient of 0.7 gPHA/gSCCA for each SCCA, while all acetate and propionate are completely consumed by the PHA producing culture, and 85% of the butyrate and valerate are consumed.

Exercise 8.9: SCCA are an excellent feedstock for PHA bioplastic production. Considering the supply chain for PHA production, approximately how much organic waste (in COD) would be needed to make an industrially-relevant 5000 tonnes PHA per year. Assume an SCCA yield of 0.34 $gCOD_{SCCA}/gCOD_{fed}$, an overall PHA yield of 0.5 g biomass/$gCOD_{SCCA}$, a PHA content of 70% and a total extraction efficiency of 63%.

Exercise 8.10: A WWTP with a capacity of 80 000 PE produces around 60 gTS/PE d with a VS/TS ratio of 78%. All the produced sludge is fermented to produce SCCA with an observed fermentation yield of 0.21 $gCOD_{SCCA}/gVS_{fed}$. The SCCA-rich stream (7500 $mgCOD_{SCCA}$/L) is then used for PHA production with anaerobic feast/famine process. The PHA-production line is composed of two different SBRs reactors:

(a) Selection reactor (S-SBR);
(b) Accumulation reactor (A-SBR).

A vOLR of 4.0 $kgCOD_{SCCA}/m^3$ d is applied to the S-SBR, which is operated at a feast/famine ratio of 0.1 min/min. The Y_{PHA} under feast conditions is 0.71 $gCOD_{PHA}/gCOD_{SCCA}$, while the Y_{obs} is 0.33 gXa/$gCOD_{PHA}$. The SRT is set at 2.5 days and the SBR works with 3.5 cycles per day. The excess sludge produced from the S-SBR is then fed to the A-SBR for the accumulation test to maximize PHA production.

In the A-SBR, the SCCA-rich stream is added seven times (once per hour) to achieve an initial COD concentration of around 1 $gCOD_{SCCA}$/L. The volume of each addition is 0.56 m^3. The PHA yield is about 0.58 $gCOD_{PHA}/gCOD_{SCCA}$ and the growth yield is 0.24 gXa/ $gCOD_{SCCA}$.

Conversion factor for PHA: 1.67 gCOD/gPHA

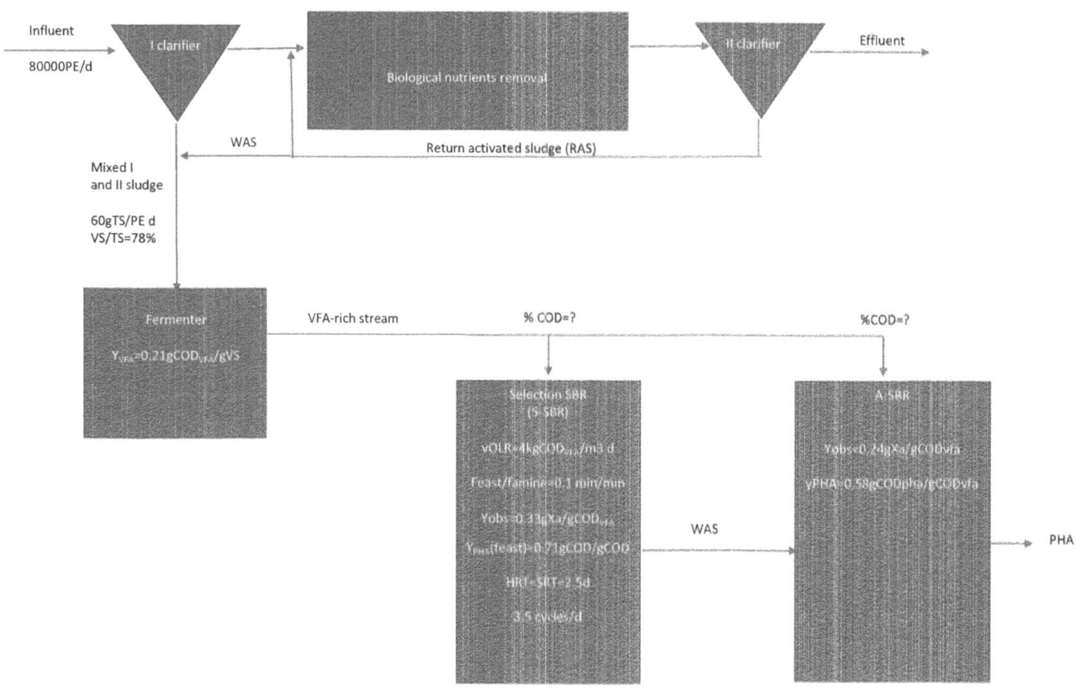

Calculate:

(1) The daily SCCA amount produced from the whole WWTP;
(2) The length of both the SBR cycle and feast/famine of the S-SBR;
(3) The volume of the S-SBR to sustain a production of 10 kgPHA/d, assuming a PHA content in the biomass of 40% (gPHA/gTS);
(4) The biomass concentration of S-SBR;
(5) The flowrate of carbon source to S-SBR and A-SBR, respectively (for the production of 10 kg PHA);
(6) The volume of the A-SBR (considering seven carbon source additions);
(7) The maximum annual PHA production potential of the whole WWTP.

8.8 DISCUSSION QUESTIONS

Question 8.1 (*technology and economy*): As the innovation manager of a large water utility, you are in charge of reorganizing the existing water infrastructure from its current situation to a more circular approach within a timeframe of 25 years. The current wastewater treatment infrastructure comprises three large-scale wastewater treatment plants with capacities in the order of 500 000 PE each. In addition, it comprises more than 20 small scale WWTPs with a capacity of around 10 000 PE. You are asked to give a presentation to the board of directors in which you evaluate the current status and justify your masterplan. Where would you focus on in terms chosen technology, location (i.e. which WWTP to focus on) and which resources discussed in this chapter would you target? What are your key considerations/motivations?

Question 8.2 (*social license to operate, risks and market analysis*): A wastewater utility operating a WWTP with a capacity of 1 000 000 PE is ending its service life. Assume a hypothetical situation in which the WWTP will be completely rebuilt from scratch and you are the business development manager of the wastewater utility in charge of selecting the most appropriate resource to recover. You are asked to give a presentation to the board of directors in which you evaluate the economic potential of the recovery of cellulose, SCCA or energy recovery through anaerobic digestion. Based on your assessment, from an economic perspective, recovery of cellulose is the most promising followed by SCCA recovery. Nevertheless, in your presentation to the board of directors you recommended the implementation of an anaerobic digester for the production of biogas. Discuss what the key criteria and considerations the business development manager most likely considered as critical?

Question 8.3 (*market analysis, risks and economy-of-scale*): In developed countries, the wastewater infrastructure is aging and would require major upgrades in the coming decades. In less developed countries and emerging economies a significant amount of new infrastructure will need to be realized in the coming decades. Considering the above, there is an ongoing debate regarding how our future urban waster infrastructure will look like, that is more centralized or more modular and decentralized. Evaluate the general characteristics of centralized and decentralized systems and provide advantages and disadvantages for each of these scenarios in the context of cellulose and PHA recovery, taking into account economy of scale, practical feasibility and maintenance, monitoring and quality control and market requirements of the two resources.

Question 8.4 (*product quality, chemical industry and economy-of-scale*): Plastics are currently produced from oil and other fossile resources. PHA could be one biodegradable alternative to replace them. Considering that petrochemical plastic production can ensure large scale manufacturing at constant quality, discuss the potential impact of the latter in terms of market potential for PHA. In your answer, include the following aspects: (i) quality and quantity aspects; (ii) logistics (i.e. means of storage, transportation and distribution); and (iii) ultimate end-user.

Question 8.5 (*process performance, capital costs*): The implementation of new resource recovery schemes and technologies is (partly) dependent on their capital and operational costs and the return on investment period. What are the key operational factors that affect the CAPEX and OPEX of cellulose recovery and PHA production, respectively, and how do they affect the process economics.

Question 8.6 (*design considerations, process stability, robustness and operator confidence*): Discuss the potential advantages and/or disadvantages of cellulose recovery and PHA production from a process stability/robustness and wastewater operator confidence and overall willingness of wastewater treatment operators to introduce these recovery approaches.

Question 8.7 (*economic considerations, market demands, decision-making*): You are the Innovation Manager of a large wastewater utility that needs to prioritize the resources to be recovered from your largest wastewater treatment facilities. The resources discussed in this chapter (i.e. cellulose and PHA) cannot be recovered simultaneously. Which of the two would be your priority as an innovation manager? Justify your choice.

Question 8.8 (*social license to operate, reduce not recover*): According to the waste management hierarchy, prevention and reduction of the use of resource are more favorable options compered to resource recovery. Within this context, discuss options to prevent the use of toilet paper and its potential impact on municipal wastewater treatment. Also, discuss the importance of social acceptance of finding alternatives for the use of toilet paper. In your discussion, specifically mention the situation in Japan (and other Asian countries).

Question 8.9 (*process stability, robustness and operator confidence*): Only in recent years has the recovery of cellulose gained a lot of interest. However, the use of micro-sieves as a pre-treatment step for the removal of suspended solids from municipal wastewater were developed and implemented in Norway as a means to comply with EU regulation with respect to the removal of suspended solids during primary treatment rather than being developed with the purpose of recovering cellulose. The practical feasibility of microsieves as a viable alternative technology for pre-treatment of municipal wastewater treatment was thus already demonstrated prior to discovering its additional benefits of enabling the recovery of cellulose. Discuss the importance of the latter in the context of market uptake and replication.

Question 8.10 (*public acceptance and market uptake*): One of the biggest challenges of resource recovery is to develop new markets and applications for the recovered products. Changing the opinion and practices of stakeholders is an important component of the resource recovery concept. Producers need to rethink the composition and design of their products with respect to the targeted information as well as to the stakeholders' involvement. What is the importance and influence of stakeholders and/ or policymakers on helping/hampering the development of markets for recovered organics materials?

FURTHER READING MATERIALS

Lema, J. M. and Suarez, S. (eds), Innovative Wastewater Treatment & Resource Recovery Technologies: Impacts on Energy, Economy and Environment. IWA Publishing, London, UK. https://doi.org/10.2166/9781780407876

Puyol, D., Batsone, D. J., Hülsen, T., Astals, S., Peces, M. and Krömer, J. O. (2017). Resource recovery from wastewater by biological technologies: Opportunities, challenges, and prospects. *Frontiers in Microbiology*, **7**, 1–23. https://doi.org/10.3389/fmicb.2016.02106

REFERENCES

Ahmed A. S., Bahreini G., Ho D., Sridhar G., Gupta M., Wessels C., Marcelis P., Elbeshbishy E., Rosso D., Santoro D. and Nakhla G. (2019). Fate of cellulose in primary and secondary treatment at municipal water resource recovery facilities. *Water Environment Research*, **91**, 1479–1489. https://doi.org/10.1002/wer.1145

Akyol Ç., Foglia A., Ozbayram E. G., Frison N., Katsou E., Eusebi A. L. and Fatone F. (2020). Validated innovative approaches for energy-efficient resource recovery and re-use from municipal wastewater: From anaerobic treatment systems to a biorefinery concept. *Critical Reviews in Environmental Science and Technology*, **50**, 869–902. https://doi.org/10.1080/10643389.2019.1634456

Albuquerque M. G. E., Eiroa M., Torres C., Nunes B. R. and Reis M. A. M. (2007). Strategies for the development of a side stream process for polyhydroxyalkanoate (PHA) production from sugar cane molasses, *Journal of Biotechnology*, **130**(4), 411–421.

Albuquerque M. G. E., Concas S., Bengtsson S. and Reis M. A. M. (2010). Mixed culture polyhydroxyalkanoates production from sugar molasses: The use of a 2-stage CSTR system for culture selection. *Bioresource Technology*, **101**, 7112–7122. https://doi.org/10.1016/j.biortech.2010.04.019

Albuquerque M. G. E., Martino V., Pollet E., Avérous L. and Reis M. A. M. (2011). Mixed culture polyhydroxyalkanoate (PHA) production from volatile fatty acid (VFA)-rich streams: Effect of substrate composition and feeding regime on PHA productivity, composition and properties, *Journal of Biotechnology*, **151**(1), 66–76.

Anjum A., Zuber M., Zia K. M., Noreen A., Anjum M. N. and Tabasum S. (2016). Microbial production of polyhydroxyalkanoates (PHAs) and its copolymers: A review of recent advancements. *International Journal of Biological Macromolecules*, **89**, 161–174. https://doi.org/10.1016/j.ijbiomac.2016.04.069

Behera C. R., Santoro D., Gernaey K. V. and Sin G. (2018). Organic carbon recovery modeling for a rotating belt filter and its impact assessment on a plant-wide scale. *Chemical Engineering Journal*, **334**, 1965–1976. https://doi.org/10.1016/j.cej.2017.11.091

Bengtsson S., Werker A., Christensson M. and Welander T. (2008). Production of polyhydroxyalkanoates by activated sludge treating a paper mill wastewater. *Bioresource Technology*, **99**(3), 509–516. https://doi.org/10.1016/j.biortech.2007.01.020

Bengtsson S., Karlsson A., Alexandersson T., Quadri L., Hjort M., Johansson P., Morgan-Sagastume F., Anterrieu S., Arcos-Hernandez M., Karabegovic L., Magnusson P. and Werker A. (2017). A process for polyhydroxyalkanoate (PHA) production from municipal wastewater treatment with biological carbon and nitrogen removal demonstrated at pilot-scale. *New Biotechnology*, **35**, 42–53. https://doi.org/10.1016/j.nbt.2016.11.005

Bout A. E., Pfau S. F., van der Krabben E. and Dankbaar B. (2019). Residual biomass from Dutch riverine areas-from waste to ecosystem service. *Sustainability*, **11**, 1–14. https://doi.org/10.3390/su11020509

Campanari S., e Silva F. A., Bertin L., Villano M. and Majone M. (2014). Effect of the organic loading rate on the production of polyhydroxyalkanoates in a multi-stage process aimed at the valorization of olive oil mill wastewater. *International Journal of Biological Macromolecules*, **71**, 34–41. https://doi.org/10.1016/j.ijbiomac.2014.06.006

Carvalho G., Oehmen A., Albuquerque M. G. E. and Reis M. A. M. (2014). The relationship between mixed microbial culture composition and PHA production performance from fermented molasses. *New Biotechnology*, **31**, 257–263. https://doi.org/10.1016/j.nbt.2013.08.010

Carvalho G., Pedras I., Karst S. M., Oliveira C. S. S., Duque A. F., Nielsen P. H. and Reis M. A. M. (2018). Functional redundancy ensures performance robustness in 3-stage PHA-producing mixed cultures under variable feed operation. *New Biotechnology*, **40**, 207–217. https://doi.org/10.1016/j.nbt.2017.08.007

Cavaillé L., Grousseau E., Pocquet M., Lepeuple A. S., Uribelarrea J. L., Hernandez-Raquet G. and Paul E. (2013). Polyhydroxybutyrate production by direct use of waste activated sludge in phosphorus-limited fed-batch culture. *Bioresource Technology*, **149**, 301–309. https://doi.org/10.1016/j.biortech.2013.09.044

Chen S. S., Wang L., Yu I. K. M., Tsang D. C. W., Hunt A. J., Jérôme F., Zhang S., Ok Y. S. and Poon C. S. (2018). Valorization of lignocellulosic fibres of paper waste into levulinic acid using solid and aqueous Brønsted acid. *Bioresource Technology*, **247**, 387–394. https://doi.org/10.1016/j.biortech.2017.09.110

Conca V., Ros C., Valentino F., Eusebi A. L., Frison N. and Fatone F. (2020). Long-term validation of polyhydroxyalkanoates production potential from the sidestream of municipal wastewater treatment plant at pilot scale. *Chemical Engineering Journal*, **390**, 1–13. https://doi.org/10.1016/j.cej.2020.124627

Crutchik D., Frison N., Eusebi A. L. and Fatone F. (2018). Biorefinery of cellulosic primary sludge towards targeted short chain fatty acids, phosphorus and methane recovery. *Water Research*, **136**, 112–119. https://doi.org/10.1016/j.watres.2018.02.047

Crutchik D., Franchi O., Caminos L., Jeison D., Belmonte M., Pedrouso A., Val del Rio A., Mosquera-Corral A. and Campos J. L. (2020). Polyhydroxyalkanoates (PHAs) production: A feasible economic option for the treatment of sewage sludge in municipalwastewater treatment plants? *Water (Switzerland)*, **12**, 1–12. https://doi.org/10.3390/W12041118

Da Ros C., Conca V., Laura A., Frison N. and Fatone F. (2020). Sieving of municipal wastewater and recovery of bio-based volatile fatty acids at pilot scale. *Water Research*, **174**, 1–10. https://doi.org/10.1016/j.watres.2020.115633

Dionisi D., Majone M., Tandoi V. and Beccari M. (2001). Sequencing batch reactor: Influence of periodic operation on performance of activated sludges in biological wastewater treatment. *Industrial and Engineering Chemical Research*, **40**, 5110–5119. https://doi.org/10.1021/ie001008k

Fernández-Dacosta C., Posada J. A., Kleerebezem R., Cuellar M. C. and Ramirez A. (2015). Microbial community-based polyhydroxyalkanoates (PHAs) production from wastewater: Techno-economic analysis and ex-ante environmental assessment. *Bioresource Technology*, **185**, 368–377. https://doi.org/10.1016/j.biortech.2015.03.025

Franchi A. and Santoro D. (2015). Current status of the rotating belt filtration (RBF) technology for municipal wastewater treatment. *Water Practice and Technology*, **10**, 319–327. https://doi.org/10.2166/wpt.2015.038

Frison N., Katsou E., Malamis S., Oehmen A. and Fatone F. (2015). Development of a novel process integrating the treatment of sludge reject water and the production of polyhydroxyalkanoates (PHAs). *Environmental Science and Technology*, **49**, 10877–10885. https://doi.org/10.1021/acs.est.5b01776

Gavrilescu M. (2014). Biomass potential for sustainable environment, biorefinery products and energy. In: Visa, I. (ed.). Sustainable Energy in the Built Environment – Steps Towards NZEB. Springer International Publishing, Cham, pp. 169–194.

Huang F. (2013). What is biomass? In: Materials for Biofuels, A. J. Ragauskas (ed.), World Scientific, Singapore, pp. 1–26.

Johnson K., Jiang Y., Kleerebezem R., Muyzer G. and Van Loosdrecht M. C. M. (2009). Enrichment of a mixed bacterial culture with a high polyhydroxyalkanoate storage capacity. *Biomacromolecules*, **10**, 670–676. https://doi.org/10.1021/bm8013796

Johnson K., Kleerebezem R., and Loosdrecht M. C. M. (2010). Influence of the C/N ratio on the performance of polyhydroxybutyrate (PHB) producing sequencing batch reactors at short SRTs. *Water Research*, **44**(7), 2141–2152.

Kehrein P., van Loosdrecht M., Osseweijer P., Garfí M., Dewulf J. and Posada J. (2020). A critical review of resource recovery from municipal wastewater treatment plants – market supply potentials, technologies and bottlenecks. *Environmental Science: Water Research and Technology*, **6**, 877–910. https://doi.org/10.1039/c9ew00905a

Keijsers E. R. P., Yilmaz G. and Van Dam J. E. G. (2013). The cellulose resource matrix. *Carbohydrate Polymersm*, **93**, 9–21. https://doi.org/10.1016/j.carbpol.2012.08.110

Khatami K., Perez-Zabaleta M., Owusu-Agyeman I. and Cetecioglu Z. (2021). Waste to bioplastics: How close are we to sustainable polyhydroxyalkanoates production? *Waste Management*, **119**, 374–388. https://doi.org/10.1016/j.wasman.2020.10.008

Kim M., Lee E.-K., Choi C.-J., Lee M.-H., Kim J.-M., Ji S.-W., Jeon B.-J., Park G.-P. and Song K.-B. (2017). Brick insulation composite and method for manufacturing same. US Pat. 2017/0191264 A1. https://doi.org/10.1016/j.(73)

Kourmentza C., Plácido J., Venetsaneas N., Burniol-Figols A., Varrone C., Gavala H. N. and Reis M. A. M. (2017). Recent advances and challenges towards sustainable polyhydroxyalkanoate (PHA) production. *Bioengineering* **4**, 1–43. https://doi.org/10.3390/bioengineering4020055

Kumar M., Morya R., Gnansounou E., Larroche C. and Thakur I. S. (2017). Characterization of carbon dioxide concentrating chemolithotrophic bacterium Serratia sp. ISTD04 for production of biodiesel. *Bioresource Technology*, **243**, 893–897. https://doi.org/10.1016/j.biortech.2017.07.067

Kumar M., Ghosh P., Khosla K. and Thakur I. S. (2018). Recovery of polyhydroxyalkanoates from municipal secondary wastewater sludge. *Bioresource Technology*, **255**, 111–115. https://doi.org/10.1016/j.biortech.2018.01.031

Li S., Wu Z., Zixin Wu. and Liu G. (2020). Enhancing fiber recovery from wastewater may require toilet paper redesign. *Journal of Cleaner Production*, **261**, 121138. https://doi.org/10.1016/j.jclepro.2020.121138

Mannina G., Presti D., Montiel-Jarillo G., Carrera J. and Suárez-Ojeda M. E. (2020). Recovery of polyhydroxyalkanoates (PHAs) from wastewater: A review. *Bioresource Technology*, **297**, 1–12. https://doi.org/10.1016/j.biortech.2019.122478

Matos M., Cruz R. A. P., Cardoso P., Silva F., Freitas E. B., Carvalho G. and Reis M.A.M. (2021). Combined strategies to boost polyhydroxyalkanoate production from fruit waste in a three-stage pilot plant. *ACS Sustainable Chemical Engineering* 9(24), 8270–8279. https://doi.org/10.1021/acssuschemeng.1c02432

Morgan-Sagastume F., Valentino F., Hjort M., Cirne D., Karabegovic L., Gerardin F., Johansson P., Karlsson A., Magnusson P., Alexandersson T., Bengtsson S., Majone M. and Werker A. (2014). Polyhydroxyalkanoate (PHA) production from sludge and municipal wastewater treatment. *Water Science and Technology*, **69**, 177–184. https://doi.org/10.2166/wst.2013.643

Morgan-Sagastume F., Heimersson S., Laera G., Werker A. and Svanström M. (2016). Techno-environmental assessment of integrating polyhydroxyalkanoate (PHA) production with services of municipal wastewater treatment. *Journal of Cleaner Production*, **137**, 1368–1381. https://doi.org/10.1016/j.jclepro.2016.08.008

Możejko-Ciesielska J. and Kiewisz R. (2016). Bacterial polyhydroxyalkanoates: Still fabulous? *Microbiological Research*, **192**, 271–282. https://doi.org/10.1016/j.micres.2016.07.010

Nechyporchuk O., Belgacem M. N. and Bras J. (2016). Production of cellulose nanofibrils: A review of recent advances. *Industrial Crops and Products*, **93**, 2–25. https://doi.org/10.1016/j.indcrop.2016.02.016

Oehmen A., Pinto F. V., Silva V., Albuquerque M. G. E. and Reis M. A. M. (2014). The impact of pH control on the volumetric productivity of mixed culture PHA production from fermented molasses. *Engineering in Life Sciences*, **14**, 143–152. https://doi.org/10.1002/elsc.201200220

Oliveira C. S. S., Silva C. E., Carvalho G. and Reis M. A. (2017). Strategies for efficiently selecting PHA producing mixed microbial cultures using complex feedstocks: Feast and famine regime and uncoupled carbon and nitrogen availabilities. *New Biotechnology*, **37**, 69–79. https://doi.org/10.1016/j.nbt.2016.10.008

Pagliano G., Galletti P., Samorì C., Zaghini A. and Torri C. (2021). Recovery of polyhydroxyalkanoates from single and mixed microbial cultures: A review. *Frontiers in Bioengineering and Biotechnology*, **9**, 1–28. https://doi.org/10.3389/fbioe.2021.624021

Pakalapati H., Chang C. K., Show P. L., Arumugasamy S. K. and Lan J. C. W. (2018). Development of polyhydroxyalkanoates production from waste feedstocks and applications. *Journal of Bioscience and Bioengineering*, **126**, 282–292. https://doi.org/10.1016/j.jbiosc.2018.03.016

Palmieri S., Cipolletta G., Pastore C., Giosuè C., Akyol Ç., Eusebi A. L., Frison N., Tittarelli F. and Fatone F. (2019). Pilot scale cellulose recovery from sewage sludge and reuse in building and construction material. *Waste Management*, **100**, 208–218. https://doi.org/10.1016/j.wasman.2019.09.015

Raza Z. A., Abid S. and Banat I. M. (2018). Polyhydroxyalkanoates: Characteristics, production, recent developments and applications. *International Biodeterioration and Biodegradation*, **126**, 45–56. https://doi.org/10.1016/j.ibiod.2017.10.001

Reddy K. H. K., Patra M. R., Roy D. S. and Pradhan B. (2012). An adaptive scheduling mechanism for computational desktop grid using GridGain. *Procedia Technology*, **4**, 573–578. https://doi.org/10.1016/j.protcy.2012.05.091

Rodriguez-Perez S., Serrano A., Pantión A. A. and Alonso-Fariñas B. (2018). Challenges of scaling-up PHA production from waste streams. A review. *Journal of Environmental Management*, **205**, 215–230. https://doi.org/10.1016/j.jenvman.2017.09.083

Ruiken C. J., Breuer G., Klaversma E., Santiago T. and van Loosdrecht M. C. M. (2013). Sieving wastewater – cellulose recovery, economic and energy evaluation. *Water Research*, **47**, 43–48. https://doi.org/10.1016/j.watres.2012.08.023

Rusten B. and Odegaard H. (2006). Evaluation and testing of fine mesh sieve technologies for primary treatment of municipal wastewater. *Water Science and Technology*, **54**, 31–38.

Rusten B., Razafimanantsoa V. A., Andriamiarinjaka M. A., Otis C. L., Sahu A. K. and Bilstad T. (2016). Impact of fine mesh sieve primary treatment on nitrogen removal in moving bed biofilm reactors. *Water Science and Technology*, **73**, 337–344. https://doi.org/10.2166/wst.2015.498

Rusten B., Rathnaweera S. S., Rismyhr E., Sahu A. K. and Ntiako J. (2017). Rotating belt sieves for primary treatment, chemically enhanced primary treatment and secondary solids separation. *Water Science and Technology*, **75**, 2598–2606. https://doi.org/10.2166/wst.2017.145

Samorì C., Abbondanzi F., Galletti P., Giorgini L., Mazzocchetti L., Torri C. and Tagliavini E. (2015). Extraction of polyhydroxyalkanoates from mixed microbial cultures: Impact on polymer quality and recovery. *Bioresource Technology*, **189**, 195–202. https://doi.org/10.1016/j.biortech.2015.03.062

SMART-Plant (2020a). D4.5 Socio-economic Assessment Including Life Cycle Costing (LCC) and Cost Benefit Analysis (CBA) Reports. Available at: https://ec.europa.eu/research/participants/documents/downloadPublic?documentIds=080166e5cf44b308&appId=PPGMS (accessed 26 January 2022).

SMART-Plant (2020b). D4.4 Environmental Impact Report, incl. LCA (Life Cycle Assessment). Available at: https://publications.kompetenz-wasser.de/pdf/Remy-2020-1273.pdf (accessed 26 January 2022).

STOWA (2017). PHARIO – Stepping Stone to A Sustainable Value Chain for PHA Bioplastic Using Municipal Activated Sludge. STOWA, Amersfoot, The Netherlands.

Valentino F., Morgan-Sagastume F., Campanari S., Villano M., Werker A. and Majone M. (2017). Carbon recovery from wastewater through bioconversion into biodegradable polymers. *New Biotechnology*, **37**, 9–23. https://doi.org/10.1016/j.nbt.2016.05.007

van der Hoek J. P., Struker A. and de Danschutter J. E. M. (2015). Amsterdam as a sustainable European metropolis: Integration of water, energy and material flows. *Urban Water Journal*, **14**, 61–68. https://doi.org/10.1080/1573062X.2015.1076858

Wang X., Bengtsson S., Oehmen A., Carvalho G., Werker A. and Reis M. A. M. (2019). Application of dissolved oxygen (DO) level control for polyhydroxyalkanoate (PHA) accumulation with concurrent nitrification in surplus municipal activated sludge. *New Biotechnology*, **50**, 37–43. https://doi.org/10.1016/j.nbt.2019.01.003

Werker A., Johansson P. and Magnusson P. (2015). Process for the Extraction of Polyhydroxyalkanoates From Biomass. VEOLIA Water Solutions & Technologies, Saint-Maurice, France, US20150368393A1.

Werker A., Bengtsson S., Korving L., Hjort M., Anterrieu S., Alexandersson T., Johansson P., Karlsson A., Karabegovic L., Magnusson P., Morgan-Sagastume F., Sijstermans L., Tietema M., Visser C., Wypkema E., Van Der Kooij Y., Deeke A. and Uijterlinde C. (2018). Consistent production of high quality PHA using activated sludge harvested from full scale municipal wastewater treatment – PHARIO. *Water Science and Technology*, **78**, 2256–2269. https://doi.org/10.2166/wst.2018.502

doi: 10.2166/9781780409566

Chapter 9

Producing microbial-based protein from reactive nitrogen recovered from wastewater

Ilje Pikaar[1], Heidy Cruz[1], Tim Hülsen[2], Damien Batstone[2], Korneel Rabaey[3,4], Paul Jensen[2], Yoram Avnimelech[5] and Willy Verstraete[3]

[1]The School of Civil Engineering, The University of Queensland, St Lucia, QLD 4072, Australia
[2]The Advanced Water Management Centre, The University of Queensland, St Lucia, QLD 4072, Australia
[3]Centre for Microbial Ecology and Technology (CMET), Ghent University, 9000 Gent, Belgium
[4]Centre for Advanced Process Technology for Urban Resource Recovery, Frieda Saeysstraat 1, 9052 Gent, Belgium
[5]Israel Institute of Technology, Faculty of Civil and Environmental Engineering, Haifa, Israel

9.1 INTRODUCTION

Recent estimates of the UN indicate that the global population will grow to 9–10 billion people by the year 2050 (United Nations, 2015). For physical and mental health, it is important that on average each person can consume some 0.66 g protein/kg body weight per day (World Health Organization, 2002). The current route to provide this protein is mainly through agricultural plant production, using massive amounts of mineral nitrogen fertilizer fabricated by means of the Haber–Bosch process (Bodirsky *et al.*, 2014). This mineral nitrogen is generated as ammonium and may be converted to nitrate prior to use. These forms of reactive nitrogen are, when applied to the soil as fertilizer, taken up by the plants to produce plant proteins. The plant proteins can be directly used as food for the human population, but are, to a large extent, used as feed to produce animal protein. In the latter case, approximately 4% of the Haber–Bosch nitrogen is ultimately consumed as high-value animal proteins. The overall nitrogen efficiency for plant proteins for human consumption is substantially higher, albeit still low, with an efficiency of 14% (Galloway & Cowling, 2002).

At present, the need for animal (and hence vegetable) protein containing essential amino acids is increasing because a growing part of the world desires to, and can afford to, consume more protein products and higher quality protein products (Bodirsky *et al.*, 2015; Godfray *et al.*, 2010). The current major supply routes for high quality protein are agri-crops (for about 50% of food protein inputs, particularly wheat, rice and pulses such as soy beans), animal proteins based on agri-crops (another 40%) and finally fish (some 5–10%). These routes are facing limitations. Limitations of the agri-crop route include the massive amount of nitrogen fertilizer used worldwide, that is ~100 Mton of nitrogen fertilizer, and this is expected to increase to about 150 Mton of by 2050. Indeed the Haber–Bosch process consumes about 1–2% of the total world industrial energy (Erisman *et al.*, 2008) and moreover – due to the fact that agriculture is subjected to losses by leaching, run-off and denitrification – nitrogen pollution is of major environmental concern (Erisman *et al.*, 2013; Galloway & Leach, 2016), often referred to as the nitrogen cascade (Galloway *et al.*, 2003). The last route is very much struggling with the international limits set to prevent overfishing of the world seas.

Ultimately, a substantial fraction of proteins consumed by humans ends up as wastewater ammonium, often referred to as reactive nitrogen; currently about 20 million tons, which is expected to increase to over 35 million tons by 2050. For more than 100 years now, emphasis has been placed on removing the reactive nitrogen from wastewater by oxidizing it back into its elemental, non-reactive form (N_2) by means of biological oxidation, predominantly by the conventional activated sludge process (Jenkins & Wanner, 2014). Recent innovations such as anammox (Kartal *et al.*, 2010) and aerobic granular sludge (Pronk *et al.*, 2017) decrease the energy requirements of nitrogen removal and may even result in energy self-sufficient wastewater treatment plants. While these innovations can be considered a step forward, the working principle remains unchanged: the dissipation of reactive nitrogen to N_2. Yet generally, the concept that there is sufficient nitrogen in the air (i.e., about 80% of air is nitrogen gas – N_2) has dominated the mindset in urban water management. The requirement for 2–3 L of fossil fuel equivalent to produce 1 kg of reactive Haber–Bosch nitrogen, and another 2 L of fossil fuel per kg wastewater nitrogen is needed to subsequently dissipate that nitrogen and return it to the atmosphere, has largely been disregarded.

The time has come to recover reactive nitrogen embedded within the wastewater matrix as microbial biomass rich in proteinaceous components. This chapter examines the possibility to recover nitrogen from wastewater in its reactive form, coupled with upgrading the recovered reactive nitrogen into microbial proteins. This microbial-based protein can, depending on the quality and type of the treated wastewater, subsequently be used as a food, as a feed or as an organic nitrogen fertilizer, contributing to a more sustainable nitrogen cycle. Alternatively, the recovered nitrogen can also be used as fertilizer as an alternative to Haber–Bosch nitrogen fertilizer, or as feedstock in the chemical industry (e.g., in the Denox process).

9.2 LEARNING OBJECTIVES

At the completion of this chapter you should be able to:

- Describe the current status of nitrogen management in urban water management and its relevance in relation to the global nitrogen cycle.
- Understand the limitations of conventional nitrogen dissipation technologies in wastewater treatment.
- Explain the fundamental principles of microbial-based protein production by means of autotrophic, organotrophic, and phototrophic microbial growth.
- Describe the key design criteria and performance parameters for aerobic growth of microbial cells using recovered nitrogen in different wastewater matrices.
- Characterize the process and explain the key technological challenges and limitations for the production of microbial protein using recovered reactive nitrogen.
- Evaluate the different microbial protein production processes used to recover the nitrogen from various wastewater streams against conventional nitrogen removal processes in terms of process economy, process robustness and economy of scale of the produced microbial protein.

9.3 CONCEPTUAL OVERVIEW PRODUCTION OF MICROBIAL PROTEIN USING RECOVERED NITROGEN

9.3.1 Fundamental principles of the microbial conversion

In general, microbial growth involves chemical transformation coupled to energy generation (catabolism), with energy stored chemically (generally as ATP) to enable cell synthesis (anabolism). The dominant mode is coupled chemical oxidation and reduction with electrons transferred from one molecule (i.e., the electron donor) and this molecule becomes oxidized. These electrons are transferred to an acceptor (i.e., electron acceptor) which becomes reduced. The process of anabolism may be oxidative or reductive, depending on the substrate. Specifically, the electron state of the substrate needs to change to match that of the microbial biomass. There are different types of microorganisms that can use different types of electron donors. In this chapter, two key microbial synthesis routes most relevant to production of microbial protein using recovered nitrogen from wastewater are described in detail. These are the *organotrophic* and *autotrophic* microbial cell synthesis through aerobic fermentation.

Very importantly, independent of the synthesis route, catabolism must be reduced to the minimum necessary for energy generation, since anabolism leads to generation of the desired product and catabolism is a cost factor. At increasing sludge ages, additional oxygen consumption is required for endogenous respiration (Van Haandel & Van Der Lubbe, 2012). It is therefore vitally important that the system is operated at the shortest possible sludge age in order to minimize catabolism and maximize cell yields.

9.3.1.1 Organotrophic microbial cell synthesis through aerobic fermentation

The most straightforward and easiest way to produce microbial protein from an engineering perspective is the use of organotrophic bacteria. Organotrophic bacteria directly use soluble organic compounds present in the growth medium, in this case wastewater. To make chemistry more transparent, the conversions are written at the level of the electron. Since no bacterium can be 'charged', one must always come to electron neutrality. As an example, consider that a bacterium is supplied with acetate, a common organic compound present in wastewater, as an energy source. This gives the following equations:

The electron donor reaction:

$$CH_3COO^-(aq) + 3H_2O(aq) \rightarrow CO_2(g) + HCO_3^-(aq) + 8H^+(aq) + 8e^- \tag{9.1}$$

The electron acceptor reactions:
For energy generation:

$$2O_2(g) + 8H^+(aq) + 8e^- \rightarrow 4H_2O(aq) \tag{9.2}$$

For cell biosynthesis using ammonia as nitrogen source (Rittmann & MacCarty, 2001):

$$5CO_2(g) + 1NH_3(aq) + 20H^+(aq) + 20e^- \rightarrow C_5H_7O_2N(s) + 8H_2O(aq) \tag{9.3}$$

According to Equation (9.3), the microbial biomass formed (approximated by $C_5H_7O_2N$) has a molecular weight of 113 g. The nitrogen present in this microbial biomass is considered to be present in the form of protein. Protein contains 16% nitrogen on weight basis (known as the Jones factor). Hence if the nitrogen content is known, the protein content can be calculated: protein content = nitrogen-N × 6.25 (100 divided by 16). If one mole of microbial biomass is produced at a molecular weight of 113 g, the biomass will contain 14 g of nitrogen, therefore the amount of protein produced equals 14 × 6.25 = 87.5 g of protein. This represents 77% of the dry weight in this example. Let us consider the aerobic production of 'young' bacterial cells at very low sludge ages (i.e., 1–2 days maximum). Under these conditions, the cell maintenance metabolism becomes negligible. Consequently, under such conditions for every electron equivalent generated approximately half goes to producing energy and the other half goes to growth of cellular biomass (Park *et al.*, 2015), one then obtains the following overall balance:

$$\begin{aligned} &5CH_3COO^-(aq) + 5O_2(g) + 1NH_3(aq) \rightarrow C_5H_7O_2N(s) \\ &\quad + 5HCO_3^-(aq) + 3H_2O(aq) \end{aligned} \tag{9.4}$$

At both sides of the equation, the ions and all chemical elements are equated. In this example five moles of acetate (MW 59), representing 5 × 59 = 295 g of organic matter, are used to generate 1 mole of biomass (MW 113). Thus, a cell yield of 113/295~0.38 or about 40% on a dry matter basis is obtained. Under conditions of stress (insufficient oxygen levels, limitation of essential nutrients or at long(er) cell residence times (i.e., in wastewater treatment this is referred to as sludge age)), this yield factor may become lower. It should be noted that different values are obtained in case one uses a more oxidized substrate (e.g., formic acid) or a more reduced substrate (e.g., ethanol). As a rule of thumb however, 0.4–0.5 is a good value for produced microbial biomass at short sludge ages (i.e., 1–2 days).

Organotrophic microorganisms in general can use a wide variety of electron donors. Besides a very common product such as acetate used in the example of above, they use all kinds of carbohydrates as energy source: sugars, starches, cellulose but also lipids and hydrocarbons. They can also use proteins and resynthesize them to their own cellular proteins. The provided oxygen functions as electron acceptor. All these substrates result in a cell synthesis efficiency of about 50% under optimal growth conditions and short sludge ages. The fact that microbes can use a wide range of organic compounds is important as in wastewater a large variety of organics can be present. In this respect, microbes are extremely effective compared to, for instance, insects or higher animals that achieve only some 10–25% of conversion efficiency. Indeed, microbes have no such complex instruments such as eyes and ears to make and can focus on their rather straightforward cellular components, that is a cell wall, a cell membrane, a nucleus and the cytoplasm. The latter is the major protein component of the microbial cell. It should be emphasized that the above described cell yields considers only exogenic respiration and no endogenic respiration. In practice, this can only be achieved when the sludge age is very low (i.e., 1–2 days). At higher sludge ages, the endogenic respiration is generally greater than the exogenic respiration, which can substantially lower the overall cell yield (Van Haandel & Van Der Lubbe, 2012).

Besides using ammonia as a nitrogen source, bacteria can also use nitrogen gas. Under such conditions of nitrogen fixation, they must spend a lot of energy to convert the N_2 to ammonia. Actually, microbial nitrogen fixation is far less efficient than the Haber–Bosch process. Therefore, it makes sense to use the industrial process to fix atmospheric nitrogen to reactive nitrogen and 'keep the unused' reactive nitrogen (that ultimately ends up in wastewater) in its reactive form and use the latter in a microbial fermenter to produce microbial protein. Microbes can also use nitrate and nitrite as electron acceptors instead of oxygen. However, this form of electron acceptor is not suitable for microbial protein production due to cost, and it is only used where nitrate or nitrite are to be removed to produce nitrogen gas rather than assimilation of nitrogen into cell biomass.

9.3.1.2 Autotrophic microbial cell synthesis

Plants can fix CO_2 by using sunlight, and thereby convert the inorganic carbon to plant biomass. In Equation (9.3), it appears that the microbial cell synthesis also starts from CO_2, but this is not the case for the so-called organotrophic microorganisms. In their biochemistry, organotrophic microorganisms do not break down the organic compounds fully to CO_2 but rather they recover small units of organics and integrate them in their own complex building units. However, there is a group of microorganisms capable of growth without utilizing organic molecules. These organisms are autotrophic – that is, can generate their own organics (mainly from CO_2). Two groups of autotrophic bacteria can be distinguished, namely (i) the chemo-lithotrophic organisms and (ii) photo-lithotrophic organisms.

9.3.1.2.1 The chemo-lithotrophic organisms

Chemo-lithotrophic microorganisms use a chemical reaction to generate energy. The most prominent example is hydrogenotrophic bacteria, which utilize the reaction shown in Equation (9.5):

$$H_2(g) \rightarrow +2H^+(aq) + 2e^- \tag{9.5}$$

Hydrogenotrophic bacteria are of particular interest since the current trend is evolving towards a hydrogen driven economy. The hydrogen, for instance generated by means of water electrolysis using renewable energy (i.e., green electricity), can be used to split water by electrolysis into hydrogen and oxygen. These two gases thus represent the major part of the energy present in the electricity (at a conversion efficiency of approximately 70% of the electrolysis process using state-of-the-art fuel cell technology). Moreover, one electron equivalent in hydrogen combustion releases 143 kJ, which is about 30% higher than that of methane (i.e., 111 kJ per electron equivalent). The energy present in the combination of oxygen and hydrogen, when offered in a reactor system to hydrogenotrophic bacteria, allows the latter to grow by using CO_2 and NH_3 as given in Equation (9.6):

$$\frac{1}{4}CO_2(g) + \frac{1}{20}NH_3(aq) + H^+(aq) + e^- \rightarrow \frac{1}{20}C_5H_7O_2N(s) + \frac{2}{5}H_2O(aq) \tag{9.6}$$

Actually, they grow, expressed in terms of electron equivalents handled as indicated above, with yields, which are of the same order of organotrophs. Indeed 1 mole of hydrogen requires half a mole of oxygen:

$$H_2(g) + \frac{1}{2}O_2(g) \rightarrow H_2O(aq) \tag{9.7}$$

Thus, 2 g of hydrogen equals to 16 g of oxygen, in other words 16 g of COD equivalents. On the basis of hydrogen expressed as COD, the hydrogenotrophic bacteria, although they indeed have to build up their biomass starting from CO_2, attain yields in the order of 0.3 g of biomass per gram COD converted (expressed in 100% dry weight) (Matassa *et al.*, 2015b). The overall reaction equates to (Ishizaki & Tanaka, 1990):

$$21.36\,H_2 + 6.21\,O_2 + 4.09\,CO_2 + 0.76\,NH_3 \rightarrow C_{4.09}H_{7.13}O_{1.89}N_{0.76} + 18.70\,H_2O \tag{9.8}$$

9.3.1.2.2 Photo-lithotrophic organisms

Purple phototrophic bacteria (PPB) are a diverse group of anoxygenic phototrophs that are mostly Gram-negative facultative anaerobes spread throughout the phylogenetic tree of bacteria with many subdivisions, particularly within the *Proteobacteria*. This group of bacteria is characterized by an extremely diverse metabolism including anaerobic photoheterotrophic and photoautotrophy with light, aerobic chemoheterotrophy or chemoautotrophy in the dark as well as fermentation (Tabita, 1995). As opposed to oxygenic photosynthesis, these bacteria cannot utilize H_2O as electron donor to reduce CO_2 to cell materials. Instead, they utilize a range of lower potentials electron donors such as hydrogen sulphide (H_2S), thiosulphate ($S_2O_3^{2-}$) ferrous iron (Fe^{2+}) and several others for photoautotrophy, and organics such as volatile fatty acids (e.g., acetic, propionic, butyric acid), alcohols and some sugars for photoheterotropy. Consequently, PPB do not generate oxygen during photosynthesis and this process is therefore referred to as anoxygenic photosynthesis. This process is further differentiated from oxygenic photosynthesis by the absorption maxima at which photons from sunlight are harvested. While photosynthetic active radiation (PAR), for example trees and algae, designates the spectral range of solar radiation from 400 to 700 nm, absorbed by chlorophyll, PPB utilize bacteriochlorophyll (BChl) and can absorb photons in the near infra-red >800 nm (and at lower wavelengths, e.g. 375 and 590 nm for BChl a) (Overmann & Garcia-Pichel, 1998) to drive a proton motive force to produce energy in the form of ATP. The generation of energy via photons in combination with photoheterotrophic growth makes PPB an interesting mediator for wastewater treatment. In fact, the organics and nitrogen can in theory be completely recovered as microbial biomass in case one has optimal COD/N ratios (Suhaimi *et al.*, 1987). Due to the high yield, nitrogen and phosphorous are required as macronutrients to support growth which allows for simultaneous and non-destructive removal of organics, nitrogen and phosphorous from wastewater and the option to recover these resources as PPB biomass.

For cell biosynthesis using different organic donors and ammonia as nitrogen source the following reactions can be written for acetic (Equation (9.9)) and butyric acid (Equation (9.10)):

$$\begin{aligned}1.5\,CH_3COOH(aq) + 0.48\,NH_3(aq) &\rightarrow 2.67\,CH_{1.8}O_{0.38}N_{0.18}(s)\\ &+ 0.33\,CO_2(g) + 1.32\,H_2O(aq)\end{aligned} \tag{9.9}$$

$$\begin{aligned}1.5\,C_4H_8O_2(aq) + 0.6\,CO_2(g) + 1.19\,NH_3 &\rightarrow 6.6\,CH_{1.8}O_{0.38}N_{0.18}(s)\\ &+ 1.84\,H_2O(aq)\end{aligned} \tag{9.10}$$

Based on Equation (9.9), 1 kg of acetic acid could generate approximately 0.66 kg of PPB biomass. The formed biomass contains approximately 70% protein (~11% N on a dry weight basis), which is similar to the protein content for organotrophic microbial cell synthesis. The latter implies that more nitrogen can be recovered as microbial protein than the organotrophic route. Photoautotrophic PPB growth can occur through diverse metabolic pathways, including through inorganic electron donors such as hydrogen sulphide (H_2S) (Equation (9.11)). Elemental sulphur can also serve as electron donor with sulphate or thiosulphate as oxidized products (Equation (9.12)). In these reactions, an overall cell composition of $CH_{1.6}N_{0.2}O_{0.4}$ was assumed (Van Gemerden, 1968):

$$CO_2(g) + 2.1\,H_2S + 0.2\,NH_3 \rightarrow CH_{1.6}O_{0.4}N_{0.2} + 2.1\,S^0 + 1.6\,H_2O \tag{9.11}$$

$$CO_2(g) + 0.5\,H_2S + 0.2\,NH_3 + 0.4\,H_2O \rightarrow CH_{1.6}O_{0.4}N_{0.2} + 0.5\,H_2SO_4 \tag{9.12}$$

Oxidation of H_2S to elemental sulphur results in biomass yields of 0.31 g cell dry matter (CDM) per g H_2S (or 0.25 g COD g COD^{-1}), while oxidation of H_2S through to sulphate results in biomass yields of 1.3 g CDM per gram H_2S (or 1.0 g COD g COD^{-1}). The latter is because the oxidation of H_2S to elemental sulphur (Equation (9.13)) and sulphate (Equation (9.14)) involves two and eight electrons, respectively:

$$H_2S \rightarrow S^0 + 2H^+ + 2e^- \tag{9.13}$$

$$H_2S \rightarrow 4\,H_2O \rightarrow SO_4^{2-} + 10H^+ + 8e^- \tag{9.14}$$

The processes described above can occur simultaneously in a single environment, such as a bioreactor treating wastewater. This is one reason why anaerobic cultures with mixed culture PPB do not generate the characteristic rotten egg smell of H_2S. When diverse mixed culture microbial populations are present, such as those consortia commonly in industrial bioreactors, there are bacteria that can provide CO_2 for VFA assimilation which enables direct CO_2 fixation rather than CO_2 generation from the VFA itself, as described in Equation (9.9).

The most critical process parameter (and bottleneck hindering practical implementation) of the production of PPB is the inherent light requirement of phototrophic bioreactors. Light penetration is generally very poor through aqueous cell suspensions (particularly long wavelengths), limiting the active reactor volumes to very narrow fields around the illuminated surface. This represents a major challenge and is, for example, one of the most critical bottlenecks hindering successful commercialization of algae-based, photobioreactor wastewater treatment technologies (Posten, 2009). The same issues arise for PPBs. Another critical challenge to address is the energy requirement when applying artificially illumination, which are in the order of ~10 kWh per m^3 wastewater treated (Hülsen *et al.*, 2018). In order to become viable in the future, energy requirements need to be reduced, which in terms of radiation can potentially be achieved by the utilization of sunlight. However, natural light/dark cycles are challenging for continuous removal performance. Another major cost factor is the harvesting and drying of the biomass, which has been reported to be up to 20–30% of the operational costs for algal systems (Molina Grima *et al.*, 2003). This is mainly caused by rather diluted cultures

(~1.0 g/L) and the small size of algal cells (5–50 μm). Considering the above-described challenges, it is evident that it remains uncertain whether PPB can become a viable concept for the production of microbial protein from wastewater.

9.3.2 Application and design

The processes to produce microbial protein from recovered nitrogen has to focus on the following factors: the origin of the wastewater (i.e., is the nitrogen 'mixed' with, e.g., fecal matter, pathogens, metals or other unwanted compounds), the yield of the protein, the amount generated per unit reactor volume per day (i.e., volumetric production rate), the harvesting and downscale processing of the biomass and finally the quality and end-use of the biomass produced.

9.3.2.1 The working microbiome

Ideally, one would aim to work only under very specific conditions, for instance in a sterile reactor with sterile input materials (purchased chemical grade reactive nitrogen and electron donors) and well-defined microbial strains. Such conditions are implemented in conventional industrial microbiology. However, in the case of upgrading recovered nitrogen from wastewater at the wastewater treatment plant, at best one can work under conditions which resemble those of making artisanal cheese. Thus, the incoming material is seeded with a starter culture which has been selected over time for the fact that it provides a good end-product and subsequently the environmental conditions are controlled so that the proper mixed association of microbes (also called the microbiome) brings about the conversion in a desired way. In the following sections, the most promising microbial protein production pathways are described in more detail. For every kg of nitrogen to be upgraded, a readily biodegradable organic carbon source is needed. The overall stoichiometry is approximately 20–25 kg COD equivalent per kg nitrogen. The conversion is based on aerobic microbial metabolism (see Equations (9.1)–(9.4)). Of every kg of COD supplied, approximately half is oxidized to CO_2 and H_2O. Hence the supply of oxygen is critical (about 0.5 kg oxygen needs to be transferred to the cells per kg carbohydrate converted). The energy generated by these catabolic processes is used for anabolic processes, that is to synthesize new microbial cells which will have a typical composition of young microbial cells (i.e., a dry matter consisting of 70% protein, 5% minerals (of which some 2% are phosphate) with the remainder being exo and endo polymers with a composition relating to sugars and lipids.

Besides the supply of an appropriate electron donor and oxygen as an electron acceptor, the ambient conditions for the microbiome should also be appropriate. Normally, one operates at temperatures in the 20–40°C range; higher is possible but the biomass yields tend to be lower outside that range. Normally, the pH is controlled within the range of 5–8 but more extreme ranges can used to exclude certain microbial species. Indeed, by imposing very acidic or alkaline conditions one can (even in non-sterile open reactor systems) disfavour unwanted microbes, such as for instance enteric bacteria. Similarly, light of specific wavelength can be utilized to select for desired bacteria, for example the supply of infrared radiation to outcompete algae and favour purple phototrophic bacteria in a photobioreactor. As described above, PPB biomass yields are higher (e.g., 66% for acetic acid based on stoichiometry) compared to organotrophic organisms and the reactive nitrogen upgrade is enhanced to around 1 kg N for every 13 kg of COD.

9.3.2.2 The volumetric production rate

The key feature of industrial production is thus to achieve a maximal conversion of electron donor per unit volume and per unit time. This determines the volumetric reactor substrate loading rate and relates to the kg biomass (CDM) generated per m^3 reactor per day, that is the volumetric production rate. In conventional treatment systems, the focus is on removing the last bit of organic matter (or at least to below discharge limits) from the wastewater and to obtain a minimal amount of residual microbial biomass, that is excess waste activated sludge. Hence, in waste treatment systems, one operates at low loading rates (of the order of approximately 1 kg organic matter per m^3 per day) and long reactor cell residence times of 20 days or more. Microbial biomass which has been maintained at a low food supply level for a residence time of several weeks is characterized by old cells which are low in protein and exopolysaccharides and hence can be harvested relatively easily because they dewater quite well. However, in the case of nutrient recovery by means of microbial protein production, high volumetric loading rates of at least 10 kg organic matter (or COD equivalent) per m^3 per day and cell residence times of only a few days need to be used to obtain young cells high in protein. The latter, however, can be quite difficult to harvest and to dewater. This is a key feature in the downstream processing of microbial protein, as at present it incurs high operational costs because the produced microbial biomass needs to be almost completely dewatered (i.e., >95%) in order to become a viable product with a market value.

9.3.2.3 Oxygen supply

Aerobic production is normally based upon oxygen as the electron acceptor. Indeed, oxygen is the most economical of all electron acceptors. One kilogram of oxygen dissolved in water has a cost of approximately €0.05. A principal factor in the reactor design is the transfer of oxygen to the water phase. As in used water treatment, a variety of aeration mechanisms can be used such as propellers, fine bubble distributors, jets, membranes, high-pressure reactors. According to Equation (9.4), to generate 1 kg of microbial cell dry weight, about 1 kg of oxygen input is generally required. This cost factor is not excessive, but a good supply of oxygen is connected with a second factor, namely the intensity at which the reactor is performing. It is important to realize that the aerobic oxygen uptake rates in these biomass production processes are much higher than those commonly found in wastewater treatment processes, which are aimed at 'cleaning up' and typically in the order of 40–60 mg/L.h (Van Haandel & Van Der Lubbe, 2012). In fact, in pure culture industrial fermentation often industrial grade pure oxygen is used, which incurs higher operational costs, albeit still only representing a relatively small part of the overall production cost.

9.3.2.4 Infrared irradation

Note that in case phototrophic bacteria are used, these have the benefit that they work under anaerobic conditions. However, they require substantial energy input in the form of light where the growth can be achieved with infra-red light only, but additional wavelength can be harvested, for example 375 and 590 nm for bacteriochlorophyll a (Overmann & Garcia-Pichel, 1998), which is relevant when growing cultures outside with sunlight (or filtered sunlight). A maximum light conversion efficiency (LCE) between 3.7 and 25.6% has been given by several authors and summarized by Adessi and De Philippis (2014), where LCE tend to be at the higher end when calculated using a specific wavelength, for example 860 nm, rather than integrating spectral ranges. This rather broad range shows that the real efficiency has to be determined specifically for given conditions, substrates and wavelengths. In any case the illuminated surface to volume ration is a crucial design parameter where typical volume ratios are 80–100 m^2/m^3 for algal systems (Posten, 2009), and in order to guarantee economic production estimated capital and operational costs may not exceed €40/m^2, which is currently several times higher.

9.3.2.5 Pumping and mixing

Fermentation reactors require energy for sufficient mixing and pumping as well as to dissolve the oxygen (and hydrogen and methane in case these gaseous compounds are used as electron donors). The use of these gaseous substrate requires substantial mixing in order to create high gas-to-liquid transfer conditions, which is a crucial factor in obtaining high volumetric productivity and also increases the absorption of these gaseous compounds. The latter is very important in terms of process economics considering that they represent a large fraction of the operational expenditure. Energy for pumping and mixing can be assumed to be on the order of 0.2–3 kWh per cubic meter of water handled (Niazi & Brown, 2015).

9.3.2.6 Dewatering, drying and conservation

After the fermentation step, the produced microbial biomass can undergo heat treatment, in case it is destined to be used as animal feed. Heat treatment can be carried out for multiple reasons, namely: (i) lyse cells thereby increasing the protein accessibility; (ii) decrease the nucleic acid content of the microbial biomass; and (iii) obtain a dry and quality assured end-product (Anupama & Ravindra, 2000). Prior to this, the microbial biomass is first dewatered to lower the water content by centrifugation (or other conventional dewatering methods) to lower the energy consumption for the heat-treatment step. Normally, dry solids contents of maximum 25 wt% can be achieved in this way. Spray-drying with integrated fluidized bed, a common practice in the food processing industry (Chen & Mujumdar, 2009), can be used for the drying step. The energy costs for this step are quite substantial, that is about 3500 MJ per cubic meter of water (Chen & Mujumdar, 2009).

9.3.3 Direct assimilation of recovered nitrogen into microbial biomass

In most practical situations, the recovered nitrogen is upgraded by supplying the bacteria with low value organic carbon sources like starch and organic acids that are present in various wastewater streams from the food and beverage industry and distillery side-streams. It is of crucial importance to select for a 'microbial team' (i.e., a microbiome). To the degree possible, rather than processing one type of molecule, a suitable microbiome converts the overall assembly of organics present in the wastewater to cellular biomass. The microbiome must consist of several members that complement one another so that the metabolites of one species are further used by the other species. Indeed, the final mixed liquor, when subjected to cell harvesting, should give rise to water which is low in residual nutrients (i.e., N, P) and organic carbon (cells and soluble organics) on the one hand and a harvest of biomass rich in nutritious microbial protein on the other hand.

Today, the major challenges in this respect are issues relating to the type of microbiome needed and the ways to maintain and support the microbiome in order to produce a microbial biomass of consistent quality and composition. Indeed, first a selection of species must be assembled. The focus must be on the fact that the team members should grow at high rates (with doubling times of only a couple of hours) and consume the N and electron donor down to low residual levels. The residual level of mineral nitrogen in the medium is easy to monitor; also the amount of residual soluble organics can be easily quantified (through COD or solids analysis). In addition, the oxygen uptake rate is a helpful tool to control the oxidation capacity of the microbiome. The constitution of the generated microbial biomass is more difficult to select for. It is highly advantageous for the further processing of the biomass if one can obtain a coagulation/flocculation of the single microbial cells into flocs that readily separate by sedimentation, filtration, or centrifugation forces. Indeed, one should try to obtain a harvested biomass at a level of some 5–10% dry matter content to achieve further handling in the form of a thick slurry of biomass. The latter can then be either preserved as such (e.g., by bringing the pH below 4.0 as done in the pickling of foods or ensiling of feeds) or by further drying to 85% dry matter or more. In this respect, the current R&D is focusing on the development of proper coagulating agents which are effective in the harvesting process and, moreover, which are of food/feed grade quality.

Wastewater from the food and beverage industry has one distinct advantage over domestic wastewater in terms of production of microbial protein, namely, it is free of fecal matter, pathogens and toxic metals. Consequently, there is no need for a nitrogen separation and purification step, allowing direct assimilation into microbial biomass. Moreover, wastewater streams from the food and beverage industry usually also have higher concentrations of ammonium and organics. The engineering is thus less challenging. Crucial will be to find the optimum COD/N/P ratio in order to stimulate microbial biomass high in protein. In fact, it turns out that in many cases, many waste streams are ammonium deficient, as shown in Table 9.1 below. As such, although it seems counterintuitive, one needs to add nitrogen to the wastewater!

Table 9.1 Overview of industrial wastewater streams free of fecal matter from the food and beverage industry and their COD to nitrogen ratio, adapted from Pikaar *et al.* (2017b).

Type of Industrial Wastewater	COD (mg/L)	Total N (mg/L)	COD/N*	Maximum Attainable Protein Content[a] (wt%)	Ref.
Dairy	4000	55	100:1.37	17	Kasapgil *et al.* (1994)
Dairy	4500	60	100:1.33	16	Koyuncu *et al.* (2000)
Dairy	4000	60	100:1.50	18	Koyuncu *et al.* (2000)
Dairy	1745	75	100:4.30	53	Koyuncu *et al.* (2000)
Dairy	18 045	329	100:1.82	22	Arbeli *et al.* (2006)
Dairy	4000	55	100:1.37	17	Ince (1996)
Dairy	2800	140	100:5	62	Schwarzenbeck *et al.* (2005)
Cheese	4430	18	100:0.41	5	Monroy *et al.* (1996)
Yoghurt and buttermilk	1500	63	100:4.2	52	Koyuncu *et al.* (2000)
Beverage	1750	28.4	100:1.62	20	Amuda and Amoo (2007)
Distillery	150 000	6000	100:4	50	Mohana *et al.* (2009)
Distillery (raisins)	57 500	750	100:1.30	16	Vlissidis and Zouboulis (1993)
Distillery (wines)	27 500	650	100:2.36	29	Vlissidis and Zouboulis (1993)
Distillery (figs)	35 400	880	100:2.48	31	Vlissidis and Zouboulis (1993)
Brewery	4000	52.5	100:1.31	16	Driessen and Vereijken (2003)
Sugar industry (beet)	6300	53.23	100:0.84	10	Güven *et al.* (2009)
Olive oil mill	40 000–220 000	300–1200	100:0.54–100:0.75	6–9	Azbar *et al.* (2004)
Olive oil mill	40 000–195 000	500–15 000	100:0.77–100:1.25	9–15	Sierra *et al.* (2001)
Palm oil mill	50 000	750	100:1.5	18	Ahmad *et al.* (2006)

[a]COD/N ratio where no external nitrogen is added to the wastewater.

9.3.4 Case studies and implementation

9.3.4.1 Industrial production of microbial protein

Microorganisms have always been central in basic feed and food processing techniques, for instance converting fibres into edible food when fermenting dough to produce bread, or milk into cheese, allowing its long-term preservation. They have often been used as a direct food source, as is the case for yeast or algae. The latter, together with bacteria, constitute the microbial actors involved in processing food. They can also be used directly as a feed or food source. The term microbe is used here in the broad connotation of bacteria, fungi, yeast and algae. Some 40 years ago, Imperial Chemical Industries (London, UK) developed a single cell protein (SCP) generated from methanol using the bacterium *Methylophilus methylotrophus*. The product received the name Pruteen and contained up to 70% protein and was used in pig feed. Pruteen, however, could not compete with cheaper animal feeds that were available at the end of the 1970s and production was discontinued. In recent years, methane is gaining interest as a substrate for the producing of microbial biomass. UniBio A/S (utilizing knowledge gained by Dansk BioProtein A/S (Odense, Denmark)) has developed fermentation technology to convert natural gas to animal feed protein by using methanotrophic bacteria. UniBio A/S UniBio A/S uses a U-loop fermenter to achieve a productivity of 4 kg m^{-3} h^{-1}, producing UniProtein® with ~70% protein, which has been approved for use in animal feed. The U-loop fermenter is designed to enhance mass transfer rates of methane from the gas to the liquid phase, making more methane available for the biomass. Another industrial producer using methane is Calysta Inc. (California, USA). It recently opened a production facility for their product, FeedKind®, in the UK in 2016 and is partnering with Cargill (Minnesota, USA) to build a larger production facility in the USA. The fact that natural gas is used means that ultimately such a production pathway does not provide a long-term sustainable solution. However, the use of methane generated in digesters at WWTPs seems a promising route that warrants further exploration. Clearly, in recent years, research and development around microbial protein is regaining momentum both in the scientific and industrial domains. The steep increase in the prices of fishmeal (from about €500 per ton in the 1990s to €1500–2500 in recent years), together with the environmental pressure of soybean production on land and water use in the tropical areas of the globe, justify the examination of the production of microbial protein in general and, in particular, in combination with using recovered nitrogen as a more sustainable and climate independent alternative to fishmeal. Indeed, at present, there are several ongoing initiatives at pilot and demonstration, which are described in more detail below.

9.3.4.2 Microbial protein production from used water

There are two main lines of emerging concepts that produce microbial protein from water containing organic waste materials enriched in respect to nitrogen:

(1) The direct assimilation of inorganic nitrogen into microbial protein and its on-site subsequent use as feed in aquaculture ponds, so-called the Biofloc technology (Avnimelech, 2015) or the ValPromic concept (http://avecom.be/product/promic-microbial-protein).
(2) The selective stripping of ammonia from the water and upgrading the latter to microbial biomass in a separate reactor. The latter, so-called power-to-protein, is a recent initiative for production of microbial protein from domestic wastewater (https://www.powertoprotein.eu/).

Biofloc technology – Water quality control and direct assimilation of nitrogen in aquaculture. Intensive, high-density fish (shrimp, other animals) production leads to the release of high concentrations of waste products (i.e., feed residues, fish excretions and dead algae). Unlike terrestrial animals, aquatic animals are totally immersed in this waste stream that often contains toxic components, a situation leading to the collapse of the system. One means to solve this situation included the replacement of the used water by fresh clean water. However, this solution is presently not acceptable due to the waste of clean water, introduction of pathogens and environmental pollution by the disposed water. The biofloc technology, in a way, can be seen as the ultimate circular nitrogen approach (Figure 9.1). This technology is based on a minimal to zero water exchange which leads to the build-up of organic substrates in the water and intensive development of microbial population (10^6–10^9 cells/mL). This population develops in a way that degrades and metabolizes the organic matter, including deleterious organic metabolytes (Avnimelech, 2015). Under aerobic conditions, approximately 50% of organic carbon metabolyzed is converted to CO_2 and eventually dissipated to the atmosphere. However, most of the mineralized nitrogen accumulate in the water, since under conditions existing in the pond, only a small portion of this nitrogen is converted into volatile nitrogenous species and released to the atmosphere. A large percentage of the excess nitrogen left in the water is made of ammonium, a highly toxic component to animals in the pond, nitrite, another toxic component, and nitrate. The biofloc process further deals with the use of the excess nitrogen to produce microbial proteins. The inorganic nitrogen excreted by the fish (or, e.g., shrimps) are directly assimilated into microbial protein by heterotrophic bacteria and are being used in-situ as protein rich feed for the fish, thereby creating a closed-loop system. This is controlled through the addition of carbohydrates (molasses, flour of

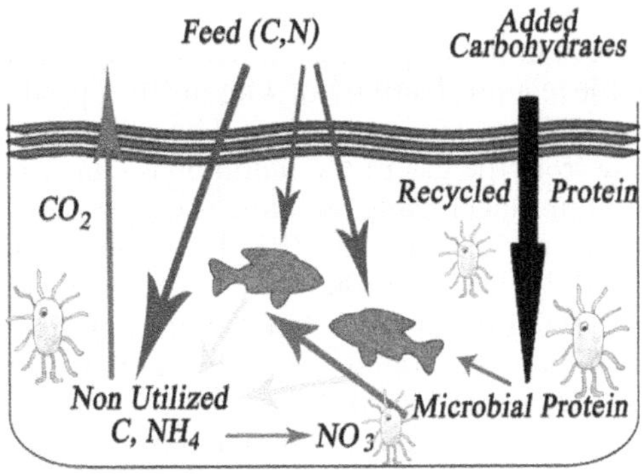

Figure 9.1 Schematic overview of the working principle of the biofloc technology.

wheat, rice or corn, and others) to the water. About 20 g carbohydrate are needed for the assimilation of 1 g nitrogen in protein (about 6.25 g carbohydrate to produce 1 g protein) (Avnimelech, 1999). The carbohydrates can be added separately to conventional feed provided to the pond (traditionally in the range of 30–50% protein), or by using a feed containing less protein. It is possible to reliably control the level of inorganic nitrogen through the control of the C/N ratio of feed and feed additives added to the pond (normally to a C/N ratio of 15–20). The microbial protein produced, present in the water in the form of bioflocs that can be easily harvested, serves as a good source of protein to important commercially-grown fishes (tilapia, carps, catfish) and shrimps. Using ^{15}N as a tracer to the microbial protein, it was found that tilapia and shrimp can consume up to 50 and 25% of their protein requirement, respectively (Avnimelech & Kochba, 2009; Burford *et al.*, 2004). This can be an important means to reduce production costs to the farmers, by saving both protein and feed amounts.

The assimilation of nitrogen into microbial protein, essentially representing a recycling of non-utilized feed protein, are being practiced in a commercial scale in ponds of varying sizes (ranging from 100 to ~10 000 m^2). The present development of biofloc technology systems all over the world may contribute to very significant nitrogen recycling in aquatic feed production (Avnimelech, 2015).

ValPromic Technology. The promic concept, a patented concept by Avecom (Ghent, Belgium) (http://avecom.be/product/promic-microbial-protein), is based on direct assimilation of nitrogen (and phosphate) present in the wastewater by a mixed culture of heterotrophic bacteria (Figure 9.2). A key difference with the biofloc concept is that wastewater from the food and beverage industry is treated without the opportunity to directly use the produced microbial protein on-site. As such, as explained in more detail in section 3.3, it requires a dewatering, drying and sterilization step. Moreover, it requires the use of a fermentation reactor, further adding complexity and capital investment. By the end of 2018, the first full-size demonstration plant of Valpromic is expected to commence operation at a total capacity of about 5000 ton microbial protein per year from potato process water (see Figure 9.3).

Power-to-Protein. For treatment of domestic wastewater, a very interesting route for upgrading recovered nitrogen is to combine it with hydrogen and oxygen generated by water electrolysis and with CO_2 from biogas, the so-called power to protein concept (https://www.powertoprotein.eu/). A schematic overview of the concept is provided in the figure below.

The key difference is the need for a 'bullet-proof' barrier between the domestic wastewater (and associated pathogens), from which the reactive nitrogen is recovered, and the microbial production step. The latter can be achieved by ammonia stripping, and in some instances one could even add a membrane step to further increase regulator and consumer confidence. The organics present in the wastewater are used to produce biogas. The produced biogas can subsequently be used to produce a clean hydrogen gas streams by means of a Combined Cooling Heating and Power unit (CCHP) fed to the fermentation reactor. As the amount of hydrogen gas produced in this way at best can only account

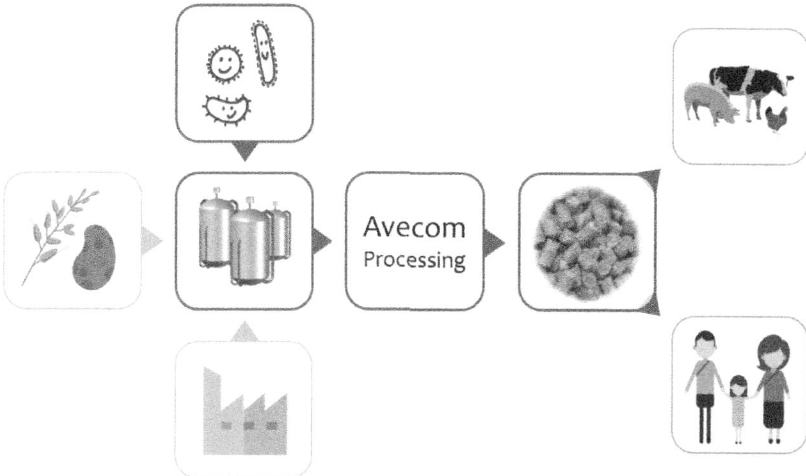

Figure 9.2 Schematic representation of Valpromic (Source: Personal communication Avecom).

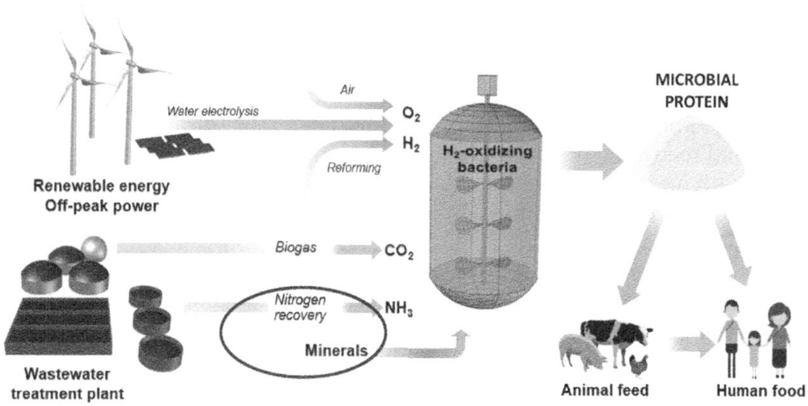

Figure 9.3 Schematic representation of the production of microbial protein using hydrogenotrophic bacteria from domestic wastewater using recovered nitrogen.

for some 12% of the total hydrogen requirements, in order to upgrade all the nitrogen that theoretically can be recovered from the wastewater, additional hydrogen is produced by means of water electrolysis driven by renewable energy. The upgrading of stripped nitrogen is currently at pilot scale explored by the association of KWR (Nieuwegein, The Netherlands) and Avecom (Ghent, Belgium) (https://www.kwrwater.nl/en/projecten/power-to-protein-pilot-phase-2/). Clearly this domain has quite a way to go before it becomes of comparative quantitative significance.

9.4 CHALLENGES, OPPORTUNITIES AND RESEARCH NEEDS

Processes to recover nitrogen from wastewater are not new at all. In fact, several mature technologies with a proven track record have been available on the market for several decades, such as struvite precipitation (http://ostara.com/nutrient-management-solutions/) and ammonia stripping (http://

www.nijhuisindustries.com). However, the economic competitiveness of such technologies are restricted to higher ammonium concentrations such as side-stream process treating digestate. Technologies that can recover reactive nitrogen from domestic wastewater directly from the main-stream at an economic competitive price are not available yet. Moreover, it is evident that technologies to recover nitrogen should always have energy requirements that are lower than the energy requirement to generate the reactive nitrogen via the Haber–Bosch process. The latter, however, is in most circumstances not the case with the energy consumption of current and emerging concepts being higher compared to the Haber–Bosch process (Matassa *et al.*, 2015a). As mentioned before, production of microbial protein from wastewater free of fecal matter does not require a separation step and the challenges are mainly related to the microbial production step, which is described in more detail below.

9.4.1 The microbial production step

Two aspects deserve special attention, namely: (i) the management of the functional microbiome; and (ii) the harvesting and downstream processing of the biomass.

The management of the functional microbiome. With respect to the management of the functional microbiome, the following aspects are considered important (Pikaar *et al.*, 2017b):

- *Steering the process towards creating suitable microbiomes*: the key feature in producing microbial protein from recovered nitrogen from wastewater is the use of proper microbiomes, in contrast to using 'pure' cultures as practiced in industrial biotechnology. While pure cultures are ideal from a product stability and quality point of view, such conditions are very hard to maintain. Therefore, it is important to gain a fundamental understanding about the dynamics and relative abundances of the different species within the microbiome.
- *Overcoming the process limitations of microbiomes*: the disadvantages of working with microbiomes and recovered nitrogen that originates from wastewater are considerable and must be addressed with an open mindset. First, the operator must create a set of conditions in terms of pH, dissolved oxygen, temperature, hydraulic and cell residence time, and specific biomass loading rate so that the microbiome achieves good conversion yields.
- *Creating a stable end-product of high quality*: it is of crucial importance that the end-product, that is the microbial biomass which is harvested, has a constant composition in terms of microbial protein content, crude amino acid composition and quality (i.e., no unwanted microbes present). Albeit the microbiome does not have to be homogenous and constant in terms of microbial composition, the overall quality of the end-product in terms of parameters such as percentage of protein, digestibility, amino acid composition and nucleic acid content has to be stable. At present, there is a lack of fundamental understanding on how to steer the microbiome in time so that the mixed culture aerobic fermentation reactors can operate in such a stable way that the overall quality of the end-product is stable. Obviously, this quality also implies that the final end-product microbiome should have metabolites (e.g., colour, odour, taste) that are attractive to the consumer, yet free of allergens at all times. Moreover, at all times, the microbiome needs to contain only species that comply with the status of GRAS (generally regarded as safe). Normally, the regulator will impose a set of detection methods for unwanted microorganisms, which are based on specific targeting by plate cultivation or quantitative PCR. For more insight, in-depth genomic analysis is also possible. However, currently, there is very little knowledge on how to install advanced and rapid monitoring of such dynamic microbiomes.

9.4.2 The harvesting of the microbial produced microbial cells

Microbial biomass needs optimal growth conditions for rapid growth, that is doubling times of hours to a couple of days. Only under such conditions is the cellular biomass very rich in digestible protein. The matrix in which the microbes are suspended has to have a low density and salt content (i.e., conductivity of 10–25 mS/cm). This results in quite dilute suspensions in the order of 5–10 kg CDM per m^3. Therefore, the cells have to concentrate with a factor of at least 10 and the water that is separated should be of such quality that it can be re-used or discharged at very low cost. Clearly, the separation is a critical factor and requires considerable development, for instance applying food grade flocculants can improve this step. Once the thick slurry is obtained, it can be further processed in several ways. It can be stabilized by decreasing the pH and used in liquid feeding. It can also be dried to a final product to be used for food, feed or as an organic fertilizer. Yet, the handling of the 5–10% slurry should be of such a nature that the material is not allowed to deteriorate in quality and it must furthermore preserve the quality of the proteinaceous compounds. Drying can cause the conversion of the proteins to Maillard products which can alter the taste, odour and digestibility. The major challenge is to remove the water at low costs. The use of techniques to rapidly and safely dry the harvested microbial biomass by using low value (waste) heat is an issue of critical importance.

9.4.3 Economic and environmental competitiveness

The aerobic fermentation process used to recover the nitrogen has to be evaluated against other processes in terms of economy and environmental impact. Normally, to remove 1 kg of reactive nitrogen from domestic wastewater by means of the conventional nitrification, the overall costs for aeration and electron donor to denitrify are of the order of €2–4 per kg reactive nitrogen. In the case of microbial production, it is clear that the overall costs will be substantially higher. Hence, the key feature for microbial protein to become a viable alternative in the future is to produce an end-product of high-quality that has a high market value similar to fish meal. Due to the microbial growth, 1 kg of nitrogen recovered from the wastewater becomes about 6.25 kg of microbial protein (or 8.9 kg CDM at a crude protein content of 70%).

Considering a WWTP with a capacity of 250 000 PE, this equals to almost 30 tons of microbial protein per day. This microbial biomass can subsequently enter the value chain. The value of food protein (100% active substance such as Quorn(TM)) is at present of the order of €3–30 per kg. That of feed protein in the form of fish meal is of the order of €1–2 Euro per kg. That of organic fertilizer is of the order of some €0.2 per kg. Hence, in the case of food and feed, upgrading of nitrogen could potentially be rewarding. However, as the costs for reactive nitrogen only represent a small fraction in the overall production costs for microbial protein, one needs to consider very carefully in which case the implementation of recovered nitrogen will be beneficial. Another key feature is of course that the microbial-based protein is indeed fully accepted in the market and can be valorized at such prices relative to other sources of quality protein. In the case of slow-release organic fertilizer, the overall economy appears negative. Yet, the use of slow-release organic fertilizer is expected to represent environmental benefits compared to the use of inorganic fertilizer such as a reduction in nitrogen emissions, increased soil organic carbon content and water holding capacity. In case the microbial is used as animal feed, it is expected to come with several environmental benefits, reduction in greenhouse gas emissions due to land use change (LULUC), pesticides, water use, soil erosion, nitrogen and phosphorus pollution (Pikaar *et al.*, 2017a). Life Cycle Analysis allows to scout for these potential unexpected environmental impacts.

9.5 CHAPTER SUMMARY

In this chapter, we first described the environmental limitations of nitrogen dissipation pathways in wastewater management. Subsequently, we indicated the need to recover the ammonium from wastewater in its reactive form. In particular, we have examined the opportunities to recover ammonium from industrial and domestic wastewater streams and to upgrade it to microbial biomass, rich in protein (i.e., microbial protein) through the implementation of an aerobic fermentation process. The microbial biomass generated in such aerobic conversion can, depending on the quality, be used as a protein rich supplement in livestock husbandry as well as aquaculture or as slow-release organic fertilizer. Such forms of upgraded mineral N have can help to alleviate the global nitrogen pollution burden. It should be noted that provided the microbial protein is produced from clean food grade inputs, it can be used as a meat replacement in human consumption, similar to other microbial biomass products such as Quorn™. Clearly, such upgrade requires a careful dialogue with the consumer and regulator in order to create sufficient consumer confidence and regulatory acceptance.

There are different aerobic fermentation approaches that can be used using different types of microorganisms. In this chapter, we have described aerobic microbial fermentation using organotrophic, autotrophic bacteria, chemo-lithotrophic and anaerobic fermentation using photo-lithotrophic bacteria, all of which are relevant for various used water treatment applications. We have also discussed the key differences in terms of design criteria, operation and maintenance and challenges for each of these microbial protein production pathways. An important difference between the various approaches described is the wastewater origin in terms of presence or absence of fecal matter, pathogens and waterborne viruses. In case the wastewater matrix is considered 'clean', which is the case for specific process waters originating from the food and beverage industry, one can adopt direct assimilation approaches. In case of fecal contaminated wastewater like domestic waster, a strict barrier, for example via a gas phase, between the wastewater from which the reactive nitrogen is recovered and the microbial production step needs to be implemented. Finally, we have described the main challenges, opportunities and research needs to achieve a widespread implementation of microbial protein production from various process and used waters.

9.6 EXERCISES

Exercise 9.1: Consider the following stoichiometry for the oxidation of methane:

$$CH_4 + 2H_2O \rightarrow CO_2 + 8H^+ + 8e^-$$

Methane is a very stable molecule that provides little energy for cell growth; typically only 20% of available energy is utilized for cell growth. Develop overall mass balance equations for a process where methane is used as an electron donor for biomass production and anomia is used as a nitrogen source and a negligible decay rat. Note: assume biomass stoichiometry of $C_5H_7O_2$ N (s).

Exercise 9.2: Consider the example where bacteria grow using starch as an electron donor and ammonium as a nitrogen source. The cell yield is 0.5 in terms of electrons going respectively to energy (i.e., catabolism) and to cell synthesis (i.e., anabolism). Develop overall mass balance equations for this process. Given the mass balance equations, calculate the amount of starch needed in order to produce 10 tons per day of microbial protein (on a dry solids basis). Note: assume biomass stoichiometry of $C_5H_7O_2$ N (s).

Exercise 9.3: Consider the following:

(a) If the typical composition of hydrogenotrophic biomass is $CH_{1.74}O_{0.46}N_{0.19}$. Present the elemental compositions as mass fractions. If all nitrogen is present as protein, determine the protein content as a fraction of CDM.
(b) Hydrogen gas is to be generated using electrolysis at a rate of 20 kg H_2 gas per MWh electricity. If a 50 MW hydrolysis cell is used to generate feed for the production of protein biomass, determine the maximum rate of biomass production per hour for a cell yield of 0.3 kg CDM per kg COD-H_2.
(c) For the process in (b), determine the mass of nitrogen that must be supplied for complete uptake of H_2 gas.
(d) If (i) hydrogen at a rate of 20 kg H_2 gas per MWh electricity and nitrogen is produced using Haber–Bosch at an energy cost of 10 MWh/ton N and energy costs are €50/MWh, determine if this process is feasible when bulk microbial protein is valued at €1/kg CDM.

Exercise 9.4: Consider the following:

(a) The typical composition of PPB is given as $CH_{1.8}O_{0.38}N_{0.18}$. Write a balanced chemical equation for the oxidation of PPB and use this equation to determine the COD to CDM ratio of PPB.
(b) If the typical composition of purple phototrophic biomass is $CH_{1.8}N_{0.18}O_{0.38}P_{0.02}$. Present the elemental compositions as mass fractions. If all nitrogen is present as protein, determine the protein content as a fraction of CDM.
(c) Research shows that a photobioreactor with a light intensity of 50 W m^{-2} will produce purple phototrophic bacteria at an average areal productivity of 20 g CDM/m^2.d. Estimate the illuminated surface of the photobioreactor required to produce enough protein to sustain 600 average people (assume 70 kg/person). What is the energy requirement to light this reactor?

Exercise 9.5: Assume that 1 kg of organic matter dry weight represents approximately 1 kg chemical oxygen demand (COD) (in reality this differs somewhat depending on the type of organics, e.g., fats or sugars), which represents the amount of oxygen needed to oxidize the organic matter present. Generally 1 kg of COD, such as sugar or starch, when consumed by rapidly growing microbial cells, gives rise to 0.4 kg CDM. The latter, when consisting out of young cells (1–5 days cell residence time), contains approximately 70% protein. Considering the above, how much carbohydrates are required to tie up 1 kg of reactive nitrogen in microbial biomass.

Exercise 9.6: Assume that 1 kg of organic matter dry weight represents approximately 1 kg COD (in reality this differs somewhat depending on the type of organics, e.g., fats, acetate or sugars), which represents the amount of oxygen needed to oxidize the organic matter present. Considering that 1 kg of acetate gives rise to 0.66 kg CDM using PPB. The latter contains approximately 70% protein. Considering the above, how much acetate is required to tie up 1 kg of reactive nitrogen in microbial biomass using PPB.

Exercise 9.7: A soft drink manufacturing process produces 0.5 ML/day of concentrated industrial wastewater with the composition shown in the table below:

Component	Concentration
COD	120 g/L
VFA	70 g/L
Alcohols	20 g/L
TKN	94 mg/L
NH_4-N	30 mg/L
Total P	104 mg/L
PO_4-P	92 mg/L

(a) If no external nitrogen is added to the wastewater (i.e., only nitrogen in the wastewater can be used for growth), calculate: (i) the maximum rate of biomass production; and (ii) the maximum rate of protein production that can be generated each day using fast growing organotrophic biomass (cell yield of 0.38 kg CDM per kg COD and protein content of 70%).
(b) Estimate the dose rate of external nitrogen that must be supplied in order to convert all COD in the wastewater into fast growing organ tropic biomass. Calculate the biomass production under these conditions.
(c) If no external nitrogen is added to the wastewater, calculate: (i) the maximum rate of biomass production; and (ii) the maximum rate of protein production that can be generated each day using purple phototrophic biomass (cell yield of 0.62 kg CDM per kg COD and protein content of 70%).
(d) Estimate the dose rate of external nitrogen that must be supplied in order to convert all COD in the wastewater into PPB. Calculate the biomass production under these conditions.

Exercise 9.8: A meat processing plant produces 3 ML/day of industrial wastewater with an average daily composition shown in the table below:

Component	Concentration
COD	5500 mg/L
VFA	1200 g/L
TKN	460 mg/L
NH_4-N	230 mg/L
TP	120 mg/L
PO_4-P	80 mg/L

(a) If wastewater discharge costs are €1.0/kg COD, €2.0/kg N and €2.5/kg P, calculate the discharge costs without wastewater treatment.
(b) If the wastewater is to be treated using PPB, calculate: (i) the maximum rate of biomass production; and (ii) the maximum rate of protein production that can be generated each day (cell yield of 0.62 kg CDM per kg COD and protein content of 70%).
(c) Using a biomass composition of $CH_{1.8}N_{0.18}O_{0.38}P_{0.02}$, assume complete conversion of the limiting component and estimate the COD, N and P composition of treated wastewater after production and harvesting of PPB.
(d) Using an areal productivity of 20 g CDM/m^2.d, calculate the required photobioreactor area for this process.
(e) Initial cost benefit assessments estimate the capital costs for the PPB process at €40/m^2, biomass production costs are estimated at €1/kg CDM and costs to harvest and process the biomass are estimated at €2/kg CDM. The biomass product is valued at €2/kg CDM. If the plant life is 20 years, and the hurdle return on investment is 15%, determine the preliminary economic feasibility of the process.

Exercise 9.9: The wastewater generated in a potato factory has the following typical composition: 10 g/L of starch, NH_4-N = 0.5 g/L and ortho-P = 0.1 g/L. This water, subjected to inoculation with a proper seed culture and aeration in a reactor will in a time period of 2–3 days convert the starch, ammonium and phosphate present in the water to microbial biomass. Actually, some 4 g CDM per L will be obtained (yield: 0.4 kg CDM per kg starch converted). Assume a starch removal efficiency of the process of 96%. Considering the above-described COD/N/P ratio, yield and typical biomass composition of young cells, calculate the expected effluent concentrations in terms of N and P that can be achieved through assimilation of the N and P into microbial biomass.

9.7 DISCUSSION QUESTIONS

Question 9.1: What is the large scale potential of microbial protein? (palm oil industry): Consider the case for which palm oil waste is upgraded by adding recovered ammonium sulphate N. The latter is obtained at zero cost because it is the result of a stripping process of gases and the level of the nitrogen in the recovered product is variable which makes it difficult to directly apply as a crop fertilizer. Consider that the reactor operates at a loading rate of 10 kg COD per m^3 reactor per day. Also take into account that the palm oil waste has a tipping fee of €100 per ton. What is the cost of microbial biomass-based fertilizer under conditions where downstream processing of the biomass (separation/drying) can be considered to be covered by heat generated by other processes within the factory.

Question 9.2: The public acceptance of microbial protein from recovered nitrogen? (public acceptance, cultural differences): Recovered nitrogen will with some part of the public raise questions in terms of acceptability, depending on its final use, food, animal food or slow release fertilizer. To convince the consumer, discuss the following number of other processes in which used nitrogen is recycled in the food chain. Examples are crop fertilized with manure or compost, mushrooms grown on chicken manure, the biofloc technology used to upgrade fish feces to microbial protein, for example.

Question 9.3: Worth the trouble? (sustainability on a planetary level). Of the total amount of nitrogen entering the biosphere (soil/water) every year by means of the Haber–Bosch process, percentage wise only a relative small fraction ultimately ends up in domestic wastewater. We can develop technology to recover this amount and upgrade it into a valuable protein source or slow release fertilizer, but considering the complexity of this concept compared to the conventional activated sludge process and/or recent emerging technologies, such as anammox and Nereda, is it worthwhile to invest in creating such a paradigm shift in urban wastewater management? In your discussion, take into account that worldwide a substantial part of the wastewater is discharged untreated back into the environment. Should we invest in those problems first?

Question 9.4: Worth the trouble of taking such a risk? (utility management, technology): As the innovation manager of a large water utility, you are in charge of reorganizing the existing water infrastructure from its current situation to a more circular approach within a timeframe of 15 years. You have heard of the production of microbial protein using recovered nitrogen (as schematized in Figure 9.1). You are enthusiastic about such an approach but you are wondering why you would take such a risk? Your job is to ensure that the wastewater is treated to below discharge limits, thereby protecting the environment and the community, which is difficult enough. You are asked to give a presentation to the board of directors in which you evaluate the current status and justify your masterplan. Would you focus on the microbial production as the key innovative process or would you focus on less break-through technology concepts. What are your key considerations/motivations with respect to your decision?

Question 9.5: What is the economic potential (economics, market potential): As described in this chapter, a key factor governing production performance is the rate by which the bacteria consume the electron donor and use it to grow. The microbiome must at all instances assure a good rate of conversion because the capital investment on the hardware is quite high. If, for instance, a reactor costs in capital investment some €5000 per m^3 installed reactor capacity, and it generates some 10 kg of end-product (microbial protein cell dry weight) per day at a gross margin of €1 per kg protein, then the payback for the capex at 220 production days per year is in the order of 2200: $(220 \times 10 \times 1) = 2.3$ year. Industrial investors often prefer payback periods of that order or less than that. Note that this is a rough estimate, not taking into account the operational cost and also assuming a very high gross margin per kg protein produced. Assume you are the business developer of an international company in the food and beverage industry that wants to improve the sustainability of their wastewater facilities worldwide. Evaluate the economic potential and practical feasibility of the full-scale implementation of: (i) organotrophic production; (ii) photo-lithotrophic; or (iii) chemo-lithotrophic microorganisms. Clearly indicate the key design criteria, the limitations of each of the routes, and estimate the expected operational costs.

FURTHER READING

Hülsen T., Barry E. M. and Lu Y. (2016). Low temperature treatment of domestic wastewater by purple phototrophic bacteria: performance, activity, and community. *Water Research*, **100**, 537–545.

Liu B., Song J. and Li Y. (2013). Towards industrially feasible treatment of potato starch processing waste by mixed cultures. *Applied Biochemistry and Biotechnology*, **171**(4), 1001–1010.

Matassa S., Batstone D. J. and Hülsen T. (2015a). Can direct conversion of used nitrogen to new feed and protein help feed the world? *Environmental Science & Technology*, **49**(9), 5247–5254.

Matassa S., Boon N. and Verstraete, W. (2015b). Resource recovery from used water: the manufacturing abilities of hydrogen-oxidizing bacteria. *Water Research*, **68**, 467–478.

Matassa S., Boon N. and Pikaar I. (2016a). Microbial protein: future sustainable food supply route with low environmental footprint. *Microbial Biotechnology*, **9**(5), 568–575.

Matassa S., Verstraete W. and Pikaar I. (2016b). Autotrophic nitrogen assimilation and carbon capture for microbial protein production by a novel enrichment of hydrogen-oxidizing bacteria. *Water Research*, **101**, 137–146.

Pikaar I., Matassa S. and Rabaey K. (2017a). Microbes and the next nitrogen revolution. *Environmental Science & Technology*, **51**(13), 7297–7303.

Pikaar I., Matassa S. and Rabaey K. (2017b). *The Urgent Need to Re-Engineer Nitrogen-Efficient Food Production for the Planet*. DNC2017 Position Paper. UNU-FLORES, Dresden.

Ritala A., Häkkinen S. T., Toivari M. and Wiebe M. G. (2017). Single cell protein-state-of-the-art, industrial landscape and patents 2001–2016. *Frontiers in Microbiology*, **8**, 1–18.

Verstraete W., Clauwaert P. and Vlaeminck S. E. (2016). Used water and nutrients: Recovery perspectives in a 'panta rhei' context. *Bioresource Technology*, **215**, 199–208.

REFERENCES

Adessi A. and De Philippis R. (2014). Photobioreactor design and illumination systems for H2 production with anoxygenic photosynthetic bacteria: a review. *International Journal of Hydrogen Energy*, **39**(7), 3127–3141.

Ahmad A. L., Sumathi S. and Hameed B. H. (2006). Coagulation of residue oil and suspended solid in palm oil mill effluent by chitosan, alum and PAC. *Chemical Engineering Journal*, **118**(1–2), 99–105.

Amuda O. S. and Amoo I. A. (2007). Coagulation/flocculation process and sludge conditioning in beverage industrial wastewater treatment. *Journal of Hazardous Materials*, **141**(3), 778–783.

Anupama and Ravindra P. (2000). Value-added food: single cell protein. *Biotechnology Advances*, **18**(6), 459–479.

Arbeli Z., Brenner A. and Abeliovich A. (2006). Treatment of high-strength dairy wastewater in an anaerobic deep reservoir: analysis of the methanogenic fermentation pathway and the rate-limiting step. *Water Research*, **40**(19), 3653–3659.

Avnimelech Y. (1999). Carbon/nitrogen ratio as a control element in aquaculture systems. *Aquaculture*, **176**(3–4), 227–235.

Avnimelech Y. (2015). *Biofloc Technology: A Practical Guidebook*. World Aquaculture Society, Baton Rouge, LA, USA.

Avnimelech Y. and Kochba M. (2009). Evaluation of nitrogen uptake and excretion by tilapia in bio floc tanks, using 15N tracing. *Aquaculture*, **287**(1–2), 163–168.

Azbar N., Bayram A., Filibeli A., Muezzinoglu A., Sengul F. and Ozer A. (2004). A review of waste management options in olive oil production. *Critical Reviews in Environmental Science and Technology*, **34**(3), 209–247.

Bodirsky B. L., Popp A., Lotze-Campen H., Dietrich J. P., Rolinski S., Weindl I., Schmitz C., Müller C., Bonsch M., Humpenöder F., Biewald A. and Stevanovic M. (2014). Reactive nitrogen requirements to feed the world in 2050 and potential to mitigate nitrogen pollution. *Nature Communications*, **5**, 1–7.

Bodirsky B. L., Rolinski S., Biewald A., Weindl I., Popp A. and Lotze-Campen H. (2015). Global food demand scenarios for the 21th century. *PLoS ONE*, **10**(11), 1–27.

Burford M. A., Thompson P. J., McIntosh R. P., Bauman R. H. and Pearson D. C. (2004). The contribution of flocculated material to shrimp (litopenaeus vannamei) nutrition in a high-intensity, zero-exchange system. *Aquaculture*, **232**(1–4), 525–537.

Chen X. D. and Mujumdar A. S. (2009). *Drying Technologies in Food Processing*. Wiley, Hoboken.

Driessen W. and Vereijken T. (2003). *Recent Developments in Biological Treatment of Brewery Effluent*. Livingstone, Zambia.

Erisman J. W., Sutton M. A., Galloway J., Klimont Z. and Winiwarter W. (2008). How a century of ammonia synthesis changed the world. *Nature Geoscience*, **1**(10), 636–639.

Erisman J. W., Galloway J. N., Seitzinger S., Bleeker A., Dise N. B., Petrescu A. M., Leach A. M. and de Vries W. (2013). Consequences of human modification of the global nitrogen cycle. *Philosophical Transactions of the Royal Society of London. Series B, Biological Sciences*, **368**(1621), 1–9.

Galloway J. N. and Cowling E. B. (2002). Reactive nitrogen and the world: 200 years of change. *Ambio*, **31**(2), 64–71.

Galloway J. N. and Leach A. M. (2016). Your feet's too big. *Nature Geoscience*, **9**, 97–98.

Galloway J. N., Aber J. D., Erisman J. W., Seitzinger S. P., Howarth R. W., Cowling E. B. and Cosby B. J. (2003). The nitrogen cascade. *BioScience*, **53**(4), 341–356.

Godfray H. C. J., Beddington J. R., Crute I. R., Haddad L., Lawrence D., Muir J. F., Pretty J., Robinson S., Thomas S. M. and Toulmin C. (2010). Food security: the challenge of feeding 9 billion people. *Science*, **327**(5967), 812–818.

Güven G., Perendeci A. and Tanyolaç A. (2009). Electrochemical treatment of simulated beet sugar factory wastewater. *Chemical Engineering Journal*, **151**(1–3), 149–159.

Hülsen T., Hsieh K., Tait S., Barry E. M., Puyol D. and Batstone D. J. (2018). White and infrared light continuous photobioreactors for resource recovery from poultry processing wastewater – a comparison. *Water Research*, **144**, 665–676.

Ince O. (1996). Performance of a two-phase anaerobic digestion system when treating dairy wastewater. *Water Research*, **32**(9), 2707–2713.

Ishizaki A. and Tanaka K. (1990). Batch culture of Alcaligenes eutrophus ATCC 17697T using recycled gas closed circuit culture system. *Journal of Fermentation and Bioengineering*, **69**(3), 170–174.

Jenkins D. and Wanner J. (2014). *Activated Sludge – 100 Years and Counting*. IWA Publishing, London, UK.

Kartal B., Kuenen J. G. and Van Loosdrecht M. C. M. (2010). Sewage treatment with anammox. *Science*, **328**(5979), 702–703.

Kasapgil B., Anderson G. K. and Ince O. (1994). An investigation into the pre-treatment of dairy wastewater prior to aerobic biological treatment. *Water Science and Technology*, **29**(9), 205–212.

Koyuncu I., Turan M., Topacik D. and Ates A. (2000). Application of low pressure nanofiltration membranes for the recovery and reuse of dairy industry effluents. *Water Science and Technology*, **41**(1), 213–221.

Matassa S., Batstone D. J., Huelsen T., Schnoor J. L. and Verstraete W. (2015a). Can direct conversion of used nitrogen to new feed and protein help feed the world? *Environmental Science & Technology*, **49**(9), 5247–5254.

Matassa S., Boon N. and Verstraete W. (2015b). Resource recovery from used water: the manufacturing abilities of hydrogen-oxidizing bacteria. *Water Research*, **68**, 467–478.

Mohana S., Acharya B. K. and Madamwar D. (2009). Distillery spent wash: treatment technologies and potential applications. *Journal of Hazardous Materials*, **163**(1), 12–25.

Molina Grima E., Belarbi E. H., Acién Fernández F. G., Robles Medina A. and Chisti Y. (2003). Recovery of microalgal biomass and metabolites: process options and economics. *Biotechnology Advances*, **20**(7–8), 491–515.

Monroy O. H., Vazquez M. F., Derramadero J. C. and Guyot J. P. (1996). Anaerobic-aerobic treatment of cheese wastewater with national technology in Mexico: the case of 'El Sauz'. *Water Science and Technology*, **32**(12), 149–156.

Niazi S. K. and Brown J. L. (2015). *Fundamentals of Modern Bioprocessing*. CRC Press, Boca Raton, Florida, USA.

Overmann J. and Garcia-Pichel F. (1998). The phototrophic way of life. In: *The Prokaryotes: An Evolving Electronic Resource for the Microbiological Community*, E. Rosenberg, E. F. DeLong, S. Lory, E. Stackebrandt & F. Thompson (eds), Springer, New York, pp. 203–257.

Park D. H.-D., Chang I.-S. and Lee K.-W. (2015). *Principles of Membrane Bioreactors for Wastewater Treatment*. CRC Press, Boca Raton, Florida, USA.

Pikaar I., Matassa S., Rabaey K., Bodirsky B. L., Popp A., Herrero M. and Verstraete W. (2017a). Microbes and the next nitrogen revolution. *Environmental Science & Technology*, **51**(13), 7297–7303.

Pikaar I., Matassa S., Rabaey K., Laycock B., Boon N. and Verstraete W. (2017b). *The Urgent Need to Re-Engineer Nitrogen-Efficient Food Production for the Planet*. DNC2017 Position Paper. UNU-FLORES, Dresden.

Posten C. (2009). Design principles of photo-bioreactors for cultivation of microalgae. *Engineering in Life Sciences*, **9**(3), 165–177.

Pronk M., Giesen A., Thompson A., Robertson S. and Van Loosdrecht M. (2017). Aerobic granular biomass technology: advancements in design, applications and further developments. *Water Practice and Technology*, **12**(4), 987–996.

Rittmann B. E. and MacCarty P. L. (2001). *Environmental Biotechnology: Principles and Applications*. McGraw-Hill, Boston, Mass.

Schwarzenbeck N., Borges J. M. and Wilderer P. A. (2005). Treatment of dairy effluents in an aerobic granular sludge sequencing batch reactor. *Applied Microbiology and Biotechnology*, **66**(6), 711–718.

Sierra J., Martı E., Montserrat G., Cruanas R. and Garau M. A. (2001). Characterization and evolution of a soil affected by olive oil mill wastewater disposal. *The Science of the Total Environment*, **279**, 207–214.

Suhaimi M., Liessens J. and Verstraete W. (1987). NH+/4-N assimilation by Rhodobacter capsulatus ATCC 23782 grown axenically and non-axenically in N and C rich media. *Journal of Applied Microbiology*, **62**(1), 53–64.

Tabita F. R. (1995). The biochemistry and metabolic regulation of carbon metabolism and CO_2 fixation in purple bacteria. In: *Anoxygenic Photosynthetic Bacteria*, R. Blankenship, M. Madigan and C. Bauer (eds), Springer, the Netherlands, pp. 885–914.

United Nations. (2015). *World Population Prospects: The 2015 Revision, Key Findings and Advance Tables*. Department of Economic and Social Affairs, The United Nations, New York.

Van Gemerden H. (1968). Utilization of reducing power in growing cultures of Chromatium. *Archiv für Mikrobiologie*, **64**(2), 111–117.

Van Haandel A. and Van Der Lubbe J. (2012). *Handbook of Biological Wastewater Treatment*. IWA Publishing, London, UK.

Vlissidis A. and Zouboulis A. I. (1993). Thermophilic anaerobic digestion of alcohol distillery wastewaters. *Bioresource Technology*, **43**, 131–140.

World Health Organization. (2002). *Report of A Joint WHO/FAO/UNU Expert Consultation. Protein and Amino Acid Requirements in Human Nutrition*. United Nations, Geneva, Switzerland.

doi: 10.2166/9781780409566

Chapter 10

Nutrient recovery from water and wastewater

Robert Colston[1], *Stephan Tait*[2], *Céline Vaneeckhaute*[3], *Heidy Cruz*[4], *Ilje Pikaar*[5], *Thomas Seviour*[6], *Johannes B. M. Klok*[7,8], *Jan Weijma*[9], *Henk Dijkman*[10], *Cees J. N. Buisman*[7,9], *Sasha Scattergood*[11], *Ana A. Robles-Aguilar*[12], *Erik Meers*[1], *Fabrice Béline*[13] *and Ana Soares*[1]

[1]*Cranfield Water Science Institute, Vincent Building, Cranfield University, Cranfield, Beds, MK43 0AL, UK*
[2]*Centre for Agricultural Engineering, University of Southern Queensland, Toowoomba, Queensland, 4350, Australia*
[3]*Research team on green process engineering and biorefineries (BioEngine), Chemical Engineering Department, Université Laval, 1065 avenue de la Médecine, Québec, QC, Canada*
[4]*Civil and Environmental Engineering Department, National University of Singapore, 1 Engineering Drive 2, #07-03 E1A, 117576, Singapore*
[5]*The School of Civil Engineering, The University of Queensland, St Lucia, QLD 4072, Australia*
[6]*WATEC Aarhus University Centre for Water Technology, Nørrebrogade 44, Bldg 1783, 8000 Aarhus, Denmark*
[7]*Wetsus, the Netherlands*
[8]*Paqell B.V., the Netherlands*
[9]*Wageningen University & Research, the Netherlands*
[10]*Paques B.V., the Netherlands*
[11]*Anaergia Inc., Burlington, ON, Canada*
[12]*Faculty of Bioscience Engineering, Ghent University, Belgium*
[13]*Irstea – National Institute for Environmental and Agricultural Science and Research, Rennes, France*

10.1 INTRODUCTION

Nitrogen (N), phosphorus (P) and sulfur (S) are essential nutrients and key to food production security, and primary components of fertilizers, for example for agriculture, energy crops (e.g., bioethanol) and aquaculture. They are also used in the pharmaceutical, chemical, automotive and aerospace industries to produce chemicals, medicines, smart devices, aeroplanes parts and car batteries and so on. Hence, nutrients are critical to a number of industries. N, P and S production is costly and resource intensive. For example, the Haber–Bosch process for N fertilizer production represents about 1–2% of the total world energy consumption (Matassa *et al.*, 2015). It uses non-renewable natural gas, and is responsible for an estimated 1.2% of greenhouse emissions worldwide. Moreover, P is a 'disappearing nutrient' and was placed on the Critical Raw Material (CRM) list in 2014 by the European Commission. This is a critical problem in the agricultural sector since P, together with N and potassium, are the main macronutrients for inorganic fertilizers. S is similarly an important agricultural nutrient, atmospheric S deposits continue to decline, and intensification of agriculture has taken up high amounts of S in soils (Hinckley *et al.*, 2020).

Wastewater is a rich source of N, P and S. N enters as ammonium and organic N from urine and industrial discharges, P from phosphates from human and animal wastes, detergents, and food wastes, and S as sulfate or sulfide from chemical coagulants in drinking water production, human wastes, and industrial discharges (Egle *et al.*, 2015; Galloway *et al.*, 2014; Pikaar *et al.*, 2014). The traditional focus of wastewater treatment processes has been to remove nutrients in order to protect public health, assets, and the environment. This is because excessive N and P in water leads to eutrophication which damages aquatic ecosystems (Robles *et al.*, 2020). Moreover, high S loadings in wastewater treatment plants (WWTPs) lead to noxious gaseous emissions, corrosion (e.g., pipes), inhibition of wastewater microbial activities, and acid rain if biosolids or biogas are combusted. Sulfide-induced concrete corrosion has been recognized as the main cause of sewer system deterioration (Pikaar *et al.*, 2014).

WWTPs have typically employed biological approaches to remove N via destructive dissipative processes and a combination of chemical and biological treatments to remove P from municipal wastewaters (McCarty, 2018; Oehmen *et al.*, 2007). S removal from municipal wastewaters is not currently widely implemented. Given the increasing demand in fertilizers to meet global food demands, however, the importance and urgency of recovering nutrients continues to grow. Consequently there has been much interest in technologies that can recover N, P and S from wastewater.

Technologies used to recover N, P and S from wastewater will be discussed in this chapter and include:

- P recovery, specifically by formation of the mineral struvite, from centrate or P-rich agricultural/ industrial wastewaters; and
- N recovery via ammonia removal by stripping and subsequent recovery, as applied to high strength industrial waters (e.g., landfill leachates) and digestate from anaerobic digestion;
- Elemental S recovery by bio-conversion to H_2S, and subsequent absorption and/or oxidation. The same process can be applied to treat industrial off-gases rich in H_2S.

The principles, technologies and case studies for these nutrient recovery processes are presented here, along with relevant challenges and opportunities for N, S and P recovery.

10.2 LEARNING OBJECTIVES

At the completion of this chapter you should be able to:

- Explain the fundamental principles of struvite formation, ammonia stripping/absorption, and sulfur conversion and recovery processes.
- Describe the key design criteria and performance parameters for relevant nutrient recovery processes.
- Characterize and examine specific challenges for struvite precipitation, ammonia stripping and sulfur recovery processes, given specific case studies.
- Evaluate resource recovery options for struvite precipitation, ammonia stripping/absorption and sulfur recovery processes, given a specific application scenario and water composition.

10.3 PHOSPHORUS RECOVERY – THE ESSENTIAL AND DWINDLING NUTRIENT

P is an essential nutrient, and has no substitute in biological functions. Shortages threaten global food security, which are exacerbated by rising population. Moreover, the movement from fossil fuels to biofuels increases P demand, as the growing market for fuel crops further increases the demand for fertilizers (Reijnders, 2014). Technological development has also found use for P in pharmaceuticals, household and energy products, placing further pressure on P supply (Chen *et al.*, 2019; Reijnders, 2014; Yan *et al.*, 2018). It is estimated that the current use of P globally is >30 million tons year^{-1} (Reijnders, 2014).

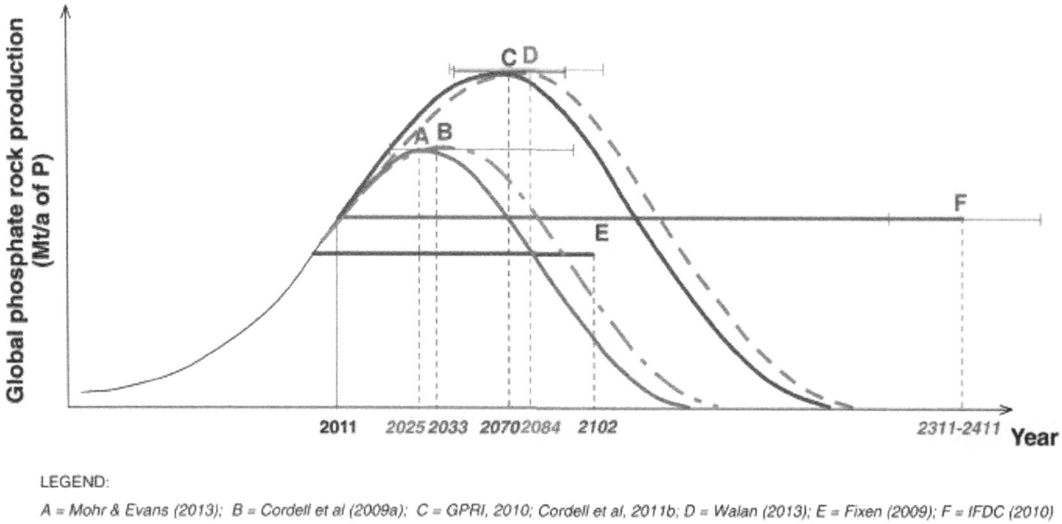

Figure 10.1 Various rock phosphate peak production estimates and depletion rates (with permission from Cordell and White (2015)).

Increased use of fertilizers has led to widespread environmental problems, such as enhanced eutrophication effects which has caused widespread algal blooms. This is detrimental to ecosystems as algal blooms reduce the concentration of nutrients available and the penetration of sunlight into the photic zone. The reduced primary productivity and anoxia that ensue threaten the local food chain. Furthermore, algal blooms can be toxic (Sekula-Wood *et al.*, 2009), and impact on utilities and services, such as bathing water quality control and tourism.

Non-renewable P is predominately refined from the apatite group ($Ca_5(PO_4)_3(OH, F, Cl)$). Apatite deposits are heterogeneous occurring in many geological settings. This poses problems when estimating global P reserves. Current estimates for when P reserves will be depleted and thus vary widely from 100 to 400 years (Figure 10.1). Estimates in phosphate ore reserves range from 70 to 300 billion tons (U.S. Geological Survey, 2019) and research models estimate that there are between 67 and 110 billion tons of refined phosphate (25–30 wt% P_2O_5) (Springer, 2017). The uncertainty about availability contributes to a very volatile ore-born P market and pricing.

A recent census showed EU members import nearly 100% of their phosphate fertilizers (Figure 10.2). To cope with increasing fertilizer demand, the EU have put in place several protocols to reduce the use of P where possible, such as reducing the concentrations of P in washing detergents (De Ridder *et al.*, 2012), thereby relieving pressures on P demand and reducing P input from grey wastewater. Regardless, alternative strategies for reducing dependence on P imports and recycling nutrients are required to ensure food security.

10.3.1 Conceptual overview – struvite precipitation

Active P recovery from wastewater has been implemented for the past 20 years or so in various different forms. The most common P-rich material recovered is struvite. It has been estimated that in 2020 between 990 and 1250 ton P was recovered as struvite in the EU (Muys *et al.*, 2021).

Struvite precipitation occurs by a phase change where the constituent ions, magnesium (Mg^{2+}), ammonium (NH_4^+) and phosphate (PO_4^{3-}), previously dissolved in wastewater, precipitate out to form

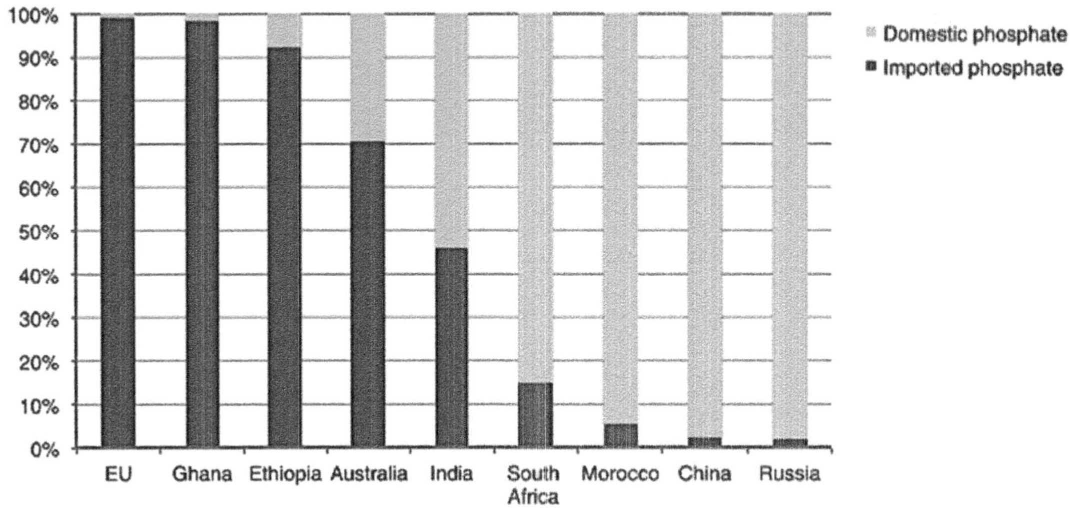

Figure 10.2 Nations' dependence on imported phosphate fertilizers as of 2010 (with permission from Cordell and White (2015)).

a crystalline mineral called struvite ($NH_4MgPO_4 \cdot 6H_2O(s)$). The struvite precipitation reaction is as follows:

$$Mg^{2+}(aq) + NH_4^{+}(aq) + PO_4^{3-}(aq) + 6H_2O \leftrightarrow NH_4MgPO_4 \cdot 6H_2O(s) \quad (10.1)$$

Six water molecules are bound in the crystal structure per mole of struvite formed. The struvite crystals can then be separated from the wastewater, enabling the recovery of N as ammonia and P as phosphate. The usefulness of the struvite process lies in:

- struvite formation upconcentrating N and P from the wastewater liquid phase by as much as 2–3 orders of magnitude into the mineral solid phase; and
- the struvite crystal being highly selective for Mg^{2+}, NH_4^{+} and PO_4^{3-}, resulting in a high struvite purity and excluding contaminants in the bulk wastewater phase.

To form struvite, N and P have to be in mineralized form (i.e., P as phosphate and N as ammonium). Within the water/wastewater context, organic P and organic N are biologically hydrolyzed into phosphate and ammoniacal N, respectively. This often causes maintenance challenges around WWTPs, when conditions become conducive to struvite forming in unwanted places in pipes, pumps and equipment, causing inorganic scale issues requiring frequent maintenance (Figure 10.3) (Le Corre, 2006; Shaddel *et al.*, 2019). However, the intentional minerals precipitation process, as outlined in this chapter, aims to promote struvite formation in a controlled way in a reaction vessel. This has the added benefit of preventing its unwanted formation in other parts of a WWTP.

10.3.1.1 Factors affecting struvite precipitation – equilibrium and solubility

Struvite precipitation is bi-directional reaction (indicated by a bi-directional arrow in Equation (10.2)). This means that the reaction occurs significantly and simultaneously in both the forwards and the backwards directions, resulting in simultaneous formation and dissolution of struvite, respectively. Bi-directional reactions eventually attain equilibrium and then satisfy an equilibrium relationship.

Figure 10.3 Unwanted inorganic scale within water pipework.

The equilibrium relationship for struvite is typically written for the reaction with the mineral phase on the left hand side as follows (the reverse of Equation (10.1) above):

$$NH_4MgPO_4 \cdot 6H_2O(s) \rightleftarrows Mg^{2+}(aq) + NH_4^+(aq) + PO_4^{3-}(aq) + 6H_2O \tag{10.2}$$

with the corresponding equilibrium relationship being:

$$K_{sp,struv} = \left(Mg^{2+}(aq)\right)_{eq} \times \left(NH_4^+(aq)\right)_{eq} \times \left(PO_4^{3-}(aq)\right)_{eq} \tag{10.3}$$

where $K_{sp,struv}$ is the equilibrium constant ($10^{-13.26}$ (Ohlinger *et al.*, 1998)), dependent only on temperature (and to a lesser extent pressure), and $(Mg^{2+}(aq))_{eq}$, $(NH_4^+(aq))_{eq}$, $(PO_4^{3-}(aq))_{eq}$ are chemical activities of the participating ions representing their 'reactive concentrations'. Note that the equilibrium relationship does not include the chemical activities of struvite and water, because these are equal to 1 with wastewater being mostly water and struvite being considered a mostly pure crystalline substance.

The equilibrium relationship formulated in terms of concentrations of ions in solution applies a multiplier-correction as follows:

$$(X) = [X] \times \gamma_i \tag{10.4}$$

where $[X]$ is the concentration of the hypothetical compound X, γ_i is its multiplier correction factor commonly termed an activity coefficient, and (X) is the chemical activity as above. The equilibrium relationship for struvite is then expanded as follows:

$$K_{sp,struv} = \left(\gamma_{Mg} \times \left[Mg^{2+}(aq)\right]_{eq}\right) \times \left(\gamma_{NH_4} \times \left[NH_4^+(aq)\right]_{eq}\right) \times \left(\gamma_{PO_4} \times \left[PO_4^{3-}(aq)\right]_{eq}\right) \tag{10.5}$$

where $[Mg^{2+}(aq)]_{eq}$, $[NH_4^+(aq)]_{eq}$, $[PO_4^{3-}(aq)]_{eq}$ are the concentrations of the respective ions in the wastewater at equilibrium and γ_{Mg}, γ_{NH4} and γ_{PO4} are their respective activity coefficients.

The value of the activity coefficient is commonly estimated by an empirical correlation. One such correlation which is commonly used is the Davies equation, which is affected by the charge or valency (z_x) of a respective constituent (X) as follows:

$$\log_{10}(\gamma_i) = -Az_x^2(I^{1/2}/(1+I^{1/2}) - 0.24I) \tag{10.6}$$

where A is a constant (= 0.5085 $M^{-1/2}$ at 25°C) that depends on the dielectric constant and temperature, and I is the ionic strength of the wastewater liquid solution, calculated as follows:

$$I = 1/2\Sigma_i m_i z_i^2 \tag{10.7}$$

where m_x is the molal concentration (mol/kg) of the respective constituent, which for a typical wastewater can be approximated as its molar concentration (mol/L). The Davies equation appears to be comparable to other more complex activity coefficient models up to $I = 0.2$ molal (Tait *et al.*, 2012). Activity coefficients of neutral ion pairs (see further below) can also be estimated as follows:

$$\log_{10}(\gamma_i) = \alpha_1 I \tag{10.8}$$

where α_1 is 0.1. Following the form of the equilibrium relationship in Equation (10.6), it is common to calculate an ion activity product (IAP) as follows:

$$\begin{aligned} \mathrm{IAP_{struv}} &= \left(\gamma_{Mg} \times [Mg^{2+}(aq)]\right) \times \left(\gamma_{NH_4} \times [NH_4^+(aq)]\right) \\ &\quad \times \left(\gamma_{PO_4} \times [PO_4^{3-}(aq)]\right) \end{aligned} \tag{10.9}$$

where in this case the concentrations of the respective ions may not be at equilibrium.

A mathematical variant of Equation (10.9), commonly termed the saturation index (SI) value, is calculated as follows:

$$\mathrm{SI_{struv}} = \log_{10}\left(\frac{\mathrm{IAP_{struv}}}{K_{sp,struv}}\right) \tag{10.10}$$

SI_{struv} and IAP_{struv} are very important for design and operation of struvite precipitation processes and are also used to predict and prevent inorganic scale formation. If $IAP_{struv} > K_{sp,struv}$ (i.e., $SI_{struv} > 0$), then struvite will precipitate to attain equilibrium (Equation (10.3)). Such conditions are commonly referred to as 'supersaturated'. In contrast, if $IAP_{struv} < K_{sp,struv}$ (i.e., $SI_{struv} < 0$) and if struvite mineral phase is present, then the struvite will want to dissolve to attain equilibrium (Equation (10.1)). Such conditions are commonly referred to as 'unsaturated'.

When $IAP_{struv} = K_{SP,struv}$, then $SI_{struv} = 0$ and conditions are at equilibrium, representing the solubility conditions of struvite in the wastewater.

Equation (10.9) importantly shows that struvite precipitation can be induced by adding any of the participating ions as a chemical reagent. For example, to induce struvite formation to remove and recover PO_4^{3-} and NH_4^+, Mg^{2+} can be added as magnesium chloride, magnesium oxide, magnesium hydroxide or other, to increase IAP_{struv} so that $SI_{struv} > 0$. This is common practice with struvite reactors.

10.3.1.2 Factors affecting struvite precipitation – Ion pairing

The struvite precipitation reaction competes with ion pairing reactions for ions in the wastewater liquid phase. These reactions occur in the wastewater liquid phase and form ion pairs without undergoing a phase change. Some relevant ion pairs that compete with struvite include the reactions:

$$Mg^{2+}(aq) + PO_4^{3-}(aq) \leftrightarrow MgPO_4^-(aq) \tag{10.11}$$

and

$$H^+(aq) + PO_4^{-3}(aq) \leftrightarrow HPO_4^{2-}(aq) \quad (10.12)$$

which compete with struvite precipitation for $Mg^{2+}(aq)$ (Equation (10.11)) and $PO_4^{-3}(aq)$ (Equations (10.11) and (10.12)). This must be considered in the design and operation of struvite precipitation processes. Some ion pairs carry a charge ($MgPO_4^-(aq)$), while others are uncharged ($MgHPO_4(aq)$), and this influences the activity coefficient calculations (Equations (10.6) and (10.8)). Some ion pairing reactions are weak acid-base reactions (Equation (10.12)) and this influences struvite precipitation via pH of a wastewater, by increasing or decreasing the competition with a weak acid-base ion pair. This is illustrated in Figure 10.4 which shows a phase diagram for struvite in a wastewater with phosphate as the limiting reagent. Generally, higher pH tends to favor the formation of struvite because it reduces competition with the weak acid-base phosphate system, specifically for $PO_4^{-3}(aq)$ (Figure 10.4). However, high pH conditions can inadvertently exist in filters and turbulent flow areas, when localized aeration causes CO_2 stripping and a pH increase, amd this is a cause for unwanted struvite scale formation in such locations (Le Corre, 2006).

Fortunately, ion pairing reactions are rapid when compared to the struvite precipitation reaction. Consequently, ion pairing reactions are usually at equilibrium and their impact on struvite precipitation can be readily estimated using well-established aqueous equilibrium models. Such models are available as standalone software packages (e.g., PhreeqC, introduced in this chapter) or have been incorporated into commercial wastewater modelling packages (e.g., BioWin, Sumo19). These models calculate the resulting free ion concentrations $[Mg^{2+}(aq)]_{eq}$, $[NH_4^+(aq)]_{eq}$ and $[PO_4^{3-}(aq)]_{eq}$ at equilibrium with the various relevant ion pairs, and then these concentrations can be used to calculate IAP_{struv} or SI_{struv} via Equations (10.9) and (10.10). These can then be used to determine if a wastewater is supersaturated or undersaturated, and thus whether struvite will precipitate or dissolve.

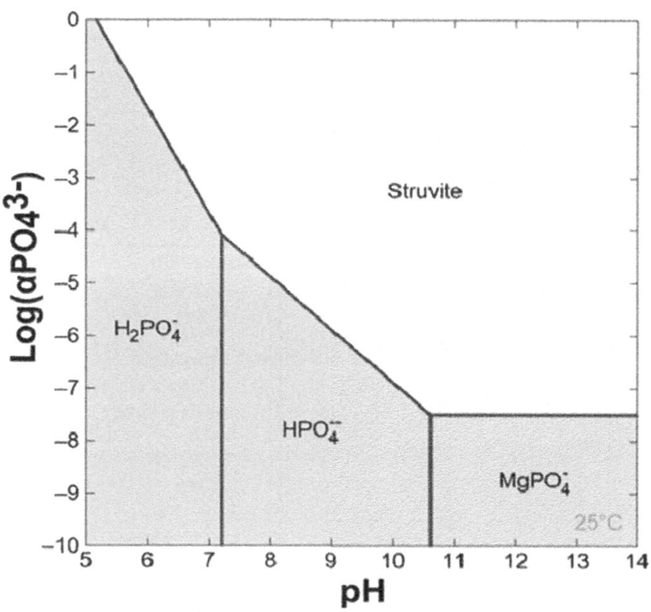

Figure 10.4 Phase diagram for ion pairing and struvite in water, with phosphate as limiting reagent. The diagram illustrates the effect of the phosphate weak acid-base system via pH and the influence of an ion pair with magnesium.

10.3.1.3 Worked exercise 1a

As an illustrated example consider a wastewater containing 200 mg P/L as phosphate, 1000 mg N/L as ammoniacal N (i.e., NH_4^+(aq) + NH_3(aq)), 100 mg Mg/L and 2550 mg/L of choride as measured by an analytical laboratory (these are total concentrations and include all ion pairs and free ions), at pH 6.12 and 25°C. We will use the model software package PhreeqC with the Minteq.dat model database to determine whether this wastewater is supersaturated or undersaturated with respect to struvite.

PhreeqC is available for download at https://www.usgs.gov/software/phreeqc-version-3. PhreeqC is free to use within the constraints of its User Rights Notice, available at https://water.usgs.gov/water-resources/software/PHREEQC/Phreeqc_UserRightsNotice.txt. The wastewater composition above is entered into PhreeqC, and the thermodynamic database used by the model is set to Minteq.dat. Figure 10.5 presents a screenshot of the text-based coding window corresponding to the wastewater composition input.

When the model is run, the following concentrations are calculated:

[Total PO_4*] = 200 mg P/L = 6.482×10^{-3} molal;
[Total Mg*] = 100 mg Mg/L = 4.129×10^{-3} molal;
[Total NH_4*] = 1000 mg N/L = 7.167×10^{-2} molal;

*again, these are total concentrations, including the free-form ion and all ion pairs.

Ion pair and free form ion concentrations are determined by the model, as shown in Figure 10.6.

Figure 10.6 shows that a large proportion of the phosphate and a moderate proportion of the magnesium in the wastewater is paired with other ions as follows:

[PO_4^{-3}(aq)]/[Total PO_4*] = $1.54 \times 10^{-9}/6.482 \times 10^{-3} = 2.376 \times 10^{-7}$

with $H_2PO_4^-$(aq) representing the majority ion pair with phosphate, and

[Mg^{2+}(aq)]/[Total Mg*] = $3.49 \times 10^{-3}/4.129 \times 10^{-3} = 0.845$

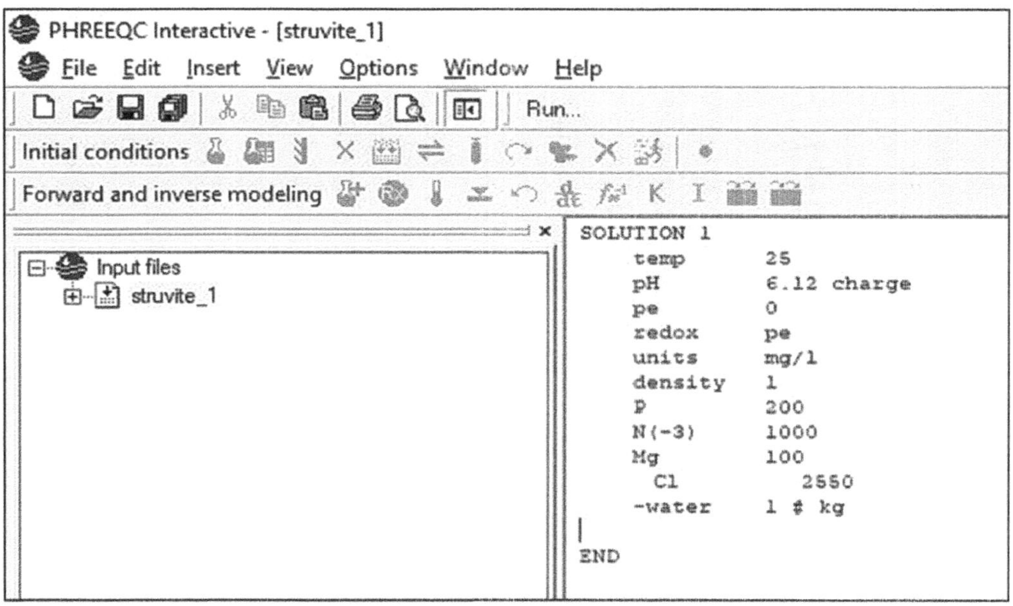

Figure 10.5 Simulation coding window (input) for PhreeqC simulation – Step 1a.

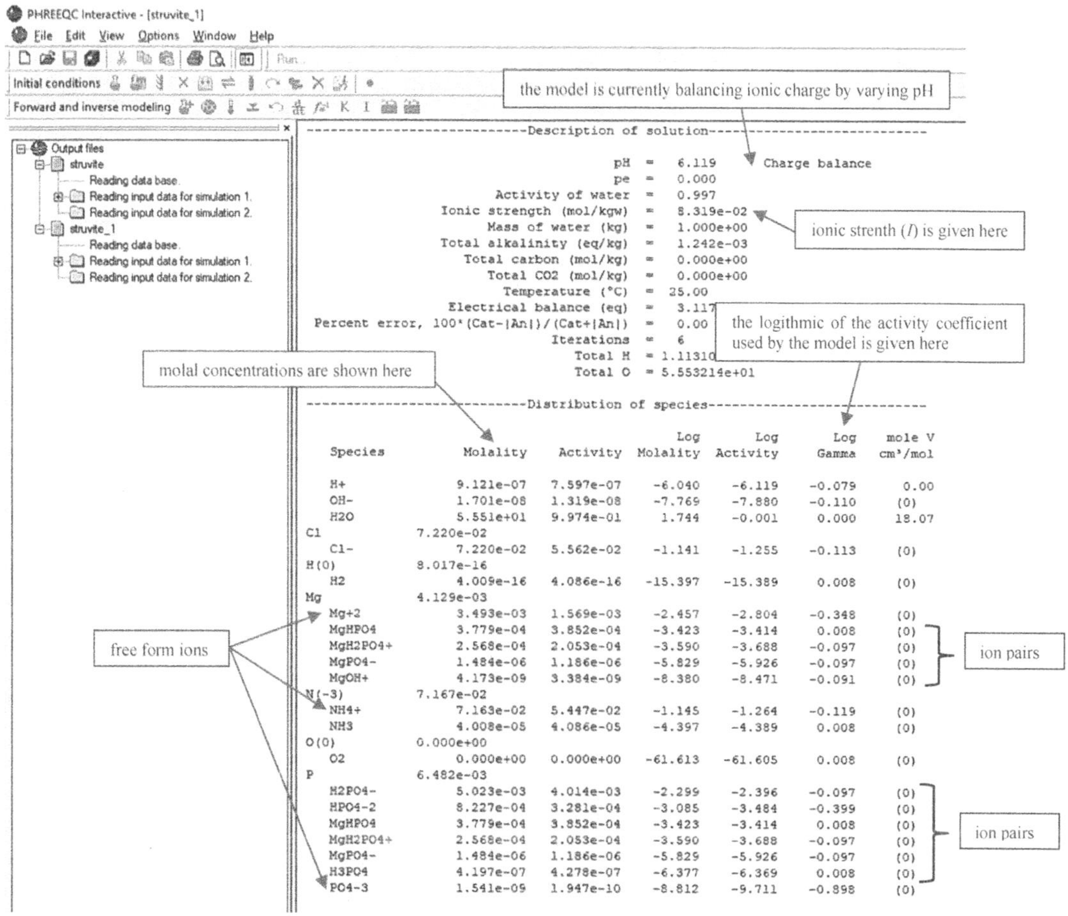

Figure 10.6 Simulation results window (output) for PhreeqC simulation – Step 1a.

with $MgHPO_4$ representing the majority ion pair with magnesium, and as a result the concentrations of the free ions that participate in struvite formation are $[PO_4^{-3}(aq)] = 1.54 \times 10^{-9}$ molal; $[Mg^{+2}(aq)] = 3.49 \times 10^{-3}$ molal; and $[NH_4^{+}(aq)] = 7.16 \times 10^{-2}$ molal. PhreeqC also calculates activity coefficients (e.g., Davies equation, Equation (10.6)) and estimates the ionic strength ($I = 0.08319$ molal) for this purpose.

Figure 10.6 shows the following activity coefficients: $\log_{10}(\gamma_{Mg2+}) = -0.348$ so $\gamma_{Mg2+} = 0.449$; $\log_{10}(\gamma_{NH4+}) = -0.119$ so $\gamma_{NH4+} = 0.760$; $\log_{10}(\gamma_{PO4\text{-}3}) = -0.898$ so $\gamma_{PO4\text{-}3} = 0.126$.

IAP_{struv} and SI_{struv} at these wastewater conditions can then be calculated as follows:

$$\begin{aligned} IAP_{struv} &= \left(\gamma_{Mg} \times \left[Mg^{2+}(aq)\right]\right) \times \left(\gamma_{NH_4} \times \left[NH_4^{+}(aq)\right]\right) \times \left(\gamma_{PO_4} \times \left[PO_4^{3-}(aq)\right]\right) \\ &= (0.449 \times 3.49 \times 10^{-3} \times 0.760 \times 7.16 \times 10^{-2} \times 0.126 \times 1.54 \times 10^{-9}) \\ &= 1.655 \times 10^{-14} \text{ or } 10^{-13.78} \quad (K_{sp,struv} = 10^{-13.26}) \end{aligned}$$

and

$$SI_{struv} = \log_{10}\left(\frac{IAP_{struv}}{K_{sp,struv}}\right) = \log_{10}\left(\frac{10^{-13.78}}{10^{-13.26}}\right) = -0.521$$

which shows that $IAP_{struv} < K_{sp,struv}$ and $SI_{struv} < 0$ and that the wastewater conditions are undersaturated with respect to struvite. Hence, struvite will not form, and if the mineral phase was present it will instead dissolve to attain equilibrium.

10.3.1.4 Worked exercise 1b

If sodium hydroxide (NaOH) was added to the wastewater in exercise 1a above, so that the amount of sodium in the wastewater composition is now instead 184 mg/L, the PhreeqC model can calculate the resulting increase in pH using an ionic charge balance. The modelling input and outputs are shown in Figures 10.7 and 10.8, respectively. Note how the equilibrium data for struvite precipitation has now been added into the model (Figure 10.7) because it was not previously in the model database. In the coding, log_k refers to the common logarithm of the solubility product constant, which in the case of struvite is -13.26 ($K_{sp,struv} = 10^{-13.26}$).

The results show that pH of the wastewater will rise to 7.98. Moreover, the simulation now gives free ions concentrations of $[PO_4^{-3}(aq)] = 6.098 \times 10^{-7}$ molal; $[Mg^{+2}(aq)] = 2.402 \times 10^{-3}$ molal; $[NH_4^{+}(aq)] = 6.890 \times 10^{-2}$ molal; and $I = 0.08849$ molal, from which $IAP_{struv} = 10^{-11.40} > K_{sp,struv}$ and $SI_{struv} = 1.86$. These show that by adding NaOH to increase the wastewater pH, conditions have been

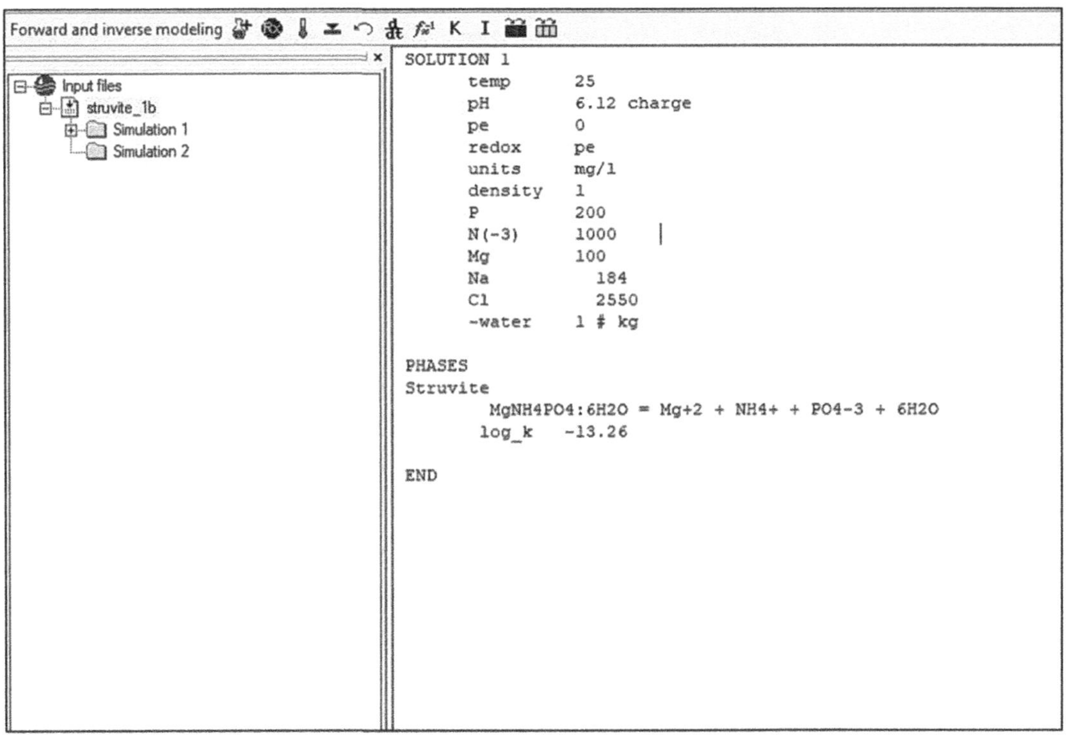

Figure 10.7 Simulation coding window (input) for PhreeqC simulation – Step 1b.

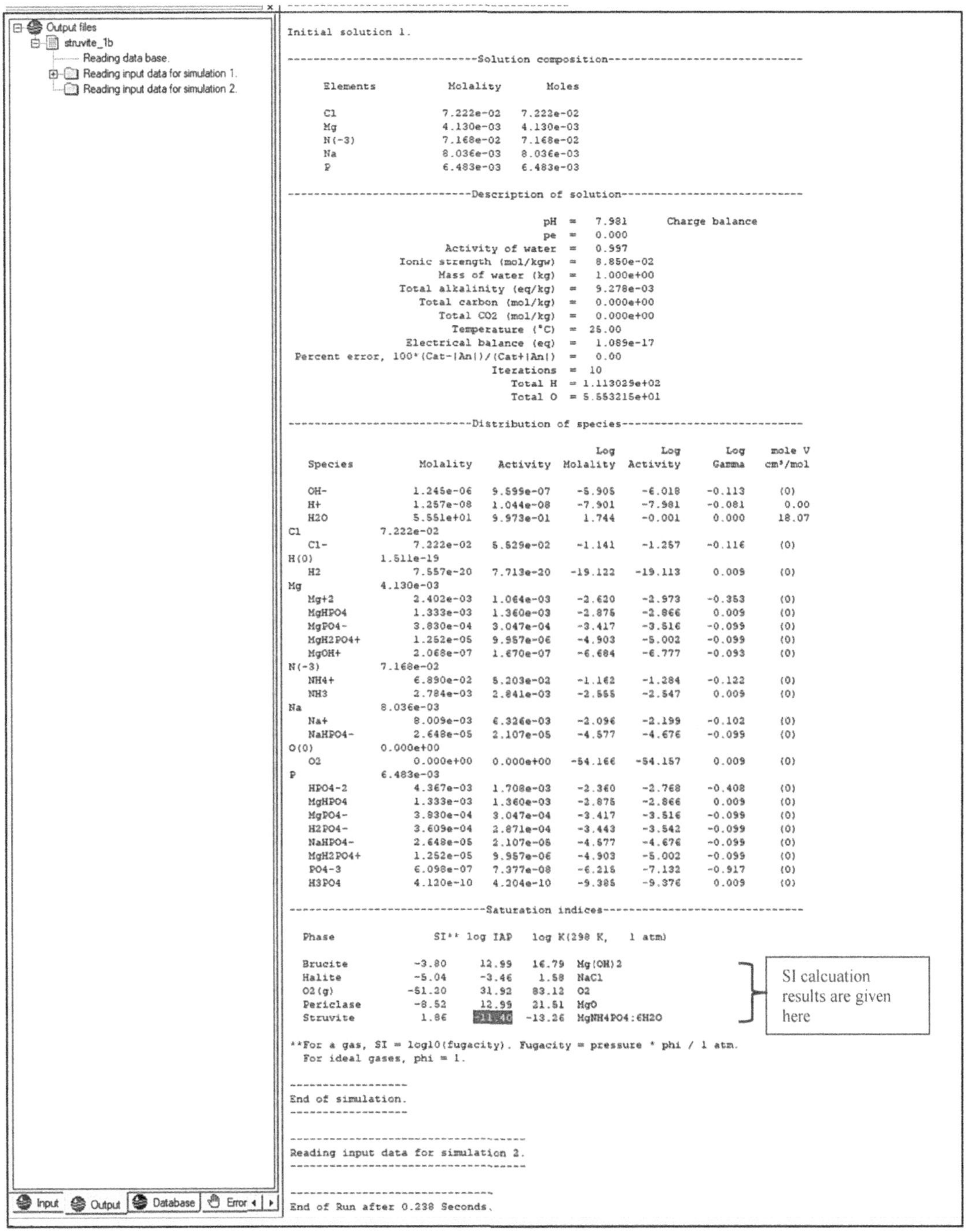

Figure 10.8 Simulation results window (output) for PhreeqC simulation – Step 1b.

changed from being undersaturated to being supersaturated, and instead struvite can now precipitate to attain equilibrium. This is primarily because the proportion of $[PO_4^{-3}(aq)]/[\text{Total } PO_4^{*}]$ has increased by decreasing the level of competition with the phosphate weak acid-base system.

10.3.1.5 Struvite nucleation, crystal growth and crystal morphology

Struvite precipitation occurs by new crystals nucleating and then growing and/or existing struvite crystals growing (Figure 10.9). Nucleation generally requires higher levels of supersaturation than for growth because, for nucleation to occur, a chemical energy barrier has to be overcome to produce nuclei (Mersmann, 2001). There are two main types of nucleation, primary nucleation which occurs in the absence of any pre-existing crystals, or secondary nucleation which occurs in the presence of at least one crystal (Mersmann, 2001).

The level of supersaturation required for primary nucleation to occur is usually greater than for secondary nucleation to occur because the presence of a crystal catalyzes secondary nucleation (Mersmann, 2001). Levels of supersaturation required for primary nucleation can be estimated from measurements of an induction time, which is the time taken for a supersaturated wastewater void of any struvite crystals to form a measurable number of nuclei (Mehta & Batstone, 2013). In such cases, a higher level of supersaturation generally decreases the measured induction time. However, the presence of struvite crystals is often ubiquitous around WWTPs, especially in the sludge line (see further below). Thus, secondary nucleation is expected to be the most important nucleation mechanism in industrial struvite reactors.

Once crystals have nucleated they can grow, either as single crystals by layered growth, or by aggregation with multiple crystals then forming interparticle bridges and growing together as combined crystal masses, termed agglomeration growth. Agglomeration growth has been suggested to be important for struvite precipitation in commercial systems (Galbraith *et al.*, 2014). Crystal growth rate generally increases with an increasing level of supersaturation (Mersmann, 2001).

Struvite crystal morphology is controlled by crystal growth which is predominantly influenced by the extent of supersaturation. Specifically, dendritic, fibrous and irregular crystals are favored under conditions of high supersaturation (Figure 10.10). The concentration of ammoniacal N also has an effect with more irregular crystal morphologies resulting from higher ammoniacal N concentrations with increased supersaturation (Figure 10.10) (Shaddel *et al.*, 2019). An increase in pH promotes a shift from Zone 1 to Zone 3-type crystals (Figure 10.10) (Shaddel *et al.*, 2019). Factors affecting struvite aggregation have also been researched due to their impact on struvite yield (Le Corre, 2006; Shaddel *et al.*, 2019).

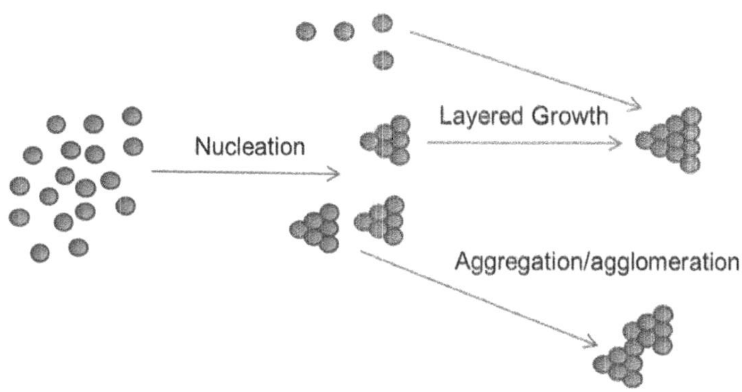

Figure 10.9 Crystal nucleation and growth processes.

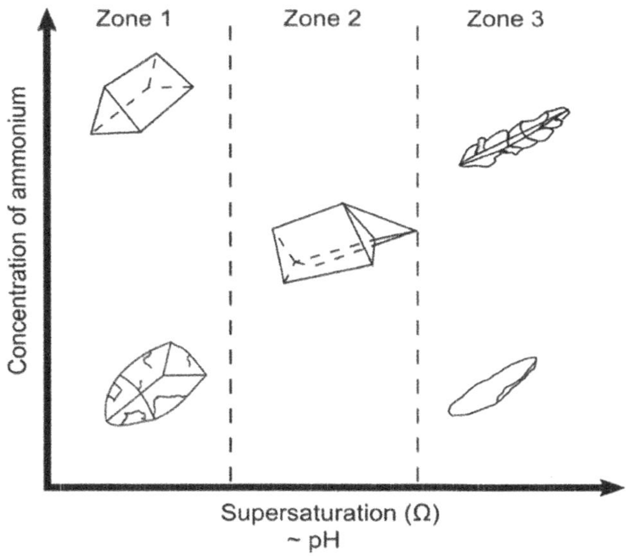

Figure 10.10 Impacts of varying supersaturation, ammonia and pH on struvite crystal morphology (adapted from Shaddel *et al.* (2019)).

Struvite crystal size is also important from an application perspective. Specifically, larger struvite crystals may be favorable for fertilizers due to ease of handling, transportation, and application (Li *et al.*, 2019a). However, the surface area/volume ratio may also be important as this would influence dissolution in soils and thus nutrient release properties (Li *et al.*, 2019a, 2019b) in fertilizer applications.

Overall, to control struvite crystal size, nucleation and particle growth processes have to be balanced by controlling levels of supersaturation in industrial processes. Specifically, a higher initial supersaturation may increase nucleation rate and produce several small particles that grow rapidly to quench the remaining supersaturation (Shaddel *et al.*, 2019). In contrast, a slow nucleation rate combined with a moderate crystal growth rate may produce a crystal product with a larger size. Some struvite precipitation processes use seed materials to promote struvite precipitation to increase particle size and improve the product purity (Le Corre *et al.*, 2009; Li *et al.*, 2019a, 2019b). Other reactor systems are designed to control supersaturation in various compartments of the precipitation reactor (see further below), to control nucleation and growth processes. The mass-based struvite formation rate (r_{cryst}, moles per L per h) which includes nucleation and growth, has been reliably represented in wastewater treatment process models by a rate expression (Mbamba *et al.*, 2015, Equation (10.13)):

$$r_{\text{cryst}} = k_{\text{cryst}} \cdot X_{\text{cryst}} \cdot \left(\left(\frac{\text{IAP}_{\text{struv}}}{K_{\text{sp,struv}}} \right)^{1/3} - 1 \right)^{3} \tag{10.13}$$

where k_{cryst} is an empirical rate coefficient (with units of per h) and X_{cryst} is struvite crystal concentration (with units of moles/L). This shows that struvite precipitation rate is catalyzed by the presence of struvite mineral phase. This can be used favorably for struvite reactor design and operation. For example, if magnesium/alkali reagent and/or raw water feed can be added in regions of a struvite reactor with high struvite crystal content (e.g., a settled crystal bed at the base of a reactor, or in locations where seed crystals are added), then the supersaturation that would result can be rapidly consumed by a fast struvite precipitation rate catalyzed by a high crystal content. This then prevents high supersaturation hotspots to minimize nucleation of poorly settleable fine crystals that would wash out and be lost with the reactor effluent.

10.3.2 Struvite precipitation – applications and design

Struvite recovery can be implemented on the sludge line of a WWTP to treat digestate or centrate produced from anaerobic digestion of sewage sludge with $P > 50$ mg/L (Figure 10.11). Within this context, N and P are accumulated in sludge which is subsequently released by hydrolysis within the anaerobic digestion stage. This results in phosphate and ammonia at concentrations well in excess of that in the water mainline of the WWTP, which is then more suitable for struvite precipitation (Sena & Hicks, 2018). Other industrial and agricultural wastewater streams have also been previously explored for P recovery by struvite precipitation, including food processing and rare-earth processing wastewaters with typically high phosphate contents (Kataki *et al.*, 2016; Ma *et al.*, 2018). High phosphate in such cases usually results from biological decay processes (e.g., manure storage) or from the addition of industrial phosphate-based chemicals.

The influent is commonly dosed with magnesium compounds, typically $MgCl_2$, $Mg(OH)_2$ or MgO, to increase supersaturation and induce struvite precipitation. The concentration of magnesium can be dosed to ensure that, stoichiometrically, to exceed a minimum molar ratio of Mg:P $\geq$ 1:1 (Le Corre *et al.*, 2009). Additionally, pH must be adjusted for most wastewater influents because municipal wastewater pH varies between 7 and 8 while optimum struvite precipitation occurs at pH $\geq$ 8.5 (Le Corre *et al.*, 2009). For increasing pH in struvite recovery processes, either the wastewater is aerated, which strips CO_2, or an alkali reagent can be dosed (i.e., NaOH or $Mg(OH)_2$). Adding NaOH to raise pH and $MgCl_2$ as magnesium source may be preferred from an operational perspective so that pH and magnesium dose can be varied independently. However, NaOH is more costly and currently less economical than alternatives for pH adjustments (Muhmood *et al.*, 2019).

Reagent dissolution of $Mg(OH)_2$ or MgO can be limited by high pH, or by the precipitation of struvite that can coat magnesium reagent particles added as a slurry to the struvite process, thereby preventing further dissolution and reaction of the magnesium reagent (Romero-Guiza *et al.*, 2015). This may increase the requirement for magnesium reagent when using $Mg(OH)_2$ or MgO. Calcium (e.g., from $Ca(OH)_2$) is an impurity that can negatively influence struvite formation and co-precipitates with phosphate as calcium phosphate minerals competing with struvite formation (Le Corre *et al.*, 2009).

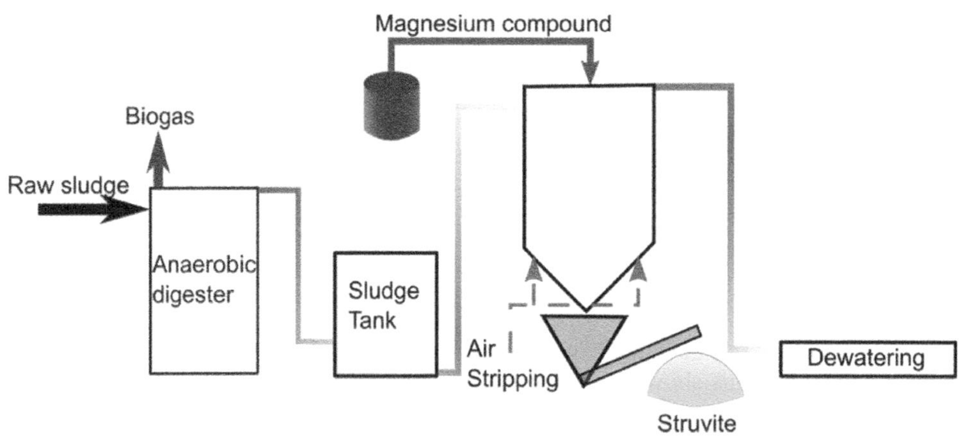

Figure 10.11 Simplified schematic of a sludge treatment section of a wastewater treatment plant with struvite precipitation.

This can especially be the case when molar ratios Mg:Ca $\geq$ 1:1, at which struvite formation can be limited and even inhibited (Le Corre *et al.*, 2009).

10.3.2.1 Worked exercise 1c – magnesium reagent requirements

As an illustrated example of how to determine magnesium requirements to achieve a certain desired P removal, consider the wastewater from exercise 1b at pH 7.98. The PhreeqC simulation will now precipitate struvite from this wastewater until equilibrium is attained (i.e., until $SI = 0$). To do this, struvite mineral phase is added to the model input. For this, a nominal amount of 0.01 moles of struvite is added to 1 L of wastewater in the simulation to allow PhreeqC to precipitate the struvite. Figure 10.12 shows the resulting simulation input window.

The results of the simulation are presented in Figure 10.13. Approximately 3.465×10^{-3} moles of struvite precipitated from 1 L of the wastewater (See 'Delta' for struvite under 'Phase assemblage') until equilibrium was attained. As a result of struvite precipitation, wastewater pH had also decreased from 7.98 to 7.14. This is expected and is a result of the removal of NH_4^+ and PO_4^{-3} which initiates a flow-on effect on ion pairs in the wastewater to overall liberate a net amount of H+ into the wastewater and thereby decrease the wastewater pH. Consequently, struvite processes may have an on-going alkali demand to maintain an elevated pH for struvite precipitation to continue. The results also show that total P content had decreased from 6.483×10^{-3} molal (200 mg/L) to 3.020×10^{-3} molal (~94 mg/L), or a P removal/recovery of 53% was achieved. It is also apparent that the residual magnesium concentration after struvite precipitation has now decreased to a negligible

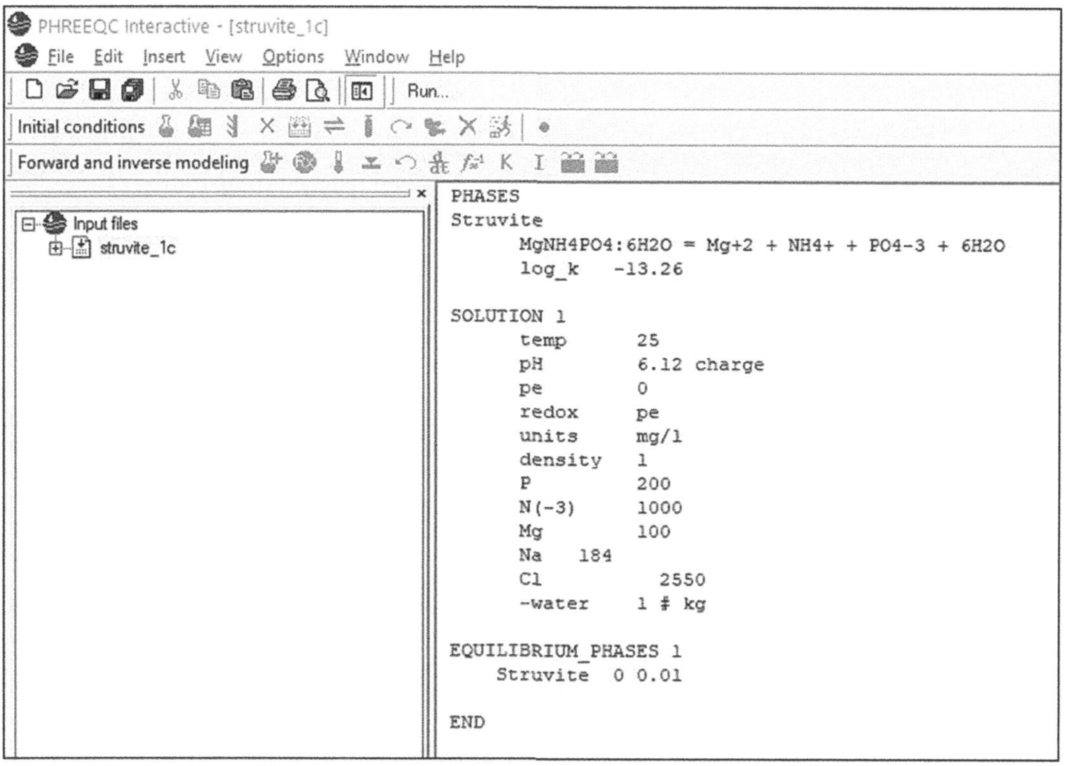

Figure 10.12 Simulation coding window (input) for PhreeqC simulation – Exercise 1c.

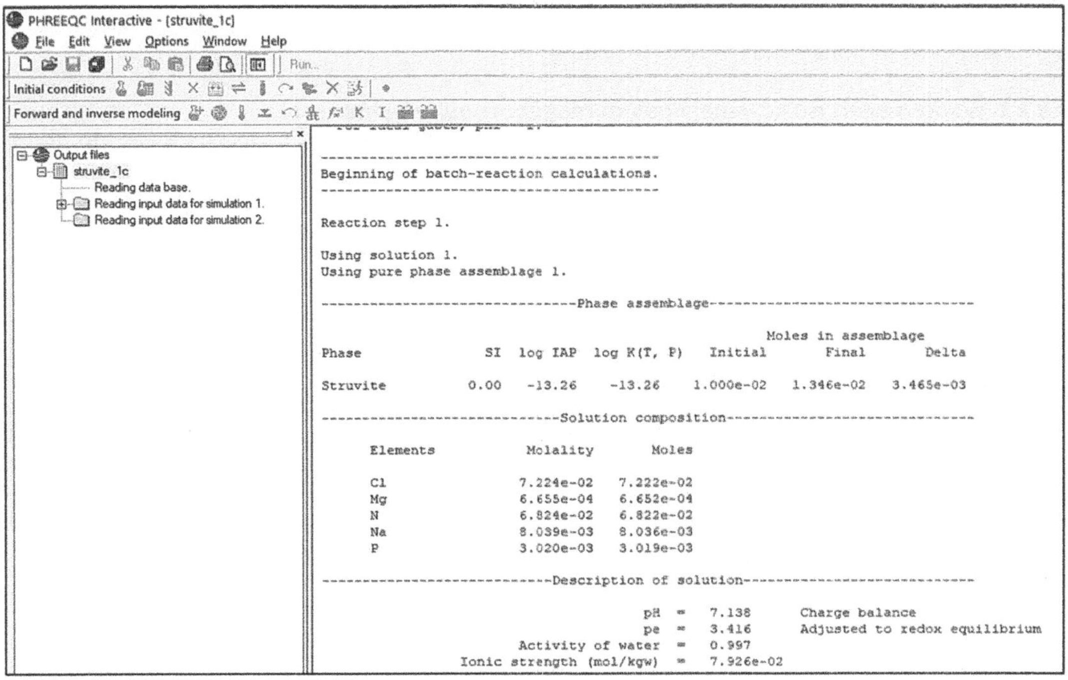

Figure 10.13 Simulation results for PhreeqC simulation – Exercise 1c.

6.655×10^{-4} molal (~16 mg/L), whilst ammoniacal N content had reduced comparatively little, down to 6.824×10^{-2} molal (~955 mg/L). This indicates that magnesium is the limiting reagent and that ammoniacal N was in large molar stoichiometric excess for struvite formation, as is typical for many wastewaters, including in the sludge line of a WWTP. This is why a magnesium reagent is often needed to increase P recovery as struvite, and also why the extent of N removal by struvite precipitation is typically low (Mehta *et al.*, 2015) unless a phosphate source is added to instead target N removal (Li *et al.*, 2019a, 2019b).

To achieve the desired >85% P recovery, the mineral Brucite ($Mg(OH)_2$) could be added to the simulation together with struvite. This can be done under the Phases 'Equilibrium_Phases' tab, or can be typed directly into the simulation command window as follows:

```
EQUILIBRIUM_PHASES 1
   Struvite 0 0.01
   Brucite 0 0.002
```

When the simulation is then rerun with this amount of brucite added, the final pH after struvite precipitation is 7.66 and the final P is 8.36×10^{-4} molal (~26 mg/L). Alternatively, magnesium chloride could be set up as a highly soluble phase (a nominal very large log_k value) as follows:

```
PHASES
Magnesium_chloride
      MgCl2 = Mg+2 + 2Cl-
log_k 100 000 000
```

and a separate alkali such as sodium hydroxide set-up:

```
PHASES
Sodium_hydroxide
        NaOH = Na+ + OH-
log_k 100 000 000
```

and a combination of these added to the simulation, instead of Brucite, until the desired 85% P removal is achieved.

In addition to the chemical requirements to control the stiochiometry and yield of struvite recovery, other process design features are implemented to achieve better control of formation and separation of struvite crystals. For this, it is important to prevent supersaturation hotspots and excessive nucleation while still promoting precipitation growth. Chemical reagents responsible for increasing/maintaining supersaturation may hence be added:

- directly into a well-mixed reaction zone together with fresh influent and a high struvite crystal concentration; or
- into areas of a reactor where there are low residual phosphate concentrations with a subsequent recycle to a well-mixed reaction zone where fresh influent is added and a high crystal content is present.

In addition, commercial designs incorporate these important features of precipitation/reaction, and also settling separation in different ways by altering:

- the hydraulic design of the reactor, for example to promote high flow in certain parts of a reactor with homogeneous mixing, and to slow down flow in other parts of the reactor to allow settling-separation of struvite crystals;
- the operational parameters, such as agitation speed, flowrate of a fluidization medium (gas or liquid), or influent flow rate;
- by varying the location and amount of chemicals added; and
- controlling pH.

These features would influence the construction costs of the struvite reactor, and also operational costs, such as energy costs and chemical costs.

Particle settling separation aims to retain fine struvite crystals so that they do not wash out with the effluent, and grow into a desired coarse crystal product. In struvite crystallizers, this settling can be achieved either with an inbuilt settling zone or an external settling tank that is separate to the struvite reactor. It is important to note that settling of particles in a bed of crystals can transition from free settling at a low solids concentration (e.g., Stoke's law) to hindered settling at a high solids concentration. Fluid up-flow velocity is also important and the various fluidization regimes are illustrated below (Figure 10.14).

The most common configurations of struvite recovery systems are: (i) fluidized bed reactors (FBR); and (ii) continuously stirred tank reactors (CSTR) (Li *et al.*, 2019a, 2019b). FBR is a continuous process where fresh influent is constantly pumped into the reactor to fluidize the struvite crystals to grow (Figure 10.15), whereas CSTR are typically operated in batch or semi-continuous mode with mechanical stirring (Figure 10.16). Reported phosphate recovery processes using struvite are summarized in Table 10.1.

The basic principles of the FBR design and operation include: constant influent flow, sometimes including return effluent to maintain fluid velocity when influent rate fluctuates (Figure 10.15) to maintain up-flow velocity in a reaction zone; and establishing supersaturated conditions within the reaction zone to promote struvite precipitation (Figure 10.15). Crystals will continue to grow until their settling velocity exceeds the up-flow velocity of the FBR (Figure 10.15). The precipitate can then

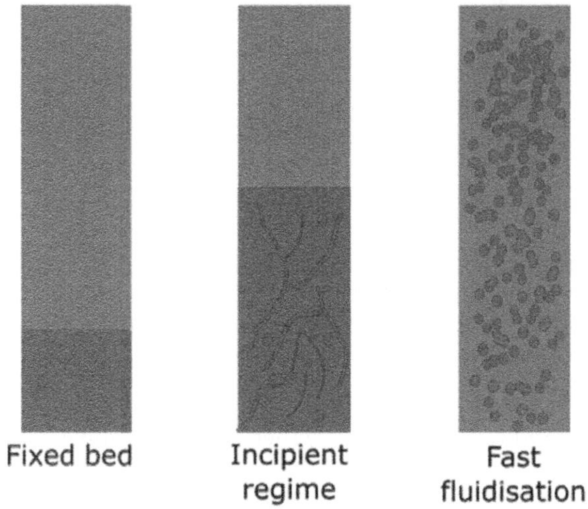

Figure 10.14 Fluidization regimes as a function of fluid velocity (adapted from University of Florida (2015)).

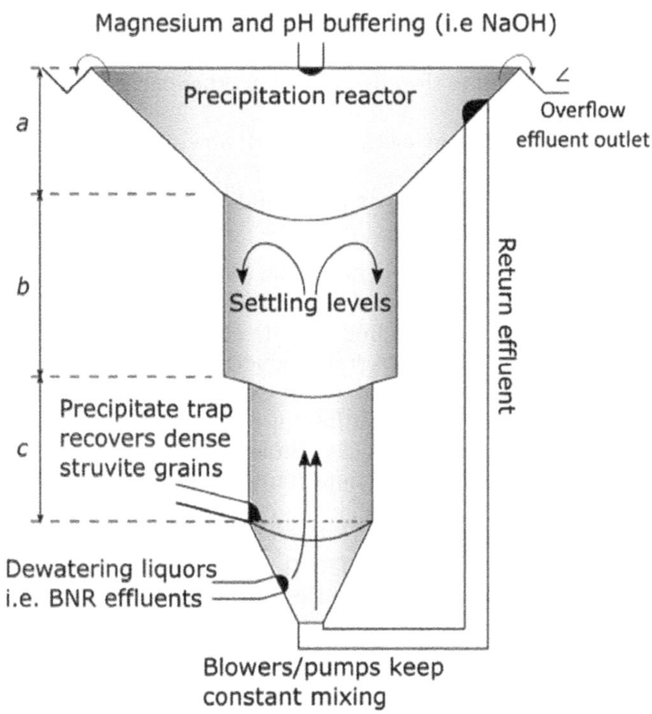

Figure 10.15 Schematic representation of a fluidized bed reactor (FBR), where dewatering liquors from a Biological Nutrient Removal (BNR) wastewater treatment plant is pumped at a constant flow into the bottom of the reactor and mixed with return effluent. Zone a = mixing and struvite reaction zone, nucleated struvite settles into zones b and c = struvite growth/aggregation zone and collection.

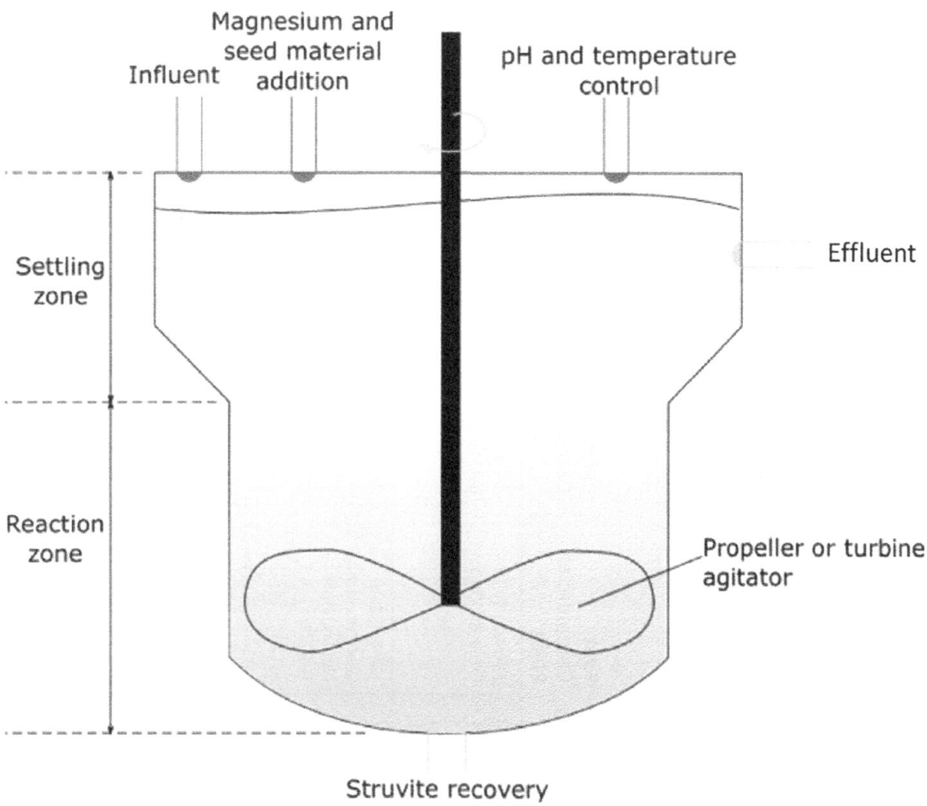

Figure 10.16 Schematic representation of a continuous stirred tank reactor (CSTR) design for struvite recovery in WWTPs.

be recovered from the base of the FBR using a mesh guard, dewatered, and dried for reuse (Figure 10.15). Ostara's Pearl© technology and Royal Haskoning DHV's Crystalactor® technology utilizes FBR struvite recovery at municipal WWTPs across the globe (Table 10.1). The Pearl© process also uses sand grains to seed struvite nucleation and ensure concentric growth to produce more aesthetic struvite that are easier to separate via settling and have properties that allow for direct application after drying of the recovered product.

The settling of a bed of struvite crystals in a FBR can help detach the solids/crystal retention time from the hydraulic retention time, so that crystals can be harvested in batch, whilst the liquid throughput occurs continuously (Le Corre *et al.*, 2009). The time constant for struvite precipitation in terms of phosphate removal appears to be in the order of hours (Mbamba *et al.*, 2015) rather than days as for biological processes. Hydraulic retention times can be as short as 1–2 hours (Munch & Barr, 2001). Simoes *et al.* (2018) suggested retention times for chemical struvite precipitation in FBRs to be in the range of 0.5–9 hours based on the cited literature studies of Ueno and Fujii (2001) and Le Corre *et al.* (2009). This indicates that a reasonable hydraulic retention time could be around 2 hours. The crystal retention time, in contrast, has to be much longer, in the order of days (Le Corre *et al.*, 2009). For example, around 10 days was required to allow sufficient time for growth of 0.5–1.0 mm struvite pellets (Ueno & Fujii, 2001). Similar results were noted by Le Corre *et al.* (2009) from other literature studies.

Table 10.1 Examples of commercially available or demonstration-scale phosphate recovery processes at WWTP (adapted from Kataki *et al.* (2016) and European phosphate platform success stories (2019)).

Technology	Influent	Reactor Type	Treatment Process	Recovery Efficiency	First Commercial Plant	Plant Size/ Capacity	Yields	Website (Where Applicable)
Pearl© Technology	Municipal wastewater	FBR	Side stream treatment of dewatering liquors via $MgCl_2$ and NaOH addition	80% $P\text{-}PO_4$, 10–15% N	2009, WWTP Oregon, USA	80 mg P/L, 600 000 PE	600 ton *Crystal Green®* struvite/ year	http://www.ostara.com
Crystalactor©	Municipal and industrial wastewater	FBR	Side stream treatment of dewatering liquors via $MgCl_2$, Aeration for stripping CO_2, or lime addition. Quartz sand for seeding	70–80% $P\text{-}PO_4$	1994, Edam Geestmerambacht, Netherlands WWTP	230 000 PE		https://www.royalhaskoningdhv.com/en/crystalactor/about-crystalactor
Phospaq™ and Anammox®	Food processing and municipal wastewater	CSTR	Aeration, stripping CO_2, increasing pH and adding MgO	80% $P\text{-}PO_4$, 90% $N\text{-}NH_4$	2011, Waterstromen-Aviko	260 kg P/L 160 000 PE	400 ton struvite/ year	https://en.paques.nl/
Struvia™	Agricultural and municipal wastewater	CSTR Turbomix™	Dosing of Mg-salts, NaOH buffers pH 8–9. Rapid mixing	80–90% $P\text{-}PO_4$	2016, Helsingør Denmark	250 mg P/L 72 000 PE	110 kg struvite/day	https://www.veoliawatertechnologies.com/en/solutions/products/struvia
AirPrex™	Sewage sludge (SS)	FBR air lift reactor	Aeration of digested sludge removes CO_2 and increases pH. Dosing of Mg-salts	90–95% P removal	2009, MG-Neuwerk, Niersverband, Germany	995 000	1000 kg struvite/day	www.p-rex.eu

Multiform	AD sludge dewatering liquors	FBR	Dosing of $MgCl_2$ and NaOH	80% PO_4, 20% N	2012, Idaho USA			www.multiformharvest.com
Polonite® and Sorbulite® filter beds	Agricultural wastewater	Granulated beds	Filtration over CaO. SiO_2, $CaSiO_3$ (polonite®) granules, P-precipitation nucleates around granules	100 mg P/g of Polonite®	4000 + farmsteads in Sweden since 2010		Reclaimed granules are slow release fertilizers	
Seaborne	AD effluents/SS	CSTR	Acidification to separate heavy metals and P, struvite precipitation using $Mg(OH)_2$ and NaOH		2007, Gifhorn WWTP, Germany	50 000 PE	250 kg struvite/day and heavy metals	
RecoPhos (electro-chemical)	Sewage sludge ash (SSA)	Electro-thermal reactor	SSA put into electrothermal carbon packed bed at 1300°C. P is volatized and distilled into acid. Molten slag collected	Up to 100% P from ash			RecoPhos® P38fertilizer or H_2PO_3 Fe-melt	RecoPhos (2012) http://www.recophos.org/
AMS Technologies	SSA	CSTR	Acidification, leachate filtered to remove suspended solids. Nanomembranes retain heavy metals	Up to 60% P as H_3PO_4			Recycled heavy metals, phosphoric acid	Paltrinieri *et al.*, (2019), http://www.amsmembrane.com/index.php/en/applications/acids

PE – population equivalent.

In a CSTR crystallization reactor, pH modification takes place through the addition of an alkaline chemical, and propeller or turbine mixing is provided within the reactor (Figure 10.16). Baffles are added to prevent the formation of vorteces which could be detrimental to crystal growth. Turbines create a radial flow, causing intensive turbulence and increased fluid shearing to produce finer precipitates, which can be desirable in some cases (Regy *et al.*, 2001).

The Struvia™ process is an example of a CSTR based technology for struvite recovery (Table 10.1) (Brockmann, 2015). The process treats nutrient-rich effluents from industries or dewatering sludge liquors. The CSTR includes a proprietary mixing technology (Turbomix®) and a lamella settler. The influent is first mixed, dosed with magnesium salts, buffered to pH 8–9 and then passed through the enhanced mixing system. The latter encourages crystal formation and growth, reduces reaction time and improves the efficiency of chemicals added. The integrated lamella settler ensures the separation of the produced struvite and the treated effluent that is returned to the head of the WWTP works. The collected struvite is then dried and processed into organo-mineral fertilizers for agricultural use.

10.3.3 Struvite precipitation – case studies of full scale implementation

Currently, there are an estimated 80 struvite precipitation installations in operation worldwide (Muys *et al.*, 2021). This section provides three case studies, showing different reactor configurations and process features.

10.3.3.1 Case study 1 – ostara®'s pearl©

Ostara®'s Pearl© nutrient recovery system is a modular process made by Ostara Nutrient Recovery Technologies Inc. (Vancouver, Canada). The process is adaptable to different nutrient-rich streams (e.g., sewage sludge liquors, streams from food processing). The wastewater to be treated is first dosed with Mg, circulated in a fluidized crystal forming reactor, with the pH strictly controlled using NaOH (Table 10.1). To initiate crystallization, small struvite crystals act as seeds to ensure that the precipitates grow into granules, fit for distribution after drying. This can be achieved by recycling into the reactor treated effluent with uncollected struvite crystals. Fluidization of the bed is achieved by a recycle, which is responsible for maintaining a constant upward velocity in the reactor regardless of the feed flow. Ostara® markets these granules as Crystal Green® (Figure 10.17), an aesthetic product designed to improve consumer confidence in recovered nutrient fertilizers.

Figure 10.17 Ostara® struvite granules commercialized as Crystal Green® Ostara pearls (left) and Pearl© Technology reactor (right).

Ostara operates at different WWTPs in Europe, including Madrid El Canal Isabell II WWTP (FuturEnviro, 2017). The nutrient recovery facility consists of two groups of feed pumps, a crystallizing reactor and its associated pumps and valves, a draining and drying product step, a product sorting and packaging step, a programmable logic controller (PLC) cabinet, an engine control centre, an $MgCl_2$ storage tank and an NaOH storage tank. The anaerobic digested sludge and the overflows of the dehydration and the flotation tanks are fed into the lower part of the reactor where they are diluted with recirculation water and $MgCl_2$ (32%; 100 L/h) and NaOH (25%; 36 L/h) are injected. Inside the reactor, the struvite precipitates in a controlled way in small hard granules. At the top of the reactor, an integrated clarifier retains the granules inside the reactor. The effluent is sent to a tank from where it is pumped to the primary decantation of the WWTP. Struvite particles are collected from the bottom of the reactor, washed, dried, sorted by size and packaged for shipment.

Struvite formation rate and its physical properties are controlled in the process so that a high-quality product is obtained, which allows it to be marketed as a fertilizer. The struvite recovery system is designed to recover phosphate from a combined feed, in the range of 50–120 m^3/h maximum anaerobically digested sludge, dehydration overflows, and flotation overflows. The reactor is designed to treat a maximum precipitable P load of 300 kg/day. The installed reactor allows production of up to 2 tons/day of marketable final product, called Crystal Green®. Under optimum operating conditions and feed quality, 600–700 tons can be produced per year.

10.3.3.2 Case study 2 – Airprex®

A variation of the FBR design uses an airlift system to achieve air stripping of CO_2 and thus increase operational pH, and also to provide mixing. An example is the Airprex® process developed by Berliner Wasserbetriebe (BWB) in cooperation with Technische Universität Berlin, which has been installed to treat a range of anaerobic digestion effluents. Magnesium is added as $MgCl_2$ and the airlift design allows for the digestate to flow upwards in the central compartment of the reactor inside a draft tube, and down via the annulus region by the downdraft of fluid (Figure 10.18).

Pumping air has a significant energy demand and costs, but the process has been demonstrated to be attractive due to savings in pH adjustment chemicals and from not requiring dewatering,

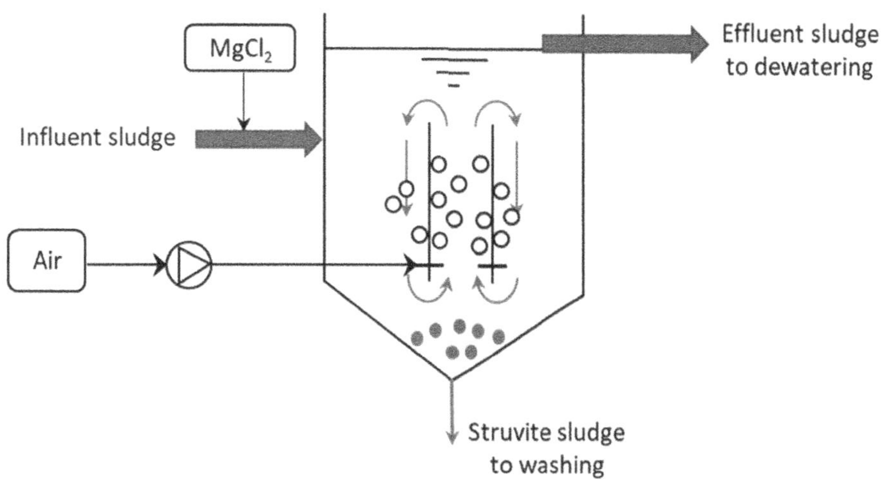

Figure 10.18 Air-lift arrangement of the Airprex® process which both provides mixing as well as CO_2 stripping to elevate pH and save on alkali dosing (adapted from P-Rex (2015)).

because P is recoverable directly from digestate with this process, and not from a centrate. The struvite product can be crystallized within the wet sludge and hence contains organic and inorganic impurities. The formed struvite crystals are harvested at the bottom of the reactor. This is done a few times per day by opening the bottom of the reactor for a minute or two and a mixture of organic material and struvite crystals is released to a pump, which delivers it to a washing system that dislodges the organics by floating it to the top of the washer to be decanted back to the treatment plant headworks.

The largest Airprex operational plant is located in Berliner Wasserbetriebe in Germany with a capacity of 2000 m^3/day of digested sludge, with struvite production of 2.3 ton/day. The AirPrex® process was said to be the first process that recovered products recognised as fertilizers, by achieving the official conformity approval and REACH registration. The application of the process is said to be limited to WWTPs with enhanced biological P removal and P concentrations >50 mg/L PO_4-P in the sludge liquor, even though the product yield can be enhanced by thermal or chemical hydrolysis prior to digestion to increase P concentration in the sludge digestate (P-Rex, 2015).

10.3.3.3 Case study 3 – Phospaq™

Stoke Bardolph WWTP in the UK serves a population equivalent of approximately 650 000 and has a unique range of technologies for sludge and centrate treatment, including the BIOPAQ® UASB for methane production combined with nutrient management, including struvite recovery, through the Phospaq™ process and N abatement using Anammox® (Table 10.1, Figure 10.19). The Anammox® promotes BNR in sludge dewatering liquors through deammonification. The deammonificaion process effluent is then treated by the Phospaq™ process.

The effluent from the Anammox® process is then aerated in two Phospaq™ struvite reactors (stripping CO_2 and increasing pH). The influent is dosed with magnesium oxide (MgO) to promote the precipitation of struvite. In addition to struvite removal, aeration increases the availability of oxygen for biological reduction of COD. The recovered struvite can be directly applied to land after drying. The Phospaq™ struvite reactors are equipped with separators that can retain the struvite to be harvested from the base of each reactor.

The Phospaq™ is said to be feasible for applications with P loadings of more than 100 kg P/d and P concentrations of 50 mg/L PO_4-P or more and N concentrations of 200 mg/L NH_4-N or more, and can achieve a removal efficiency of about 70–95% (www.paques.com).

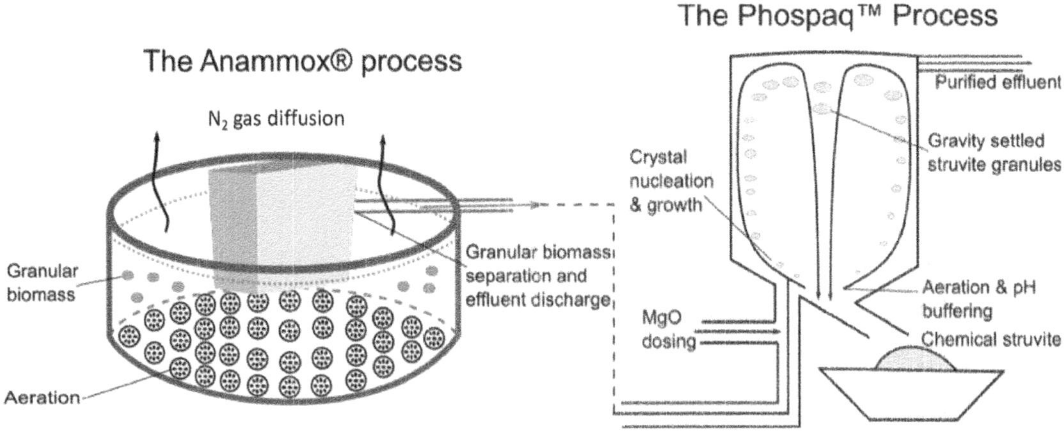

Figure 10.19 Combined Anammox® and Phospaq™ system (adapted from www.paques.com).

10.3.4 Struvite precipitation – challenges, opportunities and research needs

10.3.4.1 Struvite as a fertilizer and its value

Mined phosphate used as fertilizer is predominantly refined from the apatite group of minerals ($Ca_5(PO_4)_3(OH, F, Cl)$). In the last 10 years, almost 400 studies have been published on the effects of struvite as a fertilizer. Doyle and Parsons (2002) suggested that struvite is an excellent fertilizer, due to its low solubility and high N and P content. It has been reported that struvite is at least as effective a P source for plant nutrition as commercial mineral P fertilizers. In addition, the content of heavy metals is low compared to phosphate rock. According to Rittman (2011), struvite quality is comparable to standard fertilizers such as superphosphate and diammonium phosphate. A key factor for the management of struvite and similar recycled P sources is to increase the bioavailability and plant P-uptake efficiency (uptake per unit of root surface area) under different soil conditions. Therefore, knowing how to increase struvite P-use efficiency at different soil pH values will make its use more reliable as a substitute for rock phosphate (Robles-Aguilar *et al.*, 2019).

The current market price of struvite is estimated to be between €188 and €763 per ton. A recent study by Muys *et al.* (2021) estimated typical marketed values between €0 and €100 per ton, with some products marketed at considerably higher prices of €350 per ton (Phoshorgreen) to €1000 per ton (Pearl). Around €2–3 per kg in operational cost savings involved in P recovery can be achieved compared to conventional removal. However, Munch and Bar (2001) said that if struvite is sold as a boutique fertilizer, a conservative estimation for the selling price could be substantially higher. Chemical reagent costs strongly influence the economic feasibility of struvite recovery, which can vary significantly with wastewater composition and the type of chemicals being used (Li *et al.*, 2019a, 2019b). The cost of chemicals for pH adjustment can be greater than that for magnesium addition (Li *et al.*, 2019a, 2019b).

When considering the economic feasibility of implementing a struvite precipitation process, it is important to consider associated benefits such as: a potential to reduce the chemical dosage in traditional P removal techniques and an associated potential to reduce sludge to landfill costs (Li *et al.*, 2019a, 2019b); eutrophication prevention and an increase in the availability of a renewable P (Li *et al.*, 2019a, 2019b); and struvite deposition control around a WWTP (Li *et al.*, 2019a, 2019b). A recent review (Sena & Hicks, 2018) indicated that the environmental impacts of struvite production would be comparable to that of P removal by a conventional WWTP.

10.3.4.2 Technological and regulatory considerations

Magnesium is usually the limiting component for struvite crystallization in wastewater. The dosing of magnesium hydroxide ($Mg(OH)_2$) provides a source of magnesium and increases pH, although using $Mg(OH)_2$ can destabilize the process as discussed above (Chimenos et al., 2006; Lee *et al.*, 2003). Not withstanding process stability considerations, the most common sources of magnesium are MgO and $Mg(OH)_2$, despite their low solubility compared to $MgCl_2{\cdot}6H_2O$ and $MgSO_4{\cdot}7H_2O$. The latter are, however, more expensive.

Ammonia removal by struvite precipitation is generally low in comparison to phosphate removal because waste streams typically contain a molar excess of ammonia N (Mehta *et al.*, 2015). The molar excess ammonia then remains in soluble form and is not recovered (Mehta *et al.*, 2015), unless a phosphate source is added to specifically target ammonia removal (Li *et al.*, 2019a, 2019b), in which case ammonia removal extents of up to 99% could be achieved.

Commercialization of struvite as a fertilizer started in Japan in the year 1998 (Ueno & Fujii, 2001). Struvite is yet to be widely recognized as a fertilizer (with the exception of a few countries). However, a recent study (Muys *et al.*, 2021) sampled and analyzed struvite from 24 European production plants and showed that only three of these sampled products failed physicochemical legal limits of the revised fertilizer regulation issued by the EU in 2019. Specifically, one of the sampled products had a P content <7% and three samples exceeded the limit for organic carbon content of 3% by dry weight (Muys *et al.*, 2021). The new fertilizer regulation defines 17 physicochemical and five microbiological parameters that a product must meet to be used as a fertilizer or as a component material in N, P, K fertilizers (Muys *et al.*, 2021).

Microbiological analyzes by the same sampling and analysis study (Muys *et al.*, 2021) indicated that struvite may exceed certain legal limits, albeit that struvite crystallization is able to selectively exclude pathogens (Muys *et al.*, 2021). The study observed that struvite recovered from digestate contained more pathogens compared to struvite from the dewatering liquor of digestate (Muys *et al.*, 2021). A gentle drying process (below the decomposition temperature for struvite) may also benefit further reduction in pathogen contamination in the product (Muys *et al.*, 2021).

10.3.4.3 Other opportunities for phosphorus recovery and research needs

While struvite remains the dominant marketed fertilizer recovered from wastewaters, the search for less reactant-intensive and less energy-intense processes for P recovery continues. One possibility is to recover phosphate as vivianite, which is an iron phosphate mineral found in hydrothermal deposits, anoxic soils and sewage sludge. It can be used in lithium ion batteries, slow-release fertilizers and has magnetic properties (Wu *et al.*, 2019). Iron reduction occurs in anaerobic digesters where sulfate and methane-reducing bacteria dominate. Reduced iron reacts with phosphate ions to form vivianite ($Fe_3(PO_4)_2.8H_2O$). Vivianite crystallizes in neutral to weak-basic pH, which removes the need for dosing wastewaters with bases as currently employed in many commercial struvite removal processes. However, high concentrations of dissolved iron are required, which pose a problem when removing vivianite crystals from the sludge without incorporating organic matter into the crystals due to the high sorbent quality of organic matter to heavy metals. Possible solutions are to use magnetic plates to remove the vivianite crystals or centrifugation (Wu *et al.*, 2019). Further research is needed to elucidate the factors influencing vivianite crystallization, to enhance its removal, and to ascertain how best to incorporate it into treatment processes.

Other strategies to add magnesium to wastewater can be considered, including the use of desalination by-products (e.g., nanofiltration of sea water) that have an adequate magnesium concentration and an appropriate Mg/Ca-ratio, (Anne *et al.*, 2001), seawater (Matsumiya *et al.*, 2000) or salt concentrate (Etter *et al.*, 2011), which can be lower cost alternatives. However, these may contain other ions which could shift chemical equilibria, negatively impact on struvite formation and crystal growth, and increase the competition between struvite and calcium minerals.

Another option may be to add lightly-calcined magnesite to wastewater effluents to release magnesium through hydrolysis at pH 8 (Wei *et al.*, 2019). The released magnesium binds with ammonium and P remaining in the effluent to form struvite. Initial feasibility analysis of the process concluded that the cost of this process is just 11% of current commercial chemical struvite recovery systems (Wei *et al.*, 2019). However, the production of light calcined magnesite requires a readily available magnesium oxide source; and calcification of the magnesium oxide requires a kiln process heating the substance between 650 and 800°C for 2 hours. These processes themselves require energy and resources.

Struvite nanowires are an exciting recovery prospect as the nanowire morphology is better suited for slow phosphate release (Li *et al.*, 2019a, 2019b) and could achieve a more positive agronomic response and greater value. This controlled crystallization method regulates sodium chloride and pH to produce struvite nanowires in synthetic wastewaters. Findings showed a more pure struvite achieved in nanowires by regulating sodium chloride at 3.5–4.5 wt% and maintaining a pH of 11 (Li *et al.*, 2019a, 2019b). However, future studies into this technology would need to address application to actual wastewater at a larger scale while balancing the costs of materials to buffer the effluent and reactants to produce struvite nanowires.

Biological crystal formation of P compounds (e.g., struvite; magnesium phosphate, etc.), has been demonstrated to be a by-product of the metabolism of specific bacteria that can be found frequently in the environment. Bio-mineral formation refers to a series of processes involving selective extraction, uptake and incorporation of elements from the local environment into functional structures under strict biological control. Organisms can control construction and synthesis of minerals through changes in local chemistry with different shaping strategies. Compared with inorganic minerals, biominerals are often characterized by greater structural sophistication and morphological diversity, and biological functions such as structural support, mechanical strength, protection and storage. Biological struvite (bio-struvite) production has been investigated in sludge dewatering liquors and activated sludge liquors (Leng & Soares, 2018; Pratt *et al.*, 2012; Simoes *et al.*, 2018; Soares *et al.*, 2014). Laboratory scale experiments have shown that bacteria which cause the mineralization of struvite can function in pH from 7 to 10 (Simoes *et al.*, 2018). Furthermore, bacteria can successfully produce biostruvite in wastewater with phosphate concentrations <19.7 mg P/L, which is favorable compared to non-biological struvite precipitation processes. The biostruvite produced met fertilizer purities set by the European Union and heavy metal contaminants remained below directive levels (Leng & Soares, 2018). This research could improve the feasibility struvite recovery by reducing input and management costs, while maintaining the value of struvite as a fertilizer. However, these bacteria need to be tested in mixed cultures and at pilot plant scale to observe the impact competition has on biostruvite yields and whether controls are needed to ensure the right bacteria thrive which could incur unforeseen costs.

10.4 NITROGEN RECOVERY BY AMMONIA STRIPPING AND ABSORPTION

Nitrogen (N) in its reactive forms (ammonium, nitrate and nitrite) is essential for feed and food production for animals and humans (Matassa *et al.*, 2015). Current anthropogenic sources of N are predominately from chemical fixation using the Haber–Bosch process (100 Mt), with the remainder sourced from biological crop fixation (35 Mt) and atmospheric deposition in animal rearing (10 Mt) (Matassa *et al.*, 2015). This provides an almost endless source of N (as reactive ammonia) but the Haber–Bosch process is energy intensive and a significant source of greenhouse gas emissions.

On the other hand, N is an undesirable contaminant in many industrial, domestic and agricultural wastewaters. Current treatment technologies commonly aiming for destructive dissipation via nitrification/denitrification or deammonification using activated sludge processes (Erisman *et al.*, 2008) rather than N recovery. The loss of ammonia and nitrate to the receiving environment causes eutrophication, ammonia can be highly toxic to fish and other aquatic life, and N losses cause N_2O emissions.

Due to the growing demand for reactive N fertilizer and anticipated potential adverse effects, approaches have been proposed to more efficiently convert N into edible protein (Matassa *et al.*, 2015). However, technologies and practices are also required to progressively move wastewater management away from destructive N dissipation towards maximising N recovery and beneficial reuse.

Ammonia stripping and absorptive recovery (Figure 10.20) is among the commercially available and most established technologies for reactive N recovery from digestate, and other concentrated wastewaters and manure (Vaneeckhaute *et al.*, 2017). In ammonia stripping, ammonia-rich water is sparged with a gas (most commonly air), which causes mass transfer of ammonia from the wastewater to the gas phase. The ammonia in the extrant gas stream can then be recovered via condensation, absorption, or oxidation, to produce a concentrated fertilizer product (Mehta *et al.*, 2015).

One example of such a product is ammonium sulfate ($(NH_4)_2SO_4$) produced via the subsequent reaction of ammonia after being absorbed into a sulfuric acid solution. This $(NH_4)_2SO_4$ product, if commercialized, could largely offset the operational costs of ammonia recovery under optimal process conditions (Vaneeckhaute *et al.*, 2017).

Even though ammonia removal can be as high as 98% (Mehta *et al.*, 2015), commercially available processes are generally only operated at 80–90% ammonia removal to reduce operating costs (Vaneeckhaute *et al.*, 2017). Ammonia stripping has high energy and chemical requirements as further described below, and biological removal processes for ammonia such as deammonification via Anammox can often be less costly (Vaneeckhaute *et al.*, 2017). However, unlike biological

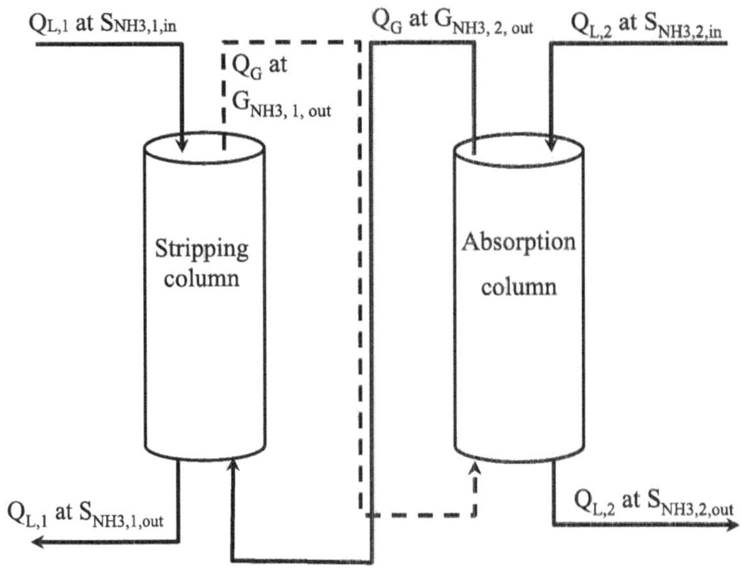

Figure 10.20 Schematic illustrating ammonia stripping and subsequent absorptive recovery, where $Q_{L,1}$ is wastewater flowrate with dissolved ammonia gas concentration of $S_{NH3,1,in}$, Q_G is gas flowrate with ammonia concentration $G_{NH3,1,out}$ (≈0).

processes, ammonia stripping recovers N, and can also remove odorous compounds and dust particles (Vaneeckhaute *et al.*, 2017). Ammonia stripping technology could thus be competitive when there is a significant commercial N-demand (Vaneeckhaute *et al.*, 2017).

In addition to producing a concentrated ammonia-based fertilizer product, ammonia recovery from wastewater and digestate can provide other benefits such as (Anaergia, 2020):

- reduced ammonia concentrations to be land-applied to meet nutrient management requirements;
- reduced potential for ammonia toxicity in anaerobic digestion of high N feedstocks; and
- reduced ammonia loads associated with returned activated sludge and centrate returned to the main water line of a WWTP for biological nutrient removal.

10.4.1 Conceptual overview – ammonia stripping and absorption

Ammonia stripping does not target N forms other than ammoniacal N, and therefore requires organic N to be mineralized into ammonia N prior to recovery. The mass transfer of ammonia from a liquid to gas (stripping), and subsequently from a gas to a liquid (absorptive recovery), is driven by a concentration driving force. This concentration driving force is the difference between dissolved (actual) ammonia concentration in the water phase ($[NH_3(aq)]$) and the solubility concentration of ammonia $[NH_3(aq)]_{eq}$in that water if it was to be at equilibrium with the contacted gas phase, as follows:

$$N'_{NH3} = K_L a\left([NH_3(aq)] - [NH_3(aq)]_{eq}\right) \tag{10.14}$$

where N'_{NH3} is the molar transfer rate of ammonia and $K_L a$ is the overall mass transfer coefficient defined in terms of liquid-phase concentrations.

An important concept in gas–liquid transfer is the two-film theory, where overall resistance to gas-liquid mass transfer could include a resistance to mass transfer on the gas-side of the transfer interface across a 'gas film' and separately a resistance to mass transfer on the liquid-side of the same transfer interface across a 'liquid film' (Treybal, 1981). This leads to a relationship between an overall mass transfer coefficient $K_L a$ (as above) and the separate liquid and gas film mass transfer coefficients ($k_L a$ and $k_G a$, respectively) as follows (Treybal, 1981):

$$\frac{1}{K_L a} = \frac{1}{k_L a} + \frac{1}{K_H \cdot k_G a} \qquad (10.15)$$

where K_H is the Henry's Law constant. This applies to relatively dilute liquid and gas phases at low to moderate pressures (as is the case with ammonia stripping from wastewater). The equilibrium solubility conditions described by Henry's law are as follows:

$$P_{NH3} = K_{H,NH3} \cdot [NH_3(aq)]_{eq} \qquad (10.16)$$

where $K_{H,NH3}$ is the Henry's law constant for ammonia gas in water at a corresponding process temperature, and P_{NH3} is the partial pressure of ammonia in the gas phase. The units of the Henry's law constant can differ from that corresponding to Equation (10.16), for example if gas composition is defined in mole fraction instead of partial pressure.

As a result of an acid-base ion pairin water, ammoniacal N is present in two forms; ammonium ions (NH_4^+), and free ammonia (NH_3, identical to dissolved ammonia gas):

$$NH_4^+(aq) \leftrightarrow NH_3(aq) + H^+(aq) \qquad (10.17)$$

Importantly, only NH_3 participates in gas–liquid mass transfer, not NH_4^+. The proportion of total ammoniacal N [that is $NH_3(aq) + NH_4^+(aq)$] that is NH_3 is influenced by pH according to the following relationship:

$$\frac{[NH_3(aq)]}{([NH_3(aq)] + [NH_4^+(aq)])} \times 100 = \frac{K_{NH4+} \cdot \gamma_{NH4+}}{(K_{NH4+} \cdot \gamma_{NH4+} + 10^{-pH} \cdot \gamma_{NH3})} \times 100 \qquad (10.18)$$

The value of K_{NH4+} ($= 10^{-9.252}$) for the reaction in Equation (10.17) is affected by temperature, and this effect of temperature can be empirically described by a van't Hoff relationship, as follows:

$$\ln\left(\frac{K(T_2)}{K(T_1)}\right) = \frac{\Delta H^0}{R}\left(\frac{1}{T_1} - \frac{1}{T_2}\right) \qquad (10.19)$$

which relates the equilibrium constants $K(T_1)$ and $K(T_2)$ at two temperatures T_1 and T_2 in degrees K, respectively, and where R is the universal gas law constant (8.3145×10^{-3} kJ per mol per K), and ΔH^0 is the standard enthalpy of reaction (52.22 kJ per mol for the reaction in Equation (10.17) at 298 K).

The resulting ion pairing effects can also be calculated at various pH and temperature values using PhreeqC (see section 10.3.1) with the following code:

```
SOLUTION 1
temp 25
pH 8
pe 4
redox pe
units mg/L
density 1
N(-3) 1000
Cl 0.01 charge
- water 1 # kg
```

Note in this code that the simulation is set to balance ionic charge in the water by altering the concentration of a nominal inert counter-ion, in this case chloride, to ensure that the simulation correctly aligns its calculations with the nominally set pH value. To visualize the effects, the values of γ_{NH4+} and γ_{NH3} were nominally set equal to unity, and the resulting proportions of NH_3(aq) calculated using Equation (10.18) with temperature corrections via Equation (10.19). The results are presented in Figure 10.21, showing that the proportion of NH_3(aq) increases at elevated temperature and elevated pH. This is important for ammonia stripping because the efficiency of stripping relies on ammoniacal N being predominately present as NH_3(aq). The stripping unit is typically operated at an elevated pH of between 10.8 and 11.5 (United States Environmental Protection Agency, 2000). While chemical pH adjustment can be used, a potentially efficient means to elevate pH of anaerobic digestate is to pre-sparge with air to strip off dissolved carbon dioxide (CO_2), and because CO_2 is a weak acid, its removal causes a pH rise. The extent to which pH can be elevated in this way depends on background alkalinity (Vaneeckhaute *et al.*, 2017). Alternatively, it may be possible to elevate temperature using waste heat if available from onsite engine generators running on biogas from anaerobic digestion.

If ammonium sulfate is produced by scrubbing, with sulfuric acid solution used as the extractant liquid, the acidic pH conditions will rapidly convert any NH_3 extracted from the gas phase into NH_4+. This then keeps the concentration of NH_3 in the liquid phase low and encourages efficient mass transfer from the gas phase to the liquid phase. The final product can be concentrated up to 38%, yielding a liquid fertilizer of 8-0-0-9 (mass%) as N-P-K-S, and can be further crystallized and dried to yield a dry prilled fertilizer with an analysis of 21-0-0-24 (mass%) as N-P-K-S. The ammonium sulfate crystallization reaction is as follows:

$$2NH_4^+(aq) + SO_4^{-2}(aq) \rightarrow (NH_4)_2SO_4(s) \tag{10.20}$$

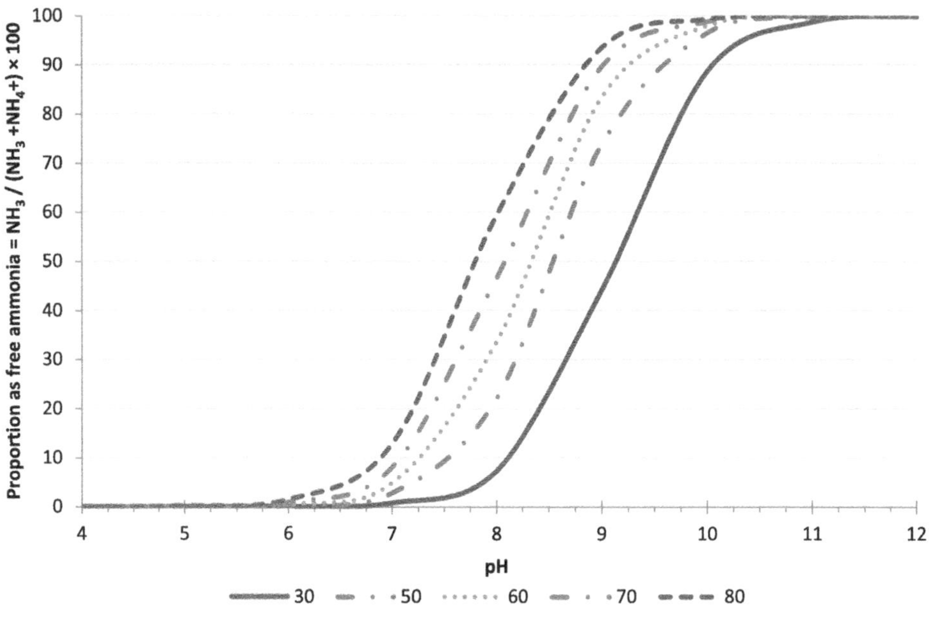

Figure 10.21 Proportion of ammoniacal nitrogen as free ammonia (NH_3), as a function of wastewater pH and operating temperature (°C).

10.4.2 Ammonia stripping and absorption – applications and design

While several possible process configurations could provide the close gas–liquid contact necessary for ammonia stripping and absorption (including plate, spray and bubble columns, and agitated vessels), ammonia stripping processes to date have mostly used packed columns (Vaneeckhaute *et al.*, 2017). Other than packed columns using structured or random packing, columns can instead be fitted with sieve plates or bubble cap trays to provide the close contact between the gas and liquid phases. In these cases the column is not liquid-filled, but is instead operated to limit pressure drop to gas flow, to prevent flooding of the column with liquid or carryover of the liquid into the gas stream (i.e., entrainment), and to ensure that liquid flow is distributed as evenly as possible across the path of the gas flow. In the case of packed columns applied to anaerobic digestate, the main technical challenge to date has been scaling and fouling of the packing material, and the resulting high energy and chemical requirements, such as for lime softening (Vaneeckhaute *et al.*, 2017). Carbonate scale formation can also be an issue, and this could be reduced with prior stripping of CO_2 (Vaneeckhaute *et al.*, 2017). This has the added benefit of saving chemicals required to raise pH to increase the fraction of NH_3 (Figure 10.21).

The tendency for minerals to precipitate (including struvite and other calcium and magnesium minerals) can be evaluated using the saturation index approaches described in section 10.3. A preceding solid–liquid separation step may be required with a packed column stripping unit to remove the majority of suspended solids that may cause clogging or fouling (Vaneeckhaute *et al.*, 2017).

Liquid-filled systems with submerged diffused gas sparging (e.g., Anaergia (2016)) on the other hand have no trays, sieves, loose media or packing that may clog or foul. As such, liquid-filled sparged reactors can be operated at higher total solids concentrations that are common in anaerobic digestion processes (e.g., 8–9% total suspended solids, (Vaneeckhaute *et al.*, 2017)) without requiring prior solids separation. However, diffused gas sparging systems tend to be less efficient than packed or low-profile tray columns (Mead & Leibbert, 1998). The height of the sparged reactor could be limited by power consumption to supply the extractant gas at a pressure above the hydraulic head at the base of the liquid-filled reactor (~1.5–2 meters). The mass transfer area in such systems would be dictated by bubble size, requiring fine bubbles to increase transfer surface area. The gas sparging system may have fine pores prone to clogging (Crittenden *et al.*, 2005) which would lead to subsequent performance decline.

For all ammonia stripping systems, demisters (mist eliminators) may be required at the top of the unit on the gas outlet to prevent liquid droplets being entrained in the exiting gas (Crittenden *et al.*, 2005).

10.4.2.1 Packed column systems

In packed columns, the gas can flow perpendicular (side-to-side) to the liquid flow (usually down), or the gas flows can flow up the column in a counter-current direction to the water or liquid extractant flowing down the column (United States Environmental Protection Agency, 2000). Commercial software packages are available for use by column suppliers to design packed, tray, or plate columns, but here the underlying analytical approaches will be presented for packed column design, because packed columns are most commonly used. The design approach for cross-flow and counter-current flow is similar, so the approach for counter-current flow will be demonstrated.

The first step in performing a preliminary design is to select an appropriate value for the stripping factor (S) defined as follows (Crittenden *et al.*, 2005):

$$S = \left(\frac{Q_G/Q_L}{1/K_H} \right) \tag{10.21}$$

with K_H in this case being dimensionless and Q_G and Q_L being the volumetric extractant gas and water/wastewater flows, each with units of m^3/s, so that S is dimensionless.

The S value indicates a minimum gas-to-water ratio for high treatment efficiencies. If $S < 1$, the stripper will be equilibrium limited and unable to achieve a high degree of removal, whilst at $S = 1$, the operation is at the minimum gas-to-water ratio required for stripping and a high degree of removal (Crittenden *et al.*, 2005). Thus, S should be >1.

A reasonable starting value is $S = 3.5$ for many systems (Crittenden *et al.*, 2005). The gas volumetric flow (Q_G) can be calculated by Equation (10.22) for the selected value of S:

$$Q_G = \frac{Q_L \cdot S}{K_H} \tag{10.22}$$

Calculations of packed column diameter can then use a generalized Eckert pressure drop correlation plot, as shown in Figure 10.22. The Eckert plot represents empirical correlations that describe the pressure drop to gas flow through the packed column as a function of gas mass flow normalized with respect to column projected surface area (G_m, with units kg/(m^2.s)) up to a limit representing the flooding point.

To size the diameter of the packed column, the value on the horizontal axis of the Eckert plot (x) is calculated as follows:

$$x = \frac{L_m}{G_m}\left(\frac{\rho_g}{\rho_l - \rho_g}\right)^{0.5} \tag{10.23}$$

where ρ_l and ρ_g are the respective gas and liquid densities with units of kg/m^3, and L_m is the water mass loading rate normalized with respect to the column projected surface area (kg/(m^2·s). The ratio of L_m and G_m can be calculated from the ratio of Q_G and Q_L as follows:

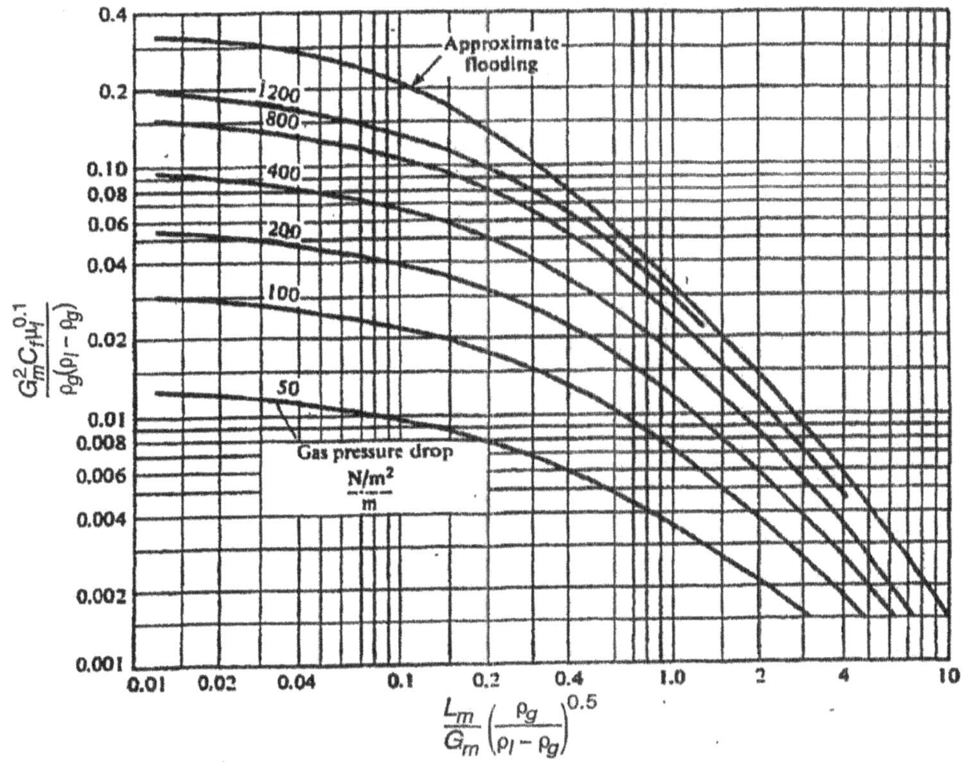

Figure 10.22 Generalised Eckert correlation plot for a random packing tower (Sourced from Crittenden *et al.* (2005) and (Eckert, 1961; Treybal, 1981) cited therein).

$$\frac{L_m}{G_m} = \left(\frac{Q_L}{Q_G}\right) \cdot \left(\frac{\rho_L}{\rho_G}\right) \tag{10.24}$$

Selecting a suitable nominal pressure drop to gas flow, the corresponding value on the vertical axis of the Eckert plot (y) is then read off and the value of G_m is calculated from the following formula:

$$G_m = \left[\frac{y \cdot \rho_G \cdot (\rho_L - \rho_G)}{C_f \cdot \mu_l^{0.1}}\right]^{0.5} \tag{10.25}$$

where μ_l is the dynamic viscosity of water/wastewater being treated by ammonia stripping (kg/(m·s) or Pa.s), and C_f is a unitless packing factor, which is a value available from packing suppliers or can be sourced for common packing types in reference texts such as Crittenden *et al.* (2005). Because the water/wastewater flow is incompressible (unlike the extractant gas flow), further calculations to determine column diameter prefers the use of L_m (instead of G_m) where L_m can be calculated as follows:

$$L_m = \left(\frac{Q_L}{Q_G}\right) \cdot \left(\frac{\rho_L}{\rho_G}\right) \cdot G_m \tag{10.26}$$

The diameter of the packed column is then calculated as follows:

$$D_t = \left(\frac{4 \cdot Q_L \cdot \rho_L}{\pi \cdot L_m}\right)^{0.5} \tag{10.27}$$

There is an important interplay between the nominally selected values of S and column gas pressure drop in the column diameter calculations, simultaneously influencing operational costs by increased gas delivery pressure to overcome in the imposed pressure drop, and influencing capital costs via increasing or decreasing column diameter. For this reason, the designer typically iterates the calculations above until a suitable compromise is reached between operational and capital costs of the design. However, column supplier design software packages incorporate these considerations and others to provide an optimized design for a particular application.

By mass balance analysis it can be shown that the required packing height (l, with units of m) can be calculated as follows (Crittenden *et al.*, 2005):

$$l = \frac{Q_L}{A \cdot K_L a} \cdot \left(\frac{S}{S-1}\right) \cdot \ln\left[\frac{1 + (C_{\text{in}}/C_{\text{out}}) \cdot (S-1)}{S}\right] \tag{10.28}$$

where C_{in} and C_{out} are the influent and effluent concentrations of the constituent undergoing mass transfer (in this case NH_3), A is the column cross-sectional area determined using the column diameter as calculated above.

A value of $K_L a$ can be estimated from Equation (10.14) once the values of $k_L a$ and $k_G a$ have been estimated using the Onda correlations, as follows (Crittenden *et al.*, 2005):

$$\begin{aligned} k_L &= 0.0051 \cdot \left(\frac{L_m}{a_w \cdot \mu_l}\right)^{2/3} \cdot \left(\frac{\mu_l}{\rho_L \cdot D_L}\right)^{-0.5} \cdot (a_t \cdot d_p)^{0.4} \cdot \left(\frac{\rho_L}{\mu_L \cdot g}\right)^{-1/3} \\ k_G &= 5.23 \cdot (a_t \cdot D_G) \cdot \left(\frac{G_m}{a_t \cdot \mu_G}\right)^{0.7} \left(\frac{\mu_G}{\rho_G \cdot D_G}\right)^{1/3} \cdot (a_t \cdot d_p)^{-2} \\ a_w &= a_t \cdot \left[1 - \exp\left[-1.45 \cdot \left(\frac{\sigma_c}{\sigma}\right)^{0.75} \cdot \left(\frac{L_m}{a_t \cdot \mu_l}\right)^{0.1} \cdot \left(\frac{(L_m)^2 \cdot a_t}{(\rho_L)^2 \cdot g}\right)^{-0.05} \cdot \left(\frac{(L_m)^2}{\rho_L \cdot a_t \cdot \sigma}\right)^{0.2}\right]\right] \end{aligned} \tag{10.29}$$

where g is acceleration due to gravity (9.81 m/s^2), D_G is gas-phase diffusivity (m^2/s), D_L is liquid-phase diffusivity (m^2/s), σ is the liquid surface tension (kg/s^2), σ_c is the critical surface tension of the selected packing (kg/s^2), a_t is the specific surface area of the packing (/m), d_p is nominal packing diameter (m), μ_G is gas viscosity (kg/(m·s)), μ_L is water viscosity (kg/(m·s)), a_w is the wetted packing surface area (/m), $k_La = k_L{\cdot}a_w$ and $k_Ga = k_L{\cdot}a_w$, and L_m is the water mass loading rate (kg per m^2 per s). The values of a_t, d_p and σ_c correspond to the selected packing type and are sourced from a reference text for common packing types (e.g., Crittenden *et al.* (2005)) or from a packing supplier for specialist packing types.

In addition to the process design component as described above, packed columns include mechanical internals such as supports to hold the wetted weight of the packing, whilst allowing a free path for the gas and liquid (Sinnott, 1999). Liquid distributors are also important to provide an even distribution of liquid across packed columns, and restributors may also be used to collect liquid to the column walls to be redistributed over the packing medium (Sinnott, 1999). Lastly, hold-down plates or bed limiters can be used to prevent top layers of packing from being fluidized if short periods of high gas flow or surging occur or to limit the expansion of a bed of packing (Sinnott, 1999). Typically, the standard fittings developed by a packing manufacturer are to be specified (Sinnott, 1999). Columns are typically tall with packing heights ranging between 6.1 and 7.6 meters (United States Environmental Protection Agency, 2000). A consideration of the material of construction is also important when using sulfuric acid solution to recover ammonia in the form of ammonium sulfate, and this may require high-grade steel (e.g., AISI 316 stainless steel or Duplex steels 2304) or suitable structural plastic components tolerant of elevated operating temperatures.

10.4.3 Liquid-filled diffused gas-sparged systems

With a liquid-filled column/reactor, small bubbles are produced via diffused gas sparging to rise through the liquid column and cause liquid–gas mass transfer (Mead & Leibbert, 1998). An example of this application is a system by Anaergia (2016). Multiple columns/reactors operating in series may also help to emulate plug-flow behavior and thereby benefit from kinetic effects of the first-order rate of gas-liquid mass transfer (Equation (10.13)). Tanks in series could be physically separated, or potentially separated by internal baffles. The design would depend on the desired extent of ammonia recovery and the operating ammonia concentrations in the wastewater, pH, and temperature. To design a liquid-filled gas-sparged column/reactor, the liquid-phase constitutive balance is solved for an elected n number of complete-mixed reactors in series (Figure 10.23), as follows:

$$V\frac{[\mathrm{NH_3(aq)}]_n}{dt} = Q_L \cdot [\mathrm{NH_3(aq)}]_{in,n} - Q_L \cdot [\mathrm{NH_3(aq)}]_n - V \cdot K_La \cdot \left([\mathrm{NH_3(aq)}]_n - \frac{P_{\mathrm{NH3}}}{K_{H,\mathrm{NH3}}}\right) \tag{10.30}$$

where V is the liquid hold-up volume of each reactor, assumed to be constant, Q_L is the wastewater or liquid extractant flow-rate, also assumed to be constant (a dilute phase), $[\mathrm{NH_3(aq)}]_n$ is the concentration of free ammonia inside the nth mixed reactor (also same as exit concentration), and K_La, $K_{H,\mathrm{NH3}}$ and P_{NH3} are as above, except for use of units consistent with the constitutive mass balance.

For the design, it is necessary to know the mass transfer coefficient K_La. Correlations are available to estimate K_La in gas-sparged reactors (e.g., Treybal (1981)).

10.4.4 Ammonia stripping and absorption – case studies of implementation

10.4.4.1 Case study 1 – anaergia – ammonium recovery from anaerobically digested chicken manure

An example of this process is shown below in Figure 10.24 at an operating facility in a large chicken manure anaerobic digestion plant. Chicken manure, largely due to the high protein content, is high in ammonia and total N. When digested, the N is largely solubilized as ammonia in the liquid phase. Treatment of the liquid was achieved at the facility using a series of four CSTR tanks providing plug flow conditions. The ammonia strippers were supplied with waste heat from a 500 kW combined heat

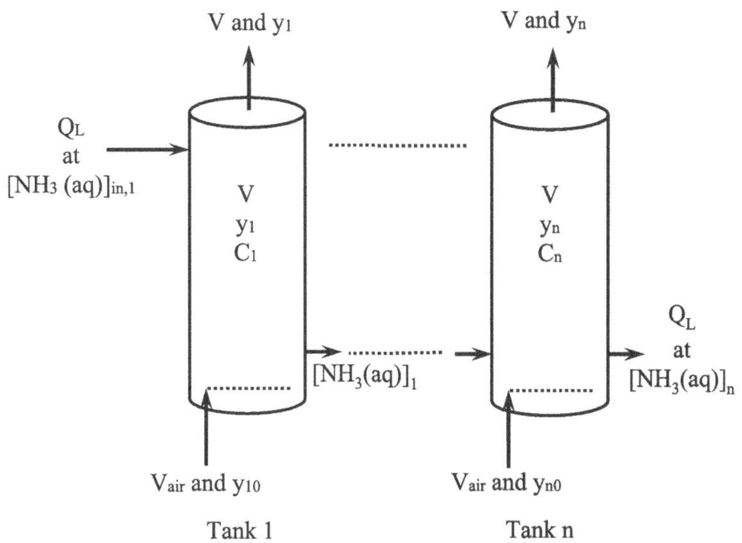

Figure 10.23 Schematic of liquid-filled columns/reactors in series to be used for gas–liquid mass transfer processes (i.e., stripping or absorption).

Figure 10.24 Operating ammonia recovery facility treating the digestate from a large poultry manure anaerobic digester (courtesy of Anaergia).

and power biogas generator. The concentrated ammonia gas was recovered using a single-stage vertical packed tower scrubber. The scrubber applied 93% sulfuric acid, thus transforming the ammonia gas into an acid salt fertilizer of ammonium sulfate.

10.4.4.2 Case study 2: nijhuis water technology B.V. – ammonia from anaerobic digestate

The N-load into an existing biological treatment plant was exceeding its capacity, with further increases forecast due to population growth in the catchment. The largest contributor to the plant, in terms of N-load, was digestate from a poultry slaughterhouse. An ammonia stripping process was installed to recover ammonia directly from the digestate and reduce the N-load into the biological process to defer the need to upgrade the biological process. The digestate flow was 75 m^3/d with a load of 2.5 g NH_4-N/L. The stripping process consisted of a heat exchanger, CO_2 stripper, ammonia stripper and a scrubber, and aimed to recover 70% of digestate ammonia. The influent was heated to 70°C, predominately through a tube-in-tube heat exchanger on the effluent flow, with additional heat input available from cogeneration plant offgas, that is to generate additional heating. Fresh air was injected into a CO_2 stripper tank to remove CO_2 and the high alkalinity in the digestate liquor ensured that the pH of the process stream could be increased to around 9.0 during CO_2 stripping without the addition of any chemicals, that is the loss of CO_2 raised pH

Offgas from both CO_2 and ammonia strippers was treated in separate ammonia scrubbers with sulfuric acid added to produce ammonium sulgate. Recirculated air from the scrubber was used as carrier gas in the ammonia stripper and a polypropylene packing material (38 mm nominal size, 150 m^2/m^3 surface area) was added to the stripper to increase the effective contact surface area between liquid and gas phases (Figure 10.25). The stripper consistently achieved recovery efficiencies of 75–80% (even without caustic addition) and reduced the effluent ammonium concentration to 0.5–1.0

Figure 10.25 High surface area packing material used in the Nijhuis ammonia recovery system.

NH_4-N g/L. There was a decrease in pH and temperature throughout the stripping column, which reduced the recovery efficiency relative to the theoretical amount, however increasing the influent pH and temperature to allow for this reduction in pH and temperature could be implemented to increase recovery efficiency even further (albeit with an increase in operating costs).

In addition to costs associated with heating (0.4 €/kg N recovered), sulfuric acid (0.4 €/kg N), maintenance (0.10 €/kg N), and installed power (0.36 €/kg N), there was a cost associated with providing anti-scalant to prevent clogging in the stripper columns. Another solution could be to encourage scaling in the CO_2 stripping tank by reducing alkalinity and removing the precipitate mechanically, which increased capital costs slightly but reduced costs associated with scale removal by 0.35 €/kg N.

Costs associated with ammonium recovery from the digestate are offset slightly by the sale of recovered ammonium sulfate to local farmers, where the availability of local consumers for this scheme increased revenue from the sale of ammonium sulfate as well as contributing to creating a circular economy regarding waste treatment. At this site, ammonium sulfate was sold at a cost of 0.08 €/kg N. The cost effectiveness of the process increased at higher total ammoniacal N as a greater fraction of ammonia can be recovered at higher concentrations. At total ammoniacal N concentrations of >2 g/L, ammonium stripping was found to become more financially attractive than side-stream biological processes commonly applied to reduce the ammonium load in digestate returned to mainline biological treatment processes.

10.4.5 Ammonia stripping and absorption – challenges, opportunities, and research needs

As resources become more and more scarce and the demand for more sustainable agricultural practices increase, the need for N recovery will increase. While the technology can be economically applied using acid scrubbing agents like sulfuric acid, these too have sustainability concerns and inherent health and safety risks. Future research needs will focus on alternate scrubbing agents which can themselves be recovered from waste products (Jamaludin *et al.*, 2018). Moreover, as for struvite, regulatory and farmer's acceptance of the fertilizer product remains a challenge. Improved methods and technologies for recovering heat from, for example, onsite digesters or neighboring infrastructure, would reduce the energy demand for shifting equilibrium towards ammonia gas. This would lead to more installations in WWTPs. Scaling of packing material can lead to forced shutdown of stripping columns and reduced throughput. Thus, new methods for controlling scaling, in terms of understanding wastewater constituents that contribute to this in stripping columns and potentially removing them in targeted pre-treatments, would reduce the need for lengthy shutdowns. Additionally, there is still scope for developing new packing materials that promote better liquid–gas interactions, and further minimize fouling and scaling.

Better integration into wastewater treatment plant processes could consider ammonia stripping together with other resource recovery processes (e.g., struvite), could optimize whole-of-plant strategies for temperature and pH control, minimize chemical salts addition (e.g., caustic), and minimize the overall cost of N recovery.

10.5 SULFUR RECOVERY

Sulfur is one of the most abundant elements on earth and is essential for life, being a key constituent of many amino acids and an essential nutrient for plant growth. Until recently, elemental sulfur was mostly obtained by the Frasch hot-water method, by mining native sulfur associated with the caprock of salt domes, which often also held oil and natural gas (U.S. Geological Survey, 2002). Currently, due to the depletion of most sub-surface deposits, global sulfur production is mainly linked to petrochemical operations and recovered via the de Claus process (Eow, 2002).

Sulfur can prevail in many oxidation states and it exists in a great number of different chemical forms within its oxidation states (Cai *et al.*, 2017). Most of them play a very important role in aqueous systems, where redox reactions occur either microbially or chemically following the thermodynamics of the system. Sulfur is highly redox sensitive and the many possible oxidation/reduction reactions make the sulfur chemistry quite complex. Sulfur species are also constituents of a large number of industrial products; for instance, sulfuric acid and sulfite and SO_2, and sulfides represent the major environmental pollutants. Additionally, hydrogen sulfide (H_2S), which is mainly found in gas form, is a significant human health hazard that causes death through asphyxiation.

This sub-section focuses on the basic principles and application of technologies for the recovery of sulfur from wastewater and biogas, which allow not only the recovery of an important resource that may become scarce when phasing out fossil resources, but also addresses possible environmental, health and economic issues derived from the presence of sulfur in those streams.

10.5.1 Conceptual overview

Sulfur can be present in the environment in many chemical forms. In what concerns sulfur recovery in the wastewater context, the focus lays mostly on hydrogen sulfide (H_2S), its most reduced form, either when: i) dissolved at high concentrations in industrial sour water, for example coming from the petrochemical sector; or ii) present in biogas produced by anaerobic digestion from organic-rich wastewater or waste activated sludge. Desulfurization technologies are, thus, key to treat those biogas/water streams so that they can be (re)used/valorized, also providing an opportunity for the recovery of sulfur. The most common approach to desulfurize biogas is to absorb H_2S in an alkaline solution in a counter-current scrubber similar to that described for ammonia removal in section 10.4. This sulfide-rich solution is regenerated in a biological oxidation step, in which sulfide is oxidized to elemental sulfur that can be recovered as a semi-solid or slurry.

10.5.2 Fundamental principles

The removal and recovery of sulfur from biogas requires two distinct steps, absorption of biogas in an aqueous solution and its biological oxidation. The removal of sulfides from liquid streams may only require the second of the two. In this sub-section, the key fundamental principles of both steps are briefly described.

10.5.2.1 Sulfide absorption

Sulfide in gaseous streams, for instance biogas, needs to be transferred into the liquid phase for treatment. Certain technologies can also treat sulfide in the gas phase, but they are either not applicable to the wastewater context or do not allow the recovery of sulfur (e.g., Clauss process, Sulfatreat process, etc).

Sulfide in aqueous solution is a weak acid and dissociates as follows:

$$H_2S_{aq} \rightleftarrows HS^- + H^+ \quad pK_1 = 7(20°C) \tag{10.31}$$

$$HS^- \rightleftharpoons S^{2-} + H^+ \quad pK_2 = 17(20°C) \tag{10.32}$$

H_2S has a high water solubility, ranging between 3000 and 4000 mg/L at atmospheric pressure and normal temperature. The H_2S solubility in water decreases with increasing temperatures.

Sulfide can be easily transferred from the gas to the liquid phases in absorption columns similar to those described in section 10.4. Since their principles design and operation are very similar, they will not be further discussed here. For H_2S absorption, alkaline solutions (pH >9–10) are needed to ensure >99% of H_2S is transferred from the gas to the liquid phase, and absorbed in their ionic forms.

10.5.2.2 Sulfide oxidation

Sulfide can be oxidized either chemically or biologically. Due to their ability to convert H_2S into elemental sulfur, sulfide oxidizing bacteria are suitable to be used in biological desulfurization processes. Chemical sulfide oxidation can also concomitantly take place. Both processes are discussed here.

Biological sulfide oxidation. A wide variety of microorganisms are known for their capability to oxidize reduced H_2S under a broad range of environmental conditions, for example Thioalkalivibrio, Thiobacillus, Thiomonas, and so on. (Kelly *et al.*, 1997; Sorokin *et al.*, 2011). The oxidation of sulfide yields energy, which enables bacterial growth. Oxidation of H_2S by sulfur oxidizing bacteria can be divided into phototrophic and chemotrophic processes. Phototrophic bacteria catalyze the so called 'van Niel' reaction under anaerobic conditions (Barton & Fauque, 2009), while chemotrophic ('colorless') sulfide oxidizing bacteria obtain their energy from the aerobic (with oxygen) or anoxic (with nitrate) oxidation of reduced sulfur compounds, for example sulfide, sulfur and sulfite.

The following general reactions describe the major phototrophic 'van Niel' biological hydrogen sulfide oxidation reactions:

$$H_2S + CO_2 + light \rightarrow (CH_2O)_n + H_2O + S^0 \quad (10.33)$$

$$H_2S + 2CO_2 + 2H_2O + light \rightarrow 2(CH_2O)_n + H_2SO_4 \quad (10.34)$$

The main chemotrophic biological hydrogen sulphide oxidation reactions are given by:

$$2H_2S + O_2 \rightarrow 2S^0 + 2H_2O \quad (10.35)$$

$$H_2S + 2O_2 \rightarrow H_2SO_4 \quad (10.36)$$

$$5H_2S + 8NO_3^- + 4H_2O \rightarrow 5H_2SO_4 + 4N_2 + 8OH^- \quad (10.37)$$

Abiotic sulfur conversions. Abiotic oxidation of H_2S with oxygen is relatively slow compared to oxidation by chemolithotrophic sulfide oxidizing microorganisms. The main oxidation product of abiotic oxidation is $S_2O_3^{2-}$ (O'Brien & Birkner, 1977):

$$2H_2S + 2O_2 \rightarrow S_2O_3^{2-} + H_2O + 2H^+ \quad (10.38)$$

Transformations of sulfur compounds can also occur as the result of chemical conversions when pH, temperature and oxidation-reduction conditions are favorable. Examples can be found in the formation of polysulfides (Kleinjan *et al.*, 2006), which form when elemental sulfur and dissolved sulfide react. Other sulfur compounds which can be chemically formed are inorganic elemental sulfur and thiosulfate. Chemically formed elemental sulfur is poorly dispersable in water, which is distinctively different when compared to sulfur formed by biological oxidation of sulfide. The origin of this hydrophilicity of biologically produced sulfur can be explained by adsorbed organic polymers such as proteins at the surface of the produced sulfur particles (Kleinjan, 2005).

Inorganic polysulfur compounds, that is polysulfide (S_x^{2-}) and their methylated form (R_2S_x), play a role in both biological and geological sulfur cycles as many biochemical mechanisms depend on this form of sulfur. In the presence of both sulfur and sulfide, polysulfides are formed according to (Roman *et al.*, 2014):

$$HS^- + (x-1)/8S_8 \rightarrow S_x^{2-} + H^+ \quad (10.39)$$

The polysulfide ions can be present in aqueous solution as complex mixtures of S_x^{2-} with $x \geq 2$. The formation of various polysulfide species is rapidly established based upon a chemical equilibrium. For biologically produced sulfur, the equilibrium constant is in on the order of $pK_x = 9.10$ (21°C) and polysulfides have an average chain length in of approximately the order of 4–5. The pK_x value for 'inorganically' produced sulfur is significantly lower ($pK_x = 8.78$; 21°C) (Kleinjan, 2005).

10.5.3 Applications and design

The basic concept of the biological desulfurization process (Thiopaq®) and recovery of elemental sulfur was first developed for the removal of H_2S from biogas in the late 1980s (Buisman *et al.*, 1990).

Biogas often contains some H_2S due to microbiological reduction of sulfur compounds, especially sulfate, present in the feed to anaerobic reactors. H_2S needs to be removed before combustion of the biogas to prevent the corrosion of the engines and their accelerated depreciation.

In the Tiopaq® process, H_2S-containing gas is humidified during bed irrigation, and subsequently contacted with microorganisms attached to a fixed bed (Graaf, 2012). One of the benefits of the process, next to the desulfurization of the gas stream, is the recovery of biologically produced elemental sulfur particles. The first full-scale facility for biogas desulfurization was built in 1993 in Eerbeek, the Netherlands (Janssen *et al.*, 2009). Since then, more than 270 installations have been built worldwide (Klok *et al.*, 2018), especially for desulfurization of biogas and several types of gas in the oil and gas industry. The technology is suitable for high pressure (up to 80 bar) as well as low pressure (below atmospheric pressure) feed gases and with a wide range of inlet H_2S concentrations. The operational parameters are shown in Table 10.2.

Biogas is typically a mixture of methane, carbon dioxide (CO_2), H_2S and higher alkanes. The composition and pressure of each gas varies widely and depends mainly on the gas source. Since H_2S and CO_2 are 'acidic gases', the Thiopaq process requires an alkaline solution to remove these compounds from a gas stream. The haloalkaline conditions which are applied in the process are established by a sodium carbonate/bicarbonate buffer solution (0.3–1.5 M) with a pH of 8.0–9.5. High CO_2 partial pressures will affect the removal of H_2S from gas streams when scrubbing the gas with an alkaline solution, and therefore the process requires elevated carbonate concentrations.

The process for desulfurization of 'sour' gas streams under haloalkaline conditions consists of three process steps (Figure 10.26): (1) absorption of H_2S in the process solution, (2) sulfide oxidation to elemental sulfur (S_8) in an aerated bioreactor and (3) removal of the formed S_8 from the suspension. In the first step, sour gas is directed to the bottom section of an absorber column, where it is put into contact with a counter current of washing solution, thereby consuming buffer capacity (see Table 10.3). The desulfurized gas, that is H_2S levels of <4 ppmv, leaves the absorber at the top for further processing.

The washing solution, containing the dissolved (bi)sulfide (HS^-), is fed to the bioreactor. Here, a mixed culture of sulfide oxidizing bacteria, dominated by *Thioalkalivibrio* spp., oxidize sulfide to primarily elemental sulfur (S_8), consuming O_2 as the final electron acceptor (Table 10.3, reaction 3). The O_2 is supplied to the bioreactor via the injection of air. In Figure 10.27, a scanning electron microscope (SEM) picture shows sulfide oxidizing bacteria excreting sulfur particles.

In addition to S_8, a fraction of sulfide is biologically oxidized to sulfate (SO_4^{2-}) (see Table 10.3, reaction 4). Besides biological oxidation, various chemical reactions can occur. Thiosulfate ($S_2O_3^{2-}$) is the main chemically formed intermediate of the oxidation of sulfide and polysulfide (Table 10.3, reactions 5 and 7). The selectivity for product formation depends on reactor conditions such as substrate levels, temperature and pH. The formation of sulfate and thiosulfate is unwanted for several

Table 10.2 General operational parameters (de Rink *et al.*, 2020; Driessen *et al.*, 2011; Echt *et al.*, 2017).

Parameter	Operational Window	Unit
Feed gas pressure	0–80	bar
H_2S levels in feed gas	0.05–100	% mol
Temperature	20–40	°C
Alkalinity/buffer capacity	0.3–1.5	$mol \cdot L^{-1}$
Conductivity of process solution	20–90	$mS \cdot cm^{-1}$
pH of process solution	8.0–9.5	–

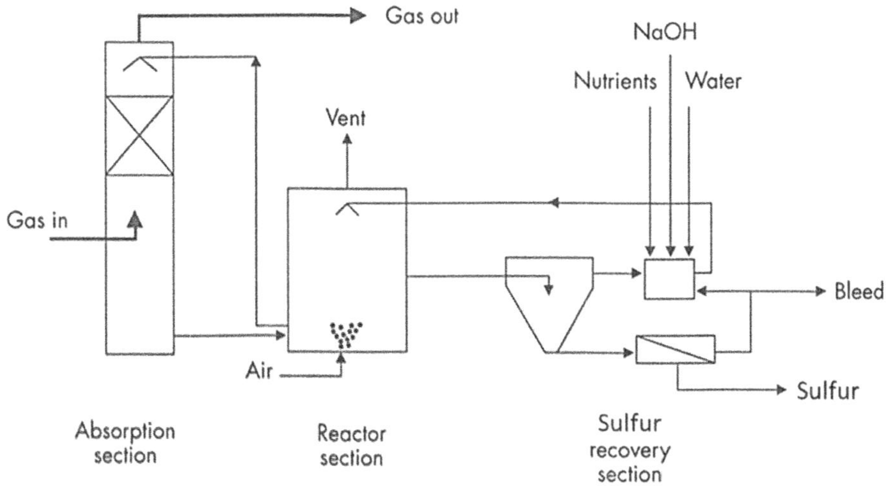

Figure 10.26 Flow scheme for the biological desulfurization process.

Table 10.3 Main reaction equations occurring in the biological desulfurization process. Bio: biological conversion. Chem: chemical conversion (from Klok (2015)).

	Reaction	Bio/Chem	Remarks
1	$H_2S_{(g)} \leftrightarrow H_2S_{(l)}$	Chem	–
2	$H_2S_{(l)} + OH^- \leftrightarrow HS^- + H_2O$	Chem	–
3	$HS^- + ½O_2 \rightarrow S^0 + OH^-$	Bio	–
4	$HS^- + 2O_2 \rightarrow SO_4^{2-} + H^+$	Bio	–
5	$2HS^- + 2O_2 \rightarrow S_2O_3^{2-} + H_2O$	Chem	–
6	$HS^- + (x-1)S^0 \leftrightarrow S_x^{2-} + H^+$	Chem	pH >8.0
7	$S_x^{2-} + 1½O_2 \rightarrow S_2O_3^{2-} + (x-2)S^0$	Chem	–

reasons. First, (thio)sulfate formation leads to the formation of protons and thus acidification of the medium, leading to more caustic consumption. Second, the addition of make-up water is required as (thio)sulfate can only be removed via a bleed stream, which is a waste stream purged from the process to control salinity levels. To optimize the formation of S_8, the bioreactor is operated at low oxygen/low oxidation-reduction potential (ORP) levels (i.e., ORP values below –200 mV) (Janssen *et al.*, 1998).

The effluent of the bioreactor i s primarily recirculated to the top of the absorber. In addition, a side flow from the reactor is directed to a gravity settler to remove sulfur from the process. From the bottom of the settler, a sulfur slurry is directed to a decanter centrifuge to further dewater the sulfur and form a sulfur cake. Nutrients to promote the growth of the bacteria, water and a concentrated caustic solution are added to the filtrate before the stream is returned to the bioreactor.

10.5.4 Case studies of implementation

The Thiopaq® process can be applied to the treatment of various H_2S-rich gaseous streams, including biogas, natural gas, and so on. An example of a full scale Thiopaq® installation can be found at

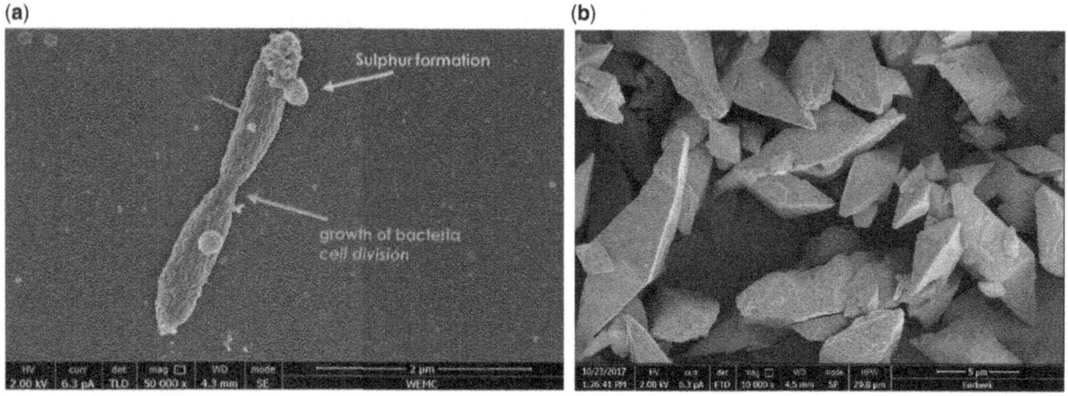

Figure 10.27 (a) Scanning electrode microscopy (SEM) picture of sulfide oxidizing bacteria excreting sulfur particles. Sulfur is artificially colored yellow for clarity. (b) Sulfur crystals in the final product.

an oil field in Southern Illinois, USA, containing over 200 smaller production wells. While not an example of desulfurization and sulfur recovery applied to the water sector, it is a very comprehensive example of S-recovery from an H_2S-rich gaseous stream and similar principles can be applied to the desulfurization of biogas.

The oil field considered here covers about 40 km^2 and produces approximately 1500 barrels of oil per day. Oil wells may contain natural gas, which is released during crude oil extraction. This so-called associated gas contains valuable condensable hydrocarbon liquids but is usually considered as waste in oil production. At the oil field in Southern Illinois, the associated gas had been flared for over 70 years, oxidizing H_2S in the feed gas and emitting substantial amounts of SO_2 to the atmosphere.

In order to reduce SO_2 emission levels for improvement of air quality, and to recover the condensable hydrocarbons, a Thiopaq and a liquefied petroleum gas (LPG; a mixture of propane and butane) recovery system were installed for gas treatment. The biological desulfurization plant (Thiopaq O&G) was commissioned in 2006 (see Figure 10.28(b)) (Lanning *et al.*, 2008). As the facility was 'skid built', all equipment was built to be road transportable, and consequently two smaller bioreactors instead of one large bioreactor were constructed (see Figure 10.28(a)).

The feed gas flow to the absorber is approximately between 800 and 1100 normal cubic meters·h^{-1} at a pressure of 4 bar. The major components in the gas are specified in Table 10.4 (Roman *et al.*, 2016). The feed gas contains concentrations of H_2S from 1 to 5 mol%, resulting in a total sulfide loading of 1–1.4 tons of sulfur per day (Figure 10.28(c)). The treated gas stream (i.e., <4 ppmv H_2S) is directed to an LPG recovery section where the LPG is recovered and stored. After LPG recovery, the remaining gas (~60 mol% CH_4 and ~30 mol% C_2H_6) is sold to the local distribution grid.

10.5.5 Challenges, opportunities and research needs

The activity and selectivity of the microorganisms are vital parameters in the design and operation of any biotechnological process. In the case of the biological desulfurization process, the fraction of sulfide which is converted into elemental sulfur should be maximized. In the current process, the selectivity of the sulfide to elemental sulfur conversion amounts to 90–94%. Hence, a significant amount of sulfuric acid is still formed, which requires the addition of caustic and results in formation of more bleed. When the selectivity towards sulfur formation is increased, the process becomes more efficient, resulting in lower caustic consumption and bleed formation. The latter is a side-stream, that is waste stream, of the process solution to control salinity levels. In full-scale facilities, two main factors determine the operational costs: (i) consumption of chemicals and (ii) formation of a

Figure 10.28 Pictures of the Thiopaq facility in Southern Illinois. (a) Overview picture of the facility with control room, absorber section, bioreactor section and sulfur recovery section. (b) Picture of the bioreactor section. (c) Picture of the recovered sulfur cake from the decanter centrifuge.

waste stream (bleed stream). Both factors are associated with the biological formation of sulfate and chemical formation of thiosulfate. Biological desulfurization plants yielding over 150 tons of sulfur per day would become economically feasible when selectivity for sulfur formation increases to >98%, which would largely reduce the largest operational costs. Improving selectivity towards elemental sulfur formation represents a challenge for future research.

Table 10.4 Inlet and outlet gas composition of the biological desulfurization unit found at an oil well in southern illinois (Roman *et al.*, 2016).

Component	Inlet Gas	Outlet Gas
H_2S	1–5 vol. %	<4 ppm(v)
CO_2	1–3 vol. %	<2 vol. %
CH_4	40–50 vol. %	40–50 vol. %
C_2H_6	15–20 vol. %	15–20 vol. %
C_3H_8	12–17 vol. %	12–17 vol. %
C_4+	10–15 vol. %	10–15 vol. %
H_2O	<2 vol. %	Saturated

10.6 CHAPTER SUMMARY

This chapter describes the fundamental principles of P, N and S recovery by three very different mixtures of physical, chemical and biological approaches; P-recovery by struvite precipitation invoking concepts of crystallization, N-recovery by stripping employing concepts of liquid–gas phase equilibrium, mass transfer and extraction, and S-recovery by an essentially biological process. While the theory underpinning each of the technologies is very different, they all provide a way to stem the loss of critical nutrients to the environment. Furthermore, they are faced by similar challenges with regards to product (i.e., purity, legislation, finding suitable end users) and operation (how best to integrate into other processes, interference from other components in wastewaters etc). Nonetheless, all technologies are being applied at full scale and have shown the potential to be economically feasible. Their economically feasibility will only increase as each nutrient becomes more limited and their value increases. Improvements in recovering energy (e.g., better heat exchanges, renewable sources etc), will further increase the uptake of ammonia recovery by stripping. Similar developments in the other technologies, coupled with increasing demand, will increase the number of their installations. This will in turn lead to greater awareness and acceptance of the final products, and ultimately the establishment of markets for these nutrients recovered from wastewaters.

10.7 EXERCISES

A landfill site produces leachate at a liquid flowrate of 5 m^3/h with a total ammoniacal nitrogen concentration of 700 mg NH_4-N/L at pH 9.5 and 80°C ($\rho_l = 971.76\ kg/m^3$; $\mu_L = 3.537 \times 10^{-4}$ Pa.s; $\sigma = 0.0626$ N/m). A stripping column is to be designed using a random packed tower containing Plastic tellerettes packing ($C_f = 20$; $a_t = 112\ m^2/m^3$; $d_p = 0.0508$ m; $\sigma_c = 0.033$ N/m; (Crittenden *et al.*, 2005)) to remove 95% of the free ammonia from the leachate using air at 80°C and 1 atm as the extractant gas ($\rho_G = 1\ kg/m^3$; $\mu_G = 20.88 \times 10^{-6}$ Pa.s). This aims to assist the final polishing of any remaining ammonia from the leachate in a subsequent biological nutrient removal plant.

Phase equilibrium for the dissolution of gaseous NH_3 in water can be described by Henry's Law, $K_{H,NH_3} = 1818$ Pa/M at 298 K (Sander, 2015), with $\Delta H^0/R$ for the dissolution of ammonia from air into water being 4100 (Sander, 2015). Diffusivity of ammonia in air is $D_G = 0.28 \times 10^{-4}\ m^2/s$ (Incropera & DeWitt, 2001), and diffusivity of ammonia in water is $D_L = 2 \times 10^{-9}\ m^2/s$ (Perry & Green, 1997; Table 2-372).

Exercise 10.1: Determine Henry's Law constant under the operating conditions.

Exercise 10.2: What is the dimensionless Henry's Law constant?

Exercise 10.3: What is the minimum gas flowrate for stripping the ammonia?

Exercise 10.4: Determine the diameter of a stripping column appropriate for the application.

Exercise 10.5: Determine the height of the stripping column.

10.8 DISCUSSION QUESTIONS

Question 10.1: What are the advantages and disadvantages of struvite precipitation on a sidestream (sludge) line treating digestate from an anaerobic digester. Can struvite precipitation be applied to the water mainline at a WWTP?

Question 10.2: What are the pollutants of concern if struvite is recovered from a municipal wastewater treatment plant and is to be used as a fertilizer in agriculture? To answer this, refer to relevant legislation.

Question 10.3: Nutrient recovery in WWTPs is slowly being implemented worldwide, please comment on the sentence: 'the bottleneck is not the technology but finding a welcoming market and suitable regulations'.

Question 10.4: How feasible would it be to implement P recovery in the following scenarios? For this, reflect on a likely amount and concentration of nutrients in the wastewater in each case, and think about a suitable technology (or technologies):

- >100 000 PE WWTP with BNR and sludge stabilization through thermal processing
- 200 000 PE WWTP with BNR and sludge stabilization through Cambi (AD)
- 20 000 PE WWTP with BNR and sludge dewatering. Sludge is stabilized in nearby WWTP
- 10 000 PE WWTP with trickling filters + chemical P removal + and sludge dewatering. Sludge is stabilized in nearby WWTP
- Decentralized system – one block of flats with yellow and black water separation

Question 10.5: Using the principles of a first-order reaction, explain why multiple tanks in series would be a more efficient process to air strip ammonia from a liquid slurry which contains 5000 mg/L NH_3-N?

Question 10.6: Select three different temperatures and pH levels which achieve an ammonia unionization of at least 90% NH_3. Discuss the implications of how to achieve the conditions from a chemical and thermodynamic perspective.

Question 10.7: Discuss consequences of choosing a higher/lower temperature for ammonia stripping in terms of effect of mass transfer coefficient and ammonia/ammonium equilibrium?

Question 10.8: Describe how fouling may affect efficacy of ammonia recovery (in terms of mass transfer, flooding, packing parameters etc)?

Question 10.9: Can stripping and struvite precipitation be implemented in the same process? Discuss.

ACKNOWLEDGEMENTS

Marcel Giesbers of the Wageningen Electron Microscopy Centre (WEMC) kindly provided the SEM pictures in Figure 10.27.

REFERENCES

Anaergia (2016). Ammonia stripper. United states patent US 2016/0130158 A1. Available at: http://www.patentsencyclopedia.com/imgfull/20160130158_01 (accessed 21 April 2021).

Anaergia (2020). Your Partner for a More Sustainable World. Turning Waste into Renewable Energy, Fertilizer, and Water for Reuse. Available at: https://www.swx.global/wp-content/uploads/2020/09/Anaergia-Product-Line.pdf (accessed 21 April 2021).

Anne C. O., Trebouet D., Jaounen P. and Quemeneur F. (2001). Nanofiltration of seawater: fractionation of mono- and multi-valent cations. *Desalination*, **140**, 67–77.

Barton L. L. and Fauque G. D. (2009). Biochemistry, physiology and biotechnology of sulfate-reducing bacteria. In: Advances in Applied Microbiology, A. J. Laskin, S. Sariaslani and G. M. Gadd (eds.), Academic Press, Amsterdam, The Netherlands, Vol. **68**, pp. 41–98.

Brockmann M. (2015). Struvia Technology for Phosphorus Recovery. Re-Water Braunschweig, 2–3 November 2015, Braunschweig, Germany.

Buisman C. J. N., Geraats B. G., Ijspeert P. and Lettinga G. (1990). Optimization of sulphur production in a biotechnological sulphide-removing reactor. *Biotechnology and Bioengineering*, **35**(1), 50–56.

Cai J., Zheng P., Quaisar M. and Zhang J. (2017). Elemental sulfur recovery of biological sulfide removal process from wastewater: a review. *Critical Reviews in Environmental Science and Technology*, **47**(21), 2079–2099.

Chen Y., Feng Y., Deveaux J. G., Masoud M. A., Chandra F. S., Chen H., Zhang D. and Feng L. (2019). Biomineralization forming process and bio-inspired nanomaterials for biomedical application: a review. *Minerals*, **9**(2), 1–21.

Chimenos J. M., Fernandez A. I., Hernandez A., Haurie L., Espiell F. and Ayora C. (2006). Optimization of phosphate removal in anodizing aluminum wastewater. *Water Research*, **40**(1), 137–143. https://doi.org/10.1016/j.watres.2005.10.033

Cordell D. and White S. (2015). Tracking phosphorus security: indicators of phosphorus vulnerability in the global food system. *Food Security*, **7**(2), 337–350.

Crittenden J. C., Trussell R. R., Hand D. W., Howe K. J. and Tchobanoglous G. (2005). Water Treatment Principles and Design, 2nd edn. John Wiley & Sons, Hoboken, New Jersey.

De Ridder M., de Jong S., Polchar J. and Lingemann S. (2012). Risks and Opportunities in the Global Phosphate Rock Market. Available at: www.hcss.nl (accessed 9 October 2019).

de Rink R., Klok J. B. M., van Heeringen G. J., Keesman K. J., Janssen A. J. H., ter Heijne A. and Buisman C. J. N. (2020). Biologically enhanced hydrogen sulfide absorption from sour gas under haloalkaline conditions. *Journal of Hazardous Materials*, **383**, 1–8.

Doyle, J. D. and Parsons, S. A. (2002). Struvite Formation, Control and Recovery. *Water Research*, **36**, 3925–3940.

Driessen W., Zessen E. and Visser M. (2011). Full-Scale Expererience with Biological Biogas Desulphurization Using the Thiopaq Technology. 16th European Biosolids and Organic Resources Conference, Leeds, UK.

Echt B., Leppin D., Mamrosh D., Mirdadian D., Seeger D. and Warren B. (2017). Low-Tonnage Sulfur Removal and Recovery. Laurance Reid Gas Conditioning Conference, Norman, Oklahoma.

Eckert J. S. (1961). Design techniques for sizing packed towers. *Chemical Engineering Progress*, **57**(9), 54–58.

Egle L., Rechberger H. and Zessner M. (2015). Overview and description of technologies for recovering phosphorus from municipal wastewater. *Resources, Conservation and Recycling*, **105**, 325–346.

Eow J. (2002). Recovery of sulfur from sour acid gas: a review of the technology. *Environmental Progress and Sustainable Energy*, **21**(3), 143–162.

Erisman J. W., Sutton M. A., Galloway J., Klimont Z. and Winiwarter W. (2008). How a century of ammonia synthesis changed the world. *Nature Geoscience*, **1**(10), 636–639.

Etter B., Tilley E., Khadka R. and Udert K. (2011). Low-cost struvite production using source-separated urine in Nepal. *Water Research*, **45**(2), 852–862.

European Sustainable Phosphorus Platform (2019). Success Stories. Available at: https://www.phosphorusplatform.eu/success-stories (accessed 4 October 2019).

FuturEnviro (2017). Spain's First Phosphorus Recovery Plant. June, 2017. Available at: https://futurenviro.es/en/spains-first-phosphorus-recovery-plant/ (accessed 12 December 2021).

Galbraith S. C., Schneider P. A. and Flood A. E. (2014). Model-driven experimental evaluation of struvite nucleation, growth and aggregation kinetics. *Water Research*, **56**, 122–132.

Galloway J. N., Winiwarter W., Leip A., Leach A. M., Bleeker A. and Erisman J. W. (2014). Nitrogen footprints: past, present and future. *Environmental Research Letters*, **9**(11), 1–11.

Graaf M. D. (2012). Biological Treatment of Spent Caustics Under Haloalkaline Conditions Using Soda Lake Bacteria. Wageningen University, the Netherlands.

Hinckley E.-L. S., Crawford J. T., Fakhraei H. and Driscoll C. T. (2020). A shift in sulfur-cycle manipulation from atmospheric emissions to agricultural additions. *Nature Geoscience*, **13**(9), 597–604.

Incropera F. and DeWitt D. (2001). Frank Incropera's Introduction to Heat Transfer, 4th edn. Wiley, Hoboken, New Jersey, USA.

Jamaludin Z., Rollings-Scattergood S., Lutes K. and Vaneeckhaute C. (2018). Evaluation of sustainable scrubbing agents for ammonia recovery from anaerobic digestate. *Bioresource Technology*, **270**, 596–602.

Janssen A. J. H., Meijer S., Bontsema J. and Lettinga G. (1998). Application of the redox potential for controling a sulfide oxidizing bioreactor. *Biotechnology and Bioengineering*, **60**(2), 147–155.

Janssen A. J. H., Lens P. N. L., Stams A. J. M., Plugge C. M., Sorokin D. Y., Muyzer G., Dijkman H., Van Zessen E., Luimes P. and Buisman C. J. N. (2009). Application of bacteria involved in the biological sulfur cycle for paper mill effluent purification. *Science of the Total Environment*, **407**(4), 1333–1343.

Kataki S., West H., Clarke M. and Baruah D. C. (2016). Phosphorus recovery as struvite: Recent concerns for use of seed, alternative Mg source, nitrogen conservation and fertilizer potential. *Resources, Conservation and Recycling*, **107**, 142–156.

Kelly D. P., Shergill J. K., Lu W.-P. and Wood A. P. (1997). Oxidative metabolism of inorganic sulfur compounds by bacteria. *Antonie van Leeuwenhoek*, **71**(1), 95–107.

Kleinjan W. (2005). Biologically Produced Sulfur Particles and Polysulfide Ions: Effects on a Biotechnological Process for the Removal of Hydrogen Sulfide from Gas Streams. PhD thesis, Wageningen University, The Netherlands.

Kleinjan W. E., Lammers J. N. J. J., de Keizer A. and Janssen A. J. H. (2006). Effect of biologically produced sulfur on gas absorption in a biotechnological hydrogen sulfide removal process. *Biotechnology and Bioengineering*, **94**(4), 633–644.

Klok J. B. M. (2015). Modeling Studies of Biological Gas Desulfurization under Haloalkaline Conditions. PhD thesis, Wageningen University, The Netherlands.

Klok J. B. M., Heeringen G. V., Rink R. D., Wijnbelt H. and Bowerbank G. (2018). Techno-Economic Impact of the Next Generation Thiopaq O&G Process for Sulfur Removal. 26th Annual Technical Conference, GPA-GCC & GPA Europe.

Leng Y. and Soares A. (2018). Understanding the Fundamentals of Bio-struvite Biomineralization in Wastewater. Cranfield University, Swindon, UK.

Lanning A., O'Brien M. and Mirdadion D. (2008). Straded no More – Turning Sour Casinghead Gas into Profits Using Shell-Paques Bio-Desulfurization. Laurance Reid Gas Conditioning Conference, Norman, Oklahoma.

Le Corre K. S. (2006). Understanding Struvite Crystallisation and Recovery. Cranfield University, Swindon, UK.

Le Corre K. S., Valsami-Jones E., Hobbs P. and Parsons S. A. (2009). Phosphorus recovery from wastewater by struvite crystallization: a review. *Critical Reviews in Environmental Science and Technology*, **39**(6), 433–477.

Lee A., Weon S., Lee C. and Koopman B. (2003). Removal of nitrogen and phosphate from wastewater by addition of bittern. *Chemosphere*, **51**, 265–271.

Li B., Boiarkina I., Yu W., Huang H. M., Munir T., Wang G. Q. and Young B. R. (2019a). Phosphorous recovery through struvite crystallization: challenges for future design. *Science of the Total Environment*, **648**, 1244–1256.

Li H., Yao Q., Dong Z., Zhao T., Zhou G. and Fu S. (2019b). Controlled synthesis of struvite nanowires in synthetic wastewater. *ACS Sustainable Chemistry and Engineering*, **7**(2), 2035–2043.

Ma H., Guo Y., Quin Y. and Li Y. (2018). Nutrient recovery technologies integrated with energy recovery by waste biomass anaerobic digestion. *Bioresource Technology*, **269**, 520–531. https://doi.org/10.1016/j.biortech.2018.08.114

Matassa S., Batstone D. J., Hulsen T., Schnoor J. and Verstraete W. (2015). Can direct conversion of used nitrogen to new feed and protein help feed the world? *Environmental Science & Technology*, **49**(9), 5247–5254.

Matsumiya Y., Yamasita T. and Nawamura Y. (2000). Phosphorus removal from sidestreams by crystallisation of magnesium-ammonium-phosphate using seawater. *Journal of the Chartered Institution of Water and Environmental Management*, **14**, 291–296.

Mbamba C. K., Tait S., Flores-Alsina F. and Batstone D. J. (2015). A systematic study of multiple minerals precipitation modelling in wastewater treatment plants, *Water Research*, **85**, 359–370.

McCarty P. L. (2018). What is the best biological process for nitrogen removal: when and why? *Environmental Science & Technology*, **52**(7), 3835–3841.

Mead E. and Leibbert J. (1998). A Comparison of Packed-Column and Low-Profile Sieve Tray Air Strippers. Conference on Hazardous Waste Research, May 18–21, Snowbird, Utah.

Mehta C. M. and Batstone D. J. (2013). Nucleation and growth kinetics of struvite crystallization. *Water Research*, **47**(8), 2890–2900.

Mehta C. M., Khunjar W. O., Nguyen V., Tait S. and Batstone D. J. (2015). Technologies to recover nutrients from waste streams: a critical review. *Critical Reviews in Environmental Science and Technology*, **45**(4), 385–427.

Mersmann A. (2001). Crystallization Technology Handbook. Marcel Dekker Inc., New York.

Muhmood A., Lu J., Dong R. and Wu S. (2019). Formation of struvite from agricultural wastewaters and its reuse on farmlands: Status and hindrances to closing the nutrient loop. *Journal of Environmental Management*, **230**, 1–13. https://doi.org/10.1016/j.jenvman.2018.09.030

Munch, E. and Barr K. (2001). Controlled struvite crystallization for removing phosphorus from anaerobic digester sidestreams. *Water Research*, **35**, 151–159.

Muys M., Phukan R., Brader G., Samad A., Moretti M., Haiden B., Pluchon S., Roest K., Vlaeminck S. E. and Spiller M. (2021). A systematic comparison of commercially produced struvite: quantities, qualities and soil-maize phosphorus availability. *Science of the Total Environment*, **756**, 1–12.

O'Brien D. J. and Birkner F. B. (1977). Kinetics of oxygenation of reduced sulfur species in aqueous solution. *Environmental Science & Technology*, **11**(12), 1114–1120.

Oehmen A., Lemos P. C., Carvalho G., Yuan Z., Keller J., Blackall L. L. and Reis M. A. M. (2007). Advances in enhanced biological phosphorus removal: from micro to macro scale. *Water Research*, **41**(11), 2271–2300.

Ohlinger K. N., Young T. M. and Schroeder E. D. (1998). Predicting struvite formation in digestion. *Water Research*, **32**(12), 3607–3614.

Paltrinieri L., Remmen K., Muller B., Chu L., Koser J., Wintgens T., Wessling M., de Smet L. and Sudholter E. (2019). Improved phosphoric acid recovery from sewage sludge ash using layer-by-layer modified membranes. *Journal of Membrane Science*. **587**, 1–9.

Perry R. and Green D. (1997). Chemical Engineers Handbook, 7th edn. McGraw-Hill, New York, USA.

Pikaar I., Sharma K. R., Hu S., Gernjak W., Keller J. and Yuan Z. (2014). Reducing sewer corrosion through integrated urban water management. *Science*, **345**(6198), 812–814.

Pratt C., Parsons S., Soares A. and Martin B. (2012). Biologically and chemically mediated adsorption and precipitation of phosphorus from wastewater. *Current Opinion in Biotechnology*, **23**(6), 890–896.

AirPrex (2015). Biosolids Treatment Optimization Process with the option of Phosphate Recoverye. MWEA Annual Biosolids Conference, Available from: https://www.mi-wea.org/docs/Forstner%20%20Biosolids%20 Treatment%20Optimization%20Process.pdf (accessed: January 25 2022).

RecoPhos (2012). RecoPhos - Recovery of Phosphorus. Available from: http://www.recophos.org/ (accessed 25 January 2022).

Regy S., Mangin D., Klein J. P., Lieto J. (2001). Phosphate recovery by struvite precipitation in a stirred reactor. Centre Europeen d'Etudes des Polyphosphates. Available from: https://phosphorusplatform.eu/images/download/Regy-Mangin-Lagep-Report-struvite-precipitation-2001.pdf (accessed 25 January 2022).

Reijnders L. (2014). Phosphorus resources, their depletion and conservation, a review. *Resources, Conservation and Recycling*, **93**, 32–49.

Rittmann B. E., Mayer B., Westerhoff P. and Edwards M. (2011). Capturing the lost phosphorus. *Chemosphere*, **84**, 846–853.

Robles Á., Aguado D., Barat R., Borrás L., Bouzas A., Giménez J. B., Martí N., Ribes J., Ruano M. V., Serralta J., Ferrer J. and Seco A. (2020). New frontiers from removal to recycling of nitrogen and phosphorus from wastewater in the circular economy. *Bioresource Technology*, **300**, 1–18.

Robles-Aguilar A. A., Schrey S. D., Postma J. A., Temperton V. M. and Jablonowski N. D. (2020). Phosphorus uptake from struvite is modulated by the nitrogen form applied.. *Journal of Plant Nutrition and Soil Science*, **18**(1), 80–90.

Roman P., Bijmans M. and Janssen A. (2014). Quantification of individual polysulfides in lab-scale and full-scale desulfurisation bioreactors. *Environmental Chemistry*, **11**, 702–708.

Roman P., Klok J. B. M., Sousa J. A. B., Broman E., Dopson M., Van Zessen E., Bijmans M. F. M., Sorokin D. Y. and Janssen A. J. H. (2016). Selection and application of sulfide oxidizing microorganisms able to withstand thiols in gas biodesulfurization systems. *Environmental Science & Technology*, **50**(23), 12808–12815.

Romero-Güiza M. S., Astals S., Mata-Alvarez J. and Chimenos J. M. (2015). Feasibility of coupling anaerobic digestion and struvite precipitation in the same reactor: evaluation of different magnesium sources. *Journal of Chemical Engineering*, **270**, 542–548.

Sander R. (2015). Compilation of henry's law constants (version 4.0) for water as solvent. *Atmospheric Chemistry and Physics*, **15**, 4399–4981. https://doi.org/10.5194/acp-15-4399-2015 (accessed 25 January 2022).

Sekula-Wood E., Schnetzer A., Benitez-Nelson C., Anderson C., Berelson W., Brzezinski M., Burns J., Caron D., Cetinic I., Ferry J., Fitzpatrick E., Jones B., Miller P., Morton S., Schaffner R., Siegel D. and Thunell R. (2009). Rapid downward transport of the neurotoxin domoic acid in coastal waters. *Nature Geoscience*, **2**(4), 272–275.

Sena M. and Hicks A. (2018). Life cycle assessment review of struvite precipitation in wastewater treatment. *Resources, Conservation and Recycling*, **139**, 194–204.

Shaddel S., Ucar S., Andreassens J. P. and Osterhus S. W. (2019). Engineering of struvite crystals by regulating supersaturation – correlation with phosphorus recovery, crystal morphology and process efficiency. *Journal of Environmental Chemical Engineering*, **7**(1), 1–9.

Simoes F., Vale P., Stephenson T. and Soares A. (2018). The role of pH on the biological struvite production in digested sludge dewatering liquors. *Scientific Reports*, **8**, 1–9.

Sinnott R. K. (1999). Coulson and Richardson's Chemical Engineering, 3rd edn. Chemical Engineering Design, Butterworth-Heinemann, Jordan Hill, Oxford, Woburn, MA, vol. **6**.

Soares A, Veesam M., Simoes F., Wood E., Parsons S. A. and Stephenson T. (2014). Bio-Struvite: A new route to recover phosphorus from waste water. *CLEAN – Soil Air Water*, **42**(7), 994–997.

Sorokin D., Kuenen J. G. and Muyzer G. (2011). The microbial sulfur cycle at extremely haloalkaline conditions of soda lakes. *Frontiers in Microbiology*, **2**(44), 1–16.

Springer N. P. (2017). Physical, technical, and economic accessibility of resources and reserves need to be distinguished by grade: application to the case of phosphorus. *Science of the Total Environment*, **577**, 319–328.

Tait S., Solon K., Volcke E. I. P. and Batstone D. J. (2012). A Unified Approach to Modelling Wastewater Chemistry: Model Corrections. In: 3rd IWA/WEF Wastewater Treatment Modelling Seminar, 26–28 February, Mont-Sainte-Anne, Quebec, Canada.

Treybal R. E. (1981). Mass-transfer Operations, 3rd edn. McGraw-Hill Chemical Engineering Series. McGraw-Hill, Singapore.

Ueno, Y. and Fuji, M. (2001). Three years experience of operating and selling recov-ered struvite from full scale plant. *Environmental Technology*, **22**, 1373–1381.

United States Environmental Protection Agency (2000). Wastewater Technology Fact Sheet. Ammonia Stripping. EPA 832-F-00-019. September 2000. Available at: https://www3.epa.gov/npdes/pubs/ammonia_stripping.pdf (accessed 21 April 2021).

University of Florida (2015). Fluidization: A Unit Operation in Chemical Engineering, pp. 1–13. Available at: https://dl.icdst.org/pdfs/files1/be196045443d68512c1e012880d22888.pdf (accessed 16 December 2021).

U.S. Geological Survey (2002). Materials Flow of Sulfur. Available at: https://pubs.usgs.gov/of/2002/of02-298/of02-298.pdf (accessed 25 January 2022).

U.S. Geological Survey (2019). Mineral Commodity Summaries 2019. Available at: https://prd-wret.s3-us-west-2.amazonaws.com/assets/palladium/production/atoms/files/mcs2019_all.pdf (accessed 25 January 2022).

Vaneeckhaute C., Lebuf V., Michels E., Belia E., Vanrolleghem P. A., Tack F. M. G. and Meers E. (2017). Nutrient recovery from digestate: systematic technology review and product classification. *Waste and Biomass Valorization*, **8**(1), 21–40.

Wei J., Ge J., Rouff A., Wen X., Meng X. and Song Y. (2019). Phosphorus recovery from wastewater using light calcined magnesite, effects of alkalinity and organic acids. *Journal of Environmental Chemical Engineering*, **7**(5), 1–8.

Wu Y., Luo J., Zhang Q., Aleem M., Fang F., Xue Z. and Cao J. (2019). Potentials and challenges of phosphorus recovery as vivianite from wastewater: a review. *Chemosphere*, **226**, 246–258.

Yan Y., Xi X. R., Miao M., He T., Dong Z. H., Zhan K., Yan J. H., Zhao B. and Xia B. Y. (2018). Bio-inspired design of hierarchical FeP nanostructure arrays for the hydrogen evolution reaction. *Nano Research*, **11**(7), 3537–3547.

doi: 10.2166/9781780409566

Chapter 11

Established technologies for metal recovery from industrial wastewater streams

Jan Weijma[1], *Johannes B.M. Klok*[2,3], *Henk Dijkman*[4], *Gijs Jansen*[5], *Irene Sànchez-Andrea*[1], *Cees J.N. Buisman*[1,2], *Eric D. van Hullebusch*[6], *Tom Hennebel*[7], *Gijs du Laing*[7,8], *Heidy Cruz*[9], *Ilje Pikaar*[10] *and Denys K. Villa Gomez*[10]

[1]*Wageningen University & Research, Wageningen, The Netherlands*
[2]*Wetsus, Leeuwarden, The Netherlands*
[3]*Paqell B.V., Balk, The Netherlands*
[4]*Paques B.V., Balk, The Netherlands*
[5]*Nyrstar Budel B.V., Budel-Dorplein, The Netherlands*
[6]*Université de Paris, Paris, France*
[7]*Ghent University, Ghent, Belgium*
[8]*Centre for Advanced Process Technology for Urban Resource Recovery, Ghent, Belgium*
[9]*National University of Singapore, Queenstown, Singapore*
[10]*University of Queensland, risbane, Australia*

11.1 INTRODUCTION

Metals are essential to global economic and social development and play an essential role in the majority of the industry value chains of the 21st century. In order to keep up with the increasing demand for metals, the mining industry has shifted to the exploitation of low(er) grade ores, as higher-grade ores become scarcer. This shift not only reduces the economic viability but also comes with increased environmental cost. Population growth, expansion of industrial infrastructures, and technological advancement have also exponentially increased the demand for metals in recent decades, leading to supply-chain risks as well as economic and geopolitical challenges [1]. Low-carbon energy technologies (e.g., solar, wind, bioenergy, carbon capture and storage), catalytic processes, and electronic equipment require large amounts of metals with low natural abundance and unequal terrestrial distribution. These challenges are particularly relevant for metals classified as critical, including cobalt, selenium, tellurium, platinum group metals (PGMs) as well as rare earth elements (REEs) such as scandium and yttrium (see Table 11.1) [2]. About 90% of the supply of these critical metals comes from only nine countries, with China providing almost half of the total supply [3]. To illustrate the scarcity and uneven resource distribution of some of these metals (e.g., the use of platinum, palladium, and rhodium in catalytic converters in automobiles) to mitigate greenhouse gas emissions has been estimated to comprise about 39, 67, and 69% of the global production capacity, respectively [2]. The majority of the supply comes only comes from two countries only, namely, South Africa and Russia [3].

Table 11.1 Metal price and demand increase from 1993 to 2013 (metal prices from [4, 5]; metal demand adapted from [6]).

Metal			Price (US$/t)	Demand (%) Increase
EU Critical raw materials	Platinum-group metals	Iridium	47 580 000	Data not available
		Palladium	42 370 000	150
		Platinum	26 060 000	100
		Rhodium	95 025 000	Data not available
		Ruthenium	11 625 000	Data not available
	Rare earth elements	Praseodymium	99 000	275
		Neodymium	59 925	275
		Cerium	5475	275
		Terbium	581 250	275
		Dysprosium	248 250	275
		Yttrium (as oxide)	34	275
		Cobalt	42 000	500
		Tantalum	186 500	350
		Gallium	154	Data not available
		Germanium	1178	300
		Tungsten	42 750	Nr
		Indium	225 750	550
EU Raw materials with high relative economic importance		Iron ore	75	250
		Aluminium	1844	200
		Nickel	11 225	300
		Zinc	2476	150
		Molybdenum	26 000	250
		Lanthanum	5625	275

To address the challenge of metal scarcity and gain long-term access to these materials at economic competitive costs and minimal environmental pollution, it is clear that a more circular management strategy for metals is needed. In this context, metal-containing wastewaters are considered an important and valuable secondary 'metal resource'. This chapter provides an introduction into the fundamentals of several established processes for metal recovery from industrial wastewater streams, with a focus on abiotic and biotic sulfur-based removal mechanisms. Specifically, guiding principles for the selection, design, and operation of sulfur-based metal recovery processes will be discussed, which are relevant to the recovery of copper, zinc, nickel, lead, iron, cobalt, cadmium, mercury, and less common metals such as thallium. Finally, an outlook for the future is provided by discussing some of the key challenges, opportunities, and research needs.

11.2 LEARNING OBJECTIVES

At the completion of this chapter you should be able to:

- Understand the importance of metals within the industry value chains of the 21st century in general and the global low carbon economy in particular.
- Describe some of the key metals that can be recovered from wastewater streams originating from various industrial activities including the oil and gas, mining, and metallurgical sectors.

- Describe and apply the fundamental principles and commonly applied technologies to recover metals from various wastewater streams and typical application area of each technology.
- Set a design basis for sulfide-based metal recovery, conduct preliminary reactor sizing, and calculate performance predictions.
- Calculate the size of reactors for sulfate reduction based on their influent characteristics.
- Calculate the pH at which one metal can be selectively precipitated as metal sulfide from a solution containing two metals, based on a difference in solubility product.
- Identify opportunities, challenges, and research needs for metal recovery from other streams in terms of environmental and economic benefits.

11.3 CONCEPTUAL OVERVIEW OF SULFIDE-BASED METHODS FOR METAL RECOVERY FROM WASTEWATER

Most metals are present in nature in the form of metal sulfides, which upon being (bio)leached are solubilized as metal ions in an acidic sulfate solution. Also, metals in solution can be precipitated as metal sulfides. As a result, technologies for metal recovery from aqueous streams have exploited this link between metals and sulfur, to recover metals in solution by means of sulfide precipitation. A key point here is the availability of hydrogen sulfide, which can be generated using sulfate already available in the wastewater or using externally supplied sulfur. This chapter will elaborate on both the principles of metal sulfide precipitation and biological sulfide production as key building blocks of current full-scale technologies for metal recovery. By treating effluents from the mining and metallurgical industry using these sulfur-based resource recovery technologies, the following valuable products can be recovered:

(1) *Metal sulfides*. The metal sulfides are produced at a dry matter content >90% and can be sold to smelters.
(2) *Water*. Sulfate can be reduced to concentrations below 250 $mg{\cdot}L^{-1}$, pH can be raised to circumneutral pH values, and heavy metals are removed to concentrations at ppb levels. The recovered water can be used for agricultural purposes or as process water in industrial or mining applications.
(3) *Elemental sulfur*. Elemental sulfur is an insoluble essential nutrient that can be produced by means of oxidation of excess sulfide (not discussed here, but for reference you can consult Chapter 10). The recovered elemental sulfur can be re-used as a soil fertilizer or as a feedstock for sulfuric acid production.

The ultimate removal and recovery mechanism for metals in sulfur-driven processes is precipitation as metal sulfide. This precipitation can be achieved in a stand-alone abiotic process, or in a coupled abiotic–biotic process. Here we introduce the fundamental principles governing sulfur precipitation (Section 11.4), the application and design of these processes (Section 11.5), and the biological production of reduced sulfur to support precipitation (Section 11.6).

11.4 FUNDAMENTAL PRINCIPLES OF METAL SULFIDE PRECIPITATION

11.4.1 Precipitation reactions and solubility products

Most metals can be precipitated from wastewater with hydroxides or sulfides to very low concentrations, due to the low solubility products of most metal hydroxide and metal sulfides. As a result, chemical precipitation is the most commonly applied method for metal removal from industrial wastewater (it is estimated that ~90% of treatment plants use this method) [7]. The solubility product (K_{sp}) determines whether the metal will remain in solution or will precipitate. For a metal hydroxide:

$$M^{2+} + 2OH^{-} \rightarrow M(OH)_2 \tag{11.1}$$

where M^{2+} and OH^- represent the dissolved metal ions and the precipitant, respectively, while $M(OH)_2$ is the insoluble metal hydroxide.

$$K_{sp} = \left[M^{2+}\right]^x \left[OH^-\right]^y \tag{11.2}$$

Metals such as copper, zinc, nickel, lead, iron, cobalt, cadmium, and mercury, but also less common metals such as thallium, can be precipitated as metal sulfides. For a solid precipitate of metal sulfide $M_xS_y(s)$, the following general solubility expression can be written:

$$M_xS_{y(s)} \rightarrow xM^{2+} + yS^{2-} \tag{11.3}$$

where M_xS_x is the insoluble metal sulfide, M^{2+} represents the metal and S^{2-} represents the reactant sulfide in any of its forms (HS^-, S^{2-}, H_2S) depending on the pH of the solution. As such, the solubility product (K_{sp}) of the metal sulfide can be defined as:

$$K_{sp} = \left[M^{2+}\right]^x \left[S^{2-}\right]^y \tag{11.4}$$

where K_{sp} is in $mol^2 \cdot L^{-2}$ when $x = y = 1$, $[M^{2+}]$ is the equilibrium activity of metal ion M^{2+} ($mol \cdot L^{-1}$) and $[S^{2-}]$ is the equilibrium activity of S^{2-} ($mol \cdot L^{-1}$), varying strongly as function of pH.

The solubility products of several metals are presented in Figure 11.1 as a function of the pH at an initial metal concentration of 100 mM. The lines in Figure 11.1 represent the total metal in solution in equilibrium with the precipitate. The pH value at which a minimum metal solubility is achieved (i.e., highest removal efficiency) varies, depending on the metal species. The latter is important as this allows for the separation of metals by sequential precipitation at different pH values, thereby enabling selective recovery of the target metal(s). For example, copper sulfide still precipitates at pH 1.5 while iron sulfide precipitates only at pH values above 5.0. It is important to remember that the values of solubility will vary depending on the constituents in the wastewater, including complexing agents (e.g., cyanide, ethylenediaminetetra-acetic acid (EDTA), citrate) [8, 9]. Although purely abiotic processes can be operated across wide ranges of pH, processes which include biological sulfur reduction are generally operated at circumneutral pH.

11.4.2 Nucleation and crystallization

Precipitation occurs when a solution is supersaturated, meaning the solute concentration is higher than the solid–liquid equilibrium value [7, 11]. The kinetic phenomena associated with precipitation are nucleation and crystallization. Figure 11.2 shows the schematic representation of the saturation zones driven by solute concentration, where nucleation and crystallization occur. Crystallization only occurs if the system is supersaturated [12]. In general, high supersaturation levels favor nucleation, thus to produce crystals the supersaturation should be minimized. At high levels of supersaturation, both phenomena compete for the available solute [13]. Depending on the conditions, either nucleation or crystal growth may be predominant over the other, and as a result, crystals with different sizes and shapes can be obtained [14]. When one wishes to produce large crystals of high crystallinity and purity, precipitation should be conducted under low supersaturation conditions. Nevertheless, the combined processes of nucleation, crystallization, and agglomeration of the metal precipitates determine the particle size distribution (PSD), which in turn affects the feasibility of recovery of the metal precipitates through settling or flotation.

It should be noted that biological reactors are usually operated at circumneutral pH, which is optimal for sulfidogenic microorganisms. In addition, in such bioreactors an excess of ~5–10 mM total sulfide (the sum of $[H_2S]$, $[HS^-]$, and $[S^{2-}]$) is maintained to keep the ORP (oxidation reduction potential) low, which is also a requirement for optimal microbial growth. The circumneutral pH, the presence of a few mM of total sulfide, and the extremely low solubility of relevant metal sulfides

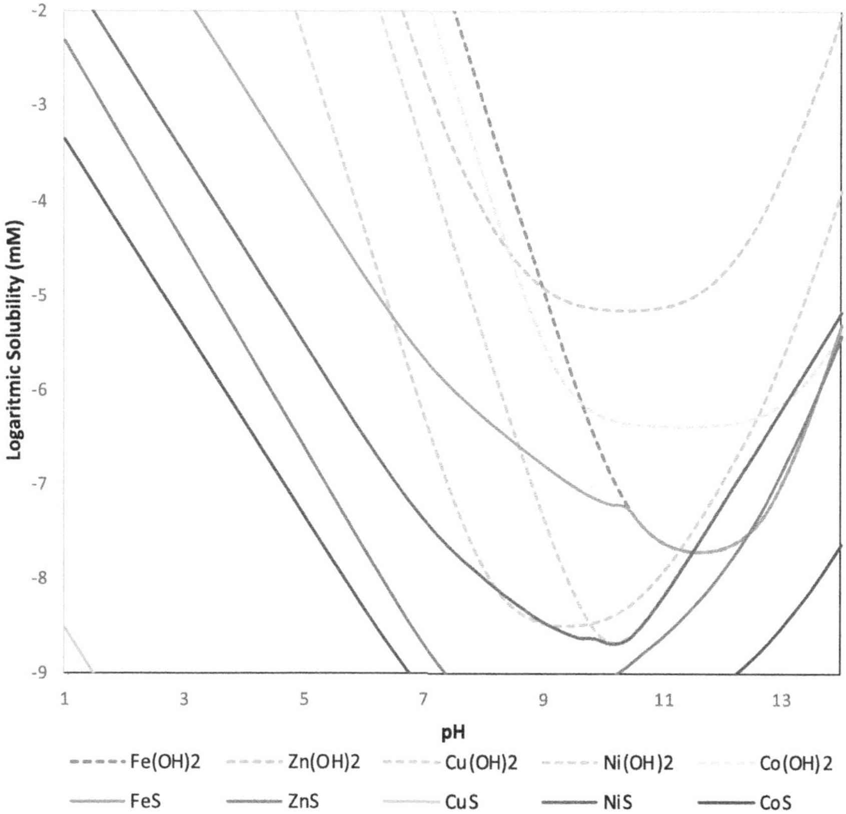

Figure 11.1 Solubility of metals at different pH values. Diagram elaborated with the MEDUSA software [10]. Equilibrium values were taken from the HYDRA database at pKa 25°C, based on initial metal concentrations of 100 mM.

(especially CuS, ZnS, NiS) implies that bioreactor solutions are highly oversaturated, with little control over the supersaturation. Therefore, for current applications of biological metal sulfide precipitation, particle size distributions cannot be controlled according to the crystallization theory explained above.

11.4.3 The impact of pH on sulfide precipitation

As already hinted in Figure 11.1, the calculation of the solubility of a metal sulfide is rather complex, due to the effect of pH, occurrence of metal sulfide complexes, and the wide range of (and sometimes conflicting) solubility products reported in the available literature. It would reach beyond the purpose of this textbook to elucidate all these complexities. Among all factors, pH has the most prominent effect on sulfide precipitation, as it strongly governs the availability of S^{2-}. Figure 11.3 depicts the speciation of the different sulfide fractions as a function of the pH.

Sulfide is in equilibrium with bisulfide (HS^-) and hydrogen sulfide (H_2S). The thermodynamic equilibrium of hydrogen sulfide can be expressed as follows [11]:

$$H_2S \xleftrightarrow{Ksp1} HS^- + H^+ \tag{11.5}$$

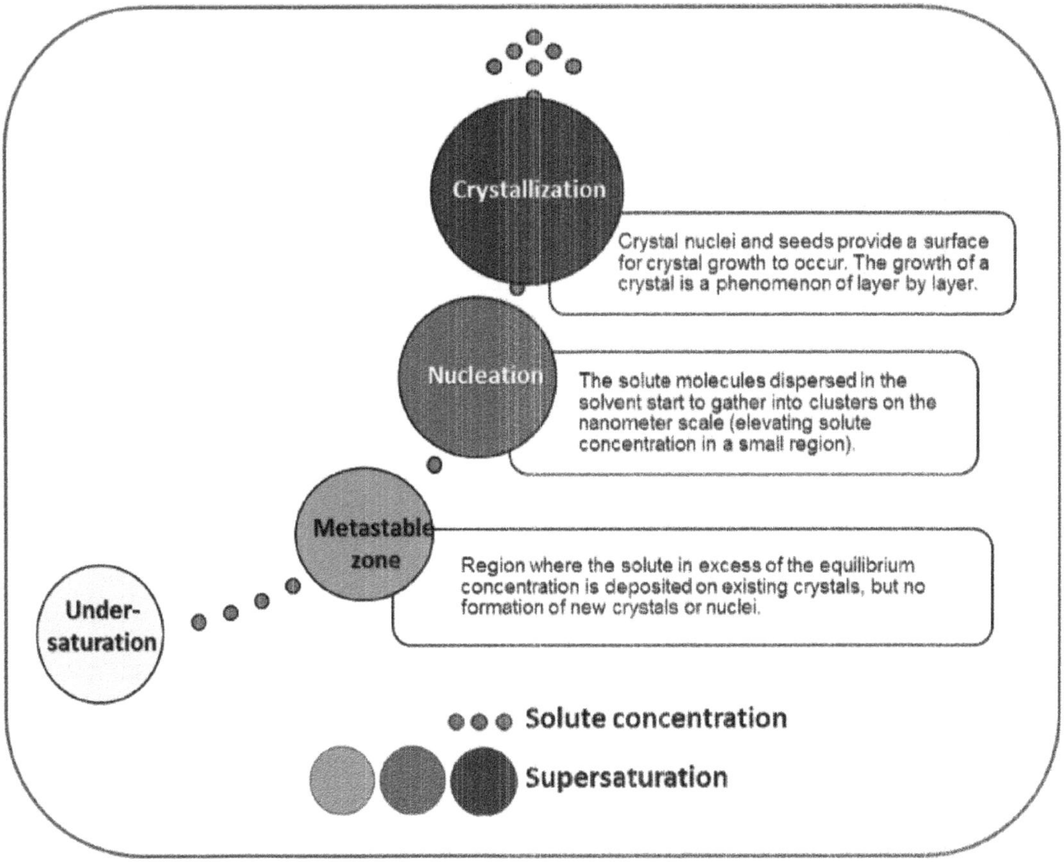

Figure 11.2 Schematic representation of the saturation zones driven by the solute concentration, where nucleation and crystallization occur (adapted from [15]).

$$\frac{[H^+]\cdot[HS^-]}{[H_2S]} = 10^{-7.0} = K_{a1}\left(pK_{a1} = 6.99\right) \tag{11.6}$$

$$HS^- \xleftrightarrow{Ksp2} S^{2-} + H^+ \tag{11.7}$$

$$\frac{[H^+]\cdot[S^{2-}]}{[HS^-]} = 10^{-17.4} = K_{a2}\left(pK_{a2} = 17.4\right) \tag{11.8}$$

in which S^{2-} represents the sulfide concentration. The pK_{sp} is defined as $-{}^{10}\log K_{sp}$, analogous to the definition of pH as measure for the proton concentration. The total dissolved sulfide (S_{tot}) concentration is given as:

$$S_{tot} = [H_2S] + [HS^-] + [S^{2-}] \tag{11.9}$$

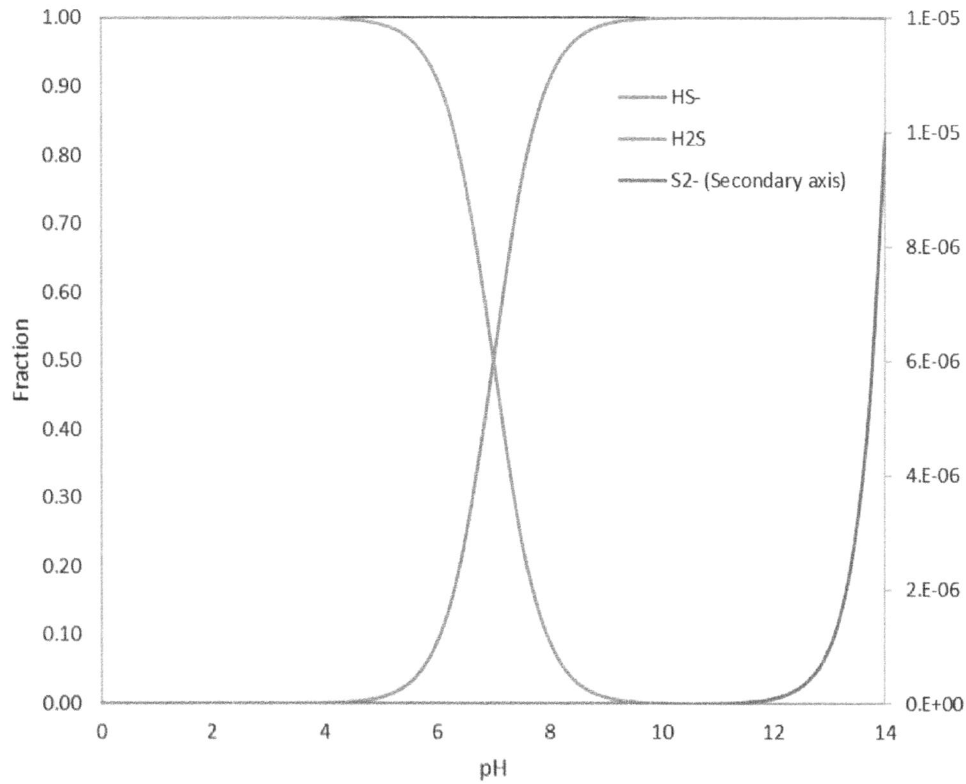

Figure 11.3 Sulfide solubility chart showing the relative fraction of each sulfide species at different pH values; H_2S, HS^-, S^{2-}. Diagram generated with the MEDUSA software [10].

The total dissolved sulfide concentration can be measured with chemical analysis. Therefore, it is useful to express $[S^{2-}]$ as a function of the total sulfide concentration and the proton concentration (given by the pH, Equation (11.10)), derived from the above equations:

$$\left[S^{2-}\right] = \frac{\left[S_{tot}\right]}{1 + \left([H^+]/K_{a2}\right) + \left([H^+]^2/K_{a1} * K_{a2}\right)} \tag{11.10}$$

A simplified model of metal sulfide precipitation is presented below, with zinc as example. For simplicity, it is assumed that the activity coefficients are equal to 1. The precipitation of zinc sulfide can be represented as:

$$Zn^{2+} + S^{2-} \leftrightarrow ZnS \tag{11.11}$$

At equilibrium:

$$\left[Zn^{2+}\right] \cdot [S^{2-}] = 10^{-24.7} = K_{sp}\left(pK_{sp} = 24.7\right) \tag{11.12}$$

Thus, at a pH of 7.0, and a total sulfide concentration of 6 mM, the sulfide concentration $[S^{2-}]$ is then 8.0×10^{-14} M. The $[Zn^{2+}]$ at equilibrium is 2.5×10^{-12} M.

11.5 APPLICATION AND DESIGN OF SULFIDE-BASED METAL PRECIPITATION

In abiotic precipitation processes, a critical design parameter relates to achieving sufficient mixing as a means to minimize supersaturation conditions by dispersing the reactants to promote crystallization over nucleation [16]. Various designs have been proposed and can be implemented, albeit the most commonly used reactor configurations for sulfide-based metal precipitation are: (i) completely stirred tank reactor (CSTR); (ii) pipe reactor; and (iii) fluidized bed reactor (FBR) [7]. Figure 11.4 provides a simplified schematic representation of each of these types of reactor designs.

Among these three configurations, the most common reactor used by industry is the CSTR, with sulfide being added at a single point and mixed into the reactor bulk solution using a central impeller. This method leads to poor crystallization, since mixing may not be adequately fast to avoid conditions that support nucleation. Pipe reactors have a pipe junction design where dosing of the precipitant agent (most commonly sulfide) is controlled, so crystallization can be favored. The advantages of pipe reactors include their simplicity and ease of operation. A disadvantage is that the local supersaturation induced by introducing the sulfide in this manner decreases the size of the precipitates in the contact zone. FBR systems have been identified as an effective reactor configuration for processes in which the product is a sparingly soluble species that is difficult to separate. Fluidized bed reactors provide ideal conditions for controlled precipitation because: (1) fluidization allows for good mixing on both the macro and meso-scale; (2) supersaturation can be controlled through multiple reagent inlet ports; (3) the precipitates can be separated from the treated water using gravitational separation; and (4) fines generated during precipitation can be recycled and allowed to agglomerate, thus forming large particles which can be recovered and thereby increase solid–liquid separation efficiency. In a fluidized bed reactor, the vessel is charged with a batch of pre-characterized seeds, which are subsequently fluidized by the aqueous metal stream entering the reactor from the bottom. The reagent is fed through inlet ports situated on the side of the reactor. As precipitate is deposited on the seeding material, the particle size distribution along the height of the column changes. Larger, denser particles migrate to the bottom of the reactor where they are removed as a product, while the lighter particles remain suspended higher up in the bed. In continuous processes, new seeding material is introduced at the top of the reactor, while large particles are removed at the bottom to maintain a constant bed height.

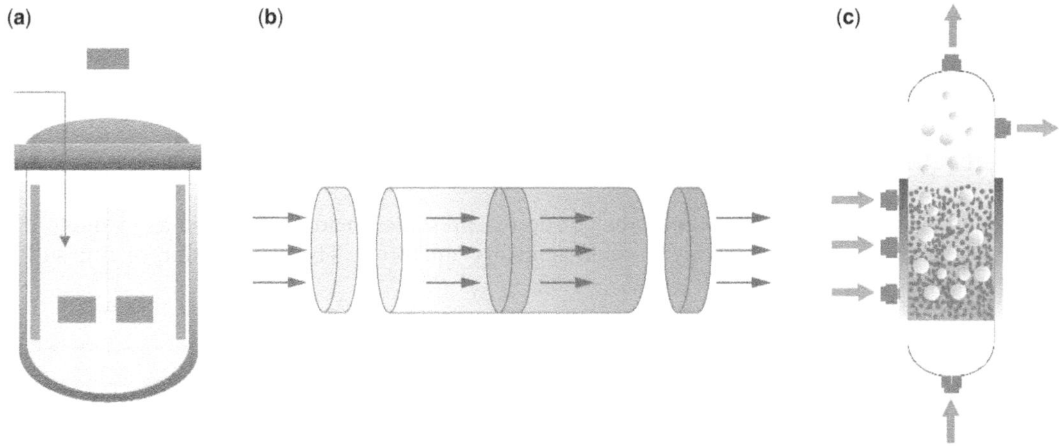

Figure 11.4 Simplified schematic representations of commonly used reactor configurations for metal precipitation using: (a) a completely stirred tank reactor (CSTR), (b) a pipe reactor, and (c) a fluidized bed reactor (adapted from [7]).

Another important design parameter is the pH value. As depicted in Figure 11.1, the solubility product of specific metal sulfide is strongly affected by the pH. This allows for selective recovery of certain metals by adjusting the pH of the solution. In practice, multiple precipitation reactors can be operated in series – each at a different pH level – as a means to sequentially recover different metals to very low concentrations to well below concentrations of 0.01–1 ppm [17].

It is important to emphasize that the use of hydrogen sulfide requires specific safety measures, as it is a toxic gas, with air concentrations of 1000 ppm leading to acute death [18]. Also, corrosion-resistant construction materials like polypropylene and high-density polyethylene are needed, as hydrogen sulfide induced corrosion may occur with concrete or steel materials [19, 20].

As discussed in Section 11.4.1. and shown in Figure 11.1, metals can also be precipitated as metal hydroxides through the addition of lime ($Ca(OH)_2$), caustic soda (NaOH), and ammonium hydroxide (NH_4OH) [8]. Of these chemicals, lime addition is the most commonly applied as it is an inexpensive, non-toxic, and non-corrosive chemical that does not require special occupational health and safety (OH&S) precautions and special storage materials. Lime is dosed in two forms, namely, in solid form as CaO (s) or as a slurry $Ca(OH)_2$. Consequently, lime addition is still commonly applied by the industry. Nevertheless, lime addition comes with disadvantages, especially in the context of metal recovery. First, lime addition results in the production of large volumes of sludge – containing a mixture of gypsum ($CaSO_4$) and metal hydroxides – which is difficult to dewater. Dewatering difficulty complicates selective metal recovery and incurs additional operational costs with respect to sludge handling and disposal. Furthermore, lime is often overdosed to ensure efficient metal removal in the presence of wastewater constituents such as complexing agents that negatively interferes with the precipitation process [8]. An important advantage of sulfide precipitation over hydroxide precipitation within a recovery context is that the sulfide precipitates that gypsum is not formed, and moreover, typically has lower levels of impurities [9]. In addition, metal sulfide precipitates can be formed in more diluted aqueous solutions due to the very low solubility product of the metal sulfides (see Figure 11.1). Equally important, the generated metal-sulfide sludge has much better dewaterability characteristics, generating a more concentrated metal sulfide product reducing the cost for transportation and further processing of the produced metal concentrates. Finally, beyond these operational benefits in terms of metal removal and lower sludge volumes, sulfide precipitates can be directly recycled within the process, as most metal refineries treat metal sulfide ores [21].

11.6 BIOLOGICAL H_2S PRODUCTION FOR METAL RECOVERY

In the section above, we have discussed the principles of sulfide-based metal precipitation. Hydrogen sulfide is commonly added in aqueous form as Na_2S, NaHS, and $(NH_4)_2S$. Alternatively, it can also be dosed in gaseous form in the form of H_2S (g). An interesting alternative that has found widespread implementation at full-scale is the onsite production of sulfide through microbial (i.e., biologically mediated) reduction of the sulfate present in the wastewater by sulfate reducing bacteria (SRB), or alternatively by reduction of added elemental sulfur by sulfur reducing bacteria. These approaches are discussed in more detail in the sub-sections below.

11.6.1 Fundamental principles of microbial sulfate and sulfur reduction

Sulfate and elemental sulfur can be reduced to hydrogen sulfide under anaerobic conditions by a diverse group of anaerobic microorganisms usually named as sulfate reducing bacteria (SRB). These microbes use oxidized sulfur compounds as electron acceptor in their metabolic pathways. These sulfide-generating bacteria are able to conserve energy by the reduction of sulfur oxyanions like sulfate, sulfite, and thiosulfate [22]. SRB are considered as a group because of their common ability to use sulfate as terminal electron acceptor for energy-generating degradation of hydrogen and/or organic substrates such as ethanol, methanol, or volatile fatty acids (VFAs). To illustrate this, for example, the reduction of sulfate with hydrogen and ethanol as electron donor proceeds as:

$$SO_4^{2-} + 4H_2 \rightarrow H_2S + 2H_2O + 2OH^- \quad (11.13)$$

$$3SO_4^{2-} + 2C_2H_5OH \rightarrow 3H_2S + 4HCO_3^- + 2OH^- \quad (11.14)$$

Besides SRB, other microorganisms carry out dissimilatory reduction of elemental sulfur, such as mycobacteria, and a variety of prokaryotes: for example, bacteria in hypersaline sediments, thermophilic archaea, and so on. The utilization of elemental sulfur as the source of sulfide needs four times less electron donor compared to sulfate, since only 2 moles electron-equivalents to 1 mol H_2 are needed for the reduction of sulfur to sulfide instead of the eight electrons equivalent to 4 moles H_2 needed for the reduction of sulfate to sulfide:

$$S^0 + H_2 \rightarrow H_2S \quad (11.15)$$

Like SRB, sulfur reducing microorganisms are highly diverse. Especially interesting is sulfur reduction at low pH, as this additionally enables selective precipitation of metals [23].

11.6.2 Applications and design of biological sulfate and/or sulfur reduction

Hydrogen sulfide is produced in biological reactors and contacted with the metals of interest for precipitation. This can be done in different configurations depending on whether or not the wastewater with metals contains (sufficient) sulfur/sulfate for the production of adequate amounts of hydrogen sulfide for subsequent metal precipitation. In general we can identify four different types of configurations, namely (Figure 11.5): (a) sulfate is present in the same stream as metals and sulfur reduction and metal sulfide precipitation occur in the same step; (b1) sulfate is present in the same stream as metals, but sulfur reduction occurs in a secondary stage, with the H_2S-rich wastewater being recirculated to the first stage for precipitation; (b2) sulfate is present in the same stream as metals, but sulfur reduction occurs in a secondary stage, with H_2S being supplied to the first stage for precipitation as a gas; and (c) the wastewater does not contain sulfur, and a source of sulfur needs to be provided in the second step, which functions analogously to (b1) and (b2). In cases where the water may be inhibitory to the microorganisms (due to pH, temperature, salinity, or other factors), it is also possible to generate H_2S in a side-stream process and recirculate it back to the metal sulfide precipitation reactor, such that the microorganisms do not come in contact with the contaminated water (configuration not shown).

Beyond the presence of sulfate or sulfur, another important aspect for biological sulfate reduction is the presence of an organic or inorganic (hydrogen) electron donor. In general, metal-rich wastewaters, especially those generated in the mining and metallurgical industries, normally do not contain sufficient electron donor to reduce the desired amount of sulfate to sulfide. Electron donors must then be supplied to the sulfate reduction process from an external source. For instance, organic waste materials such as molasses support sulfate reduction as the electron donor. However, such organic waste materials are often only available in relatively small quantities, generally limiting its applicability to streams with a low sulfate load. Furthermore, the composition of these organic waste materials is complex, which could result in less efficient availability for sulfate reduction. For instance, growth of methane producing microbial populations is more likely on such substrates. The use of organic materials may also lead to additional biochemical oxygen demand in the effluent of the process. The use of organic matter/waste is especially suitable for low-engineered systems such as permeable reactive barriers [24]. Here we will focus on the use of relatively pure bulk chemicals as electron donors for high-rate engineered processes. Hydrogen and ethanol are often used as these support high volumetric sulfate reduction rates (Table 11.2).

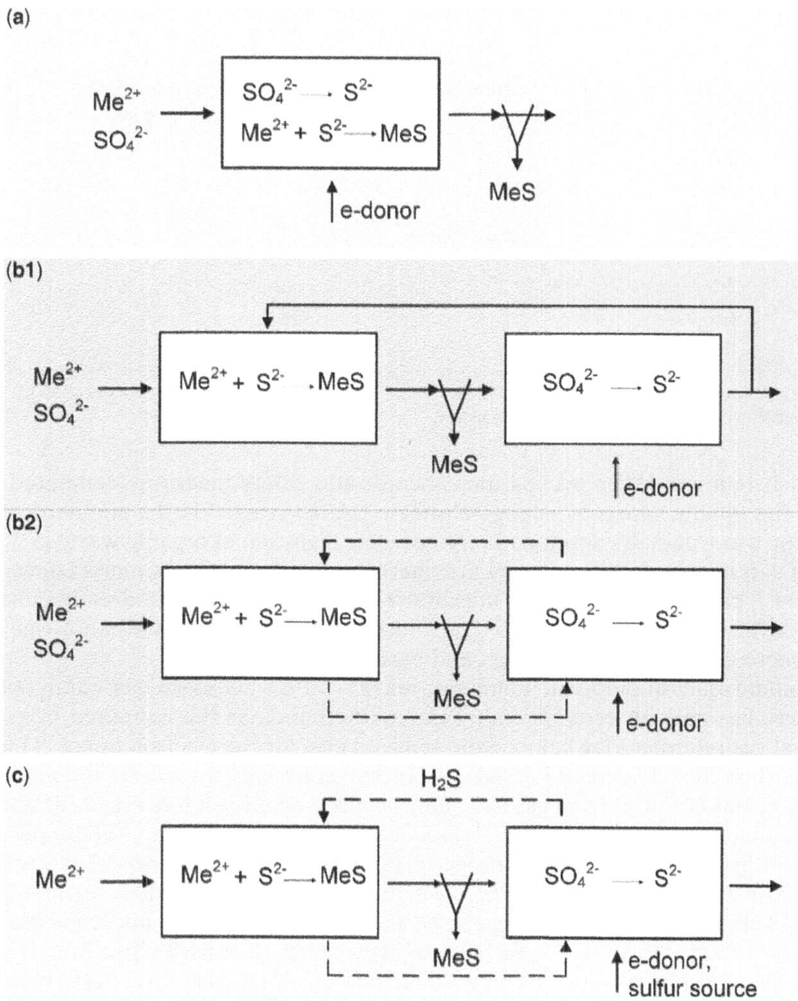

Figure 11.5 Configurations for the sulfur-associated precipitation of metals. When sulfate and metals are both in the influent stream, sulfur reduction and metal sulfide precipitation can: (a) occur in the same step; or (b) separate stages. When in separate stages, reduced sulfur can be recirculated to the first stage (where metals are present) with (b1) H_2S-rich wastewater or (b2) H_2S-containing gas; (c) When wastewater does not contain sulfur, a source of sulfur needs to be provided in the second step.

For small sulfate loads (<2–3 tons $S \cdot day^{-1}$), the use of ethanol is preferred because at such loads it is more economical than the use of hydrogen. Hydrogen gas must be produced (if not available on-site) by catalytic conversion of methanol or methane. From both methane and methanol, a mixture of H_2 and CO_2 is produced. Although the chemical costs for ethanol are somewhat higher than for hydrogen, the investment is lower since no catalytic conversion step is needed with ethanol. Ethanol can directly serve as electron donor for sulfate reducing bacteria, whereas methane and methanol support only low rates of sulfate reduction. There is an exception, however, as methanol is an effective direct electron donor under thermophilic (65°C) conditions [31]. Another cost-reducing factor for

Table 11.2 Sulfate and sulfite conversion rates as demonstrated in laboratory bioreactors with various electron donors.

Electron Donor	T (°C)	SO_4^{2-} Conversion Rate $kg \cdot m^3{}_{reactor}{}^{-3} \cdot day^{-1}$	SO_3^{2-} Conversion Rate $kg \cdot m^3{}_{reactor}{}^{-3} \cdot day^{-1}$	Reference
Molasses	31	2.4	n.a.	[26]
Synthesis gas	30	10	n.a.	[27]
H_2/CO_2	30	30	n.a.	[28]
Ethanol	30–35	29	n.a.	[29]
Acetate	33	9.4	n.a.	[30]
Methanol	65	14	18	[31]
H_2/CO_2	55	7.5	9.3	[32]

n.a. = no sulfite added, Note that the data represents a selection of the most common electron donors used for full-scale applications. A broader overview can be found elsewhere [25].

ethanol is that it requires fewer occupational health and safety measures compared to hydrogen. With ethanol, the upflow anaerobic sludge blanket (UASB) reactor is the standard reactor design. This reactor type was originally developed for anaerobic digestion of organic waste [33], and typically has a height of 4–5 meters. UASB reactors are characterized by a high biomass concentration of up to 50 kg VSS $\cdot$ m^{-3}, present as granular or floccular suspended biomass. As already indicated in Table 11.2, sulfate conversion rates of 29 kg $\cdot$ $m_{reactor}{}^{-3}$ $\cdot$ day^{-1} are possible with ethanol (see references in Table 11.2 for more details of reactor design and performance).

For larger sulfate loads at moderate temperatures (25–40°C), hydrogen gas, catalytically produced from natural gas, becomes an economic alternative to ethanol, as the increased investment for the catalytic natural gas reformer may balance the reduced cost for the electron donor. The resulting gas also contains carbon dioxide, which is used as a carbon source by the sulfate reducing bacteria. For optimal hydrogen mass transfer from gas to liquid, gas-lift loop bioreactors are used. With this design, sulfate conversion rates of 30 kg $\cdot$ $m_{reactor}{}^{-3}$ $\cdot$ day^{-1} can be achieved. Under thermophilic conditions (65°C), the use of hydrogen is less efficient due to the formation of the by-product methane from H_2 and CO_2 [32]. A relatively small part of the electron donor, typically 5–10% of the total consumed electron donor, is used for anabolic pathways to produce sulfate reducing microbial biomass.

The total sulfide concentration (i.e., the sum of hydrogen sulfide [H_2S], bisulfide [HS^-] and sulfide [S^{2-}]) in sulfate reducing reactors is kept below about 600 mg $\cdot$ L^{-1} at pH 7, in order to prevent inhibition of the sulfate reducing bacteria by the end product hydrogen sulfide. The sulfide concentration can be kept lower when metal sulfides are precipitated simultaneously. With respect to process economics (minimal electron donor consumption), it would be optimal to only produce the stoichiometric amount of sulfide needed for metals precipitation. However, in order to avoid the presence of free metals, which would severely inhibit the microbial population, sulfide is typically kept in excess of about 200 mg $\cdot$ L^{-1}. In this way, temporary decreases in sulfide production (e.g., as a result of temperature variations) or unintentional increases of the metal load can be dealt with, without compromising process stability. It does, however, imply that a post-treatment step is needed in which the excess sulfide is removed. The latter can be done by means of aeration in order to produce elemental sulfur (see Chapter 10 for more detail).

11.6.3 Case studies of implementation

11.6.3.1 Sulfate reduction for metal removal at the Nyrstar Budel zinc production facility

Nyrstar operates a zinc (Zn) smelter in the Netherlands, which produced 268 000 tons of zinc in 2018. Historically, zinc slags and polluted gypsum were stored in ponds at the site, which led to groundwater contamination. Groundwater contamination was also caused by the old pyrolytic zinc production process, which was stopped in 1973 and replaced by a roast-leach-electrolysis (RLE) process. The conventional RLE zinc production process used at the Nyrstar site produced various wastewater streams containing sulfate, chloride, fluoride, and metals (mainly zinc). Until the year 2000, the chloride and fluoride bleed streams were treated with the conventional process by neutralization with lime to produce a mixture of gypsum, metal hydroxides, and calcium fluoride. After new governmental regulations prohibited this approach, a wastewater treatment process was designed based on biological sulfate reduction, without generation of solid waste. More specifically, two streams generated at the site needed treatment, first, the scrubber discharge from the roaster gas cleaning plant, the so-called Wash Tower Acid (WTA). This wastewater stream has a typical flow of $25\ m^3 \cdot h^{-1}$ containing $15\ g \cdot L^{-1}\ H_2SO_4$, $3\ g \cdot L^{-1}$ HF, $4\ g \cdot L^{-1}$ HCl, and $1.5\ g \cdot L^{-1}\ Zn^{2+}$. The second wastewater stream requiring treatment is the Magnesium bleed. This bleed is necessary to prevent build-up of magnesium in the leaching and electrolysis circuit (which would cause operational issues due to the formation of inorganic scaling). Typically, $0.5\ m^3 \cdot h^{-1}$ of purified solution and/or spent acid is bled from the circuit. A fraction of this Mg bleed is used to make maximum use of the capacity of the sulfate reducing bioreactor. Figure 11.6 provides a simplified process flow sheet consisting of seven biological, chemical and physical stages: (i) Neutralization of Wash Tower Acid with calcine and sodium hydroxide; (ii) Fluoride removal as CaF_2 together with oxidation of As and Fe; (iii) Mixing with

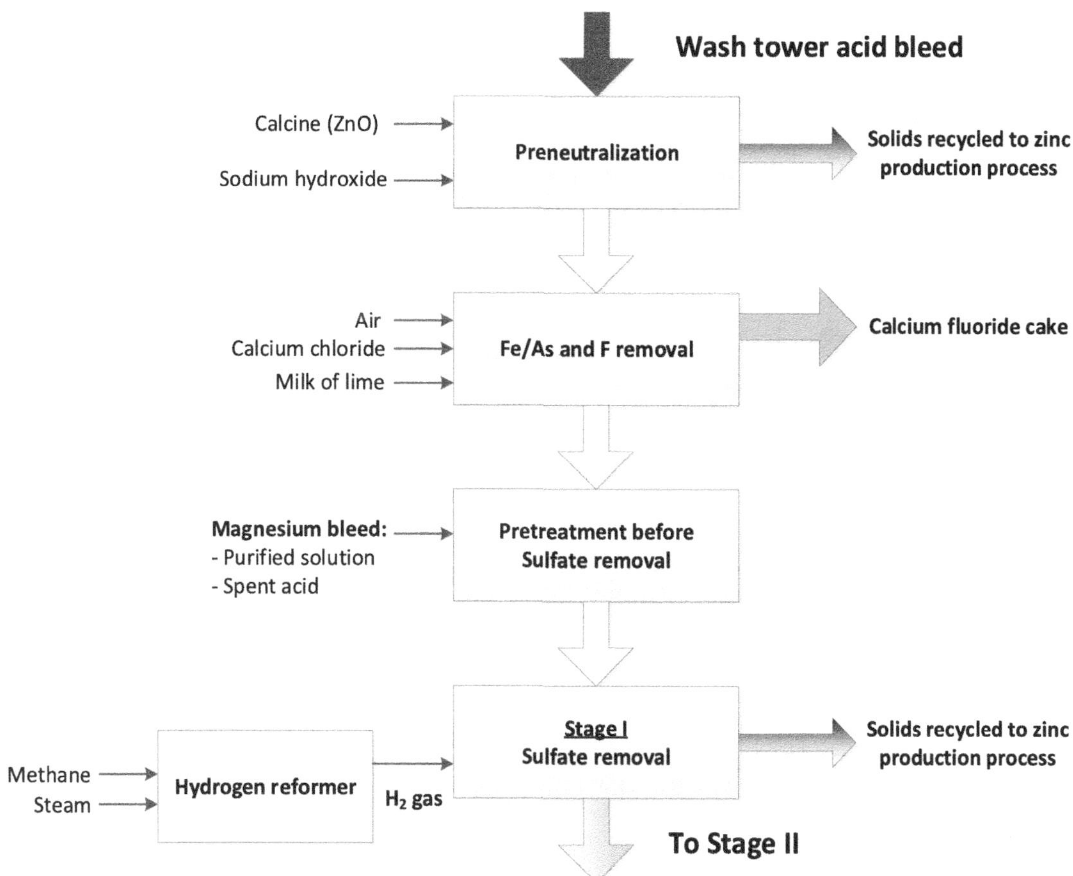

Figure 11.6 Process design for the full-scale installation for integrated sulfate reduction and zinc precipitation.

magnesium bleed (Zinc electrolyte), nutrient addition, pH, and temperature control; (iv) biological conversion of $ZnSO_4$ to ZnS, with H_2 as an electron donor; (v) separation of the formed ZnS; (vi) de-watering of the formed ZnS; and (vii) effluent of the bioreactor is further treated in the ethanol-fed bioreactor where metal traces are removed and excess sulfide is converted into elemental sulfur.

In the bioreactor, sulfate reducing bacteria convert the $ZnSO_{4(aq)}$ into $ZnS_{(s)}$ using hydrogen as electron donor:

$$ZnSO_{4(aq)} + 4\,H_2 \rightarrow ZnS_{(s)} + 4\,H_2O \tag{11.16}$$

Hydrogen gas is produced on-site from natural gas in a reformer. In the reformer unit, natural gas is converted with steam into a mixture of mainly H_2 (80%) and CO_2 which is fed to the bioreactor. CO_2 is used as a carbon source for the bacteria. In order to achieve good mixing without introducing high shear forces, a gas-lift loop type reactor is used (Figure 11.7). A small amount of the gas recirculation flow is discharged in order to bleed inert components (mainly nitrogen and methane, the latter either coming from the product gas or from methanogenic activity). This bleed gas is recycled to the reformer burner. About 15 kg of dry biomass is produced per ton of sulfate converted. The precipitated ZnS is separated in a thickener, combined with solids from the groundwater treatment and sulfide precipitation and subsequently sent to a decanter centrifuge for dewatering. Flocculant is added to obtain a clear decantate and a manageable cake. ZnS cake is recycled to the zinc production process.

The 500 m^3 bioreactor (Figure 11.7) was started up with 20 m^3 of sludge containing sulfate-reducing bacteria. Within 2 weeks, the biological activity was sufficient to treat the entire WTA flow. In 2018, the average flow to the reactor was 25 $m^3 \cdot h^{-1}$ with 3.8 $g \cdot L^{-1}$ Zn in the influent. The removal efficiency for zinc was 99.8%, while for Cd (32 $mg \cdot L^{-1}$ in the influent) removal efficiency was 99.9%. More data on long-term performance of the bioreactor can be found elsewhere [34]. The precipitated ZnS,

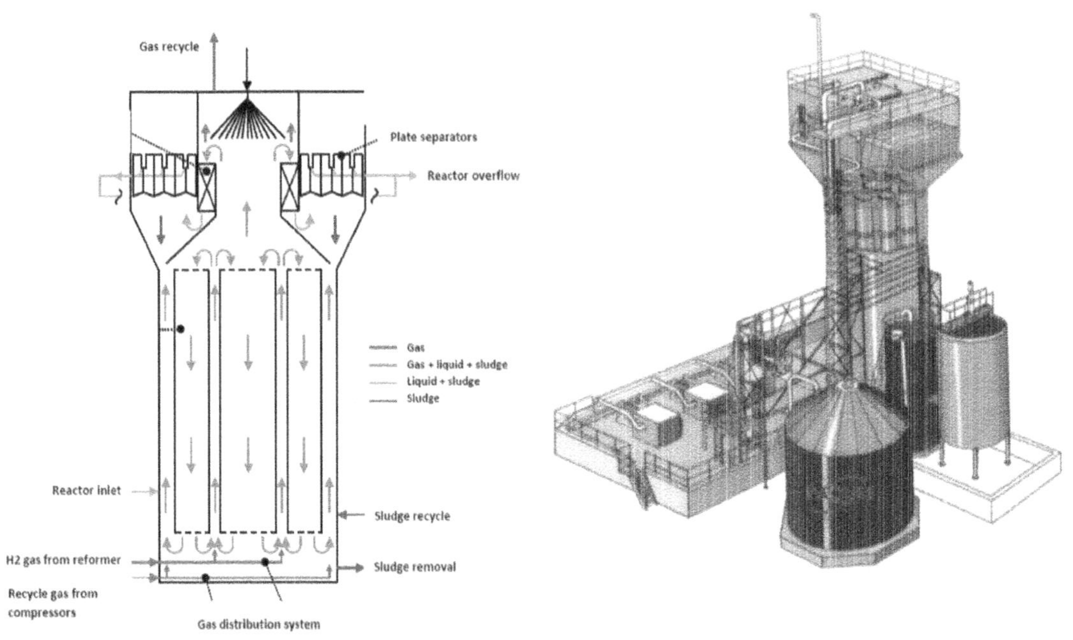

Figure 11.7 (Left) Simplified representation of a hydrogen fed SRB Gas-Lift Loop Type Reactor and (right) Full-scale sulfate reducing bioreactor for zinc and sulfate removal at Nyrstar, the Netherlands.

5 tons · day^{-1}, is returned to the zinc smelter as feedstock, such that this process produces no solid waste. For comparison, removing the zinc and sulfate with the conventional chemical lime process would result in 9 tons · day^{-1} of metal contaminated gypsum waste which is unsuitable for reuse or recycling, creating a long-term legacy as it has to be stored.

In addition to the bioreactor for integrated sulfate reduction and zinc precipitation, a second bioreactor has been in operation at the Nyrstar site in the Netherlands since 1992 for treatment of 300 m^3 · h^{-1} metal-contaminated groundwater at maximum capacity. In this bioreactor, SO_4^{2-} is reduced to H_2S with ethanol as an electron donor, followed by precipitation of metal sulfides, thereby removing sulfate, zinc, and other metals to concentrations below discharge consent values. From the year 2000 onwards, process wastewater has also been treated in the installation, and the hydrogen bioreactor installation described above also feeds into the bioreactor. Expansion of the groundwater treatment to a maximum of 400 m^3 · h^{-1} was also executed. Until 2009, sulfate reduction and metal sulfide precipitation were carried out simultaneously in the same reactor. In 2011, a separate metal sulfide precipitation was installed to optimize bioreactor performance. Excess sulfide in the metal-depleted effluent is biologically oxidized to elemental sulfur with the Thiopaq™ technology (see Chapter 10 for more details). The product of this process, a metal sulfide sludge (mainly consisting of ZnS), is recycled to the main zinc production process. The current process flow scheme is shown in Figure 11.8. The system has consistently met the required discharge limits with concentrations for sulfate, zinc, and cadmium of 450, <0.05, and <0.001 mg · L^{-1}, respectively.

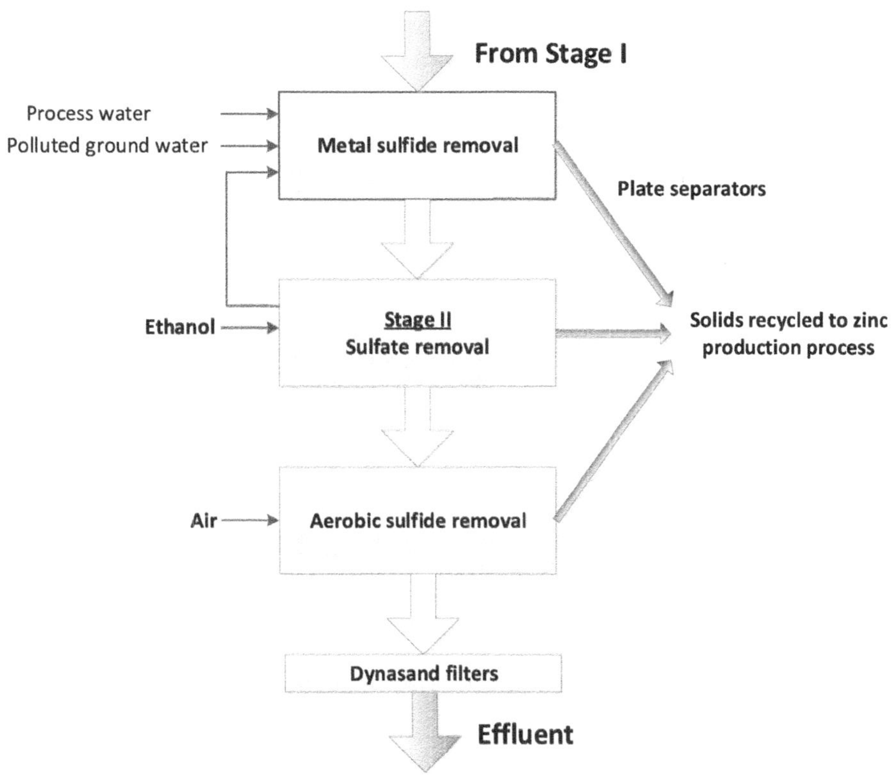

Figure 11.8 Process flow scheme of the sulfate reducing bioreactor for treatment of metal-contaminated groundwater at Nyrstar, the Netherlands.

11.6.3.2 Copper recovery at the Pueblo Viejo gold mine, Dominican Republic

Soluble copper can be found in hydrometallurgical streams resulting from gold/copper mineral processing sites. Pueblo Viejo Dominicana Corporation (PVDC) mines ores in the Dominican Republic that contain gold, silver, and copper. These ores are processed by conventional crushing, grinding, and pressure oxidation in autoclaves, cyanidation, and refining. After autoclaving, the remaining solids are washed using counter current decantation (CCD). The CCD thickeners separate the slurry (which continues to the gold/silver circuit) from the clarified liquid, which contains soluble copper at concentrations between 100 and 400 $mg \cdot L^{-1}$ and high concentrations of iron and zinc. In the past, this stream was neutralized before being recirculated to the CCD thickeners (without recovery of copper). The soluble copper was precipitated during the neutralization stage and was disposed in the tailings with the precipitated gypsum. In 2014, a THIOTEQ™ plant was started to recover the copper upstream of the neutralization plant by using biological sulfur reduction technology to produce H_2S (Figure 11.9). The use of gaseous H_2S instead of chemical NaHS avoids sodium addition to the closed loop water system, which would hamper the plant's neutralization performance. The plant was designed and commissioned by Outotec Oyj (Finland) and Paques B.V. (the Netherlands) and consists of three copper sulfide precipitation reactors and a 2000 m^3 sulfidogenic bioreactor connected by a gas recirculation system operated at 8–10% H_2S concentration in the gas. With this technology, this copper recovery plant recovers valuable copper that would be lost in the tailings while decreasing the metal content in the tailings. The process does not affect the neutralization plant performance, enabling water to be recycled back to the CCD thickeners.

Between the CCD thickeners and the copper recovery plant, an iron removal unit removes dissolved ferric iron, which is present in high concentrations. Leaving the ferric in solution would consume H_2S and lower the copper grade in the final product, because ferric iron is chemically reduced by H_2S, with ferrous iron and elemental sulfur as products of the reaction. Ferric is precipitated at high

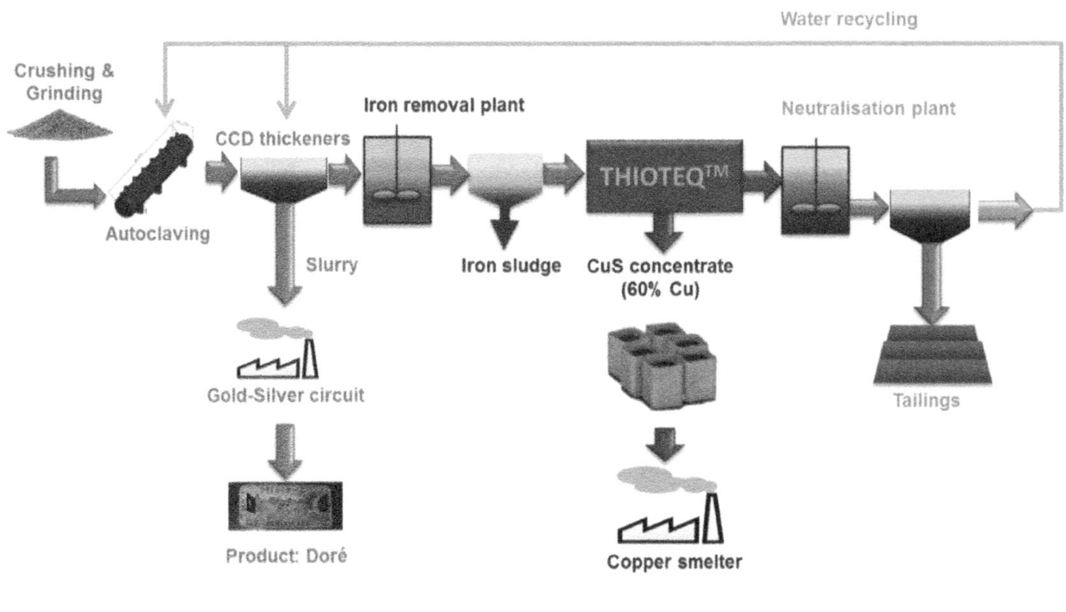

Figure 11.9 Barrick Pueblo Viejo simplified process flowsheet with copper precipitation using biogenic sulfide [35].

temperatures (>70°C) by the addition of limestone at a pH below 3, while copper and zinc remain in solution. In the copper precipitation step, the pH is not controlled and drops slightly due to the reaction with sulfide. At this low pH, zinc remains in solution, effectively making the precipitation step highly selective for copper. Although the zinc concentration in the feed is typically ten times higher than the copper concentration, the zinc content in the final product remains low at a level of around 0.2%.

The required sulfide in the form of H_2S gas is generated on-site and on-demand, at ambient temperature and pressure, in a bioreactor fed with elemental sulfur and ethanol (Figure 11.10). The reactor was seeded with naturally occurring bacteria. The bioreactor is operated 'offline' (i.e., the bacteria are not in direct contact with the metal-containing stream). The copper plant at PVDC was designed for the recovery of 12 000 tons per year of copper requiring around 20 tons of H_2S to be produced per day, making this system the largest biological H_2S generator in operation. Ground elemental sulfur is kept in excess together with a small supply of nutrients. The amount of ethanol is added based on the H_2S requirement. On average, copper removal efficiencies around 85% are achieved and are expected to increase to 95% with further optimization.

The precipitated copper settles in a 50 m diameter thickener and reaches 40% sludge density at the bottom. After dewatering in a filter press, the product is bagged and shipped to copper smelters. The produced copper concentrate contains more than 90% CuS and less than 10% gypsum, resulting in a copper grade of around 60%.

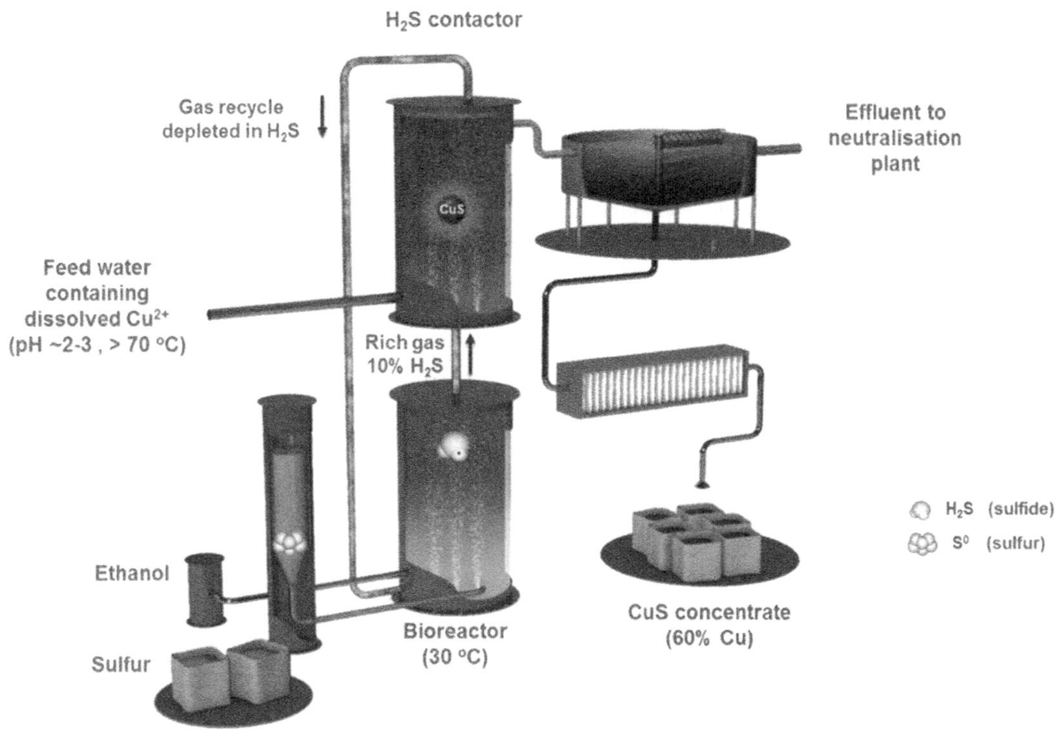

Figure 11.10 The THIOTEQ™ bioreactor produces cost-effective on-site and on-demand H_2S.

11.7 CHALLENGES, OPPORTUNITIES AND RESEARCH NEEDS

Technologies that recover metals through biological sulfur and sulfate reduction are currently applied in niche-markets. For more widespread application, the challenge is to reduce the costs of the electron donor and to widen the operational window, especially for the process parameters temperature and pH. In this respect, further development of elemental sulfur reduction technologies seems most promising. Elemental sulfur is widely available and mostly produced as a byproduct when removing sulfur-containing contaminants from natural gas and petroleum (see Chapter 10 for further details). As previously discussed, sulfide produced from elemental sulfur reduction requires four times less electron donors than from sulfate reduction. Secondly, sulfur reducers appear to span a wider range of pH (especially in the acidic range) and temperature at which they can thrive. With operation at lower pH values, sulfur reduction technologies will further generate added value for industry, as selective metal removal becomes a feasible possibility. This technology must first be investigated in more detail to obtain optimal process flowsheets.

11.8 CHAPTER SUMMARY

In this chapter, several microbial conversions of the sulfur cycle were presented, representing the basis of scalable bioreactor technologies for sulfide, sulfate, and heavy metal removal from industrial aqueous process and waste streams. These bioreactors are open, meaning that the composition of the microbial population is not controlled. However, the processes are designed to give competitive advantage to microorganisms that carry out desired conversions. For sulfate reduction, the choice of electron donor and temperature is relevant in this respect. Hydrogen and ethanol electron donors support high rates of sulfate reduction under mesophilic conditions. Sulfate reduction technologies remove sulfate and are also used for metal removal and recovery.

Although still less widespread, sulfate reduction technologies are applied in the chemical and metallurgical industries. As these sectors are not generally familiar with the use of biotechnologies, there still remains some reluctance to make full use of their potential. However, it may be expected that the technologies will increasingly penetrate the market, as they fit well with the aims of a circular economy. The recovery of reusable/recyclable products from waste streams (i.e., elemental sulfur and metal sulfides) with no or minimal solid waste production aligns especially well with this new paradigm.

11.9 EXERCISES

Exercise 11.1: A wastewater containing zinc and iron, both at a concentration 0.10 M, are aimed to be recovered through chemical sulfide precipitation. Estimate the required pH to precipitate the amount of zinc but not of iron if the initial H_2S concentration is 0.1 M. Use the equation for the metal dissolution in an acidic solution below and rearrange so the single unknown is H_3O^+. For this, calculate the solubility-product constant in acid ($K_{sp}a$). The K_{sp} values are provided in the table below and in Equation (11.5) for sulfide dissociation in acid conditions.

$$K_{spa} = \frac{k_{sp}}{k_w k_{sp1}}$$

$$MS(s) + 2\,H_3O^+(aq) \rightleftarrows M^{2+}(aq) + H_2S(aq) + 2\,H_2O(l)$$

Data for this question:

Compound	Formula	K_{sp}
Iron(II) sulfide	FeS	8×10^{-19}
Zinc sulfide (alpha)	ZnS	2×10^{-25}

Exercise 11.2: Estimate the amount of lime ($Ca(OH)_2$) required per year to remove the sulfate and metals from acid mine drainage (AMD) in the table below at a flow rate of $40\ m^3 \cdot h^{-1}$. Sulfate concentration in this AMD is $3360\ mg \cdot L^{-1}$ [36].

Wastewater Source	Acid Mine Drainage (mg/L) [24]	Wastewater Source	Acid Mine Drainage (mg/L) [24]
Magnesium	342	Nickel	3.78
Aluminum	54.3	Copper	44.9
Chromium	0.12	Zinc	5.9
Manganese	6.05	Cadmium	0.01
Iron	391	Lead	6.9
Cobalt	8.99		

(a) Estimate the amount of sludge generated per year as $CaSO_4$ and $Me(OH)_2$ and the associated costs of disposal if the cost of sludge disposal is €100/ton and the lime cost is €150/ton. Assume a solids content in the sludge of 40%.
(b) Compare this cost with the costs of using biological sulfate reduction instead with ethanol as electron donor. Consider EtOH purity 30%, density $0.95\ kg \cdot L^{-1}$, and price $100\ € \cdot m^{-3}$.
(c) Based on your calculations above, what is more attractive from an economic point of view, lime dosing or biological sulfate removal.
(d) Provide three non-economic benefits of biological sulfate removal compared with lime dosing.

Exercise 11.3: Given the operational parameters of the Thiopaq® process (Section 11.6.3) at a flowrate of $30\ m^3 \cdot h^{-1}$ and initial and final sulfate concentrations at 15 and $3\ g \cdot L^{-1}$, respectively. Calculate the consumption of H_2 in $m^3 \cdot h^{-1}$ in the biological sulfate reduction stage (Equation (11.16)).

Exercise 11.4: Given that the Thiopaq® process reported a recovery of $8.5\ ton \cdot d^{-1}$ of ZnS:

(a) How much air is used to oxidize all sulfides remaining from the calculation of exercise 11.3 in the sulfide oxidizing stage? Assume sulfides are oxidized according to:

$$H_2S + 0.5O_2 \rightarrow S^0 + H_2O$$

(b) Given air pump efficiency of 50%, energy consumption of $110\ Wh \cdot Nm^{-3}$, and electricity cost of $0.05\ € \cdot kWh^{-1}$. What is the cost of the aeration?

Exercise 11.5: In 2014, the plant from the company Paques, THIOTEQ™ Metal technology, started operation at a goldmine located in the Dominican Republic to recover copper sulfide from an acidic process water stream. The process consists of two stages: a chemical (precipitation) stage and a biological process where sulfide is produced from elemental sulfur (instead of sulfate) and ethanol. A simplified representation of the governing reactions for metal precipitation and for hydrogen sulfide production can be described as follows:

$$H_2S + M^{2+} \rightarrow MS_s + 2H^+$$

$$C_2H_5OH + 6S + 3H_2O \rightarrow 6H_2S + 2CO_2$$

The technology is able to recover 20 000 ton of Cu per year at the goldmine. Based on this amount, estimate the following:

(a) The amount of S that is being added.

(b) The amount of ethanol that is being added (ethanol: 30%, density: 0.95 kg · L^{-1}) in order to reduce sufficient sulfur to sulfide for Cu precipitation. Assume a 100% ethanol conversion efficiency.
(c) What is the advantage of using S instead of sulfate as in the Thiopaq® technology? Hint: how much ethanol would have to be added if SO_4^{2-} was the electron acceptor?
(d) Despite the advantage highlight in c), discuss a disadvantage from a process point of view.

Exercise 11.6: Give the redox equation for sulfate reduction with methanol, assuming that methanol is oxidized to HCO_3^-. Write sulfide as HS^-.

Exercise 11.7: A wastewater contains sulfate (2.5 g · L^{-1}) and zinc (1 g · L^{-1} Zn^{2+}) and will be treated biologically. Methanol (CH_3OH) is selected as electron donor for sulfate reduction. Calculate how much methanol (g · L^{-1}) needs to be added considering that:

- all zinc precipitates as zinc sulfide;
- an excess of 200 mg · L^{-1} total sulfide is targeted;
- 5% of the consumed methanol is used as carbon source;
- no hydrogen sulfide is stripped.
- How much sulfate is left in the effluent?

Exercise 11.8: Zinc needs to be removed to a level of 0.1 mg · L^{-1}. What is the theoretically required *total sulfide* level (in μg L^{-1}) to achieve this, assuming equilibrium for ZnS precipitation is reached. The pH is 7, the pK_{sp} for ZnS is 24.7.

Exercise 11.9: It turns out that besides zinc, there is also copper (1 g · L^{-1} Cu^{2+}) present in the waste stream. CuS has a pK_{sp} of 36.2:

(a) Will CuS precipitate at the pH and total sulfide concentration at which the zinc precipitates under Exercise 11.8?
(b) What consequence will the presence of 1 g · L^{-1} Cu^{2+} have on zinc precipitation?
(c) How would you solve the problem that emerged in the answer of part b?

Exercise 11.10: Zinc and copper sulfide precipitate simultaneously in the process. At which pH is selective precipitation possible?

11.10 DISCUSSION QUESTIONS

Question 11.1 (governance, market stability, self-sufficiency, political drivers). The increase in metal demand and associated supply risks have increased along with the volatility of metal prices (see Table 11.1). A good example is the significant increase in demand and price of cobalt with an increase in price of ~500 times between 1993 and 2013. The latter is related to the use of cobalt in emerging technologies, such as smartphones, electric vehicles, and the lithium-ion battery sector. Within this context, Africa has historically been the largest source of cobalt minerals. Due to a rising demand for cobalt, discuss why there is a general interest as well as political support for incentives to promote approaches that enable recovery of cobalt from wastewater. Discuss what positive impact this can have on the R&D roadmap and market implementation of technologies that enable cobalt recovery.

Question 11.2 (process control, OH&S considerations). The pH is an important driver of the biological sulfate reduction process. Discuss the implications of decreasing or increasing the pH for a one stage process, where metal precipitation and sulfate reduction occur in 1-stage, as compared to a multi-stage process (see Figure 11.5 for schematic representation of 1-stage and multi-stage configuration).

Question 11.3 (economic considerations, market demand vs. recovery potential, process limitations). In Chapter 2 of this book, a table is provided showing an overview of metals concentrations in different wastewater streams. Select two wastewaters from this table and discuss their economic potential and practical feasibility for metal recovery. In your discussion, pay particular attention to: (i) estimated volumes discharged in industry; (ii) metal concentration; (iii) market demand and price; (iv) possible process interferences/difficulties; and (v) product quality.

Question 11.4 (geopolitical considerations, non-renewable resource, political incentives). The opportunity and necessity of technologies for metal recovery is said to be evident due to the future scenarios of metal scarcity as well as potential geopolitical challenges as sources/deposits of various important metals are restricted to a limited amount of geographical locations. Based on the EU metals listed as critical metals with, moreover, relatively high economic importance, analyze which metals could be 'high on the political agenda' in the context of financial and governmental support to academia/industry to develop new technologies enabling their recovery.

Question 11.5 (ease of operation, OH&S considerations, reluctance to change, wastewater composition). As discussed in this chapter, chemical precipitation is the most common method for removal of dissolved metals from wastewater. Despite the need to change to a circular use of resources and recover metals from wastewater as well as the potential benefits of sulfide-based precipitation (both of which have been clearly identified in this chapter), at present the use of lime as a precipitation aid remains widespread. Formulate various reasons behind the latter in relation to: (i) ease of operation and OH&S considerations of the use of lime over sulfide; (ii) wastewater composition in terms of sulfate concentrations and metal concentrations and their respective discharge limits; and (iii) reluctance to change if you were in charge of the treatment facility and you have over 20 years of experience with lime dosing.

ACKNOWLEDGEMENTS

The authors are grateful for financial support from the TKI Water Technology and the Netherlands Organization of Scientific Research.

FURTHER READING MATERIALS

Barton L. L. and Fauque G. D. (2009). *Biochemistry, Physiology and Biotechnology of Sulfate-Reducing Bacteria. Advances in Applied Microbiology (Chapter 2).* Academic Press, Cambridge, Massachusetts, USA, pp. 41–98.

Hao T.-W., Xiang P.-Y., Mackey H. R., Chi K., Lu H., Chui H.-K., van Loosdrecht M. C. M. and Chen G. H. (2014). A review of biological sulfate conversions in wastewater treatment. *Water Research*, **65**, 1–21.

Muyzer G. and Stams A. J. M. (2008). The ecology and biotechnology of sulphate-reducing bacteria. *Nature Reviews Microbiology*, **6**, 441–454.

Sánchez-Andrea I., Sanz J. L., Bijmans M. F. M. and Stams A. J. M. (2014). Sulfate reduction at low pH to remediate acid mine drainage. *Journal of Hazardous Materials*, **269**, 98–109.

REFERENCES

1. Haslehner R. and Stelter B. (2015). *Mining & Metals in A Sustainable World 2050.* World Economic Forum, Switzerland, p. 43.
2. Herrero M., Bringas E., San Román M. F. and Ortiz I. (2017). Membrane operations for the recovery of valuable metals from industrial wastewater. In: Sustainable Membrane Technology for Water and Wastewater Treatment, A. Figoli and A. Criscuoli (eds), Springer, Singapore, pp. 319–348.
3. Perez J. P. H., Folens K., Leus K., Vanhaecke F., Van Der Voort P. and Du Laing G. (2019). Progress in hydrometallurgical technologies to recover critical raw materials and precious metals from low-concentrated streams. *Resources, Conservation and Recycling*, **142**, 177–188.

4. Mineral Resources Program. Cited January 2019; Available at: https://minerals.usgs.gov (accessed 22 January 2022).
5. InfoMine. Cited 2019 January; Available at: http://www.infomine.com (accessed 22 January 2022).
6. Langkau S. and Tercero Espinoza L. A. (2018). Technological change and metal demand over time: what can we learn from the past? *Sustainable Materials and Technologies*, **16**, 54–59.
7. Lewis A. (2017). Precipitation of heavy metals. In: Sustainable Heavy Metal Remediation, Volume 1: Principles and Processes, E. R. Rene, E. Sahinkaya, A. Lewis and P. N. L. Lens (eds), Springer International Publishing, Cham, pp. 101–120.
8. Cheremisinoff P. N. (2019). Handbook of Water and Wastewater Treatment Technology. CRC Press, Boca Raton, Florida, USA.
9. Tchobanoglous G., Burton F. L. and Stensel H. D. (2003). Wastewater Engineering: Treatment and Reuse, 4th edn. McGraw-Hill, Boston.
10. Puigdomènech I. (2010). Make Equilibrium Diagrams Using Sophisticated Algorithms (MEDUSA), Software Version 18 February 2004. Inorganic Chemistry Department, Royal Institute of Technology, Stockholm, Sweden.
11. Lewis A. E. (2010). Review of metal sulphide precipitation. *Hydrometallurgy*, **104**(2), 222–234.
12. Larsen P. A., Patience D. B. and Rawlings J. B. (2006). Industrial crystallization process control. *IEEE Control Systems Magazine*, **26**(4), 70–80.
13. Veeken A. H. M., Akoto L., Pol L. W. H. and Weijma J. (2003). Control of the sulfide (S2−) concentration for optimal zinc removal by sulfide precipitation in a continuously stirred tank reactor. *Water Research*, **37**(15), 3709–3717.
14. Mersmann A. (1999). Crystallization and precipitation. *Chemical Engineering and Processing*, **38**(4–6), 345–353.
15. Villa-Gomez D. K. and Lens P. N. L. (2017). Metal recovery from industrial and mining wastewaters using sulphate-reducing bioreactors. In: Sustainable Heavy Metal Remediation, Volume 2: Case Studies, E. R. Rene, E. Sahinkaya, A. Lewis and P. N. L. Lens (eds). Springer Cham, Switzerland, 81–114.
16. Giulietti M., Seckler M., Derenzo S., Ré M. and Cekinski E. (2001). Industrial crystallization and precipitation from solutions: state of the technique. *Brazilian Journal of Chemical Engineering*, **18**, 423–440.
17. Huisman J. L., Schouten G. and Schultz C. (2006). Biologically produced sulphide for purification of process streams, effluent treatment and recovery of metals in the metal and mining industry. *Hydrometallurgy*, **83**(1–4), 106–113.
18. Guidotti T. L. (2010). Hydrogen sulfide: advances in understanding human toxicity. *International Journal of Toxicology*, **29**(6), 569–581.
19. Nielsen A. H., Vollertsen J., Jensen H. S., Wium-Andersen T. and Hvitved-Jacobsen T. (2008). Influence of pipe material and surfaces on sulfide related odor and corrosion in sewers. *Water Research*, **42**(15), 4206–4214.
20. Hansson E. B., Odziemkowski M. S. and Gillham R. W. (2006). Formation of poorly crystalline iron monosulfides: surface redox reactions on high purity iron, spectroelectrochemical studies. *Corrosion Science*, **48**(11), 3767–3783.
21. Villa Gómez D. K. and Lens P. N. L. (2017). Metal recovery from industrial and mining wastewaters. In: Sustainable Heavy Metal Remediation, Volume 2: Case Studies, E. R. Rene, E. Sahinkaya, A. Lewis and P. N. L. Lens (eds), Springer International Publishing, Cham, pp. 81–114.
22. Widdel F. and Hansen T. A. (1992). The dissimilatory sulfate- and sulfur-reducing bacteria. In: The Prokaryotes, E. Rosenberg, E. F. DeLong, S. Lory, E. Stackebrandt and F. Thompson (eds), Springer-Verlag, New York, pp. 583–624.
23. Florentino A. P., Florentino A. P., Weijma J., Stams A. J. M. and Sánchez-Andrea I. (2016). Ecophysiology and application of acidophilic sulfur-reducing microorganisms. In: Biotechnology of Extremophiles: Advances and Challenges, P. H. Rampelotto (ed.), Springer International Publishing, Cham. pp. 141–175.
24. Asokbunyarat V., Lens P. N. L. and Annachhatre A. P. (2017). Permeable reactive barriers for heavy metal removal. In: Sustainable Heavy Metal Remediation, Volume 1: Principles and Processes, E. R. Rene, E. R. Rene, E. Sahinkaya, A. Lewis and P. N. L. Lens (eds), Springer International Publishing, Cham, pp. 65–100.
25. Liamleam W. and Annachhatre A. P. (2007). Electron donors for biological sulfate reduction. *Biotechnology Advances*, **25**(5), 452–463.

26. Maree J. P. and Strydom W. F. (1987). Biological sulphate removal from industrial effluent in an upflow packed bed reactor. *Water Research*, **21**(2), 141–146.
27. van Houten R. T. (1996). Biological sulfate reduction using synthesis gas as energy and carbon source. *Biotechnology and Bioengineering*, **50**(2), 136–144.
28. van Houten R. T., HulshoffPol L. W. and Lettinga G. (1994). Biological sulphate reduction using gas-lift reactors fed with hydrogen and carbon dioxide as energy and carbon source. *Biotechnology and Bioengineering*, **44**(5), 586–594.
29. De Smul A., Dries J., Goethals L., Grootaerd H. and Verstraete W. (1997). High rates of microbial sulphate reduction in a mesophilic ethanol-fed expanded-granular-sludge-blanket reactor. *Applied Microbiology and Biotechnology*, **48**(3), 297–303.
30. Dries J., De Smul A., Goethals L., Grootaerd H. and Verstraete W. (1998). High rate biological treatment of sulfate-rich wastewater in an acetate-fed EGSB reactor. *Biodegradation*, **9**(2), 103–111.
31. Weijma J., Haerkens J. P., Stams A. J. M., Hulshoff Pol L. W. and Lettinga G. (2000). Thermophilic sulfate and sulfite reduction with methanol in a high rate anaerobic reactor. *Water Science and Technology*, **42**(5–6), 251–258.
32. van Houten R. T., Yun S. Y. and Lettinga G. (1997). Thermophilic sulphate and sulphite reduction in lab-scale gas-lift reactors using H2 and CO2 as energy and carbon source. *Biotechnology and Bioengineering*, **55**(5), 807–814.
33. Lettinga G. and Hulshoff Pol LW. (1991). UASB-process design for various types of wastewaters. *Water Science and Technology*, **24**(8), 87–107.
34. van Houten B. H. G. W., van Doesburg W., Dijkman H., Copini C., Smidt H. and Stams A. J. M. (2009). Long-term performance and microbial community analysis of a full-scale synthesis gas fed reactor treating sulfate- and zinc-rich wastewater. *Applied Microbiology and Biotechnology*, **84**(3), 555–563.
35. Gonzalez Contreras P., Dijkman H., Choi Y. and Espinel R. (2016). Increasing output of metals at the pueblo vieja mine by precipitating copper sulfide using biological H_2S generation. Copper2016. Proceedings of the 9th International Copper Conference, Kobe, Japan.
36. Sahinkaya E., Uçar D. and Kaksonen A. H. (2017). Bioprecipitation of metals and metalloids. In: Sustainable Heavy Metal Remediation, Volume 1: Principles and Processes, E. R. Rene, E. Sahinkaya, A. Lewis and P. N. L. Lens (eds), Springer International Publishing, Cham, pp. 199–231.

doi: 10.2166/9781780409566

Chapter 12

Closing the loop within the water sector: circular resources

Ilje Pikaar[1], Sirajus Salehin[2], Nadine Boelee[3], Arjen van Nieuwenhuijzen[4] and Olaf van der Kolk[5]

[1]School of Civil Engineering, The University of Queensland, St Lucia, QLD 4072, Australia
[2]The Advanced Water Management Centre, The University of Queensland, St Lucia, QLD 4072, Australia
[3]Nijhuis Water Technology, 7007 CD Doetinchem, The Netherlands
[4]Witteveen+Bos, 7411 TJ Deventer, The Netherlands
[5]AquaMinerals, Groningenhaven 7, 3433 PE Nieuwegein, The Netherlands

12.1 INTRODUCTION

A crucial difference between more conventional treatment methods that are based around dissipative pathways that makes hazardous things 'disappear2 and resource recovery from used water is the need to 'do something' with the resources that are recovered. This implies that the role of utilities goes beyond that of the traditional engineering practice and (waste)water management and regulatory compliance. Provided that the water company wants to make this transition to a more circular arrangement of its processes, it also needs to include non-engineering considerations like social acceptance of the recovered resources, commercialization and marketing, cross-sectional collaboration with potential end-users in market segments and industries outside that of the water industry. The latter will most likely have important complications for management practice and required skills in the situation where resource recovery from water will become the central theme within our urban water management infrastructure. One can appreciate though that, ultimately, the most critical aspect in order for resource recovery from used water to become a central element of our urban water infrastructure in the 21st century and a driving force within the circular economy, finding an end-user that is satisfied with: (i) the quantity of the recovered materials; (ii) quality of the recovered materials; and (iii) robustness and reliability of the supply chain of the recovered materials.

There are several important aspects in relation to the quantity of the recovered materials to consider. First, what is the volume of the recovered resources/materials versus the end-user requirements? By definition there will be an imbalance between supply and demand. This imbalance is primary the challenge for the supplier to overcome, not the end-user. Second, what is the overall market size of the recovered resources/materials? Is the recovered material considered a useful resource or a replacement of a raw material for the production of other products? Third, how constant is the volume/amount of the materials that can be recovered? Is there, for instance, seasonal influence or weekend/weekday difference? There are also several important aspects in relation to the quality of the recovered materials. First, what is the achievable quality of the recovery materials compared with the quality of alternative resources and/or raw materials? End-users will only pay a premium prize when the quality of the recovered materials is equal to or exceeds that of the currently used resources and/or other available alternatives. Second, what is the reliability in terms of the quality of the recovered material? In other words, does the recovery process allow for the production of an end-product with a constant quality? Lastly, in addition to the quantity and quality, a critical aspect is the robustness and reliability of the supply chain of the recovered materials. Can a constant supply of recovered materials be guaranteed at all times?

One can imagine that the above-described characteristics could be seen as possible constraints and/or reluctance from potential end-users outside of the water industry. The latter is especially the case when the recovered materials are of low cost/value (in comparison with the overall cost of the production process of the end-user) or, for example, in situations where the recovered material is considered a bulk product that is required in large quantities. Indeed, this is often the case, for instance for recovered sand (see Chapter 1 for more details). This is a widely available bulk product with a low intrinsic value. There is no sense of scarcity for this material and possible customers have the choice between a fairly large number of potential suppliers of 'recovered sand'. The construction and demolition industry has, for instance, a large demand and supply volume, respectively, in recovered sand.

Let us imagine a situation in which one would be the person that is responsible for the process for recovery of the resources and moreover is also the person that is the end-user of the recovered materials, that is being your own customer? In other words, are there situations in which the recovered substances can be used in the same process? This could potentially substantially simplify the situation. This is in fact not fiction at all; there are various real life proven and established methods that have been implemented by the water industry. The latter has – apart from the expected positive environmental impacts (see Chapter 22) – two major advantages. First, the sector is its own customer and that saves a lot of uncertainties concerning the marketing of the recovered materials/resources. Experience has learned that the latter can really accelerate projects, investments, market uptake and replication (personal communication with representatives of the water industry). Second, it forces the water sector to think as a customer of their own materials. Being both the supplier and customer at the same time creates the notion and emphasizes on the importance of the product quality, quantity and the reliability/security of the supply-chain within the organization itself.

12.2 LEARNING OBJECTIVES

This chapter aims to define important concepts in relation to circular resources like 'cradle-to-cradle', waste management hierarchy and the precautionary principle as well as put them into a historical context. The different approaches in relation to circular resources and sustainable use of resources are also explained. This chapter will also describe and discuss the (circular) resources within the water sector and the role the water sector can play in becoming (more) circular. Lastly, various real-life full-scale examples of (recently developed) circular chains within the water sector are provided. By discussing these topics, at the completion of this chapter you should be able to:

- Define the concepts of 'cradle-to-cradle', circular resources and the Waste Management Hierarchy in your own words.
- Understand the difference between linear and circular processes.
- Understand the complications of embedding resource recovery as the central theme in urban water management on the key roles/task of water utilities.
- Explain the different approaches as it comes to sustainable use of resources.
- Explain the role the water sector can play in the transition to a more circular economy.
- Give examples of circular (re-)sources in the water sector.

12.3 CIRCULAR RESOURCES: HISTORICAL CONTEXT, CONCEPTS, AND PRINCIPLES

12.3.1 Historical background – the creation of a consumption society with linear use of resources

After World War II, economies over the world started to flourish. This looks like a long time ago, but in terms of sustainability this is actually not that long ago at all. This can be explained by the most commonly used definition of sustainability:

> 'Sustainable development is development that meets the needs of the present without compromising the ability of future generations to meet their own needs'. [1]

The word future generation refers to the need for long-term thinking, that is thousands and thousands of years. In this context, one can appreciate that a timeframe of only some 75 years is not that long at all. One of the pillars under this economic growth was 'consumption and linear use of resources'. Consumption was stimulated by a hefty increase in production volume and decreasing production costs. The relatively low consumption prices led to a less frugal use of products, products that were also less durable than those produced before this era. This altogether led to an exponential growth in volumes of waste and wastewater. In Europe this led for instance to an increase from 100 kg waste per capita per year in the early 1950s to over 480 kg waste per capita per year by the year 2016 (http://ec.europa.eu/eurostat/web/products-eurostat-news/-/DDN-20180123-1).

This increase in waste volumes was at first managed by the local municipalities. They organized waste collection and the dumping of the wastes in local landfills. This worked (and in some regions still works) for quite some time, but continued flow towards these landfills led to capacity problems. Research furthermore showed problems with the leaching of hazardous components to, for example, groundwater, often due to illegal dumping of wastes or even ignorance. In the early 1970s realization kicked in that this 'single use and throw-away' society could not last forever. For example, The Club of Rome published a very influential report entitled 'The Limits to Growth' (https://www.clubofrome.org/) [2]. This report described the use of computer simulations to prove that economic growth cannot continue forever due to the depletion of resources, and would ultimately result in the collapse of our society by the end of the 21st century. The first oil-crisis in the early 1970s further strengthened public concern about future resource scarcity and the overall sustainability of our society. Despite this increasing and more widespread concern and environmental awareness in the following decade(s), not that much changed in terms in terms of resource efficiency in the decades ahead; which was also the case for the water industry.

12.3.2 Concepts and principles

12.3.2.1 The waste management hierarchy

By the end of the 1970s it became evident that with an ever growing population, accompanied with increasing amounts of waste being produced, a paradigm shift was needed in order to deal with this enormous challenge our 'consumption society with linear use of resources' had created. This realization asked for new (decision) models in order to understand what is preferred in terms of sustainable use of resources. The most well-known model globally is known as 'the waste hierarchy', which is depicted in Table 12.1). It became a popular tool that evaluates different processes/options in order to select the most favourable in terms of sustainability. Since its conception in the early 1970s (e.g., the waste management hierarchy was embedded for the first time into European waste policy in 1975; Waste Framework Directive (1975/442/EEC)), the waste hierarchy has become one of the key pillars of environmental regulation that has been embedded in various national regulatory principles and guidelines as well international treaties and covenants.

The list provided in Table 12.1 gives a good comparison of the different routes and their hierarchy in terms of preference. However, while it is considered a very useful concept and embedded in various policies, the table also highlights the simplicity of the concept in the sense that several important

Table 12.1 The waste management hierarchy.

Waste Hierarchy – Most Favourable Solutions (from most to Least Favourable)	
1	*Quantitative prevention/Refuse* Avoid the use of a product or material
2	*Intensify use/Rethink* Use a product much more intense, for instance (1) using a MRI in an hospital 24/7 instead of the normal opening hours or (2) sharing products
3	*Qualitative prevention/Reduce* Use other or less materials giving a lower impact in the use, but also choosing those that give a lower impact in the waste phase
4	*Product re-use* Re-use a product elsewhere in its original function
5	*Repair/Refurbish* Repair a product so it can be used in its original function or restore old products and bring up to date
6	*Re-use parts/Remanufacture* Re-use parts of a disposed product in its original function
7	*Recycle parts/Repurpose* Recycle parts of a product in another function than its original
8	*Material recycling* Process materials to obtain new materials that can be used in its original or new function
9	*Use as fuel/Recover* Incineration of materials with energy recovery
10	*Incineration as disposal route with material recovery* Incineration with no/minor energy recovery but with recovery of (some) materials
11	*Incineration as disposal route without material recovery* Incineration with the sole goal to destroy the waste and/or reduce the volume
12	*Landfilling* Dumping in landfills

aspects are not included/discussed, such as various economical, technical, and environmental considerations that ultimately can result in a different waste management strategy adopted in practice. This is for example the case when:

(1) A more desired route is much more expensive than the less desired one. For instance, repairing is more expensive in labour costs than re-use parts of the product;
(2) Reusing a product is not possible in terms of the specs on hygiene and human health. For instance, the food/beverage and medical industry have evidently very high standards with respect to human health and hygiene, making re-use in some cases almost impossible;
(3) The environmental impact of a more desired route is negative (or less positive). This may be the case in situations where the materials are regenerated (material recycling) and this regeneration process consumes, for example, a lot of energy, water or chemicals.

Another important aspect, besides the multiple ways of looking at the use and pathways of resources and materials in the water sector by means of the waste management hierarchy, is what the terms and definitions used in describing them (e.g., re-use, recycle and recover) really stand for. For instance, the terms 're-use' and 'recycling' are often used to describe the 'good use' of a residual, but both terms in reality mean something completely different. Recycling refers to the process of converting waste into new materials, while re-use refers to the re-use of a material in its original function. It is therefore important to get a fundamental understanding and detailed knowledge of what these different terms really comprise. In a practical situation, it is therefore strongly advised to inquire with the client, regulator or costumer, to what they are referring to in terms of 'recycling', 're-use', 'recover' of the materials' in order to avoid any confusion and/or misunderstanding.

Various countries have adapted the waste management hierarchy in their waste-policies. As mentioned above, the European Union incorporated this waste hierarchy already in 1975 in the Waste Framework Directive (1975/442/EEC) and has after that revised its legislation on this issue several times (e.g., 1989, 1996 and 2008). Governments have, by means of implementing the Directive into national regulations, eminently the ability to influence the 'pathways' of materials/resources. Indeed, an expression often used is 'regulation and policies drive innovation'. The importance and impact of regulation/policy on practical and economic feasibility of resource recovery cannot be emphasised enough (see also Chapters 13–15). Examples of direct influence of regulation include (but are not limited to):

(1) *Stimulate desired routes by implementing subsidies* – In order to stimulate innovation and desired resource management approaches one can provide subsidies on innovation or create taxation benefits;
(2) *Discourage undesired routes by means of taxation* – With this type of taxation, the overall economics of alternative routes that are considered more environmentally friendly and sustainable (i.e., thus higher up the waste hierarchy chain), become more attractive from financial point of view. Examples of this form of taxation include for example taxation on landfilling (i.e., landfill levy) and incineration;
(3) *Prohibit undesired routes* – This can be regarded as the most powerful and direct influence of regulation on resource management. A good example for instance is the prohibition of landfilling of recyclable/reusable materials.

12.3.2.2 The 'cradle-to-cradle' concept

The phrase *cradle-to-cradle* was already introduced in the 1970s by Walter R. Stahel, but has become very popular since the publication of the book entitled 'Remaking the way we make things: Cradle to cradle' by William McDonough and Michael Braungart in 2002 [3]. The basis of the cradle-to-cradle philosophy is that processes, materials and products are designed in such a way that after their service life these materials can be used again. The basis also lies in a well thought out design of the processes where these residuals emerge. These processes are designed in such a way that the emerging residuals 'fit' a next destination or use. In the case of the urban water management, by-products are produced, and waste streams arise. So, with the ambition to 'adapt and transform' these materials into circular resources that can be completely incorporated into the cradle-to-cradle philosophy, it may thus have (substantial) complications for the design and (waste)water treatment processes. Indeed, one can appreciate that design criteria need to be extended beyond the efficiency, robustness and cost-effectiveness of the treatment process alone. Knowing and understanding what the exact specifications of these circular chains will need to look like is evidently of crucial importance in order to make this 'design' fit.

The *cradle-to-cradle* philosophy distinguishes two routes, namely, the biological cycle and the technical cycle (see Figure 12.1). It is of great importance that both cycles remain in a healthy and valuable state. This means that they these cycles remain physically apart and do not interfere with each other. Hybrid materials (materials consisting of various different elements) lead by definition to problems in the recycling phase.

The biological cycle is meant for materials that are biodegradable, thus materials that can be completely decomposed through biological processes such as composting, digestion and fermentation. The degraded materials can be used as 'food' for the soil and the crops that grow on it. For this cycle it is important to take into account the organic balance in the soil. The quantity of organic material added to the soil must ideally be in balance with the quantity that is removed in the form of crops (or other plant-based products such as timber) or lost by erosion or degradation.

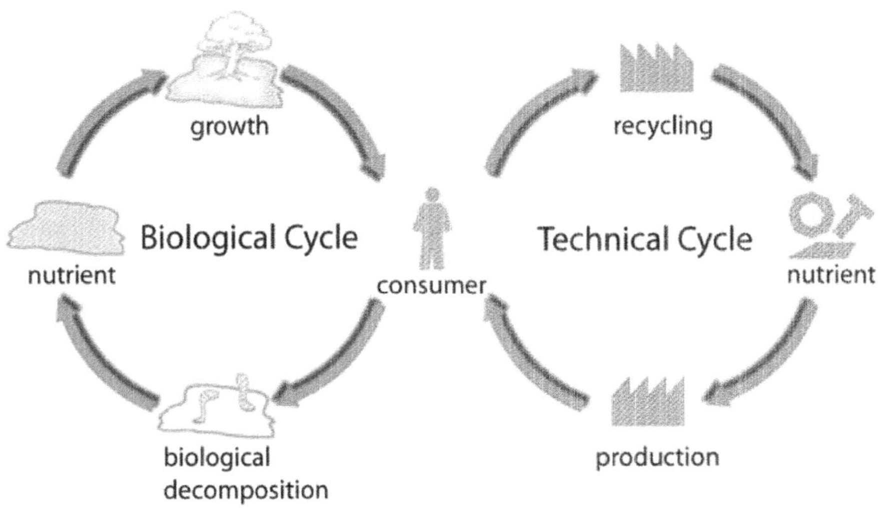

Figure 12.1 The biological and technological cycle within the cradle-to-cradle philosophy [3].

The technical cycle contains non-biological materials such as metals, sand and plastics. These materials keep circulating as valuable resources for the industry. The great challenge in this cycle is to maintain as consistent a level of quality as possible during and after use. When the quality cannot be maintained at a desirable level, reusing or recycling may not be possible anymore. In these situations, the material needs to be down-cycled to a 'lower end' product or application, or even worse, disposed of or destroyed. A frequently used term is the so-called 'closed loop' (sometimes also referred to as a 'continuous loop'). The latter term refers to a situation in which a material used in a certain product (e.g., lamp, furniture, concrete tile, and carpets) is returned back to the original manufacturer where the material is completely re-used in the same product. This concept has gained significant interest in recent years and is growing rapidly. Producers and manufacturers in fact use this as a strong marketing-tool. There are even business-concepts where the ownership of the product remains with the producer and the customer leases or rents the product (e.g., Fair phone, Philips Lightning, Technogym). This business model is often referred to as a 'product as service model'. In order to repeatedly recycle component materials from the products that contain them, it is evidently important that these individual components can easily be recovered from the product it is used in. This asks for another way of looking at designing products: *design for disassembly*. In this cycle, materials that are considered harmful/toxic in the production, use and recycle phase should be avoided. Of course, one should always aim to avoid the use of such materials at any time, but this is of special concern within the circular economy as it makes the re-use and recycling opportunities much smaller or even impossible.

12.4 CIRCULARITY WITHIN THE WATER SECTOR

When looking at the water sector one could consider three distinctive 'resource clusters' in which resources/materials can be recovered:

(1) *The (waste)water treatment process*. This process characterises itself by the use of chemicals (e.g., coagulant, metal salts, activated carbon) and the emergence of residuals/by-products (e.g., sludge, calcite, gasses). New technology increasingly leads to more recovered substances from these sludges, like cellulose, energy and struvite.

(2) *The (underground) distribution network*. This cluster contains mainly pipes, pumps and buffers. It is less (material) dynamic than the treatment process, but nevertheless contains a vast volume of materials. In fact, the distribution network comprises the largest fraction of the overall asset value of our urban water infrastructure;
(3) *The water itself*. The (drinking-) water companies use sea-, surface- or groundwater for the production process. The wastewater companies treat the 'used-water' and discharge the treated water.

While the ambition of governments and the water industry to become more 'circular' is firmly anchored in various policies and guidelines, the majority of the resources used in the water sector still originate and rely upon the use of virgin materials. There are several reasons for the latter that are worth mentioning:

(1) This choice for virgin material is often regarded as the 'safest'. Obviously, the treatment process has the highest priority and can thus not be subjected to any 'risk' of new products/materials made from non-virgin materials. However, it should be emphasised that this argument is not always valid anymore, as there are plenty of situations where the recovered materials have in fact a higher quality and spec than the virgin products/materials previously used.
(2) Various segments of the water sector have strict requirements on the use of materials and resources in terms of hygiene. The most obvious and clear-cut example would be the production of drinking water. The requirements for the chemicals used in this process (e.g., for softening or coagulation) are more difficult for non-virgin than virgin products/materials to meet, but also here recent innovations lead to new opportunities.
(3) There is often no clear incentive and/or drivers to change to different materials are lacking. So why take the risk when the current materials have proven to work just fine.
(4) There is a lack of clear guidelines and regulation on the use of resources recovered from water (see Chapter 19 for more detail).

The improvement of a more circular use of materials in the water sector can be found in the re-use/recycling of the discarded materials from the treatment process or distribution network. Most of these materials are currently landfilled, incinerated or find their way into backfills. Surprisingly, in many cases, water companies do not even know where these materials ultimately end up (personal communication water utilities and companies). However, with some extra attention to, for instance, the quality, volume, possible consumer of these materials and better management of the materials, a better use of these residuals is within reach, for example: (i) valuable materials can be recovered after incineration, such as P-recovery from the ashes of sludge incineration; (ii) residuals can be processed and recycled, such as the use of municipal wastewater sludge in agriculture or (iron-) sludge for the production of clay bricks; (iii) products such as pumps can be repaired; (iv) water can be re-used in the various processes; and (v) a better use of materials may lead to less use and therefore prevention.

However, in the above-mentioned five examples beneficial use of residuals is not 100% circular; these materials are used in processes/applications not within the water sector itself. From a more holistic point of view, the ultimate goal in the cradle-to-cradle principle would entail that resources are recovered and re-used within the water sector. Indeed, this is certainly something the water sector should pursue; would it not be great when the water sector becomes completely resource independent and self-sufficient? That is more difficult than it appears at first sight. After all, it is not only the challenge to make the water circular again, but what about the (underground) infrastructure, as well as chemicals and energy needed for the treatment processes and the residuals coming from the treatment process itself? Operating circular means also making these resources circular. Obviously, it will be very difficult for the water sector to completely achieve this goal. Nevertheless, a lot of progress can be made, and in fact there are quite a lot of examples of circular resources within the water sector, some of which are discussed in detail in the following section.

12.4.1 Circular resources within the water sector: real-life case studies

As discussed above (and in various other chapters of this book), there are various examples of resources and materials that can be re-used within the water sector. In this paragraph, we will describe four 'circular resources' within the water sector in more detail by discussing four real-life case studies within different segments of the water sector, that is drinking water production, food and beverage industry, and municipal wastewater treatment.

12.4.1.1 Case study 1. Water re-use: the Ultra-Pure Water Factory Emmen

A good example for a Water Re-use Case study is the Ultra-Pure (UPW) Water Factory Emmen (Figure 12.2), The Netherlands. This wastewater re-use plant has been in operation since early 2010 and upcycles the effluent from the wastewater treatment plant of the city of Emmen into ultra-pure water

(a)

(b)

Figure 12.2 (a) Ultra Pure Water Factory Emmen (front) with WWTP Emmen (background); (b) Interior of the UPW Factory Emmen with RO membrane unit in front (photo credits: Arjen van Nieuwenhuijzen, Witteveen+Bos).

to be used by Shell-NAM for oil extraction. The output can be as much as 10 ML/day. Shell-NAM converts the UPW into steam, which is injected into the Schoonebeek oil fields to render the thick, viscous oil more fluid. It has been estimated that by 2036 120 million barrels will be extracted using this technique.

Wastewater and industrial water process technologists and specialised engineers from Water Company Drenthe, Waterboard Velt and Vecht, WLN and Witteveen+Bos consulting engineers developed, investigated and designed the individual technologies and the complete integrated processes design. The purification process has been positively assessed by external, internationally recognised engineers and water professionals.

One can imagine that in order to produce the ultra-pure water from raw domestic wastewater requires a smart combination of various treatment methods. In fact, the innovative combination of treatment techniques utilised in the UPW Factory is unique, consisting of existing and innovative techniques comprising a sequence of five treatment and upcycling steps using the effluent of the WWTP as influent (more information on the process configuration can be found elsewhere; https://nwtr.nl/en/puurwaterfabriek.php):

(1) *Pretreatment* – Rotary screens remove the larger impurities and objects, such as leaves and hairs. A brushing mechanism ensures the screens are automatically kept clean and clear. The filtered waste together with the rinse water is collected in a buffer, from where it is returned to the wastewater treatment plant.

(2) *Ultra-filtration* – Submerged UF membranes remove the insoluble substances from the WWTP-effluent. The water is vacuum pumped from the basin through the walls of the straws. Any waste is caught by the exterior walls of these membrane straws. The wash water is returned to the Emmen WWTP. The clean water is then passed to the next step of the purification process.

(3) *Biological activated carbon filtration* – The BACF is required to prevent and control bio-fouling of the reversed osmosis membranes. By adding pure oxygen, a favourable environment for bacteria is created in the activated carbon filter, which convert organic material into water and carbon dioxide, thereby considerably reducing bio-fouling in the RO membranes. The rinse water and the biomass are returned to the WWTP. At this stage the water is almost of potable quality.

(4) *Reversed osmosis* – The RO membrane installations remove any minerals and left over impurities. The total installed RO membrane surface is 40,000 m^2. During treatment, approximately 20% of feed water is concentrated and returned to the WWTP. Adding anti-scaling prevents fouling of the membranes. Reverse osmosis happens in two stages; both installations are connected in a circuit.

(5) *Electro deionisation* – Removal of any remaining minerals (ions) takes place with the electro deionisation unit. This step is a relatively new development in the field of water purification: a combination of membrane filtration and ion exchange. After EDI the produced water is specified as Ultra Pure Water and feeds the steam production facilities of Shell-NAM. The UPW factory is the world's largest EDI (electro deionisation) facility currently in operation.

Noteworthy to mention is that also parts of the UPW Factory itself were based upon the 'cradle-to-cradle' concept, that is the concrete floor was poured using a mix which included crushed concrete from demolished structures.

12.4.1.2 Case study 2. Recovery and on-site re-use of fat as an energy source in the food and beverage industry

Many industrial wastewaters in the food and beverage industry including meat processing, candy and palm oil contain relatively high concentrations of fat. This fat can be effectively recovered using a combination of physical chemical separation techniques, as depicted in Figure 12.3.

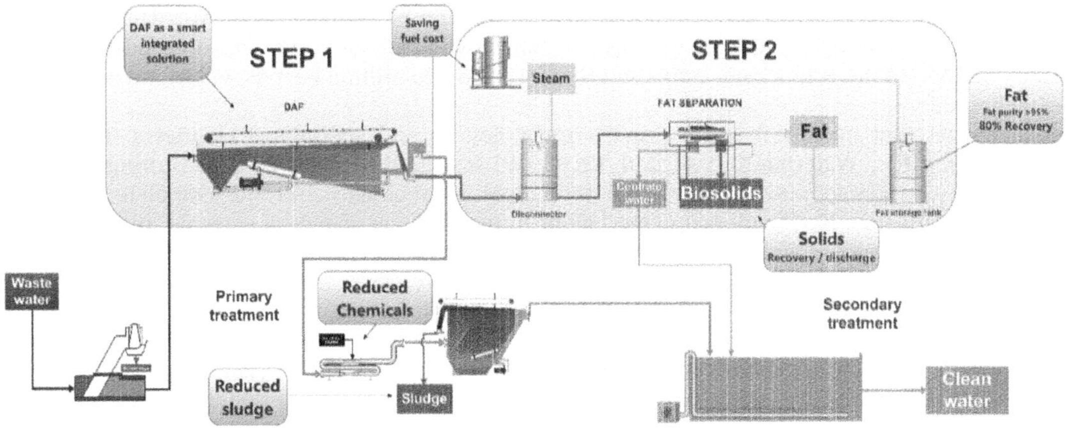

Figure 12.3 Schematic representation of the overall wastewater treatment process with Nijhuis AECO-FAT for the efficient recovery of fat in two steps.

This process has been realised for one of the leading bacon processing factories in Ireland to turn their flotation fat into recovered resource (Figure 12.4). The factory of 10,500 m^2 operates six bacon lines with a total capacity to process (i.e., slice, cook and freeze 85,000 rashers (thin slices of bacon) per hour. The latter is equals to a weekly production of about 200 tonnes of cooked strips. This production process comes with the production of substantial amount of wastewater with an average flow of 600 m^3 wastewater per day that is associated with high COD concentrations on the order of 30,000 mg/L and an oil and grease concentration of $\sim$500 mg/L.

The fat recovery process can be divided into two main steps, namely: (i) fat removal; and (ii) fat separation and re-use. In the first step, the wastewater is treated by dissolved air flotation (DAF) to remove fat and solids. In general, a removal efficiency of 45–70% total suspended solids (TSS) and 50–80% oil and grease is achieved. The produced sludge is a mixture of water, fat and solids in ratios depending on the wastewater origin and recovery method implemented. In the bacon processing wastewater more than 25% of the sludge is fat, while for instance poultry slaughterhouse wastewater sludge typically contains around 90% water and 7% fat.

The wastewater leaving the DAF is subsequently treated in a second DAF unit in combination with enhanced chemical flocculation and coagulation to remove residual solids which are not removed in the first DAF unit. This second DAF is followed by a secondary aerobic biological treatment step prior to discharge onto surface water or sewage. During the second step of the AECO-FAT system, the produced sludge is heated in a disconnector before entering the solid separation stage. In the latter, the sludge is separated into a liquid fat, a water and a solids fraction. The recovered fat is stored in heated tanks to keep it liquid for transport. The centrate water from this second step is treated in the biological reactor (see Figure 12.3). The first step of the AECO-FAT system in Ireland produces a maximum of 30 m^3 sludge per day. Around 6 m^3 of pure fat is recovered from this sludge per day. Generally, more than 80% of the fat present in the flotation sludge will be recovered. The specific recovery is dependent on the composition in the sludge (e.g., fat content, physical structure/properties etc.) which in turn depends on the wastewater origin.

In this project, daily 3 m^3 fat is used to feed a steam boiler and supply the entire factory with energy in order to become completely energy self-sufficient. The quality of the produced fat in this case is also suitable for biofuel production and the residual 3 m^3 fat per day is sold for this purpose. This 6 m^3 daily recovered fat has an energy content of 209 GJ (assuming that fat has an energy content of 40 MJ/kg and a density of 870 g/L) which is equivalent to about 58 MWh or 5500 L of diesel (assuming the energy

Figure 12.4 The case study plant overview (left top), the dissolved air flotation with 'fat' flotation sludge (right top) and the recovered fat (bottom).

content of diesel is 45.3 MJ/kg and the density of diesel is 840 g/L). In addition to becoming energy self-sufficient, the above-described fat recovery and re-use process also comes with additional benefits including a reduced chemical footprint and lower sludge production. The reduction in chemical use is due to the fact that the first DAF unit produces sludge *without* the addition of chemicals (note that chemicals cannot be used otherwise the fat could not be used in the boilers), which lowers the chemical requirement for the second DAF by ~50%. Moreover, the total amount of sludge to be disposed is decreased as fat is removed from the sludge and is no longer disposed with the sludge. The revenue of fat and the cost reduction of both chemicals and sludge discharge are depending on the local costs for the chemicals, sludge discharge and the energy price or price for which the recovered fat can be sold. It is the combination of these benefits of fat recovery that makes its business case. This makes it a recovery solution which is applicable around the globe for wastewaters containing high concentrations of fat such as wastewaters of slaughterhouses, protein processing plants (including bacon) and confectionary plants.

12.4.1.3 Case study 3. Re-use of chemical coagulants

Chemical coagulants play an important role and are dosed in enormous amounts at various places in our urban water management. They are dosed for: (i) the removal of turbidity, colour, natural organic

Figure 12.5 Application of iron-rich drinking water sludge as a means to minimise eutrophication in lakes (photo credits: Olaf van der Kolk, Aquaminerals).

matter (NOM) and pathogens during drinking water production; (ii) corrosion and odour control in sewage distribution systems; and (iii) phosphorus removal and sulphide control during anaerobic digestion at wastewater treatment plants (WWTPs). By far, aluminium sulphate (also known as alum) and iron salts (both ferrous and ferric chloride) are the most commonly used chemicals. On average, the iron content (expressed as elementary Fe) is about $27 \pm 15\%$ (personal data AquaMinerals of nine drinking water treatment plants with data gathered over a period of 10 years). With respect to beneficial re-use of these chemicals, especially iron-based coagulants are of interest as iron possesses the ability to remove both hydrogen sulphide and phosphate. Obviously, the iron content differs from one water treatment plant to another as it depends highly on the quality of the surface water. There is no need to use high-quality virgin coagulants in a sewer- or WWTP-context and the use of iron rich drinking water sludge (Figure 12.5) is a great example of circular resources in water sector and managing chemical coagulants in this way helps to close the loops within the water sector.

Another example of beneficial re-use of iron is the use of iron rich ground water sludge. A fundamental difference is that in this example the iron is not added to the water as a chemical coagulant, but naturally present in the ground water in dissolved from as ferrous iron (Fe^{2+}). In this situation, it is in fact considered a pollutant that needs to be removed from the water in order to meet drinking water regulations. The latter is normally achieved by means of aeration, thereby oxidation the ferrous iron to ferric iron (Fe^{3+}), which is non-soluble at circumneutral pH values. This process leads to a sludge which contains (very) high levels of iron (hydr-)oxide (i.e., as high as >80% $Fe_2O_3 \cdot 3H_2O$ on a dry matter basis). On average, the iron content (expressed as elementary Fe) is about $34 \pm 11\%$ (personal data AquaMinerals of 87 drinking water treatment plants with data gathered over a period of 10

Figure 12.6 Iron oxide pellets produced from iron-rich ground water sludge (photo credits: Olaf van der Kolk, AquaMinerals).

years) and can therefore be considered an excellent 'low-grade' alternative to virgin iron chloride for sewer corrosion and odour control. Since iron (hydr-)oxide is a very reactive material, this sludge is not only used in sewers for sulphide control, but for instance is also used in digesters, simply by adding it as a slurry to the anaerobic digester. Indeed, several utilities have implemented this approach with excellent results. Examples of wastewater utilities that are using iron rich drinking water sludge for sulphide control in sewer network are Hamburg Wasser, the combined drinking- and municipal waste water utility of the city of Hamburg (Germany) and the wastewater utilities Drents Overijsselse Delta and Waterschapsbedrijf Limburg (the Netherlands). Another application is the use of iron-rich drinking water sludge as a low cost 'filter' to minimise eutrophication due to excess phosphate in lakes.

Moreover, it was found that, with some processing of this sludge, the iron can be re-used in the water sector in the form of pellets (Figure 12.6). The method of producing pellets is a rather simple process that consists of the drying of the sludge, adding an organic additive and pelletising it to a granular absorbent. These pellets can be used in treating biogas from anaerobic digesters and/or for odour control in pump pits and wet wells in sewer networks. It can also be used in removing P from surface water, especially in regions where this water contains too many nutrients, leading to algal blooms. Another option to use these pellets is to remove arsenic during the production of drinking water. A key advantage of using pellets instead of sludge is that the process is better controllable and suitable for treatment of both gases and water streams, but it also gives the user the opportunity to easily change a saturated filter with a fresh one, whereas this is not possible with sludge.

12.4.1.4 Case study 4. Circular calcite

Pellet softening is often used as a step during the production of drinking water from the removal of hardness in the form of calcium (as well as magnesium). In this process, the water is fed through a column fed with sand at an up-flow velocity of 60–100 meter/hour. This applied up-flow velocity keeps the sand in suspension. Caustic soda is added at the bottom of the column to elevate the pH of the water to $\sim$8.5, causing calcium to crystalize onto the sand (which acts as a seeding material) in the form of calcium carbonate, also known as calcite. This particle starts to grow and, due to the constant shear caused by the up-flow velocity, the calcite particle is polished and rounded. When this

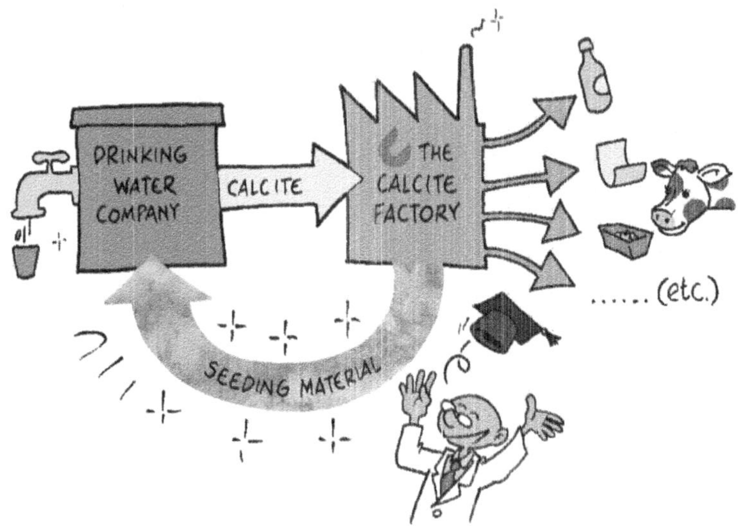

Figure 12.7 Illustration of the 100% circular calcite concept.

particle reaches a diameter of approximately 1 mm, it becomes too heavy and sinks to the bottom of the reactor and is subsequently continuously removed. The sand-calcite particle residual is among others used as resource in the steel-industry, glass-industry and as lime fertiliser in agriculture [4]. A considerable limitation of this process in terms of beneficial re-use of the calcite is actually the sand. The presence of the sand is considered a nuisance for the various re-use applications mentioned above. For example, a lot of water companies use garnet sand (due to its high specific gravity), which is rich in iron. This iron is an unwanted component for the production of white (colourless) glass, garnet-containing softening pellets are therefore not fit for this type of glass. In addition, there are applications where the calcite needs to be processed by grinding, drying, sieving and sterilisation to the desired size distribution (for instance as filler in paper, plastics, feed). The sand-core is much harder than the calcite, leading to the wearing of the mill and so a significant increase in costs.

If the calcite itself could be used as seeding material, the resulting pellet would consist of 100% calcite and would therefore become a more applicable and valuable residual with several industrial applications. Therefore, an R&D project was initiated by a collective of Dutch water companies (Brabant Water, Waternet, WML and Dunea) in the year 2014. The overall aim of this project was to develop a softening method that used 'Dutch Calcite' as 100% circular seeding material (Figure 12.7).

This 'Dutch Calcite' was made from softening pellets from the water companies themselves. After intensive research by KWR (2015; https://www.kwrwater.nl/en/actueel/high-value-re-use-of-lime-from-drinking-water-softening-is-sustainable-and-profitable/), it was found that these residuals from softening pellets can, after being ground, sieved and sterilised, be used as new seeding material within the same process. Equally important (rather surprisingly), the softening process itself became more efficient using calcite rather than garnet sand in terms of the use of seeding material and calcium removal rates. Based on these very promising results, several drinking water companies indicated that they wanted to implement this 'Dutch Calcite' in their processes. Since this is a completely innovative and novel process, no infrastructure to make this product on a rather large scale was available. Therefore, the water company of Amsterdam (Waternet) and the UK-based mineral processor Advanced Minerals took the initiative to set up a processing plant in Amsterdam. It started supplying drinking water companies with seeding material at the end of 2017 (see Figure 12.8).

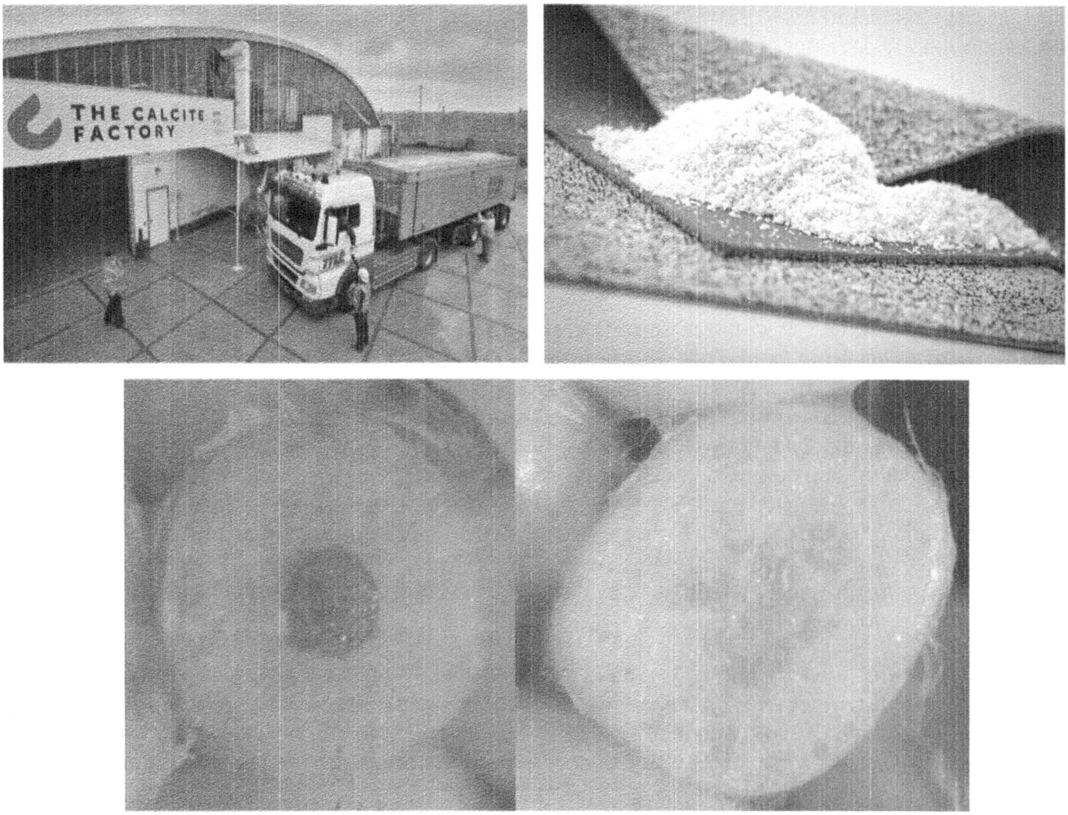

Figure 12.8 (a) The calcite factory (Amsterdam, The Netherlands); (b) Beneficial re-use of 'Dutch Calcite' for the production of carpets; and (c) comparison between 'Dutch Calcite' and conventional sand-calcite pellets.

12.5 CHAPTER SUMMARY

In this chapter, we have discussed the following concepts: (i) consumption society (with linear use of resources); (ii) 'cradle-to-cradle'; and (iii) Waste Management Hierarchy. We have described the history and development of the circular economy throughout the last decades of the 20th century with a special focus on the main opportunities and challenges for the water industry to play an important role in the transition into a circular economy through changing from linear to circular water management practices. One of the key considerations for the water industry was found to be how to embed resource recovery as the central management theme in urban water management in the context of the traditional roles/tasks of (waste)water utilities. A crucial difference between 'water management through removal processes' versus 'water management through resource recovery' identified was the need to 'do something' with the resources that are recovered. As such, the role of utilities would need to go beyond that of traditional engineering practice and regulatory compliance. The duties/considerations of utilities would also need to include commercialisation and marketing, cross-sectional collaboration with potential end-users in market segments and industries outside that of the water industry. The latter could be avoided in situations where the water industry is its own customer; the person that is responsible for the process for recovery of the resources is also the person that is the end-user of the recovered materials. Such recovered materials were defined as circular resources. In this chapter, several 'circular resources' were discussed though real-life full scale case studies of (recently developed) circular chains within the water sector.

12.6 EXERCISES

Exercise 12.1: Table 12.1 shows the waste management hierarchy. Provide at least two examples per step in the waste hierarchy ladder for the water sector as a whole. Can you provide two examples for each of the following sectors within the water industry: (i) drinking water production; (ii) municipal wastewater management and industrial (waste) water management?

Exercise 12.2: The most commonly used coagulants in drinking water treatment are iron salts (either with chloride or sulphate as counter anion) or aluminium salts in the form of aluminium sulphate (often referred to as alum) and Poly-Aluminium-Chloride (PAC). The choice differs depending on the region. For example, in Australia alum is predominantly used, whereas in countries such as the Netherlands and the US, iron-based salts are used more often. The latter is often directly related to the price of the coagulant, as both types of coagulants are capable of reaching desired water quality standards. In some regions alum is cheaper and readily available and in some iron salts are cheaper and readily available (i.e., the presence of iron and alum smelters greatly affect the price). In the context of 'circular resources', despite the fact that for your location iron salt coagulants to be used in drinking water treatment are 30% more expensive than alum-based salts, provide several arguments why iron-based salts ultimately would be considered the best choice for the urban water infrastructure at large?

Exercise 12.3: In the question above, you have provided several reasons why the choice of iron-based coagulants would be preferable for the urban water infrastructure at large. However, you may have given a different answer if you were the OH&S manager of the drinking water plant. Describe the OH&S concerns that are associated with transport, handling and storage of concentrated chemicals.

Exercise 12.4: The first implementation of the waste management hierarchy dates back to 1975 when it was included in the Waste Framework Directive (1975/442/EEC), and ever since has become one of the key pillars of environmental regulation that has been embedded in various national regulatory principles and guidelines as well international treaties and covenants. Nevertheless, as witnessed by our current way of wastewater management, in many instances the waste management hierarchy is not followed. Provide at least three reasons why this is the case.

Exercise 12.5: A drinking water utility uses alum as coagulant at a large-scale drinking water treatment plant (i.e., 500,000 PE). Currently, the produced sludge is stored on-site since regulation allows it and it is from an economic point of view (by far) the cheapest option. Moreover, there is sufficient space to store the sludge for another 30 years. Nevertheless, the utility is afraid that the regulation will change to prohibit on-site storage. Calculate the amount of sludge produced on a yearly basis used by the water utility and the financial impact it would have in case the sludge would need to be landfilled. In your answer, assume an alum dosing rate of 7 mg Al^{3+}/L (see Chapter 3 for more detail), a daily water consumption of 130 liter per person per day, a sludge moisture content of 70% and a landfill gate fees of $100 dollar per ton product (i.e., thus wet ton sludge). What about the situation where landfilling is not allowed?

12.7 DISCUSSION QUESTIONS

Question 12.1: What is the large-scale potential of iron-based coagulants from groundwater? (drinking water sector, industrial uses): Consider the production of iron-based coagulants from ground water. Look up the following information for the Dutch situation: (i) total amount of drinking water from surface water treated by means of coagulant dosing; (ii) total amount of iron recovered from treatment of ground water? Based on this, can you determine the overall market potential of iron-recovery-to-coagulant production via this route? Can you identify any potential hurdles in terms economic, social acceptance and hygiene?

Question 12.2: Worth the trouble of taking such a risk? (utility management, leadership): As the innovation manager of a large water utility, you are in charge of reorganising the existing water infrastructure from its current situation to a more circular approach within a timeframe of 15 years. You have heard of several success stories of water recycling (see Chapter 4). You are somewhat enthusiastic about such an approach but you are wondering why you would take such a risk. While successful implementations of water recycling are plenty, there is a risk of not getting sufficient support from the community as there is still scepticism from people within the community? You are asked to give a presentation to the board of directors in which you evaluate the current status and justify your masterplan. What are your key considerations/motivations with respect to your decision? Furthermore, what in your opinion will be the most important challenge of water recycling: technical, economic or social acceptance? How can you minimise the risk of failure?

Question 12.3: Who is in the lead: government, water sector or commercial market? (leadership, drivers for innovation): Resource scarcity and sustainable use of resources is a societal challenge; ultimately it affects the community at large and our future generations. Changing to a (more) circular use of recourses in general is a wise thing to do. This demands breaking with the linear economy. As described in this book, this requires a paradigm shift from the 'linear' to the 'circular' economy. The latter will not be an easy and straightforward task. You can imagine that adopting such a circular value chain will entail certain risks for the water utilities; in fact it can be argued to be a somewhat risky and challenging path. So, should the government take the initiative to stimulate or even force the creating of new circular values chains? By legislation, taxation or subsidies? Or should the initiative lie within the water sector itself? In your answer, keep in mind that for most companies in this value chain resource recovery is often not their primary focus (i.e., quite obviously, for drinking water companies this is first and foremost producing drinking water of high quality).

Question 12.4: Production of ultra-pure water from domestic wastewater effluent (industrial uses, drivers, regulation): The Water Re-use Case study the Ultra-Pure (UPW) Water Factory Emmen has been in operation since 2010 (Figure 12.2). This water reclamation plant produces up to 10 ML/day of ultra-pure water that is subsequently used by Shell-NAM for oil extraction. Discuss what might drive this type of collaboration between the water utility and Shell. In particular, focus on: (i) the drivers for the water utility to collaborate with Shell and (ii) the drivers for Shell to collaborate with the utility and use wastewater effluent? Do you think they are purely economically driven? Do you think that there also drivers for the local government and/or national government to support (financially) such an innovative project?

Question 12.5: Increasing the energy self-sufficiency (industrial uses, drivers, regulation): Biogas production by means of anaerobic digestion and subsequent beneficial on-site production of electricity (and heat) has several advantages. In some cases, applying digestion is not the most cost-effective method (under the current electricity and natural gas prices). Nevertheless, even in these situations anaerobic digestion is still applied. Discuss what the rationale for such a decision could be. In your answer include sustainability, green image, self-sufficiency, independency, regulation and potential synergy with agricultural and municipal solid waste management (e.g., co-digestion).

Question 12.6: Multiple re-use of iron in urban water management (industrial uses, drivers, utility management): Despite that direct dosing of iron-rich sludge to sewers for sulphide control is successfully applied at full-scale, some utilities have indicated that the feasibility may depend on the origin of the sludge. Provide several examples of what the reasons for the latter could be. In your answer discuss the following aspects: (i) transport distance; (ii) configuration/capacity downstream wastewater treatment plants; and (iii) presence of impurities.

Question 12.7: Putting resource recovery from water in the sustainability context (long-term thinking, global context, sustainability): It is nice to have ambitious goals as set out by various water utilities to become more sustainable. However, the term 'sustainability' is rather broad. To illustrate this, the most commonly used definition of sustainability is the so-called 'Brundtland' definition:

> 'Sustainable development is development that meets the needs of the present without compromising the ability of future generations to meet their own needs'.

How would a 'sustainable' situation with respect to wastewater management in the city you live look like by the year 2050? This sounds like an easy question, but in fact is very complex and moreover very important (i.e., you can only find the right answers/solutions if you know what you are looking for!). What does the term 'needs' mean? In your response describe the technical, social, economic, political and environmental characteristics. Finally, think about the question how resource recovery from wastewater fits in this context and what are the limiting factors (or slowing down) for the widespread adoption of resource recovery. Are these technological, economical or are these more social/political related.

Question 12.8: How to define/explain the circular economy (regulation, sustainability, critical thinking): The term circular economy has become very popular in recent years. However, many definitions and interpretations of this concept exist. Define the term circular economy in your own words, in a maximum of 250 words. After defining the 'circular economy' in your own words, read the following manuscript:

Ghisellini, P., Cialani, C. & Ulgiati, S. 2016 A review on circular economy: the expected transition to a balanced interplay of environmental and economic systems. *Journal of Cleaner Production*, **114**, 11–32. After reading this manuscript, did your understanding and/or view on the definition and the importance of the circular economy change?

Question 12.9: Policy drives innovation? (regulation, drivers for innovation). Do you think that policy makers are important stakeholders/actors in relation to the creation of a circular economy? In your opinion, how can policymakers stimulate efficient (re-)use of resources in our society? And what if you only look at the water industry, would your answer be different?

Question 12.10: How to define/explain recycling', 're-use', 'recover' of the materials' (criticial thinking): Besides the multiple ways of looking at the use and pathways of resources and materials in the water sector by means of the waste management hierarchy, the terms and definitions that describe these pathways (e.g., re-use, recycle and recover) need to be clearly understood and differentiated. For instance, the terms 're-use' and 'recycling' are often used to describe the 'good use' of a residual, but both terms in reality mean something completely different. Recycling refers to the process of converting waste into new materials, while re-use refers to the re-use of a material in its original function. It is therefore important to get a fundamental understanding and detailed knowledge of what these different terms really comprise. In a practical situation, it is therefore strongly advised to inquire with the client, regulator or costumer, to what they are referring to in terms of 'recycling', 're-use', 'recover' of the materials in order to avoid any confusion and/or misunderstanding. Define the 'recycling', 're-use', 'recover' of the materials in your own words and in particular highlight the differences between these different terms.

Question 12.11: Multiple re-use of iron in urban water management (leadership, drivers, and utility management) Hydrogen sulphide (H_2S) induced sewer concrete corrosion is a notorious and very costly issue for water utilities worldwide. Besides, the emission of H_2S from sewer manholes is a threat for sewer workers. Moreover, its obnoxious odours result in frequent complaints from the surrounding community. All of the above need to be dealt with appropriately at all times. In principle, iron-rich drinking water sludge you have produced upstream during drinking water production can theoretically be dosed to sewer networks as a means to control hydrogen sulphide. Imagine yourself being the lead engineer of the wastewater utility managing the sewer network and downstream WWTP; do you think it is better to use the drinking water sludge instead of fresh chemicals, or a combination of both? In your answer discuss the terms reliability, community engagement, potential negative side-effects and integrated urban water management.

REFERENCES

1. Keeble B. R. (1988). The Brundtland report: 'Our common future'. *Medicine and War*, **4**(1), 17–25.
2. Donella H. M. (1972). The Limits to Growth: A Report for the Club of Rome's Project on the Predicament of Mankind. Universe Books, New York.
3. McDonough W. and Braungart M. (2002). Remaking the Way We Make Things: Cradle to Cradle. North Point Press, New York, p. 104.
4. De Jong A.L. (2019) Inspiration Booklet: Reusing Water Sector Residuals. AquaMinerals. Nieuwegein, The Netherlands. Available at: https://aquaminerals.com/wp-content/uploads/2019/07/Inspiration-booklet-AquaMineralsversie-mei-2019.pdf (accessed 16 December 2021).

doi: 10.2166/9781780409566

Chapter 13

Resource recovery from used water: The (European) regulator's point of view

Ludwig Hermann and Ralf Hermann
Proman Management GmbH, Weingartenstrasse 92, 2214 Auersthal, Austria

13.1 INTRODUCTION

As discussed in the previous chapters of this book, resource recovery from wastewater (often referred to as 'used water') covers a wide scope of activities including energy, nutrients, minerals, coagulants, cellulose, metals and many more substances. Each of these product areas is governed by specific legislation that need to be considered if resource recovery activities are undertaken. Specific legislations are effective if end of waste criteria are fulfilled. The present chapter therefore goes beyond the technical and economic aspects of resource recovery and aims to provide the reader with important insight into the importance of regulation and legislation in the context of resource recovery from water. This chapters focusses in particular on the specific legislation by setting the boundaries with respect to energy and nutrient recovery within the European Union. The boundaries for energy and nutrient recovery were selected because of these sectors being the largest, best developed and most widely covered by European policies and legislation. Policies and legislation essentially aim at a sustainable use of resources, abatement of greenhouse gas emissions, preservation of soil fertility, food security and safe food production for European citizens. The legal framework consists of a mix of financial support schemes for conversion of biomass to renewable energy and restrictions to the excessive use of nutrient rich effluents on crop- and grassland. It contains a large variety of legal acts which are binding for the whole territory of the European Union, albeit in part leaving room for adaptions and interpretations when European directives are adopted in Member States' legislation. The European Court of Justice is the highest instance to decide if a national interpretation and/or adaption of a European directive corresponds to the original intention and scope of the European legal act to which the national legislation refers.

Even if the density of legal acts addressing energy and nutrient recovery activities looks overwhelming, readers will soon acknowledge that they follow a clear target and pursue a meaningful strategy. Indeed, it is expected that after reading this chapter readers will understand that this legal framework is providing a roadmap for business and revenue opportunities, albeit with some constraints inherent in every act of legislation.

Direct revenue opportunities provided by the European legal framework include feed-in tariffs, premiums and green certificates for renewable energy. Business opportunities created by the restrictions of spreading unlimited amounts of nutrient rich residues to crop- and grassland require innovation and technologies to produce more concentrated and fit for purpose fertilising products that can be shipped to destinations where nutrients are needed. New ideas for facing the existing and upcoming challenges, may be supported by funding of one of the European and/or Member State based Research and Innovation programmes. As mentioned above, this chapter covers only European legislation with the rationale that it is more comprehensive than in many other parts of the world. Consequently, it cannot give a direct guidance to legal acts governing energy and nutrient recycling and recovery in other parts of the world. However, it still can guide the reader through areas which may be governed by regulation and where attention to existing or forthcoming policy intervention should be paid. The chapter can also be taken as a reference for policy makers if they feel that certain challenges may be effectively addressed by regulatory measures.

13.2 LEARNING OBJECTIVES

This chapter aims to identify and assess regulatory support schemes and constraints that investors and operators of water facilities face when considering energy and nutrient recovery from wastewater. It covers European legislation which is probably the most comprehensive regulatory package one may encounter by undertaking that kind of activity. It gives a good example of how regulatory bodies may address challenges related to water quality, human, animal and environmental protection, renewable energy, circular economy and placing new products on the market. By studying and discussing these topics, at the completion of this chapter you should be able to:

- Understand the motivation of regulators when selecting specific topics for regulation.
- Understand which legislative bodies need to be considered if undertaking activities of recovery and recycling of energy and nutrients.
- Identify areas and activities where energy and nutrient recovery could be successfully implemented.
- Realise how to conceive projects and design processes and products in compliance with legal requirements.
- Be familiar with how to design projects and processes to make best use of the legal framework.
- Explain the governing legal framework and its motivation to colleagues and third parties.

13.3 EUROPEAN POLICIES

Legislation of the European Union consists of policies, regulations and directives which govern municipal and private holdings, energy conversion from water, wastewater and sewage sludge such as heating and cooling systems, anaerobic digestion, as well as nutrient recycling and waste management activities. Hence, investors and operators planning an activity related to energy conversion and nutrient recovery from wastewater are advised to start with considering the European regulatory framework. When it fits in this framework, one should turn to the regulatory framework of the Member State where the resource recovery facility will be located and where the recovered materials are sold and used.

13.3.1 Overview of relevant policies in the context of resource recovery from wastewater

European Policies are disclosed and implemented by a variety of acts including opinions, communications, recommendations, decisions, directives and regulations, from non-binding to binding acts. Table 13.1 provides a summary of some of the key EU policies relevant for energy and nutrient recovery from wastewater. Each of the following policies is implemented by several of these acts. Acts that are relevant to the owners and operators of resource recovery and recycling systems are briefly explained in this chapter, even if their relevance is only marginal or indirect.

Table 13.1 European policies relevant to energy and nutrient recovery from wastewater.

Policies	National Regulation Recommended	Relevance to Resource Recovery from Water
2030 Energy Strategy (see section 13.3.2)	The EU's Renewable Energy policy aims at a low carbon economy. A binding target of 20% final energy consumption from renewable sources was set for 2020. EU countries have committed to reaching their own national renewables targets. All EU countries have adopted national renewable energy action plans (NREAP). The 2030 Energy Strategy includes a new renewable energy target of at least 27% of final renewable energy consumption in the EU by 2030. On 30 November 2016, the Commission published a proposal for a revised Renewable Energy Directive to ensure that the 2030 target is met	Relevant to heating or cooling using waste- or drinking water, anaerobic digestion and combustion of sewage sludge and other organic residues from municipal and industrial wastewater treatment plants
Clean Air Package (see section 13.3.3)	The clean air package aims to substantially reduce air pollution across the EU and prevent subsequent health and environmental impacts by abating acidification, eutrophication and ground-level ozone. This is pursued by limiting the emissions of SO_2, NOx, NH_3 and volatile organic compounds by 2030 as stipulated in the Gothenburg Protocol, and has been addressed in EU legislation by the National Emission Ceilings (NEC) Directive	Relevant to storing, handling and use of nutrient containing residues from wastewater facilities
Bioeconomy Strategy (see section 13.3.4)	Europe's Bioeconomy Strategy addresses the production of renewable biological resources and their conversion into vital products and bio-energy. It aims at focusing Europe's common efforts in response to increasing populations, depletion of natural resources, impacts of increasing environmental pressures and climate change	Encourages nutrient recovery, recycling and reuse
Circular Economy Package (see section 13.3.5)	The CE package contributes to 'closing the loop' of product lifecycles through increased recycling and re-use. The plans aim at extracting the maximum value and use from all raw materials, products and waste, fostering energy savings and reducing greenhouse gas emissions. The new Fertilising Products Regulation is the first legal act coming out of the CE package	Encourages closing the material loops, particularly the nutrient loops

13.3.2 The 2030 Energy Strategy (European Commission, 2018a)

In 2018, all EU countries agreed on a new 2030 Framework for climate and energy, including EU-wide targets and policy objectives for the period between 2020 and 2030. These targets aim to help the EU achieve a more competitive, secure and sustainable energy system and to meet its long-term 2050 greenhouse gas reductions target, as committed to in the Paris Agreement in 2015. The strategy sends a strong signal to the market, encouraging private investment in new pipelines, electricity networks, and low-carbon technology. The targets are based on a thorough economic analysis that measures how to cost-effectively achieve decarbonisation by 2050. The cost of meeting the targets does not substantially differ from the price we will need to pay in any case to replace our ageing energy system. The main financial effect of decarbonisation will be to shift our spending away from fuel sources and towards renewable, low-carbon and even carbon capture and storage (CCS) and carbon capture and utilizing (CCU) technologies. The (ambitious) targets set forward for 2030 are as follows:

(1) A 40% cut in greenhouse gas emissions compared to 1990 levels;
(2) At least a 27% share of renewable energy consumption;
(3) At least 27% energy savings compared with the business-as-usual scenario.

In order to achieving these goals, the following policies will be implemented by the year 2030:

(1) A reformed EU emissions trading scheme (ETS);

(2) New indicators for the competitiveness and security of the energy system, such as price differences with major trading partners, diversification of supply, and interconnection capacity between EU countries;
(3) First ideas for a new governance system based on national plans for competitive, secure, and sustainable (renewable) energy. These plans will follow a common EU approach aiming to ensure stronger investor certainty, greater transparency, enhanced policy coherence and improved coordination across the EU.

13.3.3 The Clean Air Package (European Council, 2013)

The clean air package aims to substantially reduce air pollution across the EU. The proposed strategy sets out objectives for reducing the health and environmental impacts of air pollution by 2030 and contains legislative proposals to implement stricter standards for emissions and air pollution. The package was published by the Commission on 18 December 2013 and includes a communication (COM (2013) 918 final) on the 'Clean Air Programme for Europe', plus three legislative proposals on emissions and air pollution. The main target of the Clean Air Programme is to improve the health of European citizens and reduce health related expenses. In addition, the package aims at the protection of ecosystems, innovation and enhancing EU competitiveness in the field of green technology. Part of the Clean Air Package and relevant for stakeholders in nutrient recovery activities is the Gothenburg Protocol.

13.3.4 Gothenburg Protocol (UNECE, 1999)

Launched and adopted by the United Nations Economic Commission for Europe (UNECE) on 30 November 1999, the Gothenburg Protocol aims at abating acidification, eutrophication and excessive ground-level ozone concentrations. The Protocol is part of the Convention on Long-Range Transboundary Air Pollution (LRTAP convention) which is the main international framework for cooperation and measures to limit and gradually reduce and prevent air pollution. Fifty-one countries from the UN Economic Commission for Europe (UNECE) region are parties to the convention, including the EU Member States, Canada, the United States and several countries in Central Asia. Since its signature in 1979, the LRTAP convention has been extended by eight specific protocols, including the 1999 Protocol to stop acidification, eutrophication and ground-level ozone. To that effect, it sets national emission ceilings for each Party to be met by 2010 and thereafter for the following four air pollutants: sulphur (mainly sulphur dioxide, SO_2), nitrogen oxides (NOx), ammonia (NH_3) and volatile organic compounds other than methane (VOC). This protocol, also known as 'the Gothenburg Protocol', was approved by the Council on behalf of the EU in June 2003. It was transposed into EU law mostly through the 2016 National Emission Ceilings Directive and the 2015 Directive on emissions from large combustion plants. The focus on the abatement of acidification, eutrophication and ground-level ozone by limiting SO_2, NOx, NH_3 and volatile organic compound emissions makes the Gothenburg Protocol relevant for nutrient recovery activities, even if the emission ceilings are set for Member States.

13.3.5 The Bioeconomy Strategy (European Commission, 2012)

Launched and adopted on 13 February 2012, Europe's Bioeconomy Strategy addresses the production of renewable biological resources and their conversion into vital products and bio-energy. Under the lead of Directorate General (DG) Research and Innovation, the Strategy was co-signed by several other Commission departments, namely DG Agriculture and Rural Development, DG Environment, DG Maritime Affairs and DG Industry and Entrepreneurship.

The strategy aims to focus Europe's common efforts in response to increasing populations, depletion of natural resources, impacts of increasing environmental pressures and climate change. Its main purpose is to streamline existing policy approaches in this area. It is structured around three pillars: (i) investments in research, innovation and skills; (ii) reinforced policy interaction and stakeholder engagement; and (iii) enhancement of markets and competitiveness. The strategy is also needed to ensure that fossil fuels are replaced with sustainable and renewable alternatives as part of the shift to a post-petroleum society. The Bioeconomy Strategy manifests itself in dedicated research and innovation strategies and topics, including the Integrated EU Maritime Strategy Blue Growth covering research programmes like BONUS for the Baltic Sea and BLUEMED for the Mediterranean Basin. The strategy is relevant for potential additional research projects which may come out of nutrient recovery and recycling activities.

13.3.6 Circular Economy Package (European Commission, 2018b, 2018d)

The European Commission adopted an ambitious Circular Economy Package in December 2015 to help European businesses and consumers to engage in a transition to a stronger and more circular economy where resources are used in a more sustainable way. The European Commission started the initiative, after the withdrawal of a legislative proposal on waste, by presenting a new package in 2015 which would cover the full economic cycle (not just waste reduction targets) and is drawing on the expertise of all the Commission's services. The proposed actions contribute to 'closing the loop' of product lifecycles through increased recycling and re-use and bring benefits for both the environment and the economy. The plans aim at extracting the maximum value and use from all raw materials, products and waste, fostering energy savings and reducing greenhouse gas emissions.

The proposals cover the full life-cycle of products: from production and consumption to waste management and the market for secondary raw materials. This transition is supported financially by the European Structural & Investment Funds (ESIF), which include €5.5 billion for waste management. In addition, support is provided by €650 million under Horizon 2020 and investments in the circular economy at national level.

The new regulation of fertilising products, replacing Regulation (EC) 2003/2003 (European Commission, 2016), has been conceived as the first deliverable of the Circular Economy Package. By its holistic approach including virgin and recycled materials from mineral and organic resources and by its 'built-in' end-of-waste status for compliant products, it fully reflects the circular thinking. Other legislative proposals followed: the recast of the Waste Framework Directive including all other waste related acts, the recent EU Strategy for Plastics in the Circular Economy and the review of the list of Critical Raw Materials (European Commission, 2018d) are other examples of the highly integrated approach.

The Circular Economy Package with its new Fertilising Products Regulation is a cornerstone of research and innovation projects dealing with the globally acknowledged primary target of sustainable development: 'Decoupling', economic growth and full employment without growing use of resources. As such, it is embedded in the global framework of the Sustainable Development Goals (United Nations, 2013) and the Paris 2015 United Nations Climate Change Conference (UNFCC, 2015) agreements to confine the temperature increase to a maximum of 2°C, both adopted by virtually all countries in the United Nations by the end of 2015.

13.4 EUROPEAN REGULATIONS (EUROPA, 2018)

13.4.1 Overview

An EU Regulation is a binding legislative act. It must be applied in its entirety across the EU and inherently does not allow divergent interpretations by Member States that nonetheless exist within the activity areas covered by the present chapter. Due to frequent amendments of legislation referred to in this chapter, interested parties are advised to consult http://eur-lex.europa.eu/homepage.html (Europa, 2018) and official comments which can be easily accessed by entering the regulation denomination, as listed in Table 13.2, to an internet search engine.

13.4.2 Fertilising Products Regulation (Europa 2018)

Regulation (EC) No 2003/2003 relating to fertilisers was adopted on the 11th of December 2003 and aims at providing EU-wide rules for the type definition, nutrient content, nutrient solubility, packaging and labelling of mineral fertilisers. Important to note is that organic fertilisers are exempt from the

Table 13.2 European regulations in the context of resource recovery from wastewater.

Regulations	Binding in all Member States – Objective/Regulated Subject
New Fertilising Products Regulation – recast of Regulation (EC) 2003/2003	The current regulation defines and lists inorganic fertilisers (primary, secondary and micro-nutrients), liming materials and regulates their market placement. The draft version for the new regulation defines Component Material Categories (CMC) and Product Function Categories (PFC) for EC labelled fertilisers from organic and inorganic materials. Compliance with the new regulation acknowledged by the EC-label also determines Europe-wide end-of-waste status
Animal By-products Regulation (EC) 1069/2009	Laying down health rules regarding animal by-products (ABP) and derived products not intended for human consumption. Currently excludes ABP Cat 1 from use as raw materials for renewable fertiliser production, albeit with different approaches in certain Member States (UK, Portugal)
Plant Protection Product (PPP) Regulation (EC) 1107/2009	Regulates placing of plant protection products on the market. The Regulation on fertilising products, when enforced, will include bio-stimulants
Registration, Evaluation, Authorisation and Restriction of Chemicals (REACH) Regulation (EC) 1907/2006	Concerning the establishment of a European Chemicals Agency, amending Directive 1999/45/EC and repealing Council Regulation (EEC) No 793/93 and Commission Regulation (EC) No 1488/94 as well as Council Directive 76/769/EEC and Commission Directives 91/155/EEC, 93/67/EEC, 93/105/EC and 2000/21/EC
Classification, Labelling and Packaging (CLP) Regulation (EC) 1272/2008	The CLP Regulation contributes to the UN Globally Harmonised System (GHS) aim that the same hazards will be described and labelled in the same way all around the world
Waste Shipment regulation (EC) 1013/2006	On supervising and controlling shipments of waste within EU borders and to/from EFTA, OECD and Basel Convention countries

current regulation and governed by national legislation. In its Annex I, the regulation lists fertiliser types according to their specific characteristics. Once a fertiliser meets this type designation it may bear the letters 'EC'. The fertiliser may then be sold and used throughout the EU. This EC designation guarantees farmers that the fertilisers contain a minimum nutrient content and are safe to use. For a new type designation to be listed in the Annex I to the regulation, the manufacturer of a fertiliser corresponding to that type designation must lodge a request with a national competent authority. This request is forwarded to the European Commission, which consults the other EU countries and decides to accept or reject the application based on the advice of a technical committee set up by the regulation. To achieve the EC status, a fertiliser must provide nutrients effectively, not harm human, animal or plant health or the environment, and demonstrate it has been subject to the relevant sampling, analysis and test methods.

As the first deliverable of the Circular Economy Package, the Commission presented a draft proposal for a new Fertilising Products Regulation (COM (2016) 157 final) (European Commission, 2016) with substantial amendments: the new regulation includes amendments of Regulations (EC) 1069/2009 and (EC) 1107/2009 and covers organic and inorganic fertilising products, next to liming materials, soil amendments, growing media, bio-stimulants and blends thereof organized in so-called Product Function Categories (PFC) determined in Annex I to the proposed Fertilising Products Regulation and so-called Component Material Categories (CMC) determined in Annex II of the draft regulation (Table 13.3). EC labelled fertilising products may only be produced from designated CMC. PFCs set out the requirements related to the PFCs to which EU marked fertilising products shall belong. The requirements described in Annex I for a given PFC apply to EU marked fertilising products in all subcategories of that PFC.

CMCs set out the requirements to components of which EC marked fertilisers may be produced. An EC marked fertilising product shall consist solely of component materials complying with the

Table 13.3 Designation of product function categories (PFC).

Main Category	Sub-category 1	Sub-category 2
1. Fertiliser	A. Organic fertiliser	I. Solid organic fertiliser II. Liquid organic fertiliser
	B. Organo-mineral fertiliser	I. Solid organo-mineral fertiliser II. Liquid organo-mineral fertiliser
	C. Inorganic fertiliser	I. Inorganic macronutrient fertiliser including sub-categories II. Inorganic micronutrient fertiliser including sub-categories
2. Liming material		
3. Soil improver	A. Organic soil improver	
	B. Inorganic soil improver	
4. Growing medium		
5. Agronomic additive	A. Inhibitor	I. Nitrification inhibitor II. Urease inhibitor
	B. Chelating agent	
	C. Complexing agent	
6. Plant bio-stimulant	A. Microbial plant bio-stimulant	
	B. Non-microbial plant bio-stimulant	I. Organic non-microbial plant bio-stimulant II. Inorganic non-microbial plant bio-stimulant
7. Fertilising product blend		

requirements of one or more of the CMCs listed in Table 13.4. The component materials, or the input materials used to produce them, shall not contain one of the substances for which maximum limit values are indicated in Annex I of this Regulation in such quantities as to jeopardise the EC marked fertilising product's compliance with one of the applicable requirements of that Annex.

Table 13.4 Designation of component material categories (CMC).

CMC 1	Virgin material substances and mixtures
CMC 2	Non-processed or mechanically processed plants, plant parts or plant extracts
CMC 3	Compost
CMC 4	Energy crop digestate
CMC 5	Other digestate than energy crop digestate
CMC 6	Food industry by-products
CMC 7	Microorganisms
CMC 8	Agronomic additives
CMC 9	Nutrient polymers
CMC 10	Other polymers than nutrient polymers
CMC 11	Certain animal by-products (excluding Cat 1 animal by-products)
CMC 12	Determined industrial by-products
CMC 13	Precipitated phosphate salts and derivates (to be added after JRC study/policy decision)
CMC 14	Thermal oxidation materials and derivates (ashes) (to be added after JRC study/policy decision)
CMC 15	Pyrolysis and gasification materials and derivates (biochar) (to be added after JRC study/policy decision)

The draft EU Fertilising Products Regulation is now in the Trilogue, that is the process by which Commission, Parliament and Council elaborate an unanimously accepted final version of the new regulation that can be finally adopted. The Fertilising Products Regulation is highly relevant for nutrient recovery and recycling as it sets the rules for a common European market for the products which are developed and put to the market. It aims at superseding the highly diversified national rules including the criteria determining the end of waste status of a material and thus avoid notification between Member States (necessary if waste materials are shipped across EU borders) and create a common market for recycled fertilisers. The regulation is a form of facultative regulation. National regulation will remain in force and may be chosen by a producer not trading products cross borders. An EC marked fertilising product will qualify for free trade within the EU regardless of national regulations.

13.4.3 Animal By-product (ABP) Regulation (Europa, 2018)

Regulation (EC) No 1069/2009, adopted on 21st October 2009, lays down health rules regarding animal by-products and derived products not intended for human consumption. The regulation deals with the movement, processing and disposal of ABPs. Animal by-products (ABPs) are materials of animal origin that people do not consume. They include among others

- animal feed, for example based on fishmeal and processed animal protein;
- organic fertilisers and soil improvers, for example manure, guano, processed organic fertilisers/ soil improvers on the base of processed animal protein; and
- technical products, for example pet food, hides and skins for leather, wool, blood for producing diagnostic tools.

ABP of Category 2 (high risk) and Category 3 (low risk) may be placed on the market provided they are pressure sterilised (Art. 15), come from registered and approved plants and – in case of Category 2 material – must be prevented from use as feed by adding an appropriate component that excludes this use. Equally, digestion residues from transformation of ABP Category 2 and 3 into biogas or compost may be placed on the market and used as organic fertilisers or soil improvers. The ABP Regulation is of concern to energy and nutrient recycling from wastewater regarding two relevant aspects: (i) animal by-products include manure which is subject to use limitations in nitrate vulnerable zones designated by Member States in accordance with the Nitrate Directive; and (ii) Category 1 (very high risk) animal by-products must be incinerated and may be used as fuel and for nutrient enrichment in sludge incinerators. Since ash from Category 1 ABP is excluded from use under CMC 14 (thermal oxidation materials), this practice is not in compliance with the draft Fertilising Product Regulation. The critical question arises whether incineration ash marks the endpoint of ABP and the corresponding regulation or not. DG Environment and the British and Portuguese governments have different answers to this question. National governments consider it as an endpoint, allowing the use of ash from ABP-Category 1 incineration for use as fertilisers whereas DG Environment insists that Cat 1 ABP ash is excluded from use as a fertiliser if the use of crops or grass as animal feed cannot be ruled out.

The draft version of the new Fertilising Products Regulation (COM 2016) 157 final) (European Commission, 2016) includes provisions for use of ABP Cat 2 and 3 as CMC, provided they are free of salmonella (in a 25 g product sample) and do not contain *Escherichia coli*, or Enterococcaceae in a concentration of more than 1000 CFU/g fresh mass in the CE marked product. ABP for use as CMC must have reached the endpoint in the manufacturing chain as determined in accordance with Regulation (EC) No 1069/2009.

13.4.4 Plant Protection Products (PPP) Regulation (Europa, 2018)

Regulation (EC) 1107/2009 of the European Parliament and of the Council concerning the placing of plant protection products (pesticides) on the market. The regulation sets out rules governing the authorization of plant protection products in commercial form and their placing on the market, use and control within the European Community. It applies to products consisting of or containing active substances for: (i) protecting plants or plant products against harmful organisms; (ii) influencing the life process of plants; (iii) preserving plant products; and (iv) destroying undesired plants or their parts.

A plant protection product usually contains more than one component. The active component against pests/plant diseases is called 'active substance'. The Commission evaluates every active substance for safety before it reaches the market in a product. Substances must be proven safe for people's health, including their residues in food and effects on animal health and the environment. They may include microorganisms, pheromones and botanical extracts.

Before any PPP can be placed on the market and/or used, it must be authorised in the Member State(s) concerned. Regulation (EC) No 1107/2009 lays down the rules and procedures for authorisation of PPPs. Currently, the PPP regulation also covers plant bio-stimulants that may be produced from sewage sludge. Albeit, provisions regarding plant bio-stimulants will become subject to the new Fertilising Products Regulation and to REACH registration and thus relieved from the stricter provisions of the PPP regulation including individual authorisation by Member States. As soon as the new Fertilising Products Regulation will be enforced, the PPP Regulation will cease to be relevant to the wastewater sector.

13.4.5 REACH Regulation (Europa, 2018)

REACH Regulation (EC) No 1907/2006 aims at ensuring a high level of protection of human health and the environment by a better identification of intrinsic properties of chemical substances. It includes the promotion of alternative methods for assessment of hazards of substances, as well as the free circulation of substances on the internal market.

This regulation lays down provisions on substances and mixtures that shall apply to the manufacture, placing on the market or use of such substances on their own, in mixtures or in articles. This makes it very relevant for materials recovered from (waste-)water with the goal to bring it into the market.

REACH holds industry accountable to manage the risks from chemicals and to provide safety information on the substances. Manufacturers and importers are required to gather information on the properties of their chemical substances, which will allow their safe handling, and to register the information in a central database in the European Chemicals Agency (ECHA) in Helsinki. The Agency is the central point in the REACH system: it manages the databases necessary to operate the system, co-ordinates the in-depth evaluation of potentially risky chemicals and is building up the public database in which consumers and professionals can find hazard information. The regulation also calls for the progressive substitution of the most dangerous chemicals (referred to as 'substances of very high concern') when suitable alternatives have been identified. REACH requires increasing levels of information when larger volumes of a given substance are manufactured and placed on the market. Requirements are stipulated in the annexes and aim at reflecting the level of potential impacts from (hazardous) substances and expenses needed for compliance. Several substances are exempt from REACH: (i) radio-active substances; (ii) substances used in food; (iii) pharmaceuticals; and (iv) waste. Chemical products manufactured by resource recovery and recycling techniques are subject to registration if the end-of-waste status is achieved. Compost is explicitly exempt from REACH registration. An extension of the derogation for compost to other types of digestate is currently negotiated.

13.4.6 CLP Regulation (Europa, 2018)

Regulation (EC) 1272/2008 on Classification, Labelling and Packaging of Substances and Mixtures aims at identifying hazardous chemicals and informing users about their hazards through standard symbols and phrases. CLP is legally binding across the Member States and directly applicable to all industrial sectors. It requires manufacturers, importers or downstream users of substances or mixtures to classify, label and package their hazardous chemicals appropriately before placing them on the market and makes it therefore relevant for recovered substances from (waste-)water.

Hazard identification is the process by which information about the intrinsic properties of a substance or mixture is assessed to determine its potential to cause harm. If the nature and severity of an identified hazard meets the classification criteria in Annex I to the CLP Regulation, a certain hazard class will be assigned to the substance or mixture. There are hazard classes for physical hazards, health hazards, and environmental hazards.

For most substances and mixtures, manufacturers, importers, end users and distributors must determine the hazard classification themselves. In certain cases, the classification of a chemical is harmonised at EU level. The list of harmonised classification and labelling is included in Annex VI to the CLP Regulation.

The United Nations' Globally Harmonised System of Classification and Labelling of Chemicals (GHS) provides a harmonised basis for globally uniform physical, environmental, and health and safety information on hazardous chemical substances and mixtures. It sets up criteria for the classification of chemicals for physical-chemical, health, and environmental hazards of chemical substances and mixtures and sets up standardised hazard information to facilitate the global trade of chemicals. GHS was adopted by the United Nations in 2002 and is periodically updated. The implementation of the GHS in the EU in 2008 makes the EU a front-runner in its uptake. In line with the GHS standard, CLP allows for the identification of hazardous chemicals and the communication of these hazards to users through labelling. It also provides the basis for safety data sheets (SDS) regulated under the REACH Regulation and sets requirements for the packaging of hazardous chemicals. CLP sets general packaging standards to ensure the safe supply of hazardous substances and mixtures. In the case of certain hazards, substances and mixtures need to carry child-resistant fastenings and/or tactile warnings. Further information on CLP is available in a specific ECHA web-site called 'Understanding CLP' (https://echa.europa.eu/regulations/clp/understanding-clp).

13.4.7 Waste Shipment Regulation (Europa, 2018)

Regulation (EC) 1013/2006 deals with supervising and controlling shipments of waste within EU borders and shipments to/from EFTA, OECD and Basel Convention countries. The regulation addresses the problem of uncontrolled transport of waste and lays down procedures for the transboundary shipments (i.e., transport) of waste. It implements into EU law the provisions of the 'Basel Convention on the Control of Transboundary Movements of Hazardous Wastes and Their Disposal' as well as the OECD Decision. The Regulation includes a ban on the export of hazardous wastes to non-OECD countries ('Basel ban') as well as a ban on the export of waste for disposal.

Different regimes apply to shipments of wastes for disposal and for recovery, as well as to hazardous and 'green-listed' non-hazardous wastes. The shipment of hazardous wastes and of wastes destined for disposal is generally subject to notification procedures with the prior written consent of all relevant authorities of dispatch, transit and destination. However, as a rule, the shipment of 'green-listed' wastes for recovery within the EU and OECD does not require the consent of the authorities.

Despite the Regulation, illegal shipments of waste are still a significant problem (some estimates suggest that the overall non-compliance rate with the Regulation could be around 25%). To strengthen Member States' inspection systems, the Regulation was amended in 2014 through Regulation (EU) No 660/2014 of 15th May 2014, laying down minimum inspection requirements with a focus on problematic waste streams (such as hazardous waste and waste sent illegally for dumping or sub-standard treatment). EU countries were tasked with preparing inspection plans by 2017.

The Waste Shipment Regulation is relevant to the fertilising products from wastewater derived residues and biogas plants if they have not yet achieved the end-of-waste status, requiring mutual notification between departing and receiving states. Even if the shipment is between Member States and if none of the states rejects the shipment it may cause considerable delays.

13.5 EUROPEAN DIRECTIVES (EUROPA, 2018)

13.5.1 Overview

An EU directive is a legislative act that sets out a goal that all EU Member States must achieve (Table 13.5). However, it remains up to the individual countries to devise their own laws on how to reach these goals. Readers are advised to consult http://eur-lex.europa.eu/homepage.html (Europa, 2018) for amendments, updates and official comments, which can be easily accessed by entering the regulation denomination as listed below to an internet search engine. Any planned energy or nutrient recovery facility needs to be compliant with national legislation of the Member State where it is located.

13.5.2 Renewable Energy Directive (RED) (Europa, 2018; European Commission, 2018c)

Directive 2009/28/EC of the European Parliament and of the Council of 23rd April 2009 on the promotion of the use of energy from renewable sources. It establishes an overall policy for the production and promotion of energy from renewable sources in the EU. It requires the EU to fulfil at least 20% of its total energy needs with renewables by 2020 – to be achieved through the attainment of individual national targets. All EU countries must also ensure that at least 10% of their transport fuels come from renewable sources by 2020.

The directive specifies national renewable energy targets for each country, taking into account its starting point and overall potential for renewables. These targets range from a low of 10% in Malta to a high of 49% in Sweden. EU countries set out how they plan to meet these targets and the general course of their renewable energy policy in national renewable energy action plans (NREAP). Progress towards national targets is measured every two years when EU countries publish national renewable energy progress reports.

Biofuels and bioliquids are instrumental in helping EU countries meet their 10% renewables target in transport. The Renewable Energy Directive sets out biofuels sustainability criteria for all biofuels produced or consumed in the EU to ensure that they are produced in a sustainable and environmentally friendly manner.

Recast of the Renewable Energy Directive (RED-recast) (Europa, 2018). On the 30th November 2016, the Commission published a proposal for a revised Renewable Energy Directive to make the EU a global leader in renewable energy as well as to ensure that the target of at least 27% renewables in the energy consumption in the EU by 2030 is met. The recast of the Renewable Energy Directive (RED-recast) (COM(2016) 767), in Art. 26(7) (European Commission, 2017), specifies the minimum greenhouse gas (GHG) emissions saving thresholds that bioenergy must comply with to count towards the renewables targets and to be eligible for public support. Annex V (liquid biofuels) and Annex VI (solid and gaseous biomass) of the RED-Recast describe the methodology for GHG savings calculations needed to comply with the GHG criteria. They also provide a list of default GHG emission values, aggregated and disaggregated, that operators can use to demonstrate compliance of their product with the GHG criteria.

A corresponding JRC report (JRC Science for Policy Report, 2017) describes the input data, assumptions and methodological approach applied by the JRC when compiling the updated dataset used to calculate GHG emissions for the different biomass pathways. The GHG emissions resulting from the application of the methodology from COM (2016) 767, and presented in Annex VI of the document, are also shown. The report aims to provide operators, stakeholders, and the scientific community with all the necessary information to explain the assumptions chosen as well as to guarantee reproducibility of the results. Additional analysis to test the sensitivity of the results to various assumptions is presented in the final section of the report.

Renewable Energy Support Schemes. The Renewable Energy Directive governs grid access and national support schemes of high relevance for the revenues and the viability of energy and nutrient recovery and recycling business cases connected to biogas plants. Apart from the incentives represented by feed-in tariffs, feed-in premiums and green certificates provided by most Member States

Table 13.5 European directives.

Directives	Corresponding National Regulation Required – Objective/Regulated Subject
Renewable Energy Directive 2009/28/EC under recast (COM(2016) 767)	Establishes an overall policy for the production and promotion of energy from renewable sources in the EU
Liability of defective products Directive 85/374/EEC	Guarantees the liability of the producer in the event of damage caused by malfunctioning products
Biofuels and Indirect Land Use Change Directive (ILUC) Directive (EU) 2015/1513	Encourages advanced biofuels that do not displace food crops or cause indirect land use change (ILUC), relevant for the feedstock of anaerobic digesters
Nitrates Directive 91/676/EEC	Sets the limit of 170 kg N/ha/year from livestock manure in nitrate vulnerable zones. Member States declare designated regions as nitrate vulnerable zones (NVZ) for which National Action Programmes (NAP) are developed. Directive allows Member States to get derogations to go beyond the 170 kg limit, under strict conditions
Water framework Directive (WFD) 2000/60/EC	Establishes a comprehensive, cross-border approach to water protection organised around river basin districts (RBDs), aiming at good water quality status for European water bodies
Groundwater Directive 2006/118/EC	Complements the WFD, sets groundwater quality standards, including 50 mg/L of nitrates, and introduces measures to prevent or limit pollutant inputs to groundwater
Drinking Water Directive 98/83/EC	Lays down the essential quality standards of water intended for human consumption at EU level, requiring monitoring of 48 parameters including nitrates (limit 50 mg/L)
Bathing Water Directive 76/160/EEC amended by 2006/7/EC	It requires Members States to monitor and assess bathing water for at least two parameters of (fecal) bacteria and inform the public about water quality
Marine Strategy Framework Directive 2008/56/EC	Aims to achieve Good Environmental Status (GES) of the EU's marine waters by 2020 and to protect the resource base upon which marine-related economic and social activities depend
Urban Wastewater Directive 91/271/EEC amended by 98/15/EC	Requires collection and at least two stage treatment of wastewater for agglomerations >2000 person-equivalents (PE) and nutrient removal for plants covering >10 000 PE
Sewage Sludge Directive 86/278/ EEC	Seeks to encourage the use of sewage sludge in agriculture and to regulate its use in such a way as to prevent harmful effects on soil, vegetation, animals and man
Waste Framework Directive 2008/98/EC	Sets the basic concepts and definitions related to waste management, such as definitions of waste categories (including provisions for end-of-waste status), recycling and recovery
Landfill Directive 1999/31/EC	Defines different categories of waste and aims to prevent negative effects on the environment, on surface water, groundwater, soil, air, and on human health by introducing stringent technical requirements for waste and landfills
Hazardous Waste Directive 91/689/EEC amended by Directive 94/31/EC	Provides additional labelling, record keeping, monitoring and control obligations from 'cradle to the grave', that is from the waste producer to the final disposal or recovery
Air Quality Directive 2008/50/EC	Merges four directives and one council decision into a single directive on air quality, setting standards and target dates for reducing concentrations of fine particles ($PM_{2.5}$)
Industrial Emissions Directive (IED) 2010/75/EU – replaces IPPC Directive 96/61/EC	The IED is the main EU instrument regulating pollutant emissions from industrial installations. Sewage sludge treatment plants are subject to the IED if exceeding the threshold of 75 t/d processing capacity
EU National Emissions Ceilings Directive 2016/81/EC (under revision)	Sets national emissions ceilings for five important air pollutants: nitrogen oxide (NO_x), non-methane volatile organic compounds (NMVOCs), sulphur dioxide (SO_2), ammonia (NH_3) and fine particulate matter ($PM_{2.5}$)

to biogas-based energy carriers, policy measures such as incentivising compressed or liquified biogas as vehicle fuel, as promoted by most northern countries, may become effective tools for promoting integrated energy conversion and nutrient recycling systems.

The EU adopted guidance for EU countries (SWD (2013) 439) (Europa, 2018) when designing and reforming renewable energy support schemes. This guidance suggests that:

(1) Financial support for renewables should be limited to what is necessary and should aim to make renewables competitive in the market;
(2) Support schemes should be flexible and respond to falling production costs. As technologies mature, schemes should be gradually removed. For instance, feed in tariffs should be replaced by feed in premiums and other support instruments that incentivise producers to respond to market developments;
(3) Unannounced or retroactive changes to support schemes should be avoided as they undermine investor confidence and prevent future investment;
(4) EU countries should take advantage of the renewable energy potential in other countries via cooperation mechanisms. This would keep costs low for consumers and boost investor confidence.

Support schemes are also governed by the 'Guidelines on State aid for environmental protection and energy 2014–2020' (2014/C 200/01) (Europa, 2018) published as a Commission Communication in 2014. It aims at preventing State Aid from distorting competition in the internal market and affecting trade between Member States in a way which is contrary to the common interest, Article 107(1) of the Treaty on the Functioning of the European Union. The Treaty lays down the principle that State aid is prohibited. In certain cases, however, State aid may be compatible with the internal market under Articles 107(2) of the Treaty (Europa, 2018). The document provides guidelines on how environmental protection and renewable energy support schemes should be designed to be compatible with the Treaty.

13.5.3 Liability of Defective Products Directive (Europa, 2018)

Directive 85/374/EEC, modified by Directive 1999/34/EC, guarantees the liability of the producer in the event of damage caused by malfunctioning products. The Directive lays down a common rule for strict liability (i.e., 'liability without fault') of producers at EU level and offers a real possibility to consumers to claim financial compensation for death or personal injuries or for damage caused to an item of property intended for private use with a threshold of €500. The Directive provides the injured person with an extracontractual regime of liability. The 'strict liability' approach extends the liability of producers to injured consumers who suffer loss through defective products without the need to prove a contractual link, a duty of care or failure to take reasonable care to comply with relevant legislation.

Directive 85/374/EEC applies to all movable products, even when incorporated into other movable or immovable property, including primary agricultural products and electricity. It establishes that a product is defective when it does not provide the safety which a person is entitled to expect, taking all circumstances into account, including the presentation of the product or the time when the product was put into circulation.

13.5.4 Biofuels and Indirect Land Use Change Directive (ILUC) (Europa, 2018)

Directive (EU) 2015/1513, adopted on 9th September 2015, and amending the Fuel Quality Directive 98/70/EC and the above-mentioned Renewable Energy Directive 2009/28/EC on the promotion of the use of energy from renewable sources.

The European Union has been considering proposals to amend the Renewable Energy Directive (RED) and the Fuel Quality Directive (FQD) to encourage advanced biofuels that do not displace food crops or cause indirect land use change (ILUC).

ILUC issues have a considerable impact on nutrient recovery and recycling if energy crops are used as feedstock for biogas production, not allowed in some European countries and with much lower (50% and further decreasing) allowance in Germany, where until 2017 biogas production could be largely based on energy crops. Consequently, waste and residues from wastewater purification become more important as a resource for renewable energy conversion.

13.5.5 Nitrates Directive (Europa, 2018; European Commission, 2018e)

Directive 91/676/EEC of 12 December 1991 concerning the protection of waters against pollution caused by nitrates from agricultural sources. The Nitrates Directive aims to protect water quality across Europe by preventing nitrates from agricultural sources polluting ground and surface waters and by promoting the use of good farming practices.

The Nitrates Directive is one of the European policy instruments that plays a crucial role to the national legislations regarding the digestates' management. Under the Directive Member States must analyse the nitrate concentration in their waters and have installed close to 60 000 groundwater and surface water monitoring stations. Member States have designated territories draining into waters that are or could be affected by high nitrate levels or eutrophication as vulnerable zones. Austria, Denmark, Finland, Germany, Ireland, Lithuania, Luxembourg, Malta, the Netherlands and Slovenia have adopted a whole territory approach, meaning they have designated their whole territory as vulnerable zone.

All Member States had to establish codes of good practice for farmers, to be implemented on a voluntary basis throughout their territory and develop specific action programmes for compulsory implementation by farmers located in nitrate-vulnerable zones. Action programmes include periods when fertilisation is prohibited, minimum storage capacity for livestock manure, and rules to control the spreading of nutrients near water or on slopes, to reduce the risk of contamination. All of them include the limit of 170 kg nitrogen per hectare per year from livestock manure that is set out in the Directive. The Directive allows Member States to get derogations to go beyond the 170 kg limit, under strict conditions. Six Member States (BE, NL, DK, IE, DE, UK) had been granted derogation from the maximum allowable nitrogen application rate of 170 kg of nitrogen per hectare with animal manure, typically 250 kg N/ha (DK 230 kg N/ha). DE has had no more derogation since 2014, Italy has had one since 2009.

The Nitrates Directive's impact goes beyond its scope of protecting the water bodies. Inter alia it could support climate change abatement by improving livestock management and fertilising practices reducing nitrous oxide (N_2O) and methane emissions (CH_4) with a global warming potential of 265 times (N_2O) and 28 times (CH_4) of CO_2 (IPPC Intergovernmental Panel on Climate Change 2014). However, a holistic approach to farmyard manure and fertiliser management is necessary to protect water bodies and climate.

Fertiliser use has been and still is a major source of water pollution in Europe. Regardless of mineral fertiliser consumption having stagnated since the early 1990s, overall nitrogen consumption has increased by 6%. Farming remains responsible for over 50% of the total nitrogen discharge into surface waters.

The Nitrates Directive is only indirectly relevant to nutrient recovery from wastewater due to sewage sludge and sewage sludge ash not being constrained by the related national regulations. However, if a facility mixes manure and sewage sludge in an anaerobic digester, digestion residues become manure in many Member States, even if manure represents only 10% of the input flows. This needs to be considered when conceiving a project.

13.5.6 Water Framework Directive (WFD) (Europa, 2018)

Directive 2000/60/EC of the European Parliament and of the Council, established on 23rd October 2000, provides a framework for the Community action in the field of water policy which establishes a comprehensive, cross-border approach to water protection organised around river basin districts

(RBDs), with the aim of achieving good status for European water bodies by 2015. The WFD defines environmental objectives for surface water, groundwater and protected areas. The criteria for achieving good groundwater chemical status are defined in Annex V and require that concentrations of pollutants (in groundwater) do not exceed any quality standards applicable under other relevant Community legislation according to Article 17. Further, there is a requirement to identify and reverse any significant and sustained upward trends in the concentration of pollutants.

The WFD was the result of a multiannual stakeholder process showing that the previous European water policies were too fragmented and not effective. Essentially, they included the Urban Wastewater Treatment Directive and the Nitrates Directive. In the second half of the 1990s, the Drinking Water Directive and the Directive for Integrated Pollution Prevention and Control (IPPC) were adopted.

Expanding the scope to all water bodies, setting the targets for achieving good status, a combined approach of quality standards and emission limit values, as well as the transboundary river basin approach, were the cornerstones of the new WFD.

The WFD is currently reviewed, inter alia due to the poor results regarding the improvements of the ecological status of water bodies in the EU – the quality status is still deplorable. Frequent and non-transparent use of exemptions, lack of control and particularly agricultural practices are considered as the main challenges.

Member States' legal acts on water are typically based on the WFD. They include regulations regarding the discharge of wastewater and digester effluents, including purified liquid effluents, into water bodies and are consequently relevant to energy and nutrient recovery and recycling.

13.5.7 Groundwater Directive (Europa, 2018)

Directive 2006/118/EC, adopted on 12th December 2006, aims at the protection of groundwater against pollution and deterioration. It sets groundwater quality standards and introduces measures to prevent or limit inputs of pollutants into groundwater. The Directive establishes quality criteria that takes account of local characteristics and allows for further improvements to be made based on monitoring data and new scientific knowledge. The Directive thus represents a proportionate and scientifically sound response to the requirements of the Water Framework Directive (WFD) as it relates to assessments of chemical status of groundwater and the identification and reversal of significant and sustained upward trends in pollutant concentrations. Member States should establish standards at the most appropriate level and take into account local or regional conditions. The Groundwater Directive complements the Water Framework Directive (WFD). It requires Member States to set groundwater quality standards, perform pollution trend studies referred to a baseline level based on 2007–2008 data, reverse negative trends and prevent or limit inputs of pollutants to assure achievement of environmental objectives by 2015 (targets not achieved). Relevant to nutrient recovery is compliance with good chemical status criteria (based on EU standards of nitrates (50 mg/L) and pesticides and on threshold values established by Member States. Limit values set by Member States may include lower nitrate limits in groundwater.

13.5.8 Drinking Water Directive (Europa, 2018)

Directive 98/83/EEC lays down the essential quality standards of water intended for human consumption at EU level. Its objective is to protect human health from adverse effects of any contamination of water intended for human consumption by ensuring that it is wholesome and clean. A total of 48 microbiological, chemical and indicator parameters must be monitored and tested regularly. In general, the World Health Organization's guidelines for drinking water and the opinion of the Commission's Scientific Advisory Committee are used as the scientific basis for the quality standards in the drinking water. When adopting the Drinking Water Directive to their own national legislation, Member States of the European Union can include additional requirements but are not allowed to set lower standards as the level of protection of human health should be the same within the whole European Union. Member States may, for a limited time, depart from chemical quality standards specified in the Directive by applying for a derogation. Derogations can be granted, provided it does not constitute a potential danger to human health and provided that the supply of water intended for human consumption in the area concerned cannot be maintained by any other reasonable means. The Directive also requires providing regular information to consumers. In addition, drinking water quality must be reported to the European Commission every three years.

Of particular relevance to nutrient recovery and recycling activities is the limit concentration of nitrates in drinking water of 50 mg/L. Member States must ensure that the sum of $NO_3/50$ (nitrates) + $NO_2/_3$ (nitrites) does not exceed the value of 1 and that the value of 0.10 mg/L for nitrites is not exceeded in the effluents of water treatment works. In February 2018 the European Commission adopted the proposal for a revised Drinking Water Directive to improve the quality of drinking water and provide better access to information for citizens. The proposal updates existing safety standards in line with the latest recommendations of the World Health Organisation (WHO) and ensure our drinking water is safe to use for the decades to come.

13.5.9 Bathing Water Directive (Europa, 2018)

Directive 76/160/EEC, amended by 2006/7/EC, provides the targets and rules to safeguard public health and clean bathing waters. The revised Bathing Water Directive (BWD) of 2006 updated and simplified these rules. It requires Members States to monitor and assess the bathing water for at least two parameters of (fecal) bacteria. In addition, they must inform the public about bathing water quality and beach management, through so-called bathing water profiles. These profiles contain for instance information on the kind of pollution and sources that affect the quality of the bathing water and are a risk to bathers' health, such as wastewater discharges. The major sources of pollution are sewage and water draining from farms and farmland. Such pollution increases during heavy rains and floods due to sewage overflow and polluted drainage water being washed into rivers, lakes and seas. All EU Member States, plus Albania and Switzerland, monitor their bathing sites according to the provisions of the EU's revised Bathing Water Directive. The legislation specifies if the bathing water quality can be classified as 'excellent', 'good', 'sufficient' or 'poor', depending on the levels of fecal bacteria detected. Where water is classified as 'poor', Member States must take measures, like banning bathing or posting a notice advising against it, providing information to the public, and suitable corrective measures. Bathing water in the EU has achieved good levels in 2016 with only a few sites reporting poor conditions.

13.5.10 Marine Strategy Framework Directive (Europa, 2018)

Directive 2008/56/EC aims to achieve Good Environmental Status (GES) of the EU's marine waters by 2020 and to protect the resource base upon which marine-related economic and social activities depend. It is the first EU legislative instrument related to the protection of marine biodiversity, as it contains the explicit regulatory objective that 'biodiversity is maintained by 2020', as the cornerstone for achieving GES.

The Directive provides a legislative framework to the management of human activities having an impact on the marine environment, integrating the concepts of environmental protection and sustainable use. It establishes European marine regions and sub-regions based on geographical and environmental criteria. Environmental criteria include eleven indicators comprising biodiversity and eutrophication, which may be affected by nutrient recovery and recycling activities.

The Directive covers four European marine regions – the Baltic Sea, North-east Atlantic Ocean, Mediterranean Sea and Black Sea – located within the geographical boundaries of the existing Regional Sea Convention. To achieve GES by 2020, each Member State is required to develop a strategy for its marine waters (or Marine Strategy). In addition, because the Directive follows an adaptive management approach, the Marine Strategies must be kept up-to-date and reviewed every six years. A compliant Marine Strategy must include an initial assessment of the status and the relevant human activities, a definition of GES and targets to be achieved by 2020 and a measuring and monitoring programme.

13.5.11 Urban Wastewater Directive (Europa, 2018)

Directive 91/271/EEC, amended by Directive 98/15/EC, aims at the protection of the environment from the adverse effects of urban and certain industrial wastewater discharges.

It requires Member States to designate sensitive areas, identify catchment areas of the sensitive areas and ensure that agglomerations with more than 10 000 PE (person equivalents) within the catchment areas comply with more stringent discharge requirements than secondary treatment and establish a technical and financial programme for the implementation of the Directive for the construction of sewage treatment systems within the deadlines set up by the Directive. Furthermore, Member States must implement regulations and provisions for discharge of municipal and industrial wastewater depending on the sensitivity of catchment areas and the volume of wastewater treated. Member States must ensure that the environmentally and technically sound reuse or disposal of sewage sludge is subject to general rules, registration or authorisation and that the requirement of specific inter-linked Directives for agricultural re-use (86/278/EEC), incineration (89/429/EEC and 89/369/EEC) and landfill (99/31/EC) are respected. The disposal of sewage sludge in surface waters is banned.

The Urban Wastewater Directive is indirectly relevant for nutrient recycling and biogas activities as far as liquid effluents should be discharged. The industrial sectors covered by the Urban Wastewater Directive include a variety of food processing industries, including dairies, fruit, vegetable, meat and fish processing facilities, which are frequently cooperating with anaerobic digestion plants or even operating them.

13.5.12 Sewage Sludge Directive (Europa, 2018)

Directive 86/278/EEC seeks to encourage the use of sewage sludge in agriculture and to regulate its use in such a way as to prevent harmful effects on soil, vegetation, animals and man. To this end, it prohibits the use of untreated sludge on agricultural land unless it is injected or incorporated into the soil. Treated sludge is defined as having undergone 'biological, chemical or heat treatment, long-term storage or any other appropriate process so as significantly to reduce its fermentability and the health hazards resulting from its use'. To provide protection against potential health risks from residual pathogens, sludge must not be applied to soil in which fruit and vegetable crops are growing or grown, or less than ten months before fruit and vegetable crops are to be harvested. Grazing animals must not be allowed access to grassland or forage land less than 3 weeks after the application of sludge. The Directive also requires that sludge should be used in such a way that account is taken of the nutrient requirements of plants and that the quality of the soil and of the surface and groundwater is not impaired.

The Directive specifies rules for the sampling and analysis of sludges and soils. It sets out requirements for the keeping of detailed records of the quantities of sludge produced, the quantities used in agriculture, the composition and properties of the sludge, the type of treatment and the sites where the sludge is used. Limit values for concentrations of heavy metals in sewage sludge intended for agricultural use and in sludge-treated soils are in Annexes I A, I B and I C of the Directive. Many EU Member States have adopted stricter regulations including certification schemes for sewage sludge use.

Sewage sludge is used as a feedstock by several nutrient recycling plants and thus they are concerned by the Sewage Sludge Directive. In addition, if used in agriculture, sewage sludge is competing with agricultural digestates and/or products produced from those digestates which is partly reflected by national regulations, for instance by setting very low heavy metal limit values for agricultural use of sewage sludge in the Netherlands and, more recently, by the ban of using sewage sludge from large wastewater treatment plants (>50 000 PE) in agriculture in Germany. Such bans frequently induce the development of technical resource recovery processes to overcome the constraints. Consequently, barriers may act as stimulants and promote innovation and technology development.

13.5.13 Waste Framework Directive (Europa, 2018)

Directive 2008/98/EC on waste sets the basic concepts and definitions related to waste management, such as definitions of waste, recycling and recovery. It explains when waste ceases to be waste and becomes a secondary raw material (so called end-of-waste criteria), which waste related properties render waste hazardous and how to distinguish between waste and by-products. The Directive lays down some basic waste management principles: it requires that waste is managed without endangering human health and harming the environment, and without risk to water, air, soil, plants or animals, without causing a nuisance through noise or odours, and without adversely affecting the countryside or places of special interest. In addition, the Directive lays down the targets for recycling of certain waste streams. Waste legislation and policy of the EU Member States shall apply as a priority order the waste management hierarchy as shown in Figure 13.1.

The Directive is currently recast in the context of the Circular Economy Package. EU Communication COM (2015) 595 (Europa, 2018) final suggests an amendment of the Waste Framework Directive, stipulating, among others, to increase the recycling target for municipal waste to 65% by 2030 and new measures for preventing food waste.

13.5.14 Landfill Directive (Europa, 2018)

Directive 1999/31/EC aims at preventing or reducing negative effects of landfilling of waste on the environment, particularly on surface water, groundwater, soil, air and on human health. It introduces stringent technical requirements for waste and landfills. The Landfill Directive defines the different categories of waste (municipal waste, hazardous waste, non-hazardous waste and inert waste) and applies to all landfills, defined as waste disposal sites for the landfilling of waste onto or into land.

Among others, the Landfill Directive does not allow landfilling of untreated waste and liquid waste. It stipulates that 15 years after its implementation (i.e., 2014) biodegradable waste must not exceed 35% of the total amount (by weight) of biodegradable municipal waste produced in 1995. In addition, most national regulations implementing the Landfill Directive do not allow landfilling waste with more than 5% biodegradable carbon content. However, several Member States' waste management practices are not in compliance with the Directive.

Landfilling any organic residues from resource recovery activities is not a legal option in Europe.

Figure 13.1 Waste hierarchy as stipulated in the waste framework directive (source: European Commission, 2008).

13.5.15 Hazardous Waste Directive (Europa, 2018)

Directive 91/689/EEC, amended by Directive 94/31/EC, lays down the rules for the management of hazardous waste in the Member States. Member States must ensure that hazardous waste is recorded and identified. They must also ensure that different categories of hazardous waste are not mixed, and that hazardous waste is not mixed with non-hazardous waste.

Hazardous wastes pose a greater risk to the environment and human health than non-hazardous wastes and thus require a stricter control regime. This is laid down in Articles 17–20 of Directive 2008/98/EC. It provides additional labelling, record keeping, monitoring and control obligations from 'cradle to the grave', that is from the waste producer to the final disposal or recovery. In addition, mixing of hazardous substances is banned to prevent risks for the environment and human health. Moreover, the permit exemptions that may be granted to installations dealing with hazardous wastes are more restrictive than for installations dealing with other wastes.

The classification into hazardous and non-hazardous waste is based on the system for the classification and labelling of dangerous substances and preparations, which ensures the application of similar principles over their whole life cycle. The properties which render waste hazardous are laid down in Annex III of Directive 2008/98/EC and are further specified by the Decision 2000/532/EC establishing a List of Wastes as last amended by Decision 2001/573/EC. The List of Wastes is currently being reviewed (see information on the Waste Framework Directive pages).

The Hazardous Waste Directive may be relevant to recovery and recycling activities that produce a waste or by-product stream which due to elevated concentration of pollutants may be considered as hazardous. Occasionally, even sewage sludge ash may be characterised as hazardous due to, for instance, elevated mercury concentrations. Many Member States have legal mechanisms by which hazardous wastes may be turned into non-hazardous wastes, frequently used to de-classify sludge ash. The draft Fertilising Product Regulation allows the use of hazardous waste as input material provided the relevant pollutants are removed. Dilution of pollutants is not legal within the Hazardous Waste Directive.

13.5.16 Air Quality Directive (Europa, 2018)

Directive 2008/50/EC merges four directives and one Council decision into a single directive on air quality. It sets standards and target dates for reducing concentrations of fine particles, which together with coarser particles known as PM10 already subject to legislation, are among the most dangerous pollutants for human health.

Under the Directive Member States are required to reduce exposure to PM2.5 in urban areas by an average of 20% by 2020 based on 2010 levels. It obliges them to bring exposure levels below 20 $\mu g/m^3$ by 2015 in these areas. Throughout their territory Member States will need to respect the PM2.5 limit value set at 25 $\mu g/m^3$. This value must have been achieved by 2015 if 2010 levels were not already below the limit.

The Directive introduces objectives for fine particles PM2.5 but does not change existing air quality standards. It does, however, give Member States greater flexibility in meeting some of these standards in areas where they have difficulty complying. Meeting PM10 limit values is proving challenging for 25 of the 27 EU Member States which are exceeding these limits in at least one part of their territory (see IP/07/1537).

The deadlines for complying with the PM10 standards can be postponed for three years after the Directive's entry into force (mid-2011) or by a maximum period of five years for nitrogen dioxide and benzene (2010–2015) provided that the relevant EU legislation such as industrial pollution prevention and control (IPPC, see MEMO/07/441) is fully implemented, and that all appropriate abatement measures are being taken. The Directive provides a list of measures that need to be considered.

The Directive on air quality (see IP/07/1895 and MEMO/07/571) is one of the key measures outlined in the 2005 Thematic Strategy on air pollution adopted by the Commission in September 2005 (IP/05/1170). It establishes ambitious, cost-effective targets for improving human health and environmental quality up to 2020.

The Air Quality Directive may have an impact on permits for installations aiming at resource recovery by limitations of small particle emissions.

13.5.17 Industrial Emissions Directive (IED) (Europa, 2018)

Directive 2010/75/EU of the European Parliament and of the Council of 24th November 2010 on industrial emissions (integrated pollution prevention and control) is based on a Commission proposal recasting seven previously existing directives (including the IPPC Directive) following an extensive review of the policy. The IED entered into force on 6th January 2011 and had to be incorporated to national legislation by Member States by 7th January 2013. The IED applies to treatment of sewage sludge, particularly biological or thermal energy conversion and resource recovery, for instance nutrient recovery.

The IED aims to achieve a high level of protection of human health and the environment taken as a whole by reducing harmful industrial emissions across the EU, in particular through application of Best Available Techniques (BAT). Around 50 000 installations undertaking the industrial activities listed in Annex I of the IED are required to operate in accordance with a permit (granted by the authorities in the Member States). This permit should contain conditions set in accordance with the principles and provisions of the IED. The IED is based on five pillars: (1) an integrated approach; (2) use of best available techniques; (3) flexibility; (4) inspections; and (5) public participation:

(1) The integrated approach means that the permits must consider the whole environmental performance of the plant covering, for example, emissions to air, water and land, generation of waste, use of raw materials, energy efficiency, noise, prevention of accidents, and restoration of the site upon closure.
(2) The permit conditions including emission limit values must be based on the Best Available Techniques (BAT). To define BAT and the BAT-associated environmental performance at EU level, the Commission organises an exchange of information with experts from Member States, industry and environmental organisations. This work is co-ordinated by the European IPPC Bureau of the Institute for Prospective Technology Studies at the EU Joint Research Centre in Seville (Spain). This process results in BAT Reference Documents (BREFs); the BAT conclusions contained are adopted by the Commission as Implementing Decisions. The IED requires that these BAT conclusions are the reference for setting permit conditions.
(3) The IED allows competent authorities some flexibility to set less strict emission limit values. This is possible only in specific cases where an assessment shows that achieving the emission levels associated with BAT described in the BAT conclusions would lead to disproportionately higher costs compared to the environmental benefits due to the geographical location or the local environmental conditions or the technical characteristics of the installation. The competent authority shall always document its justification for granting such derogations.
(4) The IED contains mandatory requirements on environmental inspections. Member States shall set up a system of environmental inspections and draw up inspection plans accordingly. The IED requires a site visit to take place at least every 1–3 years, using risk-based criteria.
(5) The IED ensures that the public has a right to participate in the decision-making process, and to be informed of its consequences, by having access to permit applications, permits and the results of the monitoring of releases.

In addition, through the European Pollutant Release and Transfer Register (E-PRTR), emission data reported by Member States are made accessible in a public register, which is intended to provide environmental information on major industrial activities.

Biological treatment of sewage sludge is subject to the IED Directive if the capacity of the plant exceeds 75 t/day. If the only treatment activity is anaerobic digestion, the plant is subject to the IED if exceeding a capacity of 100 t/day.

13.5.18 National Emissions Ceilings (NEC) Directive (Europa, 2018)

Directive 2016/2284/EU about National Emission Ceilings entered into force on 31st December 2016. Replacing earlier legislation (Directive 2001/81/EC), the new NEC Directive sets 2020 and 2030 emission reduction commitments for five main air pollutants. It also ensures that the emission ceilings for 2010 set in the earlier Directive remain applicable for Member States until the end of 2019. The new Directive introduces several new reporting requirements for Member States.

Reporting requirements are defined in Annex I of the directive and include annual information on emissions of several pollutants:

(1) the five main air pollutants NOx, NMVOCs, SO_2, NH_3 and PM2.5 as well as carbon monoxide (CO);
(2) in addition to PM2.5, also PM10, particulate matter and, if available, black carbon (BC) and total suspended particulate matter (TSP);
(3) heavy metals cadmium (Cd), lead (Pb) and mercury (Hg) and, if available, the additional heavy metals arsenic, chromium, copper, nickel, selenium and zinc); and
(4) persistent organic pollutants (POPs) including selected polycyclic aromatic hydrocarbons (PAHs), dioxins and furans (PCCDs and PCDFs), polychlorinated biphenyls (PCBs) and hexachlorobenzene (HCB).

To help ensure that information on emissions reported by Member States is consistent and harmonised with international requirements, the NEC Directive requires Member States to follow the methodologies agreed upon by the UNECE LRTAP Convention (described in the policies chapter). It also requires that Member States use the EMEP/EEA air pollutant emission inventory guidebook in preparing their inventories.

Emission reporting at the national level is expected to have an impact on national legislation about industrial emissions that may affect resource recovery and recycling activities.

13.6 OUTLOOK AND RECOMMENDATIONS

(1) Harmonisation: Europe is sharing policies, regulations and objectives but how objectives are pursued is governed by each Member State with own definitions, strategies and regulations complicating the life of actors in the common market. If at least European definitions were shared by all member states, transactions between Member States would be facilitated. It is expected that the new regulation of fertilising products is only a first step in a process extending to other European areas including supporting schemes for renewable energy conversion.
(2) Level playing field for comparable products, regardless of their origin: Circular Economy is only possible if products are regulated without discriminating recycled raw materials. This approach is pursued by the new regulation of fertilising products but the trilogue tends to water down the Commission's draft. Industrial lobbying frequently prevents strict regulation of conventional products and leads to much stricter requirements for recycled products.
(3) Long-term legislative framework reliability: Harmonised (between Member States) funding schemes must provide a reliable framework within a timeframe, corresponding to the pay-back time of investments. Renewable energy and nutrient recovery projects will only happen in a market environment where business cases will not be upset by unforeseeable legislative acts.
(4) Continued support to renewable energy supplies: Surprisingly, after Europe and Member States signed and ratified the Paris (COP21) agreement, subsidies for renewable energy were reduced, discouraging investment in biogas plants that have proven to serve as a preferred hub for nutrient recovery. In general, renewable energy from wastewater residues is much less supported than agricultural residues or biowaste.
(5) Saving greenhouse gas emissions must pay back: Emissions and other externalities must be subject to sensible taxes or other mechanisms causing a relevant penalty for the polluter.

(6) Gradual shift of taxation from labour to materials use: Nutrient recovery and recycling frequently involves a higher workload – compared to producing fertilisers from fossil resources – creating (wanted) jobs in rural areas. However, renewable labour is highly charged with taxes and social expenses whereas fossil materials use is not or only marginally taxed adversely affecting recycling.

13.7 CHAPTER SUMMARY

In this chapter we have provided an overview of some of the key regulations that are important in relation to resource recovery from wastewater. The regulatory framework governing nutrient recovery and recycling activities in EU Member States is arranged in European policies, regulations and directives and by national legislation, which is frequently based on European policies and directives. Consequently, this chapter followed the same structure. Part 1 deals with European policies that are followed by the European regulations that must be enforced by all Member States in part 2. Part 3 refers to European directives which must be adopted by Member States but not literally. Directives typically stipulate a target but leave room to the Member State for selecting the strategy and pathway to achieve the target.

A second structural string follows the purpose of legislation, distinguishing between supportive acts based on European renewable energy policies, regulations and directives on one hand and constraining legislation mainly derived from environmental regulations like the Nitrate Directive and the Water Framework Directive on the other hand. Feed-in tariffs, feed-in premiums and green certificates clearly have been the drivers behind the development of anaerobic digestion in Europe. However, operators of biogas plants are increasingly facing more stringent limits to nutrient application on cropland in terms of the application periods, application methods, the acceptable amounts of nitrogen and phosphorus and requirements to storage capacities that increase the cost of digestate handling and transport but offer opportunities for innovative nutrient recycling technologies.

Due to the frequent amendments of regulative acts, this chapter can only provide a conclusive picture as of 2013. Trying to compensate for this unavoidable drawback the authors provide the links to relevant authority and service web-sites where readers can find updated information directly in the text and/or in the reference section. The chapter provides comprehensive coverage of the European Union, albeit it cannot cover the individual legislation of the Member States, which is largely governed by the European policies, regulations and directives but going sometimes beyond the European acts. However, if readers plan a project in a Member State or even in a third country, they can always take the European regulations as a guideline and look for corresponding regulation in their own jurisdiction. The main findings of the review regarding the legal framework for energy and nutrient recovery and recycling from wastewater are:

(1) Supporting schemes for renewable energy conversion (from organic and agricultural residues) in Member States are not governed by European legislation and vary from zero subsidy in Spain and – for conversion of sewage sludge – Italy to quite generous support schemes in other Member States. However, energy conversion of municipal sewage sludge is frequently exempt from support schemes that are typically addressing agricultural residues and other biowaste.

(2) Regarding restrictive regulations mainly related to the use of nutrients on grass- and cropland, many Member States have implemented similar action programmes to improve water quality as required by the Water Framework and the Nitrates Directives leading to similar regulations. However, compliance with similar regulation is much more challenging in countries with intensive and concentrated livestock rearing compared to countries with more extensive livestock farming. Countries with high livestock density tend to limit the use of sewage sludge and require processed, higher concentrated nutrients in more elaborate products.

The chapter covered only two currently most developed recovery and recycling pathways from water, namely, energy and nutrient recovery from wastewater. Recycling of water is not specifically addressed but largely covered by the same regulatory framework. After reading this chapter, the reader should understand that legislation in general aims at implementation of policies:

- Transitions towards certain global and common targets like preventing the rise of the global average temperature (UNFCCC, 2015) and the sustainable development goals (United Nations, 2013), both agreed by more than 190 United Nations in 2015;
- Protecting human health and the environment at large;
- Protecting specific natural goods like air, soil and water.

Both UN agreements aiming at a transition provide a global framework for European and national policies. Typically, these objectives and policies are closely interlinked. It is all about securing the survival of mankind and about creating basic minimum standards for all humans sharing our planet. In Europe, standards are typically higher than in the rest of the world and consequently regulations are stricter and usually better observed. Due to having chosen energy and nutrient recycling, regulations are focusing on air, soil and water protection. Nutrients are mainly used in agriculture, hence regulations regarding soil fertility, crop nutrition as well as safe feed and food are preponderant in the chapter. If we had chosen minerals for use as construction materials, the same would apply to regulations of construction works and buildings.

The chapter did not include Safety, Health and Environmental (SHE) and social/contractual regulations related to operations itself, for instance in a water treatment or recycling plant. This is a separate body of legislation that must always be considered and it is mainly governed by national regulations. The authors hope to have shown that, regardless of which material is recovered and recycled, the basic principles of protecting air, soil, water and human livelihood remains the same. Operators should always think of the goods that may be affected by his activities and check if these goods are protected by global, multilateral or national regulations. Once the basic principles are understood, identifying and respecting the specific regulatory framework should be a not too difficult task.

13.8 DISCUSSION QUESTIONS

Question 13.1: What are the predominant concerns and targets of regulators? (leadership, utility management). After having read the comprehensive body of regulations governing the use of recovered resources from water and wastewater, think of the concerns and intentions of policy makers and interpret them in your own words. Which goods are protected by regulations and which behaviour is favoured by legal incentives? As a critical citizen, do you think the regulative body is complete or do feel that important human or natural goods are not sufficiently covered? How are the regulations linked to the global agreements like the Sustainable Development Goals or the Climate Agreement?

Question 13.2: How would you explain the legislative framework to colleagues or third parties with limited specific knowledge? (operators, utility management). In your role as an operator or manager of a utility, think of having to explain the legislative framework to third parties in your own words in 15 minutes, for instance in a presentation. Think of the structural approach in Europe with policies, directives and regulations and think of the restrictions and incentives. How is the Circular Economy addressed by regulations, if at all?

Question 13.3: What is the difference between EU regulations and regulations of your country (operators, utility management). Once you have read and understood the body of legislation in the European Union, check if you have equivalent regulations in the jurisdiction where you operate. The check should include regulations that have similar intentions and cover similar subjects, even if the explicit targets and wording are quite different from what EU regulators have adopted. Find out the differences, explain them with your own words. What is your opinion about the differences – do you think that the European regulations can serve as a model for other countries?

Question 13.4: Which regulations could support resource recovery initiatives? (leadership, utility management). Imagine you want to improve the environmental performance of your water activity, you have limited financial resources and your activity should generate additional revenues or reduce the cost of operating your facility. Reviewing the regulations, which type of regulations could contribute to your financial targets? For instance, about 100 struvite plants operate globally in wastewater treatment plants, saving maintenance costs because of preventing scaling and incrustations in pipes and tanks of EHBP plants. In many cases, recycling struvite pays back even if the product is given away for free.

Question 13.5: Which areas of the water and wastewater treatment sector are preferred targets for resource recovery? (leadership, utility management). Current legislation for water and wastewater treatment focuses on removal (not re-use) of contaminants, nutrients, organic matter, and so on. If you think of recovery and recycling, how would you select and which area would you chose? To what extent would you consider legislation and to what extent would you consider other drivers, for example operating costs? Which other arguments could influence your decision?

Question 13.6: Can anaerobic digestion contribute to reduce greenhouse gas emissions and abate global warming? (leadership, utility management). You have learned that policies aim at achieving renewable energy targets and prevent emissions to air and water bodies. Renewable energy can be recovered by anaerobic digestion of sewage sludge. Do you think that anaerobic digestion can make a relevant contribution to the renewable energy targets? How could anaerobic digestions concurrently contribute to pollution prevention?

Question 13.7: Which legislative bodies need to be considered if undertaking activities of recovery and recycling of energy and nutrients? (leadership, utility management). If you conceive a resource recovery and recycling activity, you need to plan compliance with the legislative framework outlined in this chapter. However, not all regulations are relevant for your planned activity. Please specify regulations you need to comply with for: (a) an energy recovery activity; or (b) a nutrient recovery activity. Find out which regulations need to be considered for every activity, regardless of the resource to be recovered, and which regulations are only relevant for specific recycling activities?

REFERENCES

Europa. (2018). EUR-Lex Access to European Union Law. [Online] Available at: https://eur-lex.europa.eu/homepage.html?locale=en (accessed 9 August 2018).

European Commission. (2008). Waste Framework Directive. [Online] Available at: https://ec.europa.eu/environment/topics/waste-and-recycling/waste-framework-directive_en (accessed 14 December 2021).

European Commission. (2012). Bioeconomy. [Online] Available at: https://ec.europa.eu/research/bioeconomy/index.cfm?pg=policy&lib=strategy (accessed 6 August 2018).

European Commission. (2016). Proposal for a Regulation on the Making Available on the Market of CE Marked Fertilising Products and Amending Regulations (EC) No 1069/2009 and (EC) No 1107/2009. [Online] Available at: http://ec.europa.eu/DocsRoom/documents/15949 (accessed 10 July 2018).

European Commission. (2017). COM(2016) 767 final/2. [Online] Available at: https://ec.europa.eu/energy/sites/ener/files/documents/1_en_act_part1_v7_1.pdf (accessed 12 August 2018).

European Commission. (2018a). 2030 Energy Strategy. [Online] Available at: https://ec.europa.eu/energy/en/topics/energy-strategy-and-energy-union/2030-energy-strategy (accessed 22 May 2018).

European Commission. (2018b). Circular Economy. [Online] Available at: http://ec.europa.eu/environment/circular-economy/index_en.htm (accessed 7 August 2018).

European Commission. (2018c). Energy – Clean Energy for All Europeans. [Online] Available at: https://ec.europa.eu/energy/en/topics/energy-strategy-and-energy-union/clean-energy-all-europeans (accessed 10 August 2018).

European Commission. (2018d). Report on Critical Raw Materials and the Circular Economy. [Online] Available at: https://ec.europa.eu/docsroom/documents/27327 (accessed 9 August 2018).

European Commission. (2018e). The Nitrates Directive. [Online] Available at: http://ec.europa.eu/environment/water/water-nitrates/index_en.html (accessed 11 August 2018).

European Council. (2013). The Clean Air Package. [Online] Available at: http://www.consilium.europa.eu/en/policies/clean-air/ (accessed 8 August 2018).
JRC Science for Policy Report. (2017). Solid and Gaseous Bioenergy Pathways: Input Values and GHG Emissions – Calculated According to the Methodology Set in COM(2016) 767. European Union, Luxembourg.
UNECE. (1999). Gothenburg Protocol. [Online] Available at: https://unece.org/gothenburg-protocol (accessed 14 December 2021).
UNFCCC. (2015). Paris Agreement. United Nations, Paris.
United Nations. (2013). Sustainable Development Goals. [Online] Available at: https://sustainabledevelopment.un.org/sdgs (accessed 12 August 2018).

doi: 10.2166/9781780409566

Chapter 14

Economic analysis of resource recovery

Roy Brouwer

Department of Economics and the Water Institute, University of Waterloo, Ontario, Canada

14.1 INTRODUCTION

Resource recovery aims to create value from waste materials associated with the production and consumption of goods and products in modern-day society. It is an essential part of a sustainable circular economy concept focused on waste reduction, reuse and recycling. Materials of potential value contained in waste products can be extracted through further processing and recovery methods, as described in the many chapters in this book. Concepts and paradigm shifts like *circular economy* and *cradle-to-cradle* are driven by our desire to make production and consumption sustainable. Here, we define sustainable processes as those that are long-lasting without compromising future generations' resource access and welfare levels, but also have clear financial and economic drivers and motivations. Various life cycle assessment (LCA) based methods and approaches exist to assess the material flows associated with production and consumption, many of which are discussed in this book. These LCA methods are key to our understanding of how efficient current production and consumption processes are in terms of our use of available biotic and abiotic natural resources.

Resource recovery is economically of interest because many of these natural resources (e.g., minerals, energy, water) are becoming increasingly scarce worldwide. Scarcity is core to economic thinking. Goods, products and services have economic value because they are scarce, meaning that demand is higher than supply. Resource recovery is particularly of interest because the costs of recovery technologies and processes are presumably lower than their benefits. This discrepancy between costs and benefits will only increase as the development of new technologies, as seen over the past years for example with struvite recovery or the removal of fibers with fine sieve technology unburdening the rest of the wastewater treatment process, results in lower recovery costs or overall treatment costs and raises resource value over time. Hence, thereby increasing the opportunity costs or benefits forgone of not recovering the resources. The increase in value and/or the reduction in costs or cost savings drives resource recovery innovation initiatives so as to not let residual values literally go to waste. In other words, it makes economic sense to recover resources as long as the benefits outweigh their costs. This is what economists refer to as efficiency.

14.2 LEARNING OBJECTIVES

This chapter focuses on the economic analysis of resource recovery. It aims to explain and clarify basic economic concepts and methods using practical, real-world examples. Specific learning objectives are:

- Familiarize the reader with the concept of economic analysis of resource recovery.
- Explain the difference between financial and economic analysis.
- Describe the concept of Total Economic Value.
- Characterize relevant economic valuation methods.
- Increase understanding of the use and usefulness of cost-benefit analysis to support policy and decision-making regarding resource recovery.

14.3 FINANCIAL AND ECONOMIC ANALYSIS

Economics should not be confused with finance or financial analysis, which is about the money or cash flows associated with the material flows. Economic welfare analysis is broader than simply tracing money flows or managing financial budgets, and typically involves both monetary and non-monetary consequences of policy actions or investment decisions. The non-monetary side-effects of resource use are referred to in economics as externalities. They accrue, often involuntarily, to third parties that are not directly involved in the decision-making to extract or use a resource. Examples include the impacts of mining on downstream water quality or fossil fuel use on climate change. Externalities can be positive or negative and add to the complexity surrounding resource exploitation decisions. They refer to important economic trade-offs between the positive impacts (benefits) and negative impacts (costs) of resource extraction and use. For example, oil and mineral extraction generates significant regional employment benefits besides market revenues, but they may result at the same time in significant pollution on-site and downstream in rivers.

Furthermore, these resources are often finite, and their extraction and use now reduces the potential for future consumption. The latter future benefits foregone or opportunity costs are also referred to as resource costs and should theoretically be included in an economic analysis to maximize the economic gains over time of exploiting a non-renewable resource. The calculation of these resource costs are based on the so-called Hotelling's rule. Readers interested in finding out more about Hotelling's rule in non-renewable resource extraction are referred, for example, to Dasgupta (1993). Accounting for the flow of future costs and benefits is what economists refer to as dynamic efficiency (e.g., Griffin, 2016).

14.4 COST-BENEFIT ANALYSIS

Cost-benefit analysis (CBA) is carried out in order to evaluate and compare the economic efficiency of alternative resource recovery projects or technologies. The benefits from such a resource recovery project are contrasted with the associated costs within a common analytical framework. Money is used as the common denominator to allow comparison of these costs and benefits, which can be related to a wide range of productive resources, including the resource to be recovered, and are measured in differing units. Although a set of common steps can be identified to carry out a CBA, as shown in Table 14.1, a CBA is typically context specific and hence no 'standardized' CBA exists. More specifically, the CBA aims to answer the following two questions:

(1) Is the resource recovery technology project worthwhile from an economic perspective, that is, do the benefits outweigh the costs?
(2) If so, which alternative technology project yields the highest net benefit, that is, benefits minus costs?

A CBA can be used to evaluate and compare the costs and benefits of multiple alternative technologies or just one. In the latter case, the costs and benefits of a new technology are compared

Table 14.1 Steps in a cost-benefit analysis of a resource recovery technology project.

Step 1: Define the objective of the resource recovery technology or project
Step 2: Define the baseline situation or technology: What would happen if no action is taken?
Step 3: Define the alternative resource recovery options to achieve the objective
Step 4: Quantify the investment costs of each option
Step 5: Identify and quantify the positive and negative welfare effects of each alternative option compared to the baseline situation
Step 6: Value all relevant effects in monetary terms, using market prices and economic valuation methods
Step 7: Calculate the present value of costs and benefits occurring at different points in time using an appropriate discount rate
Step 8: Calculate the Net Present Value (NPV) and Benefit-Cost (B-C) ratio of each alternative option
Step 9: Perform sensitivity analysis
Step 10: Select the economically most efficient resource recovery technology or project

to the status quo, where either no resource recovery technology is in place or an existing resource recovery technology is the point of reference and the incremental costs and benefits of the new technology are compared to the existing technology.

As before, an important distinction in CBA is that between a financial and economic CBA: A financial CBA, also referred to as a financial analysis of expenditures and revenues, evaluates advantages and disadvantages of a resource recovery technology in terms of the expenditures and earnings directly associated with its implementation for the investor. Originally devised for investment decisions, the tool can also be used to assess budgetary impacts of technology development projects for resource recovery. An economic CBA evaluates the costs and benefits of a resource recovery project or technology in a broader sense, taking into account all positive and negative effects on people and the environment at the appropriate spatial and temporal scales.

The time horizon in an economic CBA is not defined by the time it takes to implement the project as is typically the case in a financial analysis where the payback period is an important evaluation criterion, but by the time over which its impacts manifest themselves. This period may extend beyond the lifetime of the project. Similarly, the appropriate spatial scale over which impacts are (expected to be) observed may go beyond the location where a new resource recovery technology is implemented. The costs and benefits addressed in an economic CBA may furthermore include indirect (second-order) effects and non-priced external effects on society and the environment. An example of an indirect effect of phosphorous recovery from wastewater may be a change in the existing market price of phosphorous in fertilizer applications, while an example of an external effect of phosphorous recovery would be the improvement of water quality of wastewater receiving water bodies.

Carrying out a CBA is a multi-disciplinary process, involving expertise from different fields and the input from technology developers and investors. In general, the steps presented in Table 14.1 are followed in a CBA. Lazurko (2018) identifies eight steps relevant for conducting a CBA for resource recovery and reuse: framing, identifying, prioritizing, quantifying, monetizing, discounting, quantifying uncertainty and presenting. Steps 2, 4, 5 and 6 correspond with steps 4–7 presented here. Step 7 in Lazurko (2018) is similar to step 9 in Table 14.2. While economists are involved in all steps, technical expertise of many kinds may also be needed, especially in steps 3, 4 and 5. In order to ensure that the various options are technically feasible, input from engineers is required especially in step 3, and often also in step 4 to specify the exact nature of the resource recovery technology and estimate the associated engineering costs. Input from technology developers and investors is essential when defining the objective of the resource recovery project and identifying the expected incremental effects from each resource recovery option compared to the relevant baseline technology that is already in place. A key role of the economist in the whole process is to frame the relevant issues in economic terms, developing the CBA framework and ensuring it includes all relevant financial and economic stakes and stakeholders. The effects of the resource recovery project have to be identified

Table 14.2 Example of a cost-benefit analysis of phosphorous recovery. The costs refer to the incremental costs of the new technology, while the benefits are in this case study the cost savings as a result of the new phosphorous recovery technology. Note that the results of a CBA are typically presented as a balance sheet, meaning that a gain or loss is presented as a balancing item, equating total costs and benefits.

Costs	€/year	Benefits	€/year
Capital cost reactor	307 000	Reduced costs dewatering	131 000
Energy costs	26 000	Reduced costs sludge processing	649 000
Labour costs	60 000	Reduced costs transportation costs sludge	32 000
Maintenance costs	92 000		
Use of magnesium chloride	182 000		
Use of ferric chloride	134 000		
Costs	801 000	Revenues	812 000
Financial gain	11 000		
Total costs	812 000	Total revenues	812 000

Source: Adapted from Veltman (2012).

and quantified, first perhaps in physical terms (e.g., kgs of a resource recovered) and then converted into monetary terms (e.g., by multiplying the recovered resource amount in kgs by its market price in $/kg to estimate the total market revenues).

It is important to point out that the CBA is just one of the various evaluation studies that is typically carried out. Besides economic efficiency, other decision criteria such as technical feasibility and acceptability, risk management, or environmental impacts may play an important role too. The CBA has to be carefully aligned with these other studies, for example to quantify the technical performance or environmental impacts of a new technology. The various steps in a CBA will be further elaborated in the next sections.

14.5 MAKING AN ECONOMIC CASE FOR RESOURCE RECOVERY

The first steps in a CBA are summarized here in general terms. The use of resources in production and consumption results in different types of solid, liquid or dissipated waste downstream of the supply chain that can have different levels of eco-toxicity and hazard levels to human and ecosystem health. Resource recovery means that parts of this waste are used and transformed into valuable resources that can be reused in the same, similar or other production and consumption processes. This reduces the negative externalities related to potential human health and ecosystem risks, while simultaneously generating new economic value by making the recovered resources available for re-use.

Examples include the use of fossil fuels and the emission of carbon dioxide into the air. Carbon captured from a fossil fuel power station can be utilized to make other substances like plastics, concrete, or biofuel. Technologies include Bio Carbon Capture and Sequestration (CCS) Algal Synthesis where CO_2 and other greenhouse gases are injected into membranes containing wastewater and select strains of algae (Stavrakas *et al.*, 2018). Using sunlight or UV light, this process produces an oil-rich biomass that can be used as oil for plastics, fuel and feed.

Another example is the removal of phosphates from wastewater. Although many countries around the world have banned phosphates from laundry detergents in the past (e.g., EU, US, Canada, Australia), it is still an important source of pollution in domestic wastewater (e.g., sewage, dish detergent). Phosphorous is an essential input in fertilizer for agricultural food production. It comes from phosphate rock, the mining of which takes place in a limited number of countries around the world, and is becoming increasingly scarce, which affects its price and that of phosphorous-containing fertilizers (Cordell & White, 2014). Current phosphorous use in fertilizers has been qualified as inefficient causing eutrophication problems worldwide. Phosphorous is removed in wastewater treatment using chemical or bio-based technologies. The removed phosphorous ends up in sewage sludge, which is then subjected to anaerobic digestion, dewatering and incineration.

Only a handful of studies exist that have looked at the costs and benefits of phosphate removal from wastewater. Most studies indicate that the costs for phosphate recovery largely exceed the costs for phosphate from rock mining (Cornel & Schaum, 2009), making it a less attractive option for the fertilizer industry. However, in wastewater treatment phosphorous recovery, in particular controlled struvite crystallization, has the potential to reduce operating and maintenance costs by reducing the reliance on chemicals (chemical precipitation) and the downtime for cleaning uncontrolled struvite formation from recycled streams (Shu *et al.*, 2006). Further cost savings may be achieved because sludge without phosphate has been shown to dewater better than sludge with phosphate, while phosphorous recovery by applying sludge on agricultural land for food production also reduces the otherwise required amount of landfill area (Evers *et al.*, 2016).

An example of a CBA of the application of the phosphorous recovery technology Airprex in the wastewater treatment facility Amsterdam West, the Netherlands, with a treatment capacity of 920 000 inhabitant equivalents, is presented in Table 14.2. The most important benefits in this case study are the cost savings as a result of the introduction of the new phosphorous recovery technology. The new technology piloted in this urban wastewater treatment facility replaces the existing Modified University of Cape Town (M-UCT) process for biological phosphorous removal, which can result in high phosphate levels in the recycled water from the fermentation-based sludge processing, and clogging of pipes and tanks. By adding magnesium-chloride ($MgCl_2$) in combination with aeration, controlled crystallization of struvite improves the dewatering of the sludge through centrifuge and consequently reduces the amount of sludge for processing and transportation. The recovered struvite is of such high quality that it can be sold to the fertilizer industry (Veltman, 2012).

14.6 COST AND EFFECTIVENESS OF RESOURCE RECOVERY TECHNOLOGIES

Resources may be recovered using different technologies, and these technologies may have different costs (step 4 in Table 14.1). Economists aim to do things at the lowest cost possible. If various options exist to recover resources, a cost-effectiveness analysis of the available options helps to answer the question which resource recovery option is cheapest, bearing in mind that the outcome of such an analysis may be highly site and context specific. The purpose of a cost-effectiveness analysis is to find out how an objective (e.g., resource recovery) or predetermined target (e.g., recovering a certain amount of resources from wastewater) can be achieved at least cost (e.g., Lise & Van der Veeren, 2002; Tietenberg, 1992). The cost-effectiveness analysis hence does not account for any benefits as in a CBA, it just focuses on the least cost way of resource recovery. In its most simple form, the analysis can be depicted as shown in Figure 14.1.

Various possible resource recovery technologies (1, 2, 3,....*n*) are ranked in increasing order of their marginal costs ($/kg), that is the costs ($) to recover one extra unit (kg) of a particular resource. Besides a different marginal cost (on the vertical axis), each technology also has a specific recovery potential or capacity (on the horizontal axis). The least cost technology is the one that is able to recover 1 kg of a resource at the lowest cost. If a predetermined target has to be reached, as illustrated in Figure 14.1 by the environmental standard, multiple complementary technologies are activities which may have to be employed, since individual technologies may have a limited recovery potential. The total costs of implementing these technologies are found under the marginal cost curve from the origin until the point where the resource is recovered (e.g., until the environmental standard). Typically, recovering the last units of a resource is most costly, for example to reach the environmental standard. Whereas marginal and total costs are often relatively low for recovering the first units of a resource, these costs can go up exponentially once the cheapest options have been exhausted.

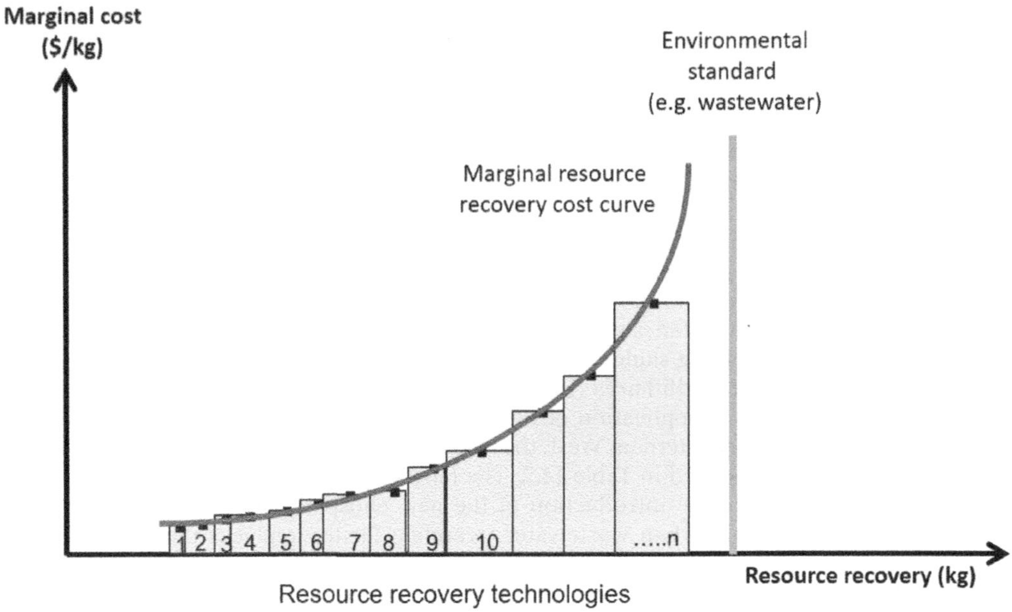

Figure 14.1 Simple form cost-effectiveness analysis.

A cost-effectiveness analysis is also a multidisciplinary exercise, requiring the input and collaboration of different scientific disciplines, such as scientists (physical effects of technologies), economists (costs of technologies), and engineers (other relevant details about the technologies such as recovery potential). Furthermore, the analysis often also needs the input and collaboration of policy and decision-makers as they determine the scope and objective of the analysis (e.g., the environmental standard).

14.7 BROADER SOCIETAL EFFECTS OF RESOURCE RECOVERY

Strictly speaking, a CBA only includes those costs and benefits that can be quantified in monetary terms. However, it will hardly ever be possible to monetize all impacts all the time. Those impacts that cannot be monetized are often left out of the analysis. Non-monetized impacts, if considered relevant, can nonetheless be included in a qualitative discussion accompanying the discussion of the CBA results. While a textbook CBA requires that all impacts be monetized, in practice different approaches exist to include non-monetized impacts in the CBA. In some approaches, they are listed as '*pro memoriam*' items on the balance sheet, expressed in qualitative or quantitative form. For example, the environmental impact of a project for which no monetary value could be calculated, like the improvement of water quality in a river or lake and the positive impact this may have on fish and other life below the water, or the spatial contours of a project like a reduction in the necessary landfill space to dump waste if it is processed and reused instead. In other cases, such impacts, if left unquantified and non-monetized, will either be ignored, left for a subsequent environmental impact analysis, or only partially monetized. The approach of monetizing impacts where possible, and including them in another form where full monetization is not possible, marks a deviation from the textbook ideal but does not discredit the CBA method as such.

In practice, investment decisions by public sector (e.g., government departments) and private sector (e.g., companies) agents are often evaluated primarily on the basis of their financial budgetary costs, as these can be assessed relatively easily. The calculation of economic costs and benefits, especially non-priced external environmental effects, is a more difficult task. An economic CBA is a more appropriate method for evaluating resource recovery technologies and projects that have significant impacts on public environmental goods, such as the improvement of water and air quality or climate change mitigation. As a result, the environmental changes induced by resource recovery projects may have an impact that goes beyond an investor or government agency and affects society as a whole. Such impacts should consequently be valued and evaluated from a societal perspective, not from the perspective of the investor or government agency alone.

An example of an economic CBA that addresses both the financial and broader economic impacts of resource recovery is given in Table 14.3. In this example, the so-called shadow price of phosphorous is also taken into account. A shadow price is the calculated or simulated price of a good or service for which no market price exists. The shadow price hence aims to reflect what this price would be if it had been traded on a market. As in this chapter's previous example (Table 14.2), the study focuses on magnesium ammonium phosphate, also known as struvite. Recovering struvite from residential wastewater reduces the chemical costs of wastewater treatment because it reduces sludge generation. Consequently, the landfill area needed for sludge disposal is also less. Struvite has been shown to be a valuable agricultural fertilizer (Evers *et al.*, 2016). In the study by Molinos-Senante *et al.* (2011), approximately 1 kg of struvite can be recovered from 100 m^3 of wastewater. The estimation of the shadow price for phosphorous recovery in this study varies between 18 and 80€/kg of recovered phosphorous based on 20 wastewater treatment plants in the coastal zone of Valencia in Spain.

Note that the costs and benefits in Table 14.3 are presented as annual values and are not discounted. The latter is preferred if the annual values occur at different points in time. For example, it may take a while before the ecosystem recovers from previous damages and hence before the avoided damage costs manifest themselves. If this is the case, the present value of the estimated damage costs would go down. If we take, for example, a discount rate of 5%, the present value of the avoided environmental

Table 14.3 Example of a financial and economic cost-benefit analysis of phosphorous recovery in 20 wastewater treatment plants in Valencia, Spain. The economic CBA includes on the benefit side the estimated avoided environmental damage costs, which were valued using a shadow price method. Note that the results of a CBA are presented as a balance sheet, meaning that a gain or loss is presented as a balancing item, equating total costs and benefits.

Costs	€/year	Benefits	€/year
Investment	70 887	Revenues from struvite sale	2378
Operation and maintenance	14 000	Savings in operation costs	10 000
Financial costs	4253		
Costs	89 140	Revenues	12 378
		Financial loss	76 762
Total costs	89 140	Total revenues	89 140
Costs	**€/year**	**Benefits**	**€/year**
Investment	70 887	Revenues from struvite sale	2378
Operation and maintenance	14 000	Savings in operation costs	10 000
Financial costs	4253	Avoided environmental damage costs	170 960
Costs	89 140	Revenues	183 338
Economic gain	94 198		
Total costs	183 338	Total revenues	183 338

Source: Adapted from Molinos-Senante *et al.* (2011).

damage costs would be around €134 000 if they would only occur after five years, €105 000 after 10 years or €82 000 if they would only occur after 15 years. A break-even point would be reached after year 16, when the present value of the avoided damage costs would dive under €76 000. This illustrates the sensitivity of the results to discounting future benefits (step 9 in Table 14.1). How to discount future costs and benefits will be addressed later in this chapter.

In a financial analysis, the upscaling of phosphorous recovery technologies in urban wastewater treatment plants stands little chance as long as the financial costs of phosphorous recovery are orders of magnitude larger than the current market price of rock phosphate, unless there are significant cost savings involved, for example in operation (e.g., energy, labor) or material (e.g., chemicals, water) costs, as shown in Table 14.2. However, accounting for the environmental externalities caused by unrecovered phosphorous discharged from wastewater treatment plants (eutrophication) and the pollution caused by phosphate mining due to the release of toxic wastewater (e.g., Reta *et al.*, 2018) and other beneficial aspects of phosphorous recovery (e.g., Mayer *et al.*, 2016) produces a different outcome. In this case study, the avoided external costs outweigh the implementation costs of phosphorous recovery.

14.8 ECONOMIC VALUATION OF RESOURCE RECOVERY EXTERNALITIES

Many environmental resources, such as water and air, are often not priced in monetary terms, creating important challenges for projects aiming to recover resources and protect the environment at the same time. For many goods and services provided by environmental resources (e.g., clean air or freshwater) there is no market where they are traded, and therefore no market price is available which reflects their economic value. There are, however, several economic valuation methods, which allow placing a monetary value on these non-market goods and services. The economic valuation of environmental resources compares people's willingness to pay for and the opportunity costs of the goods and services supplied by the resources involved. Accordingly, a wide range of environmental goods and services can be explicitly recognized in the CBA.

In economics, value is expressed as the degree to which people want to give up scarce resources, such as money or time, to acquire or retain something. Value exists in this sense only through the interaction between a subject (individual) and an object, and is therefore not considered an intrinsic quality of something (Pearce & Turner, 1990). As in other social sciences, the value people attach to something like the benefits associated with resource recovery is based upon a hypothesized positive relationship with their observed behavior or verbal responses. In economics, the value that an individual assigns to something is usually revealed through market behavior and measured in monetary terms by that individual's willingness to pay (WTP). For instance, an industry is willing to pay to some extent the market prices for the minerals it needs in its production process, a consumer is willing to pay for bottled water or a farmer is willing to pay the market prices for fertilizer to improve crop growing conditions and/or reduce the risk of crop failure. In cases where resources do not have market prices, one can ask hypothetical questions about someone's WTP. For example, visitors to recreational lakes can be asked for their WTP to improve the quality of these lakes, thereby improving their recreational experience and reducing potential health risks due, for example, to harmful algae blooms.

In their survey focusing on upgrading a local wastewater treatment plant, Paola *et al.* (2018) asked 400 residents in the city of Ferrara in north Italy's Po valley for their WTP for using reclaimed water from the local wastewater treatment plant in a constructed wetland located in a publicly-accessible green urban park surrounding the treatment plant. On average, they found a positive one-off WTP of €48 per household for the combined supply of treatment and recreational services. Aggregated across all the residents in the city of Ferrara served by the local wastewater treatment plant and living in the surroundings of the green space around the plant, the aggregated WTP provides an indicator of the total economic value (TEV).

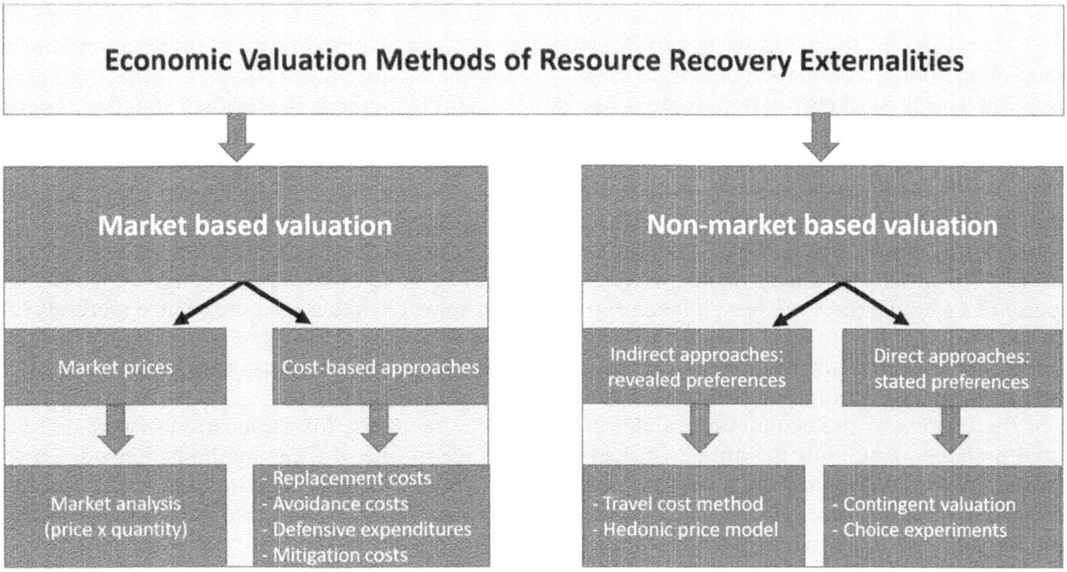

Figure 14.2 Overview of economic valuation methods for resource recovery externalities.

Environmental economists have introduced a taxonomy of this TEV, distinguishing between use and non-use values, in order to account for the various reasons and motives people may have to value environmental change. Use values are associated with the actual or potential future use of a recovered resource (e.g., fertilizer, irrigation water). In contrast, non-use values refer to values attached to resource recovery and reuse based, for example, on how the resource will benefit future generations.

A range of valuation methods exists for assessing the economic value of resource recovery externalities (Figure 14.2). The estimation of a shadow price, as presented in Molinos-Senante *et al.* (2011), is one approach (falling under the market-based valuation methods). Depending on the nature of the specific resource to be recovered, the potential externalities associated with its recovery, the presence of a market where the resource is exchanged and data availability, economic values can be estimated using various techniques, including direct and indirect market and non-market-based methods.

Market valuation means that existing market behavior and market transactions are used as the basis of the valuation exercise. Economic values are derived from existing market prices for inputs (production values) or outputs (consumption values), through more or less complex econometric modeling of dose-response or damage functions. Examples include the economic value of phosphorous in fertilizer, which is sold on a market (market analysis), the costs of replacing impaired environmental riparian functions such as nutrient retention and export through the installation of a wastewater treatment plant (replacement costs) or the costs of a water filter on tap water (avertive behavior or defensive expenditures).

Where market prices exist for the resource, these may have to be adjusted for market distortions such as taxes or subsidies in order to obtain their real or shadow prices, but otherwise they are likely to provide a relatively simple means of assessing economic market value. More advanced economic models (so-called applied general equilibrium models instead of the partial equilibrium methods discussed here) are needed if these prices are expected to change as a result of significant shifts in demand and supply of the resource involved or its alternatives.

In the absence of market prices for water or other resources, the economic value of the goods and services provided by these resources can be estimated with the help of direct and indirect non-market valuation methods. Non-market valuation means deriving economic values outside existing markets in cases where such markets are non-existent or distorted. Direct methods (also called stated preference methods) refer to contingent valuation (CV) and discrete choice experiments (DCE), where individuals are asked directly, in a social survey format, for their WTP for a specified environmental change.

WTP can also be measured indirectly through revealed preference methods by assuming that this value is reflected in the costs incurred to travel to specific sites (travel cost studies), prices paid to live in specific neighborhoods or wage differentials to account for occupational hazards (hedonic pricing studies). The latter approaches measure (environmental) use values because they typically relate to the use of a specific resource (e.g., a lake or green urban space), while stated preference methods like CV and DCE are able to also measure so-called non-use values by asking survey respondents to pay for resources without any intention to actually use them (e.g., simply because they value clean air or water or the preservation of natural resources for future generations).

In the context of the economic evaluation of resource recovery, the application of revealed and stated preference methods is limited. In theory, the choice of one of these methods depends on the relevant environmental impact of resource recovery and reuse. Travel cost studies may, for example, be appropriate if resource recovery and reuse have a positive impact on the quality of water resources that are used for recreational purposes and hence are expected to have recreational value (e.g., if phosphorous levels in wastewater effluent discharged into surrounding water bodies are reduced). Hedonic pricing studies can be applied in cases where resource recovery and reuse results in public health risks (e.g., microbial hazards or exposure to hazardous chemicals), and workers involved in resource recovery and reuse activities may only be willing to accept exposure to increased health risks in their workplace if they receive a risk premium on top of their salary. Stated preference methods such as CV and DCE have primarily been applied to measure consumer and farmer acceptance of and WTP for (partially) untreated wastewater in agriculture as an alternative water source in water scarce areas (e.g., Lienhoop *et al.*, 2014; Ndunda & Mungatana, 2013).

Wastewater reuse brings various economic and environmental benefits. Economic benefits include the supply of irrigation water and nutrients (in addition to cost savings in wastewater treatment), while environmental benefits mainly relate to reduced pollution loads into water courses that protect water quality and dependent ecosystems. Wastewater reuse is widespread worldwide due to its year-round availability and nutrient content, contributing to food security especially in those parts of the world facing serious water resource constraints (e.g., Jaramillo & Restrepo, 2017). However, it may result in a variety of negative health effects, both for those applying wastewater such as crop farmers and those handling (traders) and consuming (households) the crops grown with (partially) untreated wastewater. Partially treated and untreated wastewater reuse has furthermore been directly linked to diarrheal diseases in epidemiological studies focusing on farmers and farm workers (Helmecke *et al.*, 2020). Foodborne outbreaks due to salmonella or *Escherichia coli* related to the consumption of fresh produce like lettuce are well documented (e.g., Adegoke *et al.*, 2018). These outbreaks may negatively impact public opinion and acceptance of food products grown with (partially) untreated wastewater.

In general, public acceptance of wastewater reuse is higher for non-potable applications (e.g., irrigation of public parks or flushing toilets) (Boyer *et al.*, 2012). Baghapour *et al.* (2017) conducted a public survey among a representative sample of 562 residents living in Shiraz, one of the largest cities in Iran, and found that public tendency to accept the reuse of wastewater in various applications is greater among respondents who have a higher level of awareness of wastewater treatment processes. A possible explanation could be that better informed residents may be more capable of assessing the public health risks involved. This outcome furthermore suggests that it may pay off to invest in information and awareness raising campaigns to communicate the risks involved.

Studies applying stated preference methods like CV and DCE are summarized, for example, in Bouzit *et al.* (2018), who find an average WTP for treated wastewater and its reuse of $56 per household per year (price year 2010) based on 22 studies conducted mostly in Greece and Australia. The majority of these non-market valuation studies estimate WTP for improving the capacity or technology of wastewater treatment plants. Thirteen studies examined consumers' WTP for wastewater reuse in food production, while eight of the 22 studies focus on farmers' WTP for wastewater reuse, and one study (Menegaki *et al.*, 2007) investigated both farmers' and consumers' WTP for using and eating food crops irrigated with recycled wastewater. Those stated preference studies that included a so-called 'yuck' factor in their analysis, that is where respondents expressed a disgust for the use wastewater, yielded a significantly lower mean WTP value, while respondents' knowledge of wastewater reuse and trust in the local government significantly increased household WTP.

An example of the inclusion of the non-market benefits of wastewater reuse in CBA is given in Alcon *et al.* (2013). The costs and benefits of reclaimed water use on an experimental citrus farm in southeast Spain are compared with those of using surface water and a mixture of water sources. The non-market benefits of reclaimed wastewater were estimated based on a CV survey in the Region of Murcia, in which 352 residents were interviewed. Respondents were informed about the amount of money they currently pay for wastewater treatment and were asked if they would be willing to pay more to raise wastewater purification up to a level that would make it suitable for use in agriculture. They were told this would contribute to the preservation of the river's ecological status by reducing the pressure on the resource, while at the same time ensuring water supply for agriculture. The maximum amount respondents were willing to pay was an average of €5.13 per household per month on top of their current water rate. The CBA overview, including the estimated WTP, is presented in Table 14.4.

Table 14.4 Example of an economic cost-benefit analysis of wastewater reuse in irrigated agriculture in southern Spain.

	Surface Water	Mixed Water	Reclaimed Water
A) Annual revenues			
Yield (kg/ha)	45 625	50 566	37 955
Price (€/kg)	0.27	0.27	0.27
Total revenues (€/ha)	12 136	13 450	10 096
B) Annual costs			
Material input costs	32	33	23
Irrigation water	1927	2014	1243
Fertilizer	797	797	797
Plant protection	353	353	353
Electricity	456	456	456
Other	306	306	306
Labour costs			
Irrigation maintenance	606	606	606
Pruning	531	531	531
Phytosanitary treatment	344	344	344
Other	186	186	186
Machinery equipment	349	349	349
Tax and insurance	253	253	253
Total costs (€/ha)	6108	6195	5424
C) Financial gross margin (A-B) (€/ha)	6028	7255	4672
D) Environmental benefits (€/ha)	0	385	1927
E) Economic gross margin (C + D) (€/ha)	6028	7640	6599

Source: Alcon *et al.* (2013).

14.9 INTERNALIZING EXTERNALITIES

In some cases, governments and public agencies try to internalize the externalities of resource use, for example through the creation of a carbon tax or carbon trading scheme. By pricing externalities, investors and other decision-makers are more likely to factor them into their decisions and change their behavior accordingly. Money is a common language, often more persuasive and powerful than words or even rules and regulations. Many resources (e.g., minerals) are traded in economic markets, and their scarcity should theoretically be reflected in their market price. We would expect higher scarcity to result in a higher market price, although a straightforward relationship between geologically scarce minerals and their long-term price levels may not always exist (Henckens *et al.*, 2016). As the market price of a mineral increases, the more lucrative it becomes to develop technologies that allow us to extract it from less accessible locations or media at higher cost (e.g., oil sands). Higher prices are also expected to result in the search for alternatives or substitute resources (e.g., biofuels). In the oil industry, a distinction is therefore made between technically and economically recoverable resources (Miller & Sorrell, 2014). A similar distinction can be made for other types of resources, for example wastewater.

There are also resources, including water and wastewater, that do not necessarily have markets where they are traded and hence that do not have market prices to signal their scarcity and value. Or they have a market price (water is e.g. traded in different parts of the world where water markets were created), but this price is somehow distorted and as a result does not reflect the true value of the resource that is being used or depleted. Water pricing is a good example of such potential distortions, where water is considered an economic good (1992 Dublin Statement on Water and Sustainable Development) on the one hand, and a basic human right (to safe drinking water at an affordable price; UN Human Rights Council Resolution A/HRC/15/L.14) on the other. Economists argue that if prices do not properly reflect scarcity conditions, decision-making aimed at economic efficiency may be misinformed. These non-market resources will have no market price (i.e. are provided 'free of charge') or a very low price (e.g., due to market distortions). As a result, they may not be fully accounted for in policy and decision making and remain outside any financial analysis based on market transactions and cash flows. Regulation and standard setting may in these cases be the only option, or a combination of regulation and pricing. For example, by ensuring that every person or household is supplied a minimum amount of water for free or at an affordable low price, and a volumetric price system for every cubic meter of water consumed over and above this minimum supply standard to reconcile the views above that water is a basic human right and an economic good at the same time. Or imposing water quality standards as in the US Clean Water Act and allow water quality trading based on compliance with effluent limitations (Morgan & Wolverton, 2008).

14.10 DISCOUNTING

Costs and benefits may occur at different points in time. The expression goes that you have to invest time, effort or money before you can expect anything back. Similarly, costs and benefits associated with a project implementation often look like the flow of values in Figure 14.3. Negative values indicate costs, positive values benefits. One-off investment costs like the purchase of a new technology or the construction of a new building result in relative high negative values (costs in red) at the beginning of a project while positive annual values (benefits in blue) may only arise after a few years, after the technology has been installed and becomes operational until the end of the technology's lifetime.

In order to make costs and benefits comparable over time (e.g., the costs of a resource recovery project at the start of the project in its first year of implementation with project costs that would occur e.g. five or 10 years after the start of the project), economists discount future flows of costs

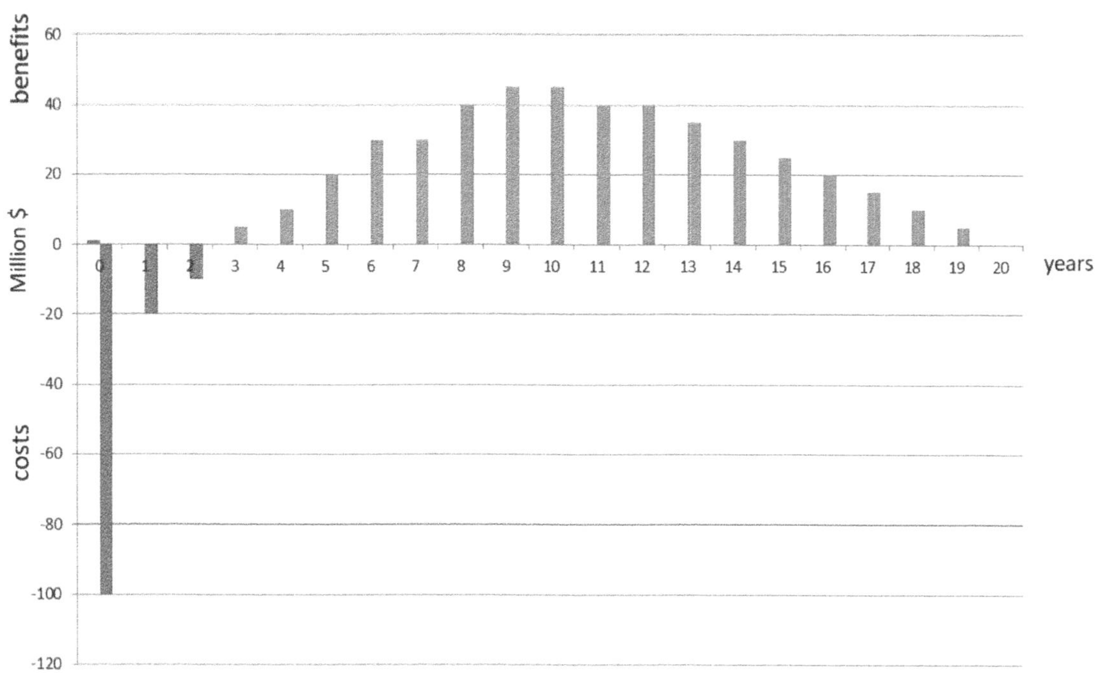

Figure 14.3 Typical flow of costs and benefits in project evaluations.

and benefits to their present value (step 7 in Table 14.1). Discounting means that the value of a future cost or benefit is weighted and modified to reflect its value at the time of starting the implementation of the project in 'year 0'. There may be various reasons for discounting future costs and benefits, including time preferences (people prefer to have money now than later in the future) and the opportunity costs of capital (investing money in one project means that it cannot be invested in another project or put in a savings account in a bank where it would generate interest revenues). Discounting is a standard economic practice in project appraisal (see e.g. the UK's HM Treasury 2018 Green Book for Central Government Guidance on Appraisal and Evaluation, https://assets.publishing.service.gov.uk/government/uploads/system/uploads/attachment_data/file/685903/The_Green_Book.pdf).

14.11 ECONOMIC DECISION CRITERIA

Ultimately, a CBA compares the costs and benefits of different technologies or resource recovery projects in monetary terms. The results of this analysis can be presented in different ways (step 8 in Table 14.1):

(i) As a net present value (NPV): the present value of the benefits minus the present value of the costs, where a positive NPV indicates that the technology or project is economically beneficial.
(ii) As a benefit-cost ratio (BCR): the present value of the total benefits divided by the present value of the total costs, where a ratio larger than 1 indicates that the technology or project is economically beneficial.

(iii) As an internal rate of return (IRR): the discount rate where the NPV is equal to zero, or the project breaks even, that is where the present value of the benefits is equal to the present value of the costs.

The choice for one or multiple decision criteria depends on the purpose of the presentation of the results. The first two criteria directly answer the first question in Section 14.4: is a project worthwhile from an economic perspective? A project is worthwhile undertaking if the NPV >0 and the BCR >1. An NPV >0 automatically also means a BCR >1. If multiple project alternatives are evaluated, these two criteria also answer the second question in Section 14.4: which project yields the highest net benefit? Here the project with the highest NPV or BCR is preferred.

The IRR is usually used to examine the opportunity costs of capital or to see how patient or impatient an investor is when discounting future flows of costs and benefits. The IRR can be compared to the rate of return when bringing the investment sum to a bank for example, or to the IRR of an alternative investment project. The IRR therefore helps to answer the second question and determine the preferred course of action by selecting the most profitable project, where the project with the highest IRR is preferred, since this is the discount rate at which the costs and benefits break even. A lower (higher) discount rate means that a future flow of net benefits breaks even more (less) rapidly. Typically, a project would have an IRR > central bank interest rate. If not, it would be more profitable to put the investment needed for the project in a bank account and earn interest on it.

A related criterion, often used in the private sector, but not based on discounting, is the payback time. This is the number of years it takes before an investment is earned back through the net benefits it generates. For example, if an investment of $200 000 generates a net benefit of $20 000 per year, the payback time is equal to 10 years ($200 000/$20 000/year = 10 years).

A pre-programmed excel sheet is included in the Appendix to this chapter which can be used to calculate the three different decision criteria mentioned above (NPV, BCR and IRR).

14.12 SUMMARY

This chapter described the most important decision-support tool in the economists' toolbox: cost-benefit analysis. Cost-benefit analysis (CBA) is an economic evaluation method, comparing in a structured and systematic way the positive and negative effects of resource recovery across one or more recovery technologies. CBA can be used to evaluate one resource recovery technology (comparing the incremental impacts of resource recovery to a situation of no resource recovery) or multiple technologies (comparing the effects of different resource recovery technologies). In the latter case, the resource recovery technology yielding the highest net benefits is preferred, while in the former case, the 'with' (resource recovery) situation is preferred to the 'without' (resource recovery) situation as long as the net benefits of resource recovery are positive.

CBA can be used before the start of a new resource recovery project, during the project implementation and afterwards. Ideally, costs and benefits are monitored throughout the lifetime of the project and thereafter, to learn and adapt based on new insights. It is usually difficult to predict all possible impacts before the start of a new resource recovery project, and costs and benefits may only become clearer and quantifiable once the project has been implemented. Insofar possible, all these effects are quantified and expressed in monetary terms, but that often proves to be challenging in practice, especially for non-priced environmental impacts.

The difference between financial and economic cost-benefit analysis was explained. Costs and benefits are ideally expressed in monetary terms, but that is not always possible. Non-monetary effects are also included in a CBA. Resource recovery is expected to have long-lasting impacts on the environment and possibly future generations, justifying the use of a broader economic or societal analysis instead of a narrow financial cash flow analysis for an individual investor. It is especially these longer-term environmental effects that are typically hard to capture in monetary terms. Moreover, using a positive discount rate implies that longer-term benefits are valued less than costs and benefits that occur nearer in the future. Systematically identifying and qualifying the different effects associated with resource recovery with the relevant decision-makers and stakeholders involved can in itself be a valuable exercise, raising awareness of the pros and cons of resource recovery. Involving and consulting different parties allows them to help identify all the relevant impacts, voice possible concerns, open up information sources, identify relevant uses and values, and ultimately make outcomes of a CBA more acceptable, because all relevant interests have been heard and are represented in the CBA.

14.13 DISCUSSION QUESTIONS

Question 14.1: Why is it important to make a distinction between financial and economic analysis of resource recovery?

Question 14.2: Discuss which internal stakeholders will be convinced by a cost-effectiveness analysis and which will be convinced by a cost-benefit analysis.

Question 14.3: Discuss why economists use non-market valuation methods in cost-benefit analysis.

Question 14.4: Discuss why economists include a project or policy intervention that fall outside existing economic markets into their cost-benefit analysis.

Question 14.5: Discuss why economists discount future costs and benefits.

Question 14.6: Discuss why and under which conditions resource recovery is an economic problem.

Question 14.7: What are important social and economic drivers and barriers to resource recovery?

Question 14.8: What are important pre-conditions to establish an economic market for resource recovery? Give examples of existing markets for marketed resources and potentially new markets for currently non-marketed resources.

Question 14.9: Discuss what role economic policy instruments like pricing, taxes or subsidies can play to incentivize resource recovery. Give practical examples where possible.

Question 14.10: Assume you are the innovation manager of a large municipal wastewater treatment plant, facing increasingly limited local government budgets. You are assigned to present a master plan to the board of directors for the coming 10 years to extend the current treatment capacity (from 400 000 to 600 000 Population Equivalents, currently with conventional activated sludge processing) and at the same time find ways to increase revenues for the different services the wastewater facility is able to provide, targeting different customers, positioning yourself strategically in an increasingly circular economy. How would you justify your master plan economically speaking? What would be key success or failure factors to implement your master plan?

Mixed water includes surface and ground water and 20% reclaimed water. Reclaimed water originates from a wastewater treatment plant with tertiary treatment. Differences in irrigation water influence irrigation water costs and yields. Using reclaimed water allows more water to remain in the river system, yielding ecological benefits. Accounting for the environmental benefits of reclaimed water, makes this economically speaking a preferred option to the use of surface water.

Appendix: Setting up your cost-benefit analysis in Excel

Year	Costs	Benefits	Net Benefits	DISCOUNT RATE	Present Value Costs	Present Value Benefits	Present Value Net Benefits
0	150	0	-150	1.10	=B2/POWER(F2,A2)	=C2/POWER(F2,A2)	=I2-H2
1	10	25	15		=B3/POWER(F2,A3)	=C3/POWER(F2,A3)	=I3-H3
2	10	50	40		=B4/POWER(F2,A4)	=C4/POWER(F2,A4)	=I4-H4
3	10	75	65		=B5/POWER(F2,A5)	=C5/POWER(F2,A5)	=I5-H5
4	10	100	90		=B6/POWER(F2,A6)	=C6/POWER(F2,A6)	=I6-H6
5	10	100	90		=B7/POWER(F2,A7)	=C7/POWER(F2,A7)	=I7-H7
TOTAL	200	350	150		=SUM(H2:H7)	=SUM(I2:I7)	=SUM(J2:J4)

Year	Costs	Benefits	Net Benefits	DISCOUNT RATE	Present Value Costs	Present Value Benefits	Present Value Net Benefits
0	150	0	-150	1.10	150	0	-150
1	10	25	15		9	23	14
2	10	50	40		8	41	33
3	10	75	65		8	56	49
4	10	100	90		7	68	61
5	10	100	90		6	62	56
TOTAL	200	350	150		188	251	63

Net Present Value (NPV)	=D12+NPV(0.1,D13:D17)	63	The NPV is positive, so the present value of the benefits exceeds the present value of the costs
Benefit-Cost Ratio (BCR)	=I18/H18	1.33	As a result, the BCR is larger than 1, indicating that the present value of the benefits is larger than the present value of the costs
Internal Rate of Return (IRR)	=IRR(D12:D17)	22%	The IRR indicates that the break-even point (NPV=0) is found at a disount rate of 22%
			22% is higher than the applied discount rate of 10%, so the project is economically efficient as long as the disount rate is less than 22%

FURTHER READING

Brouwer R. and Pearce D. W. (2005). Cost-benefit Analysis and Water Resources Management. Edward Elgar Publishing Limited, UK.

Hanley N. D. and Spash C. L. (1993). Cost-benefit Analysis and the Environment. Edward Elgar Publishing Limited, UK.

Hanley N. D. and Barbier E. B. (2009). Pricing Nature. Cost-benefit Analysis and Environmental Policy. Edward Elgar Publishing Limited, UK.

Pearce D. W. (1998). Cost-benefit analysis and environmental policy. *Oxford Review of Economic Policy*, **14**(4), 84–100.

Pearce D. W., Atkinson G. and Mourato S. (2006). Cost-benefit Analysis and the Environment: Recent Developments. Organisation for Economic Co-operation and Development, Paris, France.

REFERENCES

Adegoke A. A., Amoah I. D., Stenström T. A., Verbyla M. E., Mihelcic J. R. (2018). Epidemiological evidence and health risks associated with agricultural reuse of partially treated and untreated wastewater: A review. *Frontiers in Public Health*, **6**, 337.

Alcon F., Martin-Ortega J., Pedrero F., Alarcon J. J. and de Miguel M. D. (2013). Incorporating non-market benefits of reclaimed water into cost-benefit analysis: a case study of irrigated mandarin crops in southern Spain. *Water Resources Management*, **27**, 1809–1820.

Baghapour M. A., Shooshtarian M. R. and Djahed B. (2017). A survey of attitudes and acceptance of wastewater reuse in Iran: shiraz city as a case study. *Journal of Water Reuse and Desalination*, **7**(4), 511–519.

Bouzit M., Das S. and Cary L. (2018). Valuing treated wastewater and reuse: Preliminary implications from a meta-analysis. *Water Economics and Policy*, **5**(2), 1650044.

Boyer T. H., Overdevest C., Christiansen L. and Ishii S. K. L. (2012). Expert stakeholder attitudes and support for alternative water sources in a groundwater depleted region. *Science of the Total Environment*, **437**, 245–254.

Cordell D. and White S. (2014). Life's bottleneck: sustaining the world's phosphorus for a food secure future. *Annual Review of Environment and Resources*, **39**, 161–188.

Cornel P. and Schaum C. (2009). Phosphorus recovery from wastewater: needs, technologies and costs. *Water Science and Technology*, **59**(6), 1069–1076.

Dasgupta P. (1993). Natural resources in an age of substitutability, chapter 23. In: Handbook of Natural Resource and Energy Economics, A. V. Kneese and J. L. Sweeney (eds), Elsevier Science Publishers B.V., The Netherlands, Vol. **3**, 1111–1130.

Evers M., Vroegrijk M. and Evers T. (2016). Market Exploration and Crop Research for Struvite and Struvite-Containing Products From Municipal Wastewater. STOWA, Amersfoort, the Netherlands. Report 2016-12.

Griffin R. C. (2016). Water Resource Economics. The Analysis of Scarcity, Policies and Projects, 2nd edn. The MIT Press, Cambridge, Massachusetts, USA.

Helmecke M., Fries E. and Schulte C. (2020). Regulating water reuse for agricultural irrigation: risks related to organic micro-contaminants. *Environmental Sciences Europe*, **32**(4), 1–10.

Henckens M. L. C. M., van Ierland E. C., Driessen P. P. J. and Worrell E. (2016). Mineral resources: geological scarcity, market price trends, and future generations. *Resources Policy*, **49**, 102–111.

Jaramillo M. F. and Restrepo I. (2017). Wastewater reuse in agriculture: a review about its limitations and benefits. *Sustainability*, **9**(10), 1–19.

Lazurko A. (2018). Assessing the Value of Resource Recovery and Reuse. Social, Environmental, and Economic Costs and Benefits for Value Creation and Human Well-Being. Resource Recovery and Reuse Series 13, CGIAR Research Program on Water, Land and Ecosystems (WLE). International Water Management Institute (IWMI), Colombo, Sri Lanka. https://doi.org/10.5337/2018.229

Lienhoop N., Al-Karablieh E. K., Salman A. Z. and Cardona J. A. (2014). Environmental cost–benefit analysis of decentralised wastewater treatment and re-use: a case study of rural Jordan. *Water Policy*, **16**(1), 323–339.

Lise, W. and van der Veeren, R. J. H. M. (2002). Cost-effective nutrient emission reductions in the Rhine river basin. *Integrated Assessment*, **3**(4), 321–342.

Morgan C. and Wolverton A. (2008). Water quality trading in the United States: Trading programs and one-time offset agreements. *Water policy*, **10**(1), 73–93.

Mayer B. K., Baker L. A., Boyer T. H., Drechsel P., Gifford M., Hanjra M. A., Parameswaran P., Stoltzfus J., Westerhoff P. and Rittmann B. E. (2016). Total value of phosphorus recovery. *Environmental Science and Technology*, **50**, 6606–6620.

Menegaki A. N., Hanley N. and Tsagarakis K. P. (2007). The social acceptability and valuation of recycled water in Crete: A study of consumers' and farmers' attitudes. *Ecological Economics*, **62**(1), 7–18.

Miller R. G. and Sorrell S. R. (2014). The future of oil supply. *Philosophical Transactions of the Royal Society*, **372**, 1–27.

Molinos-Senante M., Hernandez-Sancho F., Sala-Garrido R. and Garrido-Baserba M. (2011). Economic feasibility study for phosphorus recovery processes. *Ambio*, **40**, 408–416.

Ndunda E. N. and Mungatana E. (2013). Evaluating the welfare effects of improved wastewater treatment using a discrete choice experiment. *Journal of Environmental Management*, **123**(1), 49–57.

Paola V., Mustafa A. A. and Giacomo Z. (2018). Willingness to pay for recreational benefit evaluation in a wastewater reuse project. Analysis of a case study. *Water*, **10**, 1–18. https://doi.org/10.3390/w10070922

Pearce, D. W. and Turner, R. K. (1990). Economics of Natural Resources and the Environment. Johns Hopkins University Press, Baltimore, Maryland, USA.

Reta G., Dong X., Li Z., Su B., Hu X., Bo H., Yu D., Wan H., Liu J., Li Y., Xu G., Wang K. and Xu S. (2018). Environmental impact of phosphate mining and beneficiation: review. *International Journal of Hydrology*, **2**(4), 424–431.

Shu L., Schneider P., Schneider V. and Johnson J. (2006). An economic evaluation of phosphorus recovery as struvite from digester supernatant. *Bioresource Technology*, **97**(17), 2211–2216.

Stavrakas V., Spyridaki N.-A. and Flamos A. (2018). Striving towards the deployment of bio-energy with carbon capture and storage (BECCS): a review of research priorities and assessment needs. *Sustainability*, **10**, 1–27.

Tietenberg, T. H. (1992). Environmental and Natural Resource Economics, 3rd edn. HarperCollins Publishers, New York, USA.

Veltman A. (2012). Pilot Research at the Amsterdam West Sewage Treatment Plant: Struvite Production Through the Airprex Process. STOWA, Amersfoort, the Netherlands. Report 2012-27.

doi: 10.2166/9781780409566

Chapter 15

Resource recovery from wastewater and the consumer point of view: social, cultural and economic aspects

Pay Drechsel, Miriam Otoo and Munir A. Hanjra†
International Water Management Institute, Colombo, Sri Lanka

15.1 INTRODUCTION

Between now and 2030, the sourcing of water for human needs is expected to change as the pressure on natural freshwater resources becomes more intense. This pressure is likely to come primarily from agriculture, as increasing demands for higher protein diets and biofuels will require a significant increase in agricultural output, which can only be met through greater water use. This will accelerate the over-exploitation of our freshwater resources, including a 66% increase in non-renewable groundwater withdrawals which is likely to affect millions of people by 2030, and billions by the end of the century (GWI, 2010). Under these circumstances, there will be limited alternatives to water reuse and desalination, especially where long-distance transfer is not cost-competitive. As public agencies seek economically and socially acceptable solutions to cope with increasing water demands, matching waters of different qualities with appropriate uses and implementing helpful reuse incentives will become essential for achieving the related sustainable development goals (SDGs) 6.3, 7.2 and 12.5, which directly address resource recycling, recovery and reuse.

Recovering resources such as water, energy and nutrients from wastewater has been practiced for generations in many countries (GWI, 2010; Jimenez, 2005; Lazarova *et al.*, 2013; Otoo and Drechsel, 2018; Smit and Nasr, 1992). It is expected that resource recovery and reuse (RRR) will gain more momentum where resources for agricultural production are increasingly limited under progressing climate change, competition for clean water, energy, diminishing global nutrient reserves and increasing fertilizer prices; and especially in developing countries with lower purchasing power of individual households. Wastewater not only offers opportunities to reclaim water of different quality, but also energy, nutrients and organic matter with a high application potential in agriculture (Figure 15.1). This opportunity is especially important where soils are poor and the availability of alternative inputs are physically or economically constrained. There is great potential to close the nutrient loop and to create viable businesses that make best use of what wastewater can offer.

Consumer perceptions can however limit reuse among particular clients and reuse purposes or completely undermine the potential for resource recovery and anticipated reuse impact where the

†Deceased 24 April 2019.

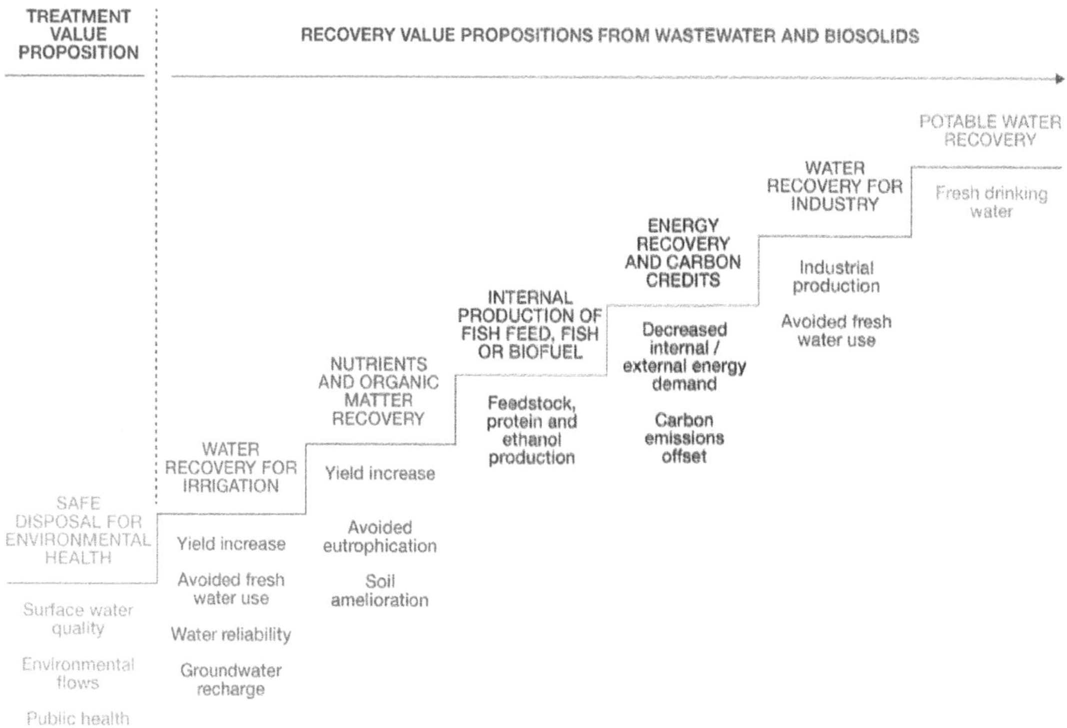

Figure 15.1 Ladder of increasing value propositions related to wastewater treatment based on increasing investment in water quality and/or the reuse value chain (Source: Drechsel *et al.*, 2015).

reuse discussion steers public perceptions and hostility. Therefore, promoting reuse in any setting requires an assessment of public views, from risk awareness to the willingness to pay for reclaimed water, under consideration of alternative options consumers may have (Drechsel *et al.*, 2015; Fielding *et al.*, 2019; Ross *et al.*, 2014).

Historically, the earliest and most commonly accepted beneficial reuse of wastewater (and the valuable resources embedded in the wastewater matrix, e.g. ammonium and phosphate) is in agriculture and landscaping, although there are numerous reuse options in both fields with different risks and demands, as well as different degrees of planning (Otoo and Drechsel, 2018). These can range from designed reuse projects to farmers unconsciously using untreated but diluted wastewater. Reuse can also take place at different scales, from household-based greywater recycling to large-scale freshwater–wastewater swaps between rural and urban areas, or it can target indirect and direct (potable) reuse through blending with freshwater, groundwater recharge and/or high-end treatment, to name just a few of the existing water reuse business models (Drechsel and Hanjra 2018; Lazarova *et al.*, 2013). Globally, Namibia, the United States of America, Australia, and Israel are among the most successful countries in introducing water reuse. Scholars and public authorities in those countries have gained substantial experience in addressing public perceptions and attitudes toward the reuse of reclaimed water, be it for direct, indirect, potable and non-potable uses (Dishman *et al.*, 1989; Dolnicar and Schafer, 2009; Higgins *et al.*, 2002; Hurlimann and McKay, 2006; Hurlimann, 2009; USEPA, 2012; Woltersdorf *et al.*, 2016). In an ideal situation, public and private concerns and benefits can be aligned and concerns about real or perceived risks weighed against the benefits of using treated (reclaimed) water. However, this requires risk education and data to make informed decisions (Prouty *et al.*,

2018); two requirements which are unavailable in the context of many low-income countries. Given the many determinants of social acceptance and the need to improve wastewater management and use in many areas, a comprehensive but location specific approach, including educational, policy, and management strategies, is needed to support public acceptance (Keremane, 2007). For example, across nine international locations of Australia, Belgium, Canada, Israel, Japan, Jordan, Mexico, Norway and Los Angeles (Hurlimann and Dolnicar, 2016) water source preference varied. Desalinated water, a purified and scientifically treated source, was preferred over rainwater for close-to-person uses. However, rainwater was preferred to recycled water that was 'purified and scientifically tested', despite the increased health risks in untreated rainwater consumption. Also, the hypothetical willingness to use alternative water sources varies between locations, and perceptions are different across locations such that context specific intelligence and interventions are required.

Especially, discussions around the introduction of direct and indirect potable reuse have sparked public interest and demand research on social acceptance. However, also recreational or agricultural reuse requires stakeholder buy-in (Marks, 2004; Marks *et al.*, 2006; McKay and Hurlimann, 2003; Po *et al.*, 2004, 2005; USEPA, 2012; Wegner-Gwidt, 1991; WHO, 2006). Failure to gain public acceptance can result in program stalling or becoming unviable (Friedler and Lahav, 2006; Keremane, 2007; Wegner-Gwidt, 1991). Depending on the region and case, cultural, religious, educational and/or socio-economic factors can support or constrain the development of wastewater use in a given location (Po *et al.*, 2004). These social acceptance challenges pertain to both the introduction of new wastewater use schemes and also to improvements in existing situations where wastewater is already (often informally) used and the recovery of resources.

Despite these challenges, the discussion on resource recovery from wastewater has garnered significant momentum, especially in water scarce regions, and focuses on water reuse for both potable and non-potable purposes with more emphasis on safety than financial aspects. The discussion on nutrient recovery from wastewater is however one step behind and so far more determined by regulatory pressure or technical opportunities for cost savings than any actual market demand for recovering nutrients. With few exceptions the demand is more theoretical, embedded in the call for a more circular economy ideally turning conventional wastewater treatment plants into resource recovery centers (Wallis-Lage 2013), alongside treating wastewater to serve the imperative of protecting public health and ecosystems. Isolating nutrient recovery from the basic function of wastewater treatment for safeguarding public health and the environment, and from the value proposition of reclaiming water or energy, appears artificial. However, depending on the local context, market demand for certain recovered resources, such as water, fertilizers, or energy can be very different and treatment operators may opt for only the one option with the highest probability of returns or social benefit. So far, phosphorus (P) and nutrient (N) recovery from waste streams is expensive but new technologies which bypass the need for costly inputs in the precipitation of phosphate are emerging. From the business perspective, this could offer a financial breakthrough even if the P price will only slowly increase. Winkler *et al.* (2013) however stresses that so far the usage of struvite in agriculture has not been well accepted in the Netherlands and expects the same in other European countries despite full compliance with required standards.

While developed countries with extensive sewer systems require advanced technology to separate nutrients from the waste stream, the low chemical and metal contamination in household-based on-site treatment facilities, such as septic tanks and latrines, makes the resulting fecal sludge (septage) a valuable soil ameliorant. The dried and composted material can for example be pelletized or blended with particular nutrients to meet farmers' needs, as shown for example in South Africa and Ghana (Harrison and Wilson, 2012; Nikiema *et al.*, 2012). So far nutrient recovery from wastewater is more driven by the treatment sector and its challenges or caused by changing regulations than by market demand for alternative fertilizers.

Market demand for recovered energy from wastewater, on the other hand, is not hardly influenced by consumer perception and inherent barriers related to safety like treated wastewater use (Otoo and Drechsel, 2018). The majority of recovered energy from wastewater currently occurs from sewage-based systems, with limited examples of non-sewered (on-site) sanitation systems. In either case, the produced energy is typically used to meet internal energy (electricity) needs and any surplus sold to households/industries via national grids. The recovered energy does not possess product attributes such as color or taste or has it any potential direct or indirect health risk to consumers from use as recovered water and nutrients may have. From that perspective, this chapter focuses on key considerations and lessons learnt mostly in the domains of water and nutrient recovery from wastewater for agricultural and potable reuse; and also differentiates between nutrient recovery options commonly seen in sewered and non-sewered (on-site) sanitation system, looking at wastewater as well as fecal sludge and biosolids.

In view of wastewater reuse markets, particularly agricultural reuse, two contrasting situations are common (Drechsel *et al.*, 2015), the one where reuse has to be marketed and the one where it is already taking place but safety options have to be marketed:

- Marketing wastewater use and its recovered resources (nutrients): An example of this refers to irrigation schemes that are planned and formally designed to use treated wastewater as a source of water. These are common in many water-scarce regions of middle- and high-income countries, where wastewater is promoted as an economic good. Wastewater is treated before being released to irrigation schemes and there are usually strict regulations guiding its use. Additionally, we can consider the case where: (a) wastewater is treated for the recovery of nutrients from sewered (e.g. phosphorus recovery) and non-sewered systems for use as an agricultural input (soil amendment); (b) treated wastewater as a medium for aquaculture; and (c) production of aquatic plants, such as algae and duckweed that grow naturally as a part of pond and lagoon treatment systems and can absorb significant amounts of nutrients and be harvested for a variety of purposes, including biofuels, or a source of protein for animal/fish feeds, and so on. (Otoo and Drechsel, 2018).
- Marketing safety: This category pertains mostly to low- or middle-income countries with limited treatment capacity, in which untreated or partially treated wastewater is polluting water bodies which are used by smallholders for irrigating crops in high market demand. The wastewater may be used either in diluted or raw form, largely due to the lack of freshwater alternatives and, where diluted, often unconsciously. In this situation, the cultural and social challenge is not the 'introduction of reuse' but to support a 'transition to safe reuse'.

While the increase of planned wastewater use receives significant policy attention through the Sustainable Development Goal 6 (SDG 6 – Ensure availability and sustainable management of water and sanitation for all), the already ongoing, but usually unsafe reuse for agricultural irrigation covers an area approximately 30 times larger than previously reported (Thebo *et al.*, 2017). Due to the significant scale of water pollution in many low-income countries, and limited capacity to monitor water quality, banning the unsafe practices is difficult to enforce as the examples of, for instance, Ghana (Drechsel and Keraita, 2014), India and Pakistan (Otoo and Drechsel, 2018) showed. In these situations, the use of polluted water remains often in a state of 'laissez-faire', without the ability of authorities to enforce restrictions or assistance to reduce potential risks (Drechsel *et al.*, 2006). Introducing risk reduction efforts would have to rely on occupational safety measures, crop restrictions, safer irrigation practices, and good post-harvest handling, following for example the WHO (2006) multi-barrier approach. In this situation, the conventional 'technical responsibility' of treatment plants to safeguard public health becomes a social responsibility task involving various stakeholders along the food chain. Thus, the challenge of 'formalizing' informal wastewater, by introducing pathogen barriers, is as much a cultural and social challenge as the introduction of reuse itself. Drawing from practical cases of project failure or success, this chapter presents a number of factors that commonly influence the introduction or improvement of wastewater use for potable and non-potable purposes, as well for nutrient and organic matter recovery. This chapter draws from Drechsel *et al.* (2015) and Otoo *et al.* (2015).

15.2 LEARNING OBJECTIVES

At the completion of this chapter you should be able to:

- Describe circumstances and understand the key factors steering public acceptance or resistance vis-à-vis wastewater reuse for potable and non-potable purposes, and for nutrient and organic matter recovery.
- Understand why stakeholder participation in planning reuse is important.
- Explain the financial consideration of agricultural producers which may favor or disfavor the acceptance of wastewater or related recovered resources.
- Understand gender roles and responsibilities, influences of culture and religion in the context of wastewater use for potable and non-potable purposes, and recovered resources (nutrient and organic matter recovery).
- Understand that besides finding acceptance for reuse, another challenge (at area-wide or larger scales) is to find acceptance for safety practices by those who have no choice but to use wastewater, or to trade or consume wastewater-grown crops, even if they are unwashed or uncooked.
- Explain options to support behavior change, like education, awareness creation, incentives, social marketing and regulations.
- Give examples of situations where cultural, political or economic reasons steered the acceptance or disapproval of reuse.
- Describe pertinent research gaps such as gender analysis vis-à-vis water reuse. There is a need for a systematic way of assessing the different impacts of actions and results on both genders. This form of analysis asks the 'who' questions – Who does what? Who has access and control over water? Who makes the decisions? Who benefits from (better) reuse implementation? Who is burdened? Many of these are questions relating to power dynamics, social roles and responsibilities and relationships (Arafa *et al.*, 2007).

15.3 FACTORS INFLUENCING ACCEPTANCE OF RESOURCES RECOVERY FROM WASTEWATER

Many studies show that across the spectrum of reuse purposes, the acceptance of wastewater as a valuable resource is influenced by many factors, ranging from expressions of disgust to calculated costs and benefits, issues of choice, trust and knowledge, attitudes toward the environment, and socio-demographic factors. These reviews also showed that water and resource scarcity itself is a good but not sufficient driver of public acceptance (Po *et al.*, 2004; USEPA, 2012). The most recent detailed literature review of common factors influencing social acceptance of (waste) water reuse was provided by Fielding *et al.* (2019). Common cross-cutting results were that

- the acceptance of recycled water decreases with increasing human contact;
- greater knowledge is related to greater acceptance of recycled water;
- health risk perceptions are consistently and negatively associated with acceptance.

While these observations appear relevant in many planned wastewater schemes, they cannot be taken for granted, and can be very different where wastewater use is already taking place, although unplanned. Depending on the context, there are a number of key factors which can play an important role, such as knowledge and risk awareness, the availability of alternative water sources, the financial implications for those directly concerned, and the need to progress in mutual agreement for the benefit of all concerned. These and other factors will be discussed in the following two sections on: (a) introducing potable and non-potable reuse; and (b) promoting safety within existing agricultural reuse operations, using examples from developed and developing countries.

15.4 ACCEPTANCE OF TREATED WASTEWATER FOR POTABLE AND NON-POTABLE PURPOSES

15.4.1 Knowledge, perceptions, and acceptability

Non-potable reuse, or in a wider sense, reuse without human contact, in general has higher acceptance than reuse with human contact. More than 45 studies assessing the level of acceptance of wastewater use, irrespective of the measure of acceptance used, show the same pattern – acceptance of recycled wastewater drops with increasing human contact, that is, lower acceptance for more personal uses. Most participants found wastewater irrigation acceptable for public parks or home garden irrigation, while only very few would accept it for drinking. Indirect contact, like via food irrigation ranges in between the extremes (Figure 15.2). The reasons can vary depending on a range of factors, such as the education and risk awareness, the degree of water scarcity or availability of alternative water sources, economic considerations, involvement in decision making, and experience with treated wastewater. Some of these factors will be looked at in more detail.

15.4.1.1 Knowledge and direct exposure

Higher levels of education are positively associated with acceptance of wastewater use. Several authors have investigated the association of socio-demographic descriptors with the acceptance of treated wastewater. The two factors that have been frequently found to be associated with the acceptance levels are the education/knowledge of the individuals expressing their opinion, and as mentioned the personal proximity or involvement in the planned reuse. In Kuwait or Greece, for example, the willingness to accept or pay for reuse increased with educational attainment (Alhumoud and Madzikanda, 2010). Positive perceptions towards reuse are usually directly the inverse of the level of physical contact with the reclaimed water (Hamilton *et al.*, 2007; Po *et al.*, 2005). For example, despite significant technical advances, potable use is usually rejected due to personal health concerns or at least avoided as long as there are alternatives (Dolnicar and Saunders, 2006; Higgins *et al.*, 2002). Wastewater use in agriculture generally is preferred to potable use, while more distant uses, such as landscape irrigation, are the most preferred (Figure 15.3). A similar perspective has been reported for Kuwait, Israel, UK, USA and Australia (Alhumoud and Madzikanda, 2010; Friedler *et al.*, 2006; Hartley, 2006; Po *et al.*, 2004; USEPA, 2012). Only two studies, in Turkey (Buyukkamaci and Alkan, 2013) and Beijing, China (Chen *et al.*, 2015) reported a negative relationship between education and support for wastewater use, which denotes a risk information-deficit. Education alone may not be a primary factor but it is correlated with other factors in influencing public acceptance of wastewater use. In peri-urban Bengaluru, India, the majority (67%) of households in apartments and gated communities appeared willing to accept and buy reclaimed wastewater for toilet flushing and landscaping, considering recycled wastewater as cost effective compared to other sources. Yet some (20%) were concerned about hygiene and others (33%) lacked trust. The main factors influencing the decision to use reclaimed wastewater in this case were: lack of trust in the public agency, health risks, quality standards and performance, psychological ('yuk') factor, affordability to buy better quality water, and alternative water availability (Ravishankar *et al.*, 2018).

15.4.1.2 Availability of alternative water sources

New reuse schemes face greater initial public resistance where freshwater or groundwater supplies are still available. Even when advanced processes are used to treat wastewater and known health risks are well managed, negative public perception can prevent well-planned projects from moving forward. The case of Singapore is such an example where the produced NEWater is technically ultra-safe but the public remains hesitant to accept it, even for indirect potable use. As a result, only a small portion (2.5% in 2011) of NEWater has been injected into Singapore's freshwater reservoirs (Lim and Seah, 2012) during dry periods. Also bottled NEWater remains more a curiosum at water conferences than a marketable product, given many competing and affordable water brands are not derived from wastewater. Thus, today most of the reclaimed NEWater is used for industrial and air-con cooling purposes.

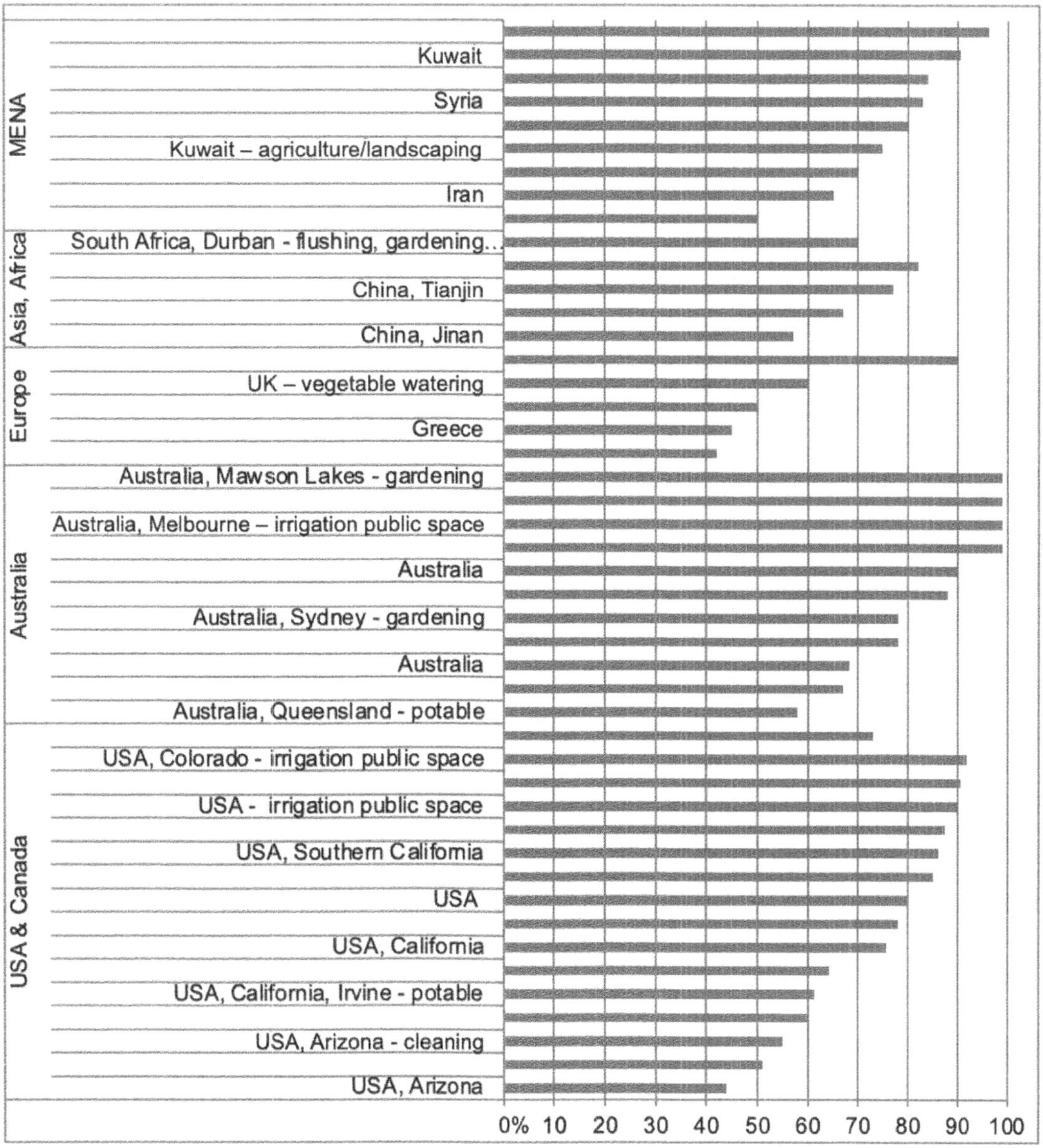

Figure 15.2 Acceptance of recycled wastewater for irrigated agriculture (or specified uses) (Source: Authors, based on a literature review).

In Windhoek, Namibia, which lacks affordable water alternatives, up to 35% of the city's wastewater is treated and blended with other potable sources to increase the drinking water supply. The success of Windhoek is supported by the fact that since the wastewater use program began in 1968, no health problems have been reported (Lahnsteiner *et al.*, 2012). The Windhoek example shows that water scarcity is an important factor in support of wastewater treatment for reuse; and that safe reuse is possible.

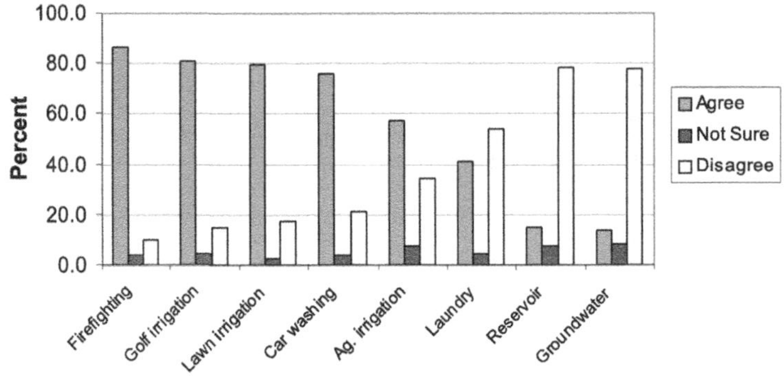

Figure 15.3 Attitudes towards wastewater use options, as expressed by 303 participants in a telephone survey in southeast United States (Source: Robinson *et al.*, 2005).

An alternative freshwater source is a crucial disincentive to the adoption of reuse, as it was reported for example for Jordan, Spain, and Tunisia (Ben Brahim and Duckstein, 2011; Molinos-Senante *et al.*, 2010; Otoo and Drechsel, 2018). In Tunisia, for example, although farmers were charged less for reclaimed water than for conventional freshwater, demand for reclaimed water remained modest, especially where farmers had access to shallow groundwater which was free, except its pumping costs. Moreover, water salinity may increase through wastewater treatment making reclaimed water less preferred than untreated wastewater or groundwater (Drechsel and Hanjra, 2018). A general recommendation in the Tunisia case was that to promote water reuse, groundwater access has to be regulated.

15.4.1.3 Public awareness and marketing challenges

In the case of wastewater irrigation, crop acceptance by the consumer remains the most critical criterion. Assuming the source of the crop is known to the consumer, his/her decision to buy or not to buy a crop produced with reclaimed water is determined by public views, knowledge and perceptions. To identify the actual consumer and to understand consumer's views, crop marketing channels need to be analyzed before assessing the perceptions (Abu-Madi *et al.*, 2008; Amoah *et al.*, 2007). In many countries, with planned reuse schemes and much more in those where informal wastewater irrigation is common reality, the existing marketing system does not differentiate between different farms or water sources, and wastewater irrigated crops are on offer together with freshwater irrigated crops. This may give an advantage to farmers using wastewater, but obviously not to consumers who can be at risk and, depending on their risk awareness, will prefer dedicated marketing channels to compare and decide. However, unless consumers clearly articulate their preference there will not be much advantage for traders to separate and display produce according to its source (Keraita and Drechsel, 2015). Dedicated marketing channels for crops produced with treated wastewater can also be a success if well branded, like in Botswana's Glen Valley, where horticultural incubators are supported with modern technologies and improved farming practices growing high value, high (off-season) demand, and high yielding crops (cherry tomatoes) in a controlled greenhouse environment with treated wastewater. Without such 'high tech' branding, reclaimed water will have a negative association which implies financial risks to farmers. The situation becomes even more challenging in view of export markets. In Northern Africa, for example, where farmers try to enter the European market, they are very reticent about using treated wastewater (Abu-Madi *et al.*, 2008; Choukr-Allah, 2013).

15.4.1.4 Public involvement, trust and buy-in

A general consensus across many reuse cases is that to achieve general acceptance of planned wastewater use schemes, especially in a social environment with the power to influence the implementation process, it is important to ensure active public involvement from the planning phase to full implementation (USEPA, 2012; WHO, 2006). Public involvement begins with early contact with potential users, and can involve the forming of an advisory committee, and public workshops on reasons, benefits and risks of reuse. The exchange of information between authorities and public representatives should ensure that concerns from perceived health or environmental impacts to lower property values have been shared and addressed (Crook *et al.*, 1992; Helmer and Hespanhol, 1997). The dialogue should build on mutual trust to provide the right climate for negotiation and conflict resolution. Timing may be an important factor. Gaining public acceptance is easier once water scarcity is affecting the public and the need to conserve high quality water sources for domestic purposes is established. In a sense, the use of wastewater becomes a solution to a problem, rather than a problem (Fawell *et al.*, 2005). In Durban, South Africa, for example, 60–70% of the surveyed consumers not only favored reuse to save water, but also the next step of dual piping systems at home (Bakare *et al.*, 2016). However, good timing alone is not a guarantee of success, as the Toowoomba example shows (Box 15.1). It will also require a sensitive approach to avoid a (e.g. politically driven) polarization of stakeholders in favor and against reuse.

Box 15.1 Public opposition to re-use

Toowoomba city in Queensland State of Australia is an often-cited case illustrating the strength of public opinion regarding wastewater use. A plan to turn wastewater into drinking water failed in Toowoomba at a referendum in 2006, although water scarcity in the community was severe, to the point that water use for gardening was completely prohibited in the 'Garden City'. With no major river nearby, the community water supply had to be pumped uphill. During several years of drought, the 140 000 residents of Toowoomba and surrounding areas endured tough water restrictions. Local officials considered that the city had no choice but to treat and use parts of its wastewater for drinking water, and given the water crisis, they expected the program would be acceptable. However, the proposal was met with fierce opposition from the community. In 2006, the residents of Toowoomba voted strongly against treating and using 25% of the city's wastewater. They relied instead on water piped from Brisbane's Wivenhoe Dam, at a cost to ratepayers of nearly $100 million more than the reuse program would have cost.

The Toowoomba proposal was an indirect wastewater use program, in which highly treated wastewater would be passed through an environmental buffer before being treated again, as part of the drinking water system. The public poll was accompanied by two dynamic campaigns building on the 'yuck' and 'fear' factors on one side, and social and financial arguments on the other. In the end, 62% of those polled opposed the project (Hurlimann and Dolnicar, 2010).

Source: Drechsel *et al.* (2015), modified.

Results from Australia indicate that actual exposure (see above) and practical experience can positively influence trust building in water authorities and community acceptance of reclaimed water, indicating also the importance of demonstration projects (Hurlimann, 2008). Dolnicar and Saunders (2006) propose reuse pilots in high-status communities first, positively influencing acceptance rates also in lower status communities.

Jordan has succeeded in informing and convincing its population about the importance of wastewater use in agriculture, by implementing an active educational campaign with strong community outreach (EMWATER, 2004). The program component included the distribution of newsletters, guidebooks, coverage of water issues in newspapers and on television and radio, websites, public educational places, and the education of land-use decision makers. Additionally, educational materials were distributed in schools, universities, and libraries (Al-Momani, 2011). However, in many low-income countries, authorities may have limited capacity to deliver on policies and enforce regulations. In such settings, local monitoring and oversight can be insufficient, and public mistrust can be fundamental (Dare and Mohtar, 2018).

In some cases, like for example Jordan, Tunisia or Kuwait, also religious concerns (Box 15.2) about water reuse were expressed in farm and household surveys (Abu-Madi *et al.*, 2008; Alhumoud and Madzikanda, 2010) while Wilson and Pfaff (2008) did not find any fundamental religious objections comparing feedback from different faith groups in South Africa.

Box 15.2 Religious concerns

In 1978, the Council of Leading Islamic Scholars (CLIS) in Saudi Arabia stated that treated wastewater can be used if its treatment included advanced technical procedures that remove impurities with regard to taste, color and smell (Faruqui *et al.*, 2001). According to Farooq and Ansari (1983), there are three ways in which impure water may be transformed into pure water:

- self-purification of the water (e.g., removal of the impurities by sedimentation);
- addition of pure water in sufficient quantity to dilute the impurities; and
- removal of the impurities by the passage of time or physical effects (e.g., sunlight and wind).

It is notable that the first and third of these transformations are essentially similar to those achieved by wastewater treatment processes.

Framing water reuse messages can strongly influence acceptance or resistance to reuse (Menegaki *et al.*, 2009; Wester *et al.*, 2016). Care must be taken that the use of negative language and images does not stigmatize wastewater use. Negative branding, especially by some media, including such headlines as 'Toilet to Tap' or 'Recycled Sewage' prevents unbiased thinking and can generate fear, stigma, and disgust (Macpherson and Slovic, 2012). Also, technical terms may not be convincing, as learned in a study in the United States (Figure 15.4).

While inadequate and negative terminology can impede clear communication, positive images and terms that enhance knowledge and understanding of water and wastewater can enhance the likelihood of success (Macpherson and Slovic, 2012). Box 15.3 presents commonly used terminology in wastewater (re)use literature.

Box 15.3 Commonly used terminology in water reuse and social acceptance modelling studies

Wastewater use: reclaimed water, reclaimed wastewater, water reuse, renovated water, treated wastewater, water recycling, recycled water, NEWater.

*Side note: technically speaking there is "water **reuse**" or "wastewater **use**", although in normal language as well as scientific papers this difference is commonly ignored.*

Social acceptance: health risk perceptions, level of support or opposition, willingness to use, willingness to pay, likelihood of use, happiness to use, feeling comfortable to use, positive attitude, satisfaction, level of assurance or trust.

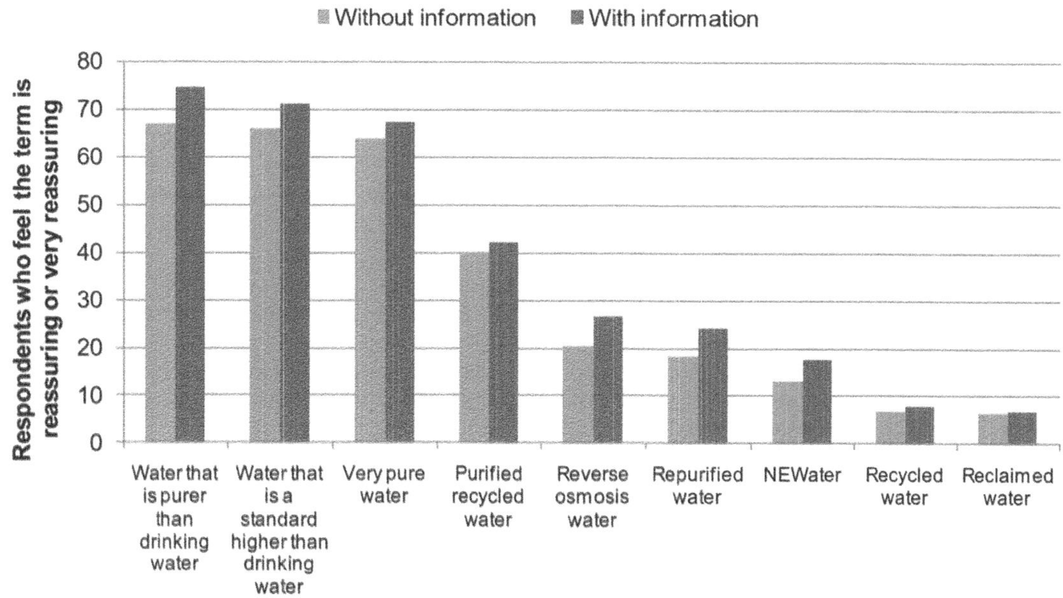

Figure 15.4 Water reclamation terms in order of declining public reassurance (USEPA 2012, based on data from the Water Reuse Association www.watereuse.org/product/07-03).

15.4.2 Accepting safety interventions for untreated or diluted wastewater use in agriculture

Highly polluted, informal wastewater use by farmers, be it raw or diluted, is far more common than planned reuse (Thebo *et al.*, 2017). Unfortunately, many countries with low treatment capacity also miss the capacity to monitor water quality for effective restrictions. The resulting situation is a widespread use of untreated or partially treated wastewater, either directly or indirectly from receiving streams, putting farmers and consumers at risk (Scott *et al.*, 2010). WHO (2006) recommends as an interim measure the adoption of safety interventions on- and off-farm to reduce health risks. These recommended practices (safer irrigation systems, produce washing or cooking, etc.) require behavior change. The adoption of these practices will largely depend on: (i) personal health risk awareness, which inter-links with educational standards, cultural and social factors; (ii) financial benefits and cost of changing current practices; and (iii) perceived water quality which may not correspond with a scientific assessment:

15.4.2.1 Risk awareness

In many low-income countries, pathogen-related risks are commonplace and many poor households face numerous risk factors daily. The risks include missing freshwater access, limited (cooling) capacity to keep food and water safe, inadequate or missing sanitation facilities, poor hygiene, and so on. In such a setting, food safety hazards (like vegetables irrigated with polluted water), which would concern consumers in developed countries, usually do not get special attention or a priority claim on the households' financial resources as experienced in a more developed environment (Whittington *et al.*, 2013). Thus, the normal living environment in large parts of Africa is characterized by several notable health hazards, such that the health risks of producing or consuming vegetables irrigated with unsafe water is usually not a primary concern of farmers, traders, or consumers, and also only one of many challenges authorities are facing.

However, not only consumers, but also farmers rank wastewater exposure commonly more like an occupational burden, which the achieved benefits from crop sales offset (see next section), and other farming practices (pest control, water carrying, weeding, etc.) as a more important health burden. In addition, whenever health risks are identified, farmers link them more to off-farm activities such as sanitation and drinking water than to farm based activities (Chaudhuri, 2008; Keraita and Drechsel, 2015; Kilelu, 2004; Ouedraogo, 2002; Weldesilassie *et al.*, 2010). Thus, there are often no significant differences in risk perception between farmers using safe and unsafe water from an outsider perspective (Gbewonyo, 2007; Gerstl, 2001), even when risk assessments predict or confirm likely health impacts (Niang, 2002; Seidu *et al.*, 2008).

Educational efforts trying to link wastewater exposure to health risk face social and scientific challenges. A limited risk awareness applies in particular to the most common situation in which wastewater is diluted (indirect use), compared to the use of raw sewage or where chemical contamination is visually evident (Binns *et al.*, 2003). The invisibility of pathogens and the lack of connection made between symptoms of potential illnesses and exposure show the need for mutually agreed on risk indicators (Box 15.4) and their use for building risk awareness.

Box 15.4 The challenge of visualizing invisible risks

Clarifying farmers and consumers' risks is normally the base for introducing health risk mitigation measures. A significant challenge for the introduction of safety options for wastewater use is getting farmers and traders to understand health risks stemming from 'invisible' contamination, such as from parasites or chemicals in water, soil and crops, and their possible transmission to consumers. Especially where farm households do not consume the (exotic) vegetables they produce, only occupational exposure problems, such as skin rashes, appear to be suitable indicators. However, the common measures to avoid skin contact, for example through the use of rubber boots, will not protect the consumer.

Studies in West Africa of traders and consumers show a generally low risk perception which is limited to visible quality characteristics, such as the produced/product's color, size and cleanliness (Acheampong *et al.*, 2012; Drechsel and Keraita, 2014; Hope *et al.*, 2008; Keraita and Drechsel, 2015). Thus it is important to also identify other risk indicators to increase awareness (Knudsen *et al.*, 2008). In Kano, Nigeria, for example, severe chemical contamination from tanneries resulted in different water colors that were well known and distinguished by local farmers in terms of possible risks (Binns *et al.*, 2003). In participatory trials with consumers and farmers, the use of UV fluorescent reactive gel (www.glitterbug.com) received positive feedback as a way to visualize potential germs (Amoah *et al.*, 2009). However, where risks remain invisible and hard to comprehend, behavior change can also be achieved without 'education' but through association of a wrong practice, for example with disgust (yuck factor), which was a successful trigger in Ghana's hand wash campaign (Scott *et al.*, 2007).

Source: Keraita *et al.* (2008), modified.

15.4.2.2 Financial benefits

Water access can have significant financial benefits for farmers, especially where it allows off-season (i.e. dry season) production of cash crops. Based on limited risk awareness or unenforced restrictions, water quality is usually a less important consideration. Studies show that farmers in West and East Africa, Southeast Asia and the MENA region generally are concerned about the quality of their irrigation water, mostly for crop growth, yet they consider the potential gains from irrigating with wastewater to be greater than any possible personal risks (Hanjra *et al.*, 2015). The common lack of safer (and at least equally beneficial) alternatives makes the use of polluted water an accepted, hardly avoidable professional trade-off (Abu-Madi *et al.*, 2008; Gbewonyo, 2007; Gerstl, 2001; Keraita *et al.*, 2008; Kilelu, 2004; Knudsen *et al.*, 2008).

A challenge related to some of the recommended safer irrigation practices, such as wastewater treatment, drip irrigation, or cessation of water application, is that these practices do not only reduce microbial contamination, but can also reduce crop yields (and income) if they are not well adapted to local conditions (Amoah *et al.*, 2011). For example, (i) drip kits with too wide inner-tube spacing in Ghana reduced plant density and yields; (ii) cessation of watering a few days before harvest made the vegetables appear dry and unattractive for traders; (iii) and wastewater treatment in ponds in Pakistan increased the salinity content of the water making it unsuitable for crops. Participatory research is required in such situations to understand farmers' constraints and adjust the technology to farmer's particular crops and farming conditions.

Health risk reduction measures will be adopted more easily if they appeal to farmers' priority challenges. For example, drip kits reduce pathogen exposure for farmers and crops, and they also enable farmers to save labor which is a high priority burden (Keraita *et al.*, 2010). In Ghana, Keraita *et al.* (2008) concluded that cost/labor savings and market incentives are the main factors which would motivate farmers to adopt best practices in the long term. However, marketing channels or an institutional framework to promote safer vegetable production and marketing are missing. To build such value chains, gender related work distribution will have to be addressed. In Ghana, for example, the marketing of most exotic vegetables is only done by women, while vegetable farming is mostly the domain of men (Drechsel *et al.*, 2013). These gender roles prevent farmers from direct marketing and result in 'safe' vegetables usually becoming mixed with unsafe vegetables in markets.

In general, the net beneficiaries of safe vegetables are the urban consumers, who may pay more for safe produce and dedicated marketing channels (Ngigi *et al.*, 2011). So far only specialist markets for more wealthy population groups show interest to pay for safety (Acheampong *et al.*, 2012; Lagerkvist *et al.*, 2013). A challenge will be how to make safe produce accessible for the most vulnerable, who have the lowest ability to pay a premium.

15.4.2.3 Water preferences

While consumers will generally prefer freshwater over (treated) wastewater, farmers' preferences can be the opposite where missing restrictions allow them to choose. In contrast to the planned introduction of reuse, where the availability of freshwater can be a strong disincentive for accepting reclaimed water, farmer preference can be very different in informal irrigation, especially if the driver of choice is income and not personal safety. Where wastewater is highly concentrated, farmers are often also aware of its fertilizer value (Van der Hoek *et al.*, 2002). There are many cases described where farmers actively seek the wastewater, and preferably untreated wastewater. In Pakistan, for example, treated wastewater did not find the same acceptance among farmers than untreated wastewater given its increase in salinity in treatment ponds (Ensink *et al.*, 2004). In Mexico, farmers protested against treatment to maintain the fertilizer value of the water (Scott *et al.*, 2000; Silva-Ochoa and Scott, 2004). In Bangladesh, farmers appeared to be well aware of actual and possible risks but still preferred wastewater for its fertilizer value or due to lack of alternative or equally (year round) reliable water sources (Mojid *et al.*, 2010), also supported from Morogoro, Tanzania (Mayilla *et al.*, 2017). A rather indifferent view was observed when reuse was indirect from streams carrying diluted wastewater. In this situation, the nutrient value of wastewater can be negligible (Erni *et al.*, 2010) while the year-round availability steers the added wastewater value. In the Mezquital Valley, Mexico, the possibility of irrigating with wastewater instead of (only) rainwater caused land rents to increase many times as the additional water enabled three crops to be harvested per year instead of one (Jimenez, 2005). Only where wastewater use is actively regulated, like in Tunisia, does its use become unattractive (Dare *et al.*, 2017).

15.4.2.4 Facilitating the adoption of safer behavior

Where authorities lack capacity to implement and monitor safety regulations, education as well as economic and/or social incentives can help to enhance food safety along the value chain. Behavior change is a particular challenge where wastewater is the only water source, and safety measures are required to facilitate a transition from informal to formal use (Prouty *et al.*, 2018). Such safety measures can be introduced from 'farm to fork', as described for example by Amoah *et al.* (2011) and WHO (2006). Where risk awareness is low and not easy to develop, research is needed to determine how best to motivate and trigger adoption of risk mitigation measures. Gender specific roles can be an important factor in this context (see below). Measures to support behavior change can include economic or social (marketing) incentives, such as access to credit, labelling/branding, dedicated marketing chains, tax exemptions, and institutional support, like the provision of extension services, awards, or tenure security, but also restrictive regulations if they can be enforced (Drechsel and Karg, 2013). Labeling of food products in a reliable manner that reveals safe or unsafe irrigation methods will be needed to support a market response to changing consumer behavior.

In many cases increased education and risk awareness will not be sufficient to motivate the desired changes in behavior. Economic incentives may be helpful in motivating wastewater farmers who are usually engaged in cash crop production, while consumers may respond better to social marketing which aims to respond to inner desires, fears and motivations (Dishman *et al.*, 1989; Scott *et al.*, 2007). Successes with social marketing have been reported from promoting latrine use and hand washing (Box 15.5). Where regulations and monitoring are weak, media publicity can encourage farmers to adopt safety practices including safer water sources, in the same way that negative media exposure can harm business activity (Drechsel and Keraita, 2014).

Box 15.5 Social-marketing studies in the West African context

'Health in your hands' refers to a marketing approach applied in a nationwide hand-washing campaign in Ghana, involving the use of professional marketing techniques facilitated through a private–public partnership to promote 'socially useful products' (in this case, hand washing with soap) through generation of demand. The underlying research revealed two main drivers for hand washing with soap: disgust of dirt (yuck factor) and caring for a child, whereas health (protection from disease) was a weak motivator. The communication campaign was thus designed to evoke the feeling of disgust without mentioning health or sickness. The campaign was fairly successful: soap use after toilet use increased by 13% and soap use before eating increased 41% (Scott *et al.*, 2007) and thus raised awareness.

'A wanted latrine is a used latrine': Many sanitation projects in developing countries have failed because they relied only on subsidized latrine construction and health education without generating demand. Thus the target community did not change established habits (like open defecation) and the latrines remained unused. In Benin, the social marketing approach was applied to improve sanitation. Research was conducted to determine what triggers people to invest in a latrine and to use it. Health benefits did not appear in the top ten triggers, whereas safety, dignity and social prestige were among the top five (Martinsen, 2008), and the social marketing approach was much more successful where local social actors such as school teachers led the campaign, like in Ethiopia (Crocker *et al.*, 2016).

15.5 ACCEPTANCE OF RECOVERED NUTRIENTS FROM WASTEWATER FROM SEWERED SYSTEMS

15.5.1 Biosolids recovery and use

The conventional view of wastewater as a public and environmental health concern still results in the linear model where large amounts of energy and chemicals are utilized to ensure that wastewater is effectively treated and/or transformed into products which meet stringent human health and environmental standards before they are released back to the environment (WRF, 2022). Where such treatment is in place, and contaminants are controlled, the beneficial uses of wastewater and sludge (biosolids) produced during the treatment process are well documented. There are various technology options to achieve Class A or Class B biosolids standards (USEPA, 2012). However, technical possibilities do not imply market demand and thus business opportunity. There are different constraints but also opportunities.

A common limitation for the biosolids market is the lack of regulatory and financial support needed to catalyze market demand for biosolids. However, in a context where landfills are filling up fast and sludge is being produced in ever greater quantities, growth in sustainable solutions for sludge treatment are on the horizon. With increasing competition for valuable landfill space and new government guidelines and compulsory policies emerging, many countries such as the UK, USA, Australia, South Africa, India, Japan and China are phasing out landfilling of sludge in favor of sludge dewatering and utilization, like land application, soil amelioration, energy and heat recovery, or the production of bricks and cement blocks (Box 15.6).

Box 15.6 Sludge management in China

In China, government policies have set national goals of treating 70% of the sludge in large cities and 50% in small cities and significant investments have been committed in sewage sludge treatment. The Ministry of Environmental Protection, together with the Ministry of Housing and Urban-Rural Development and the Ministry of Science and Technology, published the "Policy on Sludge Treatment and Pollution Prevention Technology in Urban Wastewater Treatment Plants" which aims to regulate and promote beneficial sludge utilization practices, which can have a major potential to be exploited as a Clean Development Mechanism (CDM) project. These plans show government recognition that shifting from disposal to utilization is compatible with their idea of a circular economy and are a clear signal that sludge treatment and utilization can have a future in the environmental protection industry (GTZ 2009).

Although sewage sludge has been used in agriculture in many parts of the world, its acceptability varies significantly with different cultures and beliefs among agricultural producers. Even with strict guidelines in place, in some countries like Australia and the US, agricultural producers have significant reservation in using sewage sludge (Krogmann *et al.*, 2001; Murphy, 2018). For these agricultural producers, perceived risks related to heavy metal contamination (soil build-up, crop uptake), negative food purchasers and consumers' perception, odor complaints, and potential increase of contaminants in water supply outweigh the economic incentives (nitrogen and lime value of sewage sludge) and soil improvement benefits (addition of organic matter) (Krogmann *et al.*, 2001). Farmers' reluctance to use sewage sludge is with good reason given that in China for example, 80% of the produced sludge is still transferred to landfills because industrial contamination makes the sludge unsuitable for most reuse options. On the other hand, in places like Palestine, farmers are in favor of using sewage sludge from a certified fully functioning wastewater treatment plant (Rashid *et al.*, 2017). It is important to note that the majority of the farmers were however in favor of using it for growing fruit trees, rather than growing vegetables and other plants in a greenhouse. In the case of the latter, farmers' willingness to use treated sewage sludge increased on condition that: (i) consumers are willing to buy agricultural products fertilized by sludge; (ii) sludge meets the public health requirements; and (iii) sludge is available at low costs.

Not surprisingly, public perception is a major concern of agricultural producers given that their livelihoods depend on consumer satisfaction of their produce. This has been justified by studies about consumer preferences (Krogmann *et al.*, 2001; Murphy, 2018; Rashid *et al.*, 2017; Zimmerman *et al.*, 1991); although it is important to note that limited research has been conducted on consumer preferences about food crops grown on sewage sludge applied land. As with potable vs. non-potable wastewater reuse, reuse without human contact in general has higher acceptance than reuse with human contact, as corroborated by Zimmerman *et al.*'s (1991) findings that US households have a preference for land application of sewage sludge in parks and roadways compared to flowers and food chain crops. Food purchasers and consumers' reservations are noted to be driven by key risk perceptions related to biosolids regulations, treatment, and application. Surveyed respondents believed that the benefits derived from biosolids did not offset the perceived health and safety risks. Significant gender differences were observed, with females perceiving greater risks to health and safety from biosolids use than males.

Sewage sludge for agriculture use is unlikely to grow until farmers' diverse concerns about consumer perceptions, crop and land application are addressed. Even with existing regulations and guidelines governing land application practices of sewage sludge, it is still unclear whether farmers are not knowledgeable about the regulations or are simply not convinced that the regulations are protective enough. Innovative and practical communication strategies focusing on crop or land related issues are imperative to improve farmers' adoption of sewage sludge for agriculture. Community-specific outreach programs are critical to address public risk perceptions and those driven by gender differences.

Beyond negative consumer perceptions, another common problem with biosolids-as-fertilizers in developed countries relates to the level of nutrient, in particular nitrogen (N), even in dewatered biosolids, which is oftentimes too low to support a market price that permits an independent company to be profitable; and thus actually produce it for the market. Only a fraction (5–15%) of the available nitrogen in the wastewater can be recovered through phosphate (P) based precipitates. It is more likely that P recovery will drive the process. However, making a high value biosolids-fertilizer mix can also be a viable option today because of increasing tipping fees by municipally operated wastewater treatment plants to dispose of their biosolids, especially in medium to large municipalities.

15.5.2 Phosphorus recovery and use

The currently dominant process to extract phosphorus (P) with market value from wastewater treatment streams is based on crystallization and precipitation of struvite (magnesium ammonium phosphate or MAP) (Duenas *et al.*, 2003; Rahman *et al.*, 2014). Struvite can have many uses with the most common as slow-release fertilizer or raw material for the fertilizer industry (Gaterell *et al.*, 2000). Given that the world's affordable mineable reserves of phosphorous are set to start running out in the coming decades and an expected increase of the price of high-quality rock phosphate, alternative high quality P sources, like struvite, will become more competitive than currently (Rahman *et al.*, 2014). A contemporary driver observed for example in USA, are anticipated changes to regulatory limits affecting effluent discharge permits. The result is that the number of treatment plants recovering phosphorus is continuously increasing and also the number of technologies offered for P recovery, especially in the Netherlands, Germany, Austria, Canada, and Japan.

Technology plays a significant role for P recovery as there are various options with very different costs and efficiencies. A wastewater treatment process offers several locations for phosphorus recovery.

Raw material options for phosphorus recovery processes include the sludge-free wastewater, the sludge liquid and the sludge itself, and the incinerated sludge ash, each with a different P concentration and recovery potential but also costs. Crystallization processes based on the liquid phase from sludge dewatering are so far cost and energy wise the commonly preferred option, while processes building on P recovery from sludge ash are slightly more expensive but have a significant more favorable P recovery capability. Options to recover P from sludge can extract similar amounts of P than those based on incineration, but the additional energy demand and costs makes them so far less attractive (Morf and Koch, 2009).

Limited information is available on the actual use of the recovered struvite from its registration as fertilizer under local law and how far the fertilizer sector is accepting the product at the scale of its production, or only for niche markets. So far the market value of the struvite is not a driver for phosphorus recovery and recycling (P-REX, 2013). In fact, the chemical reagents necessary for struvite production (in particular magnesium chloride) cost so far more than the market value of the produced phosphate fertilizer. However, savings in removing unwanted struvite and avoiding blocked pipes, reducing sewage sludge production and disposal, and sustainable development objectives make the innovation an appreciated and viable value proposition with payback periods of 3–7 years (Shu *et al.*, 2006).

In view of the market for recovered P, there can be a variety of challenges which differ from country to country and are still limiting the potential of P recovery despite its obvious benefits (Otoo and Drechsel, 2018). One in particular relates to the political and regulatory environment. The regulatory context in many countries still does not yet support 'secondary' phosphorus containing fertilizers and their producers as it is often classified as waste (P-REX, 2013). While stringent environmental regulations on the discharge of P effluents into water bodies are on the increase and provide an opportunity to promote recovery and reuse, and so do SDG 12.4 and 12.5, these regulations mostly favor P removal, but not yet recovery and reuse. In fact, in Europe, regulations on the reuse of waste derived resources, including urine and struvite, are often very restrictive (Winkler *et al.*, 2013). On the other hand, in many developing countries, regulations and standards may be lacking which can place resource recovery and reuse in a gray area where entrepreneurs may face few barriers, but quality control and legal security remain risk factors. However, with increasing attention to the SDGs and a circular economy the situation is changing, especially in Europe (Box 15.7).

Other limiting factors to the development of the phosphorus market include:

- In many countries a range of markets may not be accessible due to prohibitive legislations or missing legislation on the reuse of waste derived resources.
- The volumes of the recovered P are still too small compared with the market size, which increases the costs of entering the current mainstream value chain.
- Although many studies show that recovered P crystals are of high quality and are often less contaminated with metals and other micro-contaminants than natural rock phosphate, legislation and the fertilizer industry are hesitant to accept the product, be it for blending of other P sources or as stand-alone slowly-soluble fertilizer.
- More progressive legislation in support of a circular economy could help penetrate the conventional P market by demanding for a certain ratio of recovered to natural P; one example is that of the Indian Government which requires the fertilizer industry to co-sell bags of industrial fertilizer with a number of bags of waste-based compost.
- To avoid perception related risks, marketing strategies normally avoid any connection between the name of the P product and its source.
- With the never-ending generation of wastewater, also the supply of recovered P will be continuous irrespective of agricultural seasons. This will pose storage challenges unless multiple market segments next to seasonal crops are available (e.g. parks and gardens, forest or fruit plantations, year-round home gardens).

Box 15.7 P-recovery regulations and obstacles in Europe

Switzerland was the first European country to make phosphorus recovery and recycling from sewage sludge and slaughterhouse waste obligatory. The new regulation entered into force on 1.1.2016 with a transition period of 10 years. Switzerland banned direct use of sewage sludge on land in 2006, so that the new regulation will lead to obligatory technical recovery and recycling in the form of inorganic P products. Swiss sludge and slaughterhouse waste together represent an annual flow of 9100 t of phosphorus. In Germany, a new sewage sludge ordinance (AbfKlärV) is expected to enter into force early in 2018, making phosphorus recovery obligatory for larger sewage works within 12 years (>100,000 p.e.) or 15 years (>50,000 p.e.), under certain conditions. P-recovery will thus be required for around 500 sewage plants, treating around two-thirds of German sewage. Following the legislative developments in Switzerland and Germany, Austria is now also opting for mandatory P recovery from municipal sewage sludge. The draft Federal Waste Plan 2017 (Bundes-Abfallwirtschaftsplan) includes a ban of direct land application or composting for sewage sludge generated at Wastewater Treatment Plants (WWTP) with capacities of 20,000 p.e. or above within a transition phase of 10 years. Alternatively, these WWTP will have to recover the P from sludge or its ash. This regulation will cover 90% of the P contained in the Austrian municipal wastewater. However, P recovery within a Circular Economy requires reuse. Until now, struvite recovered from wastewater is only authorized for use as a fertilizer for some producers in some countries (e.g. the Netherlands, Denmark and Japan), or only on a case-by-case (e.g. Ostara plant by plant) authorization. Even in a country like the Netherlands, approval as a fertilizer does not ensure end-of-waste status for struvite. End-of-waste criteria specify when a certain material ceases to be waste and obtains a status of a product (or a secondary raw material). This current lack of clarity and disparities even between EU Member States poses a significant obstacle, also to investments in the technology as long as it cannot necessarily be sold in another country because the resulting product cannot be sold as a fertilizer. The currently (2017) discussed new EU Fertilizers Regulation will enable recycled nutrient products to be sold in any Member State, when the new Regulation comes into force. Recognized products will also be granted de-facto end-of-waste status. Composts and digestates are already included in the proposed regulation text, but struvite is not. The EU's Joint Research Centre (JRC) has been mandated to make an impact assessment and (if this concludes positively) to propose criteria to add struvite, biochars and ash-based recycled nutrient products to the new Regulation annexes.

Source: http://phosphorusplatform.eu/

- It is a significant advantage if like in the Ostara case the cost of P recovery can be (more than) absorbed by savings in conventional P removal, as the price of rock-phosphate is still too low compared with the breakeven price of recovered P, pushing recovered P into premium or niche markets which are able to pay higher-than-average prices.

15.5.3 Wastewater-based aquaculture

Greater wastewater reuse can enhance social benefits, provided health and environmental risks can be managed appropriately. The majority of wastewater reuse for aquaculture in the world today occurs in Asia, where it is a traditional practice in countries such as Vietnam and China (Edward and Pullin, 1990). Many international organizations, such as the UNDP and World Bank, are promoting the adoption of an integrated system of wastewater treatment plants and aquaculture in developing countries (Edward, 2008). These systems have the potential to improve sanitation because they provide low treatment cost options to policy makers and have the opportunity to reduce nutrients and pathogens in wastewater.

In several cities in northern Vietnam, the use of wastewater in agriculture is the only means of treatment and fertilizes about 500 ha of fish ponds (Vo, 1996; WHO, 2006). From a public sector perspective, the sale of fish and aquatic plants represent opportunities to offset the costs of the wastewater treatment. This trend is supported by national estimates of Vietnam, which indicates that wastewater-based aquatic products have the potential to generate a revenue of approximately $5760 per ha per year for vegetable production and $7200 per ha per year for fish production, which is three times higher than that of rice production, a major local staple crop (Nguyen and Leung, 2009). The food security perspective of wastewater-based aquaculture cannot be ignored, especially in view of the limited availability of reliable fresh water sources for sustainable aquaculture production. Wastewater represents an important source of nutrients and water, which can be used to increase both fish and crop production (Bunting, 2007). Wastewater-based aquaculture accrues significant social benefits via employment generation, fish for households, service providers such as producers of fish fingerlings, and economic actors involved in fish transport and marketing.

While the potential benefits of wastewater use for aquaculture are multi-fold and significant, the market prospects are limited by social barriers and consumer perceptions. A key factor that threatens wastewater-based aquaculture is the perceived consumer health risks. The World Health Organization (WHO) has developed specific guidelines for the safe use of wastewater for aquaculture to ensure public health protection (WHO, 2006). The guidelines recognize the need for public health standards to be based on epidemiological rather than microbiological guidelines, as these guidelines are context specific (WHO, 2006). Many studies show that there is no strong evidence of health risks from the consumption of wastewater-raised fish (Edward, 2008). Supporting studies from India and Egypt (Easa *et al.*, 1995; Pal and Das Gupta, 1992) suggest that fish reared in treated wastewater-raised ponds have better microbiological quality than freshwater fish cultivated in water bodies and surface waters, which may have been unintentionally polluted. A study in Vietnam corroborates this notion, where there was no significant difference found in the number of presumptive thermos-tolerant coliforms in the gut content and muscle tissue of fish raised in wastewater-based ponds and non-wastewater-based ponds (Lan *et al.*, 2007). Fattal *et al.* (1992) and Edwards and Pullin (1988) reported similar findings.

Even with the implementation of practices that satisfy health and hygiene guidelines, research shows mixed reviews on consumers' perception of wastewater-raised fish. In Ghana, for example, wastewater-raised tilapia is sold at prevailing market prices as those of freshwater systems (Murray *et al.*, 2011). Conversely, in Vietnam, although significant evidence indicates no increased human health risks from consumption of fish raised in wastewater reuse systems, concerns over toxin accumulation in edible fish has been found to significantly influence consumer demand (WHO, 2006). Mancy *et al.* (2000) found similar results in Egypt where consumers were reluctant to consume fish cultivated in wastewater although noted suitable for human consumption. This issue of food safety has been attributed to market failures related to imperfect information between households and producers with regard to product-specific attributes (Ortega *et al.*, 2011). Particularly for wastewater-raised fish, this issue also arises because of inadequate knowledge and awareness of the level of health risks associated with wastewater-raised fish, unclear policy development and regulation for fish marketing. There is limited information on the sources of water used to raise fish for the market. As a result, households' purchasing decisions are usually not based on improved fish safety and quality attributes. The assessment of fish safety attributes can restore households' confidence and provide the optimal attributes mix for potential investors to focus on for the promotion of wastewater-raised fish. If indeed consumers' purchasing decisions are influenced by source and product quality information, then understanding consumers' acceptance of fish reared in treated wastewater is critical.

Several authors have investigated the association of socio-demographic descriptors with the acceptance of fish reared in treated wastewater. The three factors that have been frequently found to be associated with acceptance levels are the education/knowledge of the individuals expressing their opinion, income, and their perception of certification by a trusted government agency. Higher levels of education are positively associated with acceptance of fish reared in treated wastewater.

Fewer studies have included informational attributes such as third-party certification. Fish safety issues often arise from lack of trust between producers and consumers with respect to product-specific attributes. Such market failures are remediable with quality certification measures using a trustworthy and independent party to test the product for its quality. Third-party certification can serve as a quality assurance indicator, which may influence household purchasing decisions (Danso *et al.*, 2017; Fonner and Sylvia, 2015; Ortega *et al.*, 2011). The existence of an independent certification can create a more competitive market to provide certified fish products and other businesses will be incentivized to pay the fees for certification, and compete to provide the given quality most effectively. Brand identity conveys a message of product quality, but this is often realized at the cost of high prices and limited supply.

Danso *et al.* (2017) found a positive correlation between consumers' willingness-to-pay (WTP) for certification of fish and information on the medium used to raise fish in Vietnam; with demand expected to increase if certification is done by a trusted government agency. This result indicates that there is a strong need for the Vietnamese government to provide adequate food safety and quality control as the households in this study preferred a government party to provide food safety certification. The results also suggest that quality assurance considerations and high incomes are factors that would increase the probability of higher expenditures on wastewater-raised fish. As the Vietnamese urban per capita income continues to rise rapidly, more people are expected to join the higher income class that is incentivized to pay a higher price for better information on food safety. This should serve as a catalyst for governmental agencies and the private sector to invest in quality control services for food safety and new wastewater-based aquaculture businesses as this approach may contribute to improve food and strengthened nutritional quality in Vietnam. While some households may be willing and able to pay for values they believe in, it must first be apparent that the market can deliver these values. Innovative marketing strategies including pricing and product promotion strategies will be required to facilitate market entry.

Beyond wastewater use for aquaculture, biological processes can be used for recovering nitrogen and phosphorus from wastewater, like through aquatic plants growing in treatment ponds, aquaculture, or wastewater irrigation. Aquatic plants, such as algae and duckweed that grow naturally as a part of pond and lagoon treatment systems, can absorb significant amounts of nutrients and be harvested for a variety of purposes, including biofuels, or a source of protein for animal/fish feeds, and so on. Ozengin and Elmaci (2007), for example, reported 83–87% total nitrogen removal and 70–85% total phosphorus removal for duckweed fed with municipal and industrial wastewater. In the US, most wastewater treatment ponds and lagoons are functionally high rate algae producers and in more recent years the systems have been explored for their potential to grow algae that can be used, for example, for biofuel production (Sturm and Lamer, 2011). Biological nutrient recovery via the production of fish food (duckweed) and/or fish has been tested successfully in different environments from Bangladesh to Peru with full cost recovery, not only of the additional costs but all operational costs of the pond-based treatment system (http://www.agriquatics.com/Case_Studies.html; Otoo and Drechsel, 2018). Although there is a potential significant market, this is a nascent sector and very limited research on end users' perceptions and the market has been conducted to date.

15.5.4 Wastewater from non-sewered systems (onsite sanitation systems) as a nutrient source

While developed countries with extensive sewer systems require advanced technology to separate nutrients from the waste stream, the low chemical and metal contamination in household based on-site treatment facilities, such as septic tanks and latrines, makes the resulting fecal sludge (septage) a valuable soil ameliorant. The dried and composted material can for example be pelletized or blended with particular nutrients to meet farmers' needs, as shown for example in South Africa and Ghana (Harrison and Wilson, 2012; Nikiema *et al.*, 2012). Fecal sludge (FS) is an abundant and valuable resource, similar to other organic manure such as farmyard manure which is used as a source of fuel and fertilizer. However, with diarrhea being the second top ranking contributor to the global burden of diseases

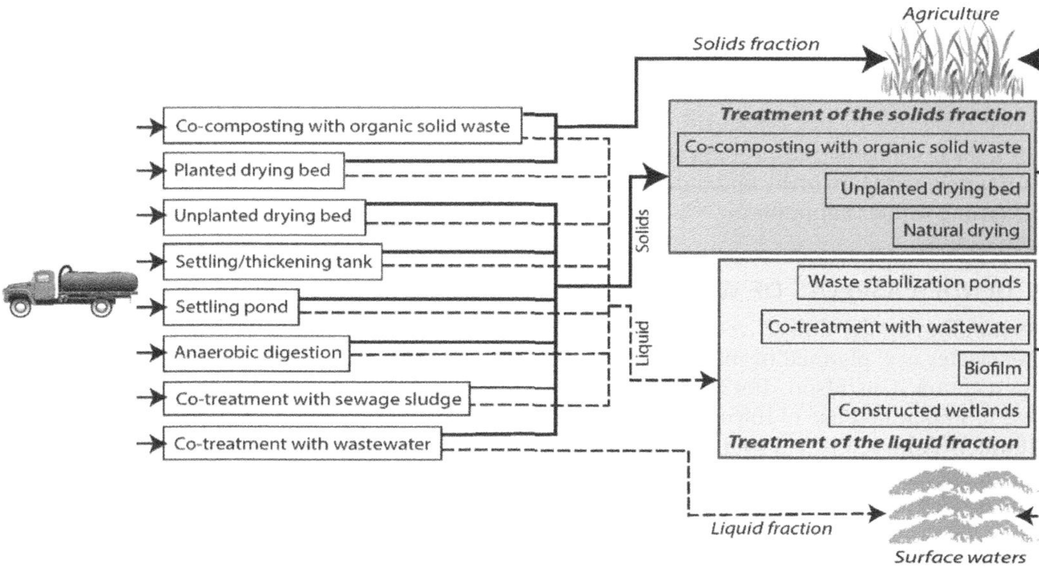

Figure 15.5 Appropriate FS treatment options in developing countries with option for nutrient and water recovery (Source: Strauss (2006), modified).

and 88% of cases of diarrhea worldwide attributed to fecal matter contamination, the management and possible reuse of human waste containing fecal matter receives priority attention across the water supply, sanitation, food and health sectors (WHO, 2006). A controlled resource recovery approach can however reduce the negative impact of fecal matter on the environment and have a positive public health impact, by turning a potential threat into an asset for the production of nutritious food. For the reuse of nutrients captured in on-site sanitation systems, a number of nutrient recovery options exists (Figure 15.5) as described, for example, by Tilley *et al.* (2008).

In areas where fertilizer production is limited and/or their prices are high, smallholder farmers may resort to the use of raw FS for fodder, tree (crop) plantation or cereal production. For example, farmers in West Africa and South India are redirecting cesspit truck operators to their fields to obtain the nutrient rich manure. Research has shown that farmers have concerns with low product nutrient content, skin diseases from product use, labor requirements and general mistrust of information on product quality, and these may significantly affect the demand for compost products (Cofie *et al.*, 2006; Murray *et al.*, 2011; Viaene *et al.*, 2015). In many situations, farmers' willingness-to-pay is either too low or farmers prefer existing substitutes for soil inputs such as cow dung, poultry manure or even dried fecal sludge. A major drawback in mitigating farmers' reservation about potential health risks is the lack of regulations and negative perceptions of the authorities themselves concerning the use of fecal matter in food production. Crop restrictions and sufficient solar drying can minimize these risks (Keraita and Drechsel, 2015; Seidu *et al.*, 2008), even where no other regulations govern the process like, for example, in the septage program of Michigan's Department of Environmental Quality, USA, where septage application in agriculture is a well-regulated option.

Formalized use of processed fecal sludge for agricultural production is still novel (Otoo and Drechsel, 2018). Few opportunities to increase the accessibility and usability of value-added FS products in agriculture are emerging, with cases identified in Nigeria, Ghana and South Africa. Different entities have adopted innovative value-addition techniques such as fortification and enrichment of FS with nutrients to boost its safety quality and fertilizer value. For the latter, a nutrient source such as 'natural' like rock-phosphate or struvite/urine or industrial fertilizer can be added as an enrichment. Another option is pelletizing composted FS, resulting in an easy to handle, safe, high-value product. These commodity-value based approaches have shown significant potential in increasing farmers' demand as has product certification (Danso *et al.*, 2017; Otoo *et al.*, 2015). Farmers' incentive to pay for certification is indicative of their need of safety assurance from the use of a fecal sludge-based product. Similar to wastewater-based aquaculture, quality certification measures using a trustworthy and independent party to test the product for its quality will be essential in increasing farmer adoption.

15.6 GENDER ASPECTS OF WASTEWATER USE

Understanding gender differences is important for examining the social aspects and public responses to wastewater use, planned or unplanned, due to common gender differences in roles, risk exposure and perception (Gustafsod, 1998). Very limited research has been conducted on gender dimensions related to unplanned use of untreated wastewater in comparison to that of planned reuse of treated wastewater for agriculture. From that perspective, this section solely focuses on the latter.

On the general issue of accepting water reuse, most gender-sensitive studies either found no significant association between gender and acceptance of wastewater use or concluded a higher acceptance of wastewater use among men than women – for example, in USA, Australia, Greece, Iran, and Bahrain (Baghapour *et al.*, 2017; Bakopoulou *et al.*, 2010; Gibson and Burton, 2014; Haddad *et al.*, 2009; Madany *et al.*, 1992; Price *et al.*, 2012; Rock *et al.*, 2012; Rozin *et al.*, 2015; Tsagarakis *et al.*, 2007), while also the opposite has been observed, like in Italy (Saliba *et al.*, 2018) or Tanzania (Mayilla *et al.*, 2017). Around Addis Ababa, Ethiopia, positive perceptions towards water reuse were higher among male farmers, while female farmers appeared more aware towards problems of eating unwashed vegetables (Woldetsadik *et al.*, 2018). However, as the reasons behind observed patterns can vary, each situation ideally requires both qualitative and quantitative analysis (Gustafsod 1998).

In many cultures, women carry the main responsibility for hygiene and health, also vis-à-vis greywater or wastewater use as reported for example from Jordan (Boufaroua *et al.*, 2013), Vietnam (Knudsen *et al.*, 2008) and Tunisia (Mahjoub, 2013). The strong connection between water use at a household level and women offers a significant potential for innovative training approaches to improve the social acceptance of safe water reuse as recently demonstrated in Jordan (Boufaroua *et al.*, 2013), as well as household livelihoods (Box 15.8). In Ghana, where households already use greywater to grow indigenous crops in backyards like sugarcane, banana/plantain, taro, sweet/wild basil, and dandelion, the major plant benefits identified were food (84% of respondents) and medicine (62% of respondents), both supporting household roles expected from women (Dwumfour-Asare *et al.*, 2018).

As shown in the example of Box 15.8, water reuse can also contribute to gender equality and shrink the income gap amid socially mediated gender roles. Perceived advantages and greater acceptance of greywater use by women than men were, for instance, reported in South Africa and India (Bakare *et al.*, 2016; Ravishankar *et al.*, 2018). Another example of gender specific responses is the acceptance and use of protective clothing to avoid health risks. In Vietnam, women were observed wearing protective gloves and boots with more consistency than men. The differences were attributed to the gender specific work separation on the farm, with men walking around the farms much more than women, where protective clothing constrained men's movements (Knudsen *et al.*, 2008). A lesson from gender analysis vis-à-vis water reuse is the need for a systematic way of assessing the different impacts of actions and results on both genders. This form of analysis asks the 'who' questions – Who does what? Who has access and control over water? Who makes the decisions? Who benefits from (better) reuse implementation? Who is burdened? Many of these are questions relating to power dynamics, social roles and responsibilities and relationships (Arafa *et al.*, 2007).

Box 15.8 Greywater reuse supporting women

In Jordan, a pilot project implemented by IDRC in partnership with the Inter-Islamic Network on Water Resources Development and Management (INWRDAM) allowed the poor in Tufileh to reuse household greywater in home gardens. The women of the community used small revolving loans to implement simple greywater recovery systems and set-up gardens. Although women were not very prominent publicly, they are actually the ones who decided what went into the system. It is they who gauge the quantity of grease or soap that enters the system and women are the 'water managers' at the household level. Their active involvement was critical for ensuring that the system worked properly. The project allowed the community to offset food purchases and generate income by selling surplus production, earning an average of 10% of their present income. Had the households used municipal sources for this supplemental irrigation, on average, they would have used 15% more water and had 27% higher water bills. Moreover, the project helped community members gain valuable gardening, irrigation, and food preservation skills. Women reported feeling more independent and proud because of the income they generated, the skills that they gained and their enhanced ability to feed their families. It is worth mentioning here that an environmental impact assessment demonstrated that the quality of the treated greywater was adequate, and the negative impacts on soil and crops were negligible.

Source: Arafa *et al.* (2007).

15.7 CHAPTER SUMMARY

While water scarcity and material value support a discussion about reuse, decisive factors may be the level of direct exposure, availability of alternative water and material sources, education levels and perceptions of health risks, extent of public participation and buy-in, religious concerns, gender specific roles and responsibilities, and the means and messages used in knowledge sharing and communication. Overall, the acceptance of (safe) wastewater and residuals use varies with and within the development stage of the society, and can be a very dynamic process which makes social feasibility studies, close participation of target groups, and trust building essential components of successful reuse programs, independently of the technical advances and geographies.

The discussion on resource recovery from wastewater has particular momentum in water scarce regions and focuses on water reuse for both potable and non-potable purposes, with more emphasis on safety than financial aspects. The discussion on nutrient recovery from wastewater is however one step behind and so far more determined by regulatory pressure or technical opportunities for cost savings than any actual market demand for recovering nutrients and farmers' and food consumers' perceptions. With few exceptions the demand is more theoretical, embedded in the call for a more circular economy, ideally turning conventional wastewater treatment plants into resource recovery centers (Wallis-Lage, 2013), alongside treating wastewater to serve the imperative of protecting public health and ecosystems.

The documented experience on consumers' perceptions related to the social and cultural dimension of marketing water reuse and related recovered resources analyzed common factors challenging the acceptance of reuse, as well as of behavior change towards increased safety where reuse is already taking place, but without enforced guidelines. Decision making appears to be influenced by context-specific factors, for example consumers' knowledge about reuse, the level of risk awareness, direct or indirect contact or exposure, the availability of alternative water options, financial benefits, buy-in through participatory planning of reuse schemes and trust building. Responses can also differ by gender and may be influenced by religious beliefs. There are geographic differences in terms of education and risk awareness, but these do not guarantee that reuse is easier promoted and/or accepted in developed countries.

For potable reuse, individual and group perceptions related to risks and disgust and the possibility of alternatives appear to be the main decisive criteria for potential users of reclaimed water. Farmers' main arguments for or against changing their water source or behavior was usually related to economic arguments, like (domestic, export) market perceptions and regulations affecting sales and revenues or cost and benefits in general (saving on fertilizer, extra harvest, reliability of supply, etc.). As a result, wastewater may be rejected in one situation, favored in another, or only be accepted if not treated in a third (to preserve nutrient while keeping the salinity level low). Even where own health impacts were experienced by farmers, these were often perceived as an acceptable challenge, balanced by financial gains. A key lesson seems to be that acceptance of reclaimed water is hard to predict and requires a careful socio-economic analysis. In fact, technical perfection is an important input, but may not be the final decision-making factor as the example of Singapore showed. This implies that the introduction of water reuse, of behavior change towards increased safety, requires a strong integration of social science research and related strategic partners and stakeholders in the strongholds of engineering and epidemiology to address possible adoption barriers and opportunities. These concern in particular:

- public perceptions and group dynamics which can easily jeopardize any reuse project;
- educational levels which may be too low to understand risks and related responsibility; which can 'favor' reuse but may also prevent it based on beliefs and myths;
- the lack of economic or social incentives for changing practices.

Compared to the significant body of references on the introduction of water reuse and its social challenges, there is comparatively little information on strategies, achievements or failures along the trajectory from unplanned to planned reuse, or informal to formal, like in Peru, Mexico or several MENA countries where both systems co-exist. The reason for this lack of information may be that there are only a few countries, like Tunisia, which began combining wastewater treatment and reuse programs early in the 1980s, resulting in significant treatment for reuse efforts supported by well-enforced regulations (Bahri, 2009; ONAS, 2012). Most other success stories derive from well-resourced developed countries with own reuse regulations. These regulations are however seldom transferable to developing countries due to differences in institutional and technical capacities.

Locally adapted and applied regulations and reuse guidelines are essential to support reuse programs in the long term. The global WHO guidelines for potable water as well as water reuse in agriculture and aquaculture provide a flexible framework for local adaptation and are particularly strong in supporting the transition from informal to formal reuse even where treatment plants are not yet able to safeguard public health. They are building on the adoption of multiple barriers (safety options) along the contamination pathway from farm to fork, similar to the well accepted Hazard Analysis and Critical Control Points (HACCP) concept of the food industry. However, the guidelines fall short in explaining how the behavior change towards their adoption could be facilitated and sustained.

The WHO 2006 guidelines face limited acceptance, probably due to their loss of simplicity by moving away from irrigation water quality thresholds to more flexible, human exposure-based targets based on local risk assessments. The more recently published Sanitation Safety Planning Manual (WHO, 2015) and Guidelines on Sanitation and Health (WHO, 2018) are addressing this challenge to increase the adoption and sustainability of safe wastewater use.

15.8 EXERCISES

Exercise 15.1: How should a campaign to introduce water reuse at household level in the MENA region be designed?

Exercise 15.2: Could you describe why it is so challenging to change behavior of, for example, African farmers who use untreated wastewater due to the lack of safer alternative water sources?

Exercise 15.3: Select a country where farmers use wastewater for agricultural production. Determine which of these measures (economic or social (marketing) incentives, such as access to credit, labelling/ branding, dedicated marketing chains, tax exemptions, and institutional support, like the provision of extension services, awards, or tenure security) would be the best in supporting behavior change and justify why?

Exercise 15.4: Even with existing regulations, can you explain the reluctance of agricultural producers in using treated wastewater for irrigation?

Exercise 15.5: What regulations need to be changed to allow/incentivize biosolids land application?

Exercise 15.6: Where risk awareness is low and not easy to develop, how best can farmers be motivated to trigger adoption of risk mitigation measures?

Exercise 15.7: Select a country and assess the potential for wastewater-based aquaculture. Determine socio-economic factors that can support and/or limit consumer demand.

Exercise 15.8: Assess the current level of formal and informal use of septage in agriculture in your country. Evaluate the factors supporting and limiting its formal use.

15.9 DISCUSSION QUESTIONS

Question 15.1: Why is wastewater use in agriculture much more common than official statistics show, and what are the related health and behavior change challenges? Consider differences between treatment capacities in developed and developing countries; consider the time frame it may take to bring treatment capacities to a similar level. What should be done in the meantime for farmers facing polluted water without choice? Consider educational levels and what it takes to change behavior. Discuss options around regulations, incentives, social marketing, education and awareness creation.

Question 15.2: Which financial considerations can influence the acceptance of water reuse by farmers, as well as of common safety interventions (like drip irrigation)? Consider chemical water quality differences (nutrients, salinity) and how treatment, in particular pond-based systems, may influence those. Consider the profit from producing cash crops (like exotic vegetables) in the dry season. Consider market expectations and regulations. How do farmers see their own exposure to polluted water assuming they are aware of it?

Question 15.3: How can we promote water reuse for potable purposes? Although water savings and reuse are very important in drier climates, many consumers do not accept water reuse despite excellent technical progress and high educational levels, like in South Africa and Australia, or at best for purposes which avoid any direct contact, like landscape irrigation. What are the reasons? What can be done to address these obstacles and promote a broader and inclusive reuse agenda?

Question 15.4: What strategies can private entities engaged in the wastewater reuse market adopt or implement to improve market demand for the recovered resources? Select 2–3 recovered resources and develop and justify strategies to be implemented.

Question 15.5: What protection or assistance do farmers need to protect themselves from liability and assist with cleanup if contamination issues are found from a wastewater treatment plant's application?

ACKNOWLEDGEMENTS

This work was supported by the CGIAR Research Program on Water, Land and Ecosystems (WLE). The findings and conclusions contained within this report are those of the authors and do not necessarily reflect positions or policies of the project funding agencies.

REFERENCES

Abu-Madi M., Al-Sa'ed R., Braadbaart O. and Alaerts G. (2008). Perceptions of farmers and public towards irrigation with reclaimed wastewater in Jordan and Tunisia. *Arab Water Council Journal*, **1**(II), 18–32. http://www.academia.edu/1152959/Public_perceptions_towards_wastewater_reuse_in_Jordan_and_Tunisia (accessed 22 January 2022)

Acheampong P. P., Braimah H., Ankomah-Danso A. and Mochiah M. B. (2012). Consumers behaviours and attitudes towards safe vegetables production in Ghana: A case study of the cities of Kumasi and Cape Coast. *Science Journal of Agricultural Research and Management*, **2012**, 1–11. Article ID sjarm-109, 11 Pages, http://doi.org/10.7237/sjarm/109

Alhumoud J. M. and Madzikanda D. (2010). Public perceptions on water reuse options: The case of Sulaibiya Wastewater Treatment Plant in Kuwait. International Business & Economics Research Journal, **9**, 141–158.

Al-Momani S. S. (2011). State of the Wastewater Management in the Arab Countries. Arab Water Council, Dubai, UAE.

Amoah P., Drechsel P., Abaidoo R. C. and Henseler M. (2007). Irrigated urban vegetable production in Ghana: Microbiological contamination in farms and markets and associated consumer risk groups. *Journal of Water and Health*, **5**(3), 455–466.

Amoah P., Schuetz T., Kranjac-Berisavjevic G., Manning-Thomas N. and Drechsel P. (2009). From world cafés to road shows: Using a mix of knowledge sharing approaches to improve wastewater use in urban agriculture. *Knowledge Management for Development Journal*, **5**(3), 246–262.

Amoah P., Keraita B., Akple M., Drechsel P., Abaidoo R. C. and Konradsen F. (2011). Low Cost Options for Health Risk Reduction Where Crops are Irrigated With Polluted Water in West Africa. IWMI Research Report Series 141, Colombo. http://www.iwmi.cgiar.org/Publications/IWMI_Research_Reports/PDF/PUB141/RR141.pdf (accessed 22 January 2022).

Arafa D., El-Fattal L. and Laamrani H. (2007). Gender & WDM in the Middle East & North Africa. Water Demand Management (WDM) Research Series, IDRC, Canada.

Baghapour M. A., Shooshtarian M. R. and Djahed B. (2017). A survey of attitudes and acceptance of wastewater reuse in Iran: Shiraz City as a case study. *Journal of Water Reuse and Desalination*, **7**, 511–519.

Bahri A. (2009). Managing the Other Side of the Water Cycle: Making Wastewater an Asset. TEC Background Paper 13. Global Water Partnership, Sweden.

Bakare B. F., Mtsweni S. and Rathilal S. (2016). A pilot study into public attitudes and perceptions towards greywater reuse in a low cost housing development in Durban, South Africa. *Journal of Water Reuse and Desalination*, **6**, 345–354.

Bakopoulou S., Polyzos S. and Kungolos A. (2010). Investigation of farmers' willingness to pay for using recycled water for irrigation in thessaly region, Greece. *Desalination*, **250**, 329–334.

Ben Brahim H. and Duckstein L. (2011). Descriptive methods and compromise programming for promoting agricultural reuse of treated wastewater. In: Computational Methods for Agricultural Research: Advances and Applications, H. A. do Prado, A. J. Barreto Luiz and H. C. Filho (eds.), IGI Global, Hershey, Pennsylvania, USA, pp. 355–388. Available at: www.igi-global.com/book/computational-methods-agricultural-research/41780 (accessed 22 January 2022).

Binns J. A., Maconachie R. A. and Tanko A. I. (2003). Water, Land and Health in Urban and Peri-urban Food Production: The Case of Kano, Nigeria. Available at: www.cityfarmer.org/WaterLandHealthKano.rtf (accessed 28 December 2013).

Boufaroua M., Albalawneh A. and Oweis T. (2013). Assessing the efficiency of grey-water reuse at household level and its suitability for sustainable rural and human development. *British Journal of Applied Science and Technology*, **3**(4), 962–972.

Bunting S. (2007). Confronting the realities of wastewater aquaculture in peri-urban Kolkata with bio economic modeling. *Water Research*, **41**, 499–505.

Buyukkamaci N. and Alkan H. S. (2013). Public acceptance potential for reuse applications in Turkey. *Resources, Conservation and Recycling*, **80**, 32–35.

Chaudhuri N. (2008). Using participatory education and action research for health risk reduction amongst farmers in Dakar, Senegal. In: Agriculture in Urban Planning: Generating Livelihoods and Food Security, M. Redwood (ed.), Earthscan/IDRC, UK, 181–199. Available at: http://kuris.ku.ac.ke/schools/human_sciences/images/stories/2016/IDL-36354.pdf#page=200 (accessed 22 January 2022).

Chen W., Bai Y., Zhang W., Lyu S. and Jiao W. (2015). Perceptions of different stakeholders on reclaimed water reuse: The case of Beijing, China. *Sustainability*, **7**, 9696–9710.

Choukr-Allah R. (2013). Wastewater Treatment and Reuse in Morocco: Situation and Perspectives. Agadir, Morocco. Available at: https://www.semanticscholar.org/paper/Wastewater-treatment-and-reuse-in-Morocco-%3A-and-Choukr-allah/7cb383a04cd695c314657ab399203fc3ac5b3ff6 (accessed 22 January 2022).

Cofie O., Agbottah S., Strauss M., Esseku H., Montangero A., Awuah E. and Kone D. (2006). Solid-liquid separation of faecal sludge using drying beds in Ghana: Implications for nutrient recycling in urban agriculture. *Water Resources*, **40**, 75–82.

Crocker J., Geremew A., Atalie F., Yetie M. and Bartram J. (2016). Teachers and sanitation promotion: An assessment of community-led total sanitation in Ethiopia. *Environmental Science & Technology*, **50**, 6517–6525.

Crook J., Ammermman D. K., Okun D. A. and Matthews R. L. (1992). *Guidelines for Water Reuse*. Camp Dresser and McKee, Inc., Cambridge, Massachusetts.

Danso G. K., Otoo M., Duy Linh N. and Madurangi G. (2017). Households' willingness-to-pay for fish product attributes and implications for market feasibility of wastewater-based aquaculture businesses in Hanoi, Vietnam. *Resources*, **6**(2), 1–22.

Dare A. and Mohtar R. H. (2018). Farmer perceptions regarding irrigation with treated wastewater in the West Bank, Tunisia, and Qatar. *Water International*, **43**, 460–471.

Dare A. E., Mohtar R. H., Jafvert C. T., Shomar B., Engel B., Boukchina R. and Rabi A. (2017). Opportunities and challenges for treated wastewater reuse in the west bank, Tunisia, and Qatar. *Transactions of the ASABE*, **60**, 1563–1574.

Dishman M., Sherrard J. H. and Rebhum M. (1989). Gaining support for direct potable water reuse. *Journal of Professional Issues in Engineering*, **115**(2), 154–161.

Dolnicar S. and Saunders C. (2006). Recycled water for consumer markets – A marketing research review and agenda. *Desalination*, **187**(1–3), 203–214.

Dolnicar S. and Schafer A. I. (2009). Desalinated versus recycled water: Public perceptions and profiles of the accepters. *Journal of Environmental Management*, **90**, 888–900.

Drechsel P. and Hanjra M. A. (2018). Wastewater for agriculture, forestry and aquaculture: An overview of presented business cases and models. In: Resource Recovery From Waste: Business Models for Energy, Nutrients and Water Reuse in Low- and Middle-Income Countries, M. Otoo and P. Drechsel (eds.), Routledge, USA, pp. 549–554.

Drechsel P. and Karg H. (2013). Motivating behaviour change for safe wastewater irrigation in urban and peri-urban Ghana. *Sustainable Sanitation Practice*, **16**, 10–20. http://www.ecosan.at/ssp/issue-16-behaviour-change/SSP-16_Jul2013_10-20.pdf/view (accessed 22 January 2022).

Drechsel, P. and Keraita, B. (eds.), (2014). Irrigated Urban Vegetable Production in Ghana: Characteristics, Benefits and Risk Mitigation, 2nd edn. International Water Management Institute (IWMI), Colombo, Sri Lanka, 247 pp.

Drechsel P., Graefe S., Sonou M. and Cofie O. (2006). Informal Irrigation in West Africa: An Overview. IWMI Research Report No. 102. International Water Management Institute, Colombo.

Drechsel P., Hope L. and Cofie O. (2013). Gender mainstreaming: Who wins? Gender and irrigated urban vegetable production in West Africa. *Journal of Gender and Water (wH2O)*, **2**, 15–17.

Drechsel P., Mahjoub O. and Keraita B. (2015). Social and cultural dimensions in wastewater use. In: Wastewater: Economic Asset in an Urbanizing World, P. Drechsel, M. Qadir and D. Wichelns (eds.), Springer, Dordrecht, The Netherlands, pp. 75–92.

Duenas J. F., Alonso J. R., Rey A. F. and Ferrer A. S. (2003). Characterisation of phosphorous forms in wastewater treatment plants. *Journal of Hazardous Materials*, **97**, 1–3.

Dwumfour-Asare B., Nyarko K. B., Awuah E., Essandoh H. M. K., Gyan B. A. and Ofori-Addo H. (2018). Indigenous plants for informal greywater treatment and reuse by some households in Ghana. *Journal of Water Reuse and Desalination*, **8**(4), 553–565.

Easa M., El-S Shereif M. M., Shaaban A. I. and Mancey K. H. (1995). Public health implications of wastewater reuse for fish production. *Water Science and Technology*, **32**, 145–152. http://doi.org/10.1016/0273-1223(96)00128-X.

Edward P. (2008). Wastewater and Excreta Use in Aquaculture, vol. **3**. Asian Institute of Technology, Bangkok, Thailand.

Edwards P. and Pullin R. S. V. (1990). Wastewater-fed aquaculture. Proceedings of the International Seminar on Wastewater Reclamation and Reuse for Aquaculture, Calcutta, India, 6–9 December 1988. Environmental Sanitation Information Center, Asian Institute of Technology, Bangkok, Thailand.

EMWATER (2004). Prospects of efficient wastewater management and water reuse in Palestine. Country study. Prepared within the framework of the EM Water Project, "Efficient management of wastewater, its treatment and reuse in the Mediterranean countries". Water Studies Institute – Birzeit University, West Bank, Palestine.

Available at: https://www.pseau.org/outils/ouvrages/enea_meda_water_iws_inwent_prospects_of_efficient_wastewater_management_and_water_reuse_in_palestine_2005.pdf (accessed 22 January 2022).

Ensink J., van der Hoek W., Matsuno Y., Munir S. and Islam M. R. (2004). Use of Untreated Wastewater in Peri-Urban Agriculture in Pakistan: Risks and Opportunities. IWMI Research Report 64. International Water Management Institute (IWMI), Colombo, Sri Lanka.

Erni M., Drechsel P., Bader H-P., Scheidegger R., Zurbruegg C. and Kipfer R. (2010). Bad for the environment, good for the farmer? Urban sanitation and nutrient flows. *Irrigation and Drainage Systems, Special Issue on Wastewater Reuse*, **24**(1–2), 113–125. http://doi.org/10.1007/s10795-009-9083-9

Farooq S. and Ansari Z. I. (1983). Water reuse in Muslim countries – an Islamic perspective. *Environmental Management*, **7**(2), 119–123.

Faruqui N. I., Biswas A. K. and Bino M. J. (eds.), (2001). Water Management in Islam. International Development Research Centre/United Nations University Press, Ottawa, Canada.

Fattal B., Doan A. and Tchorsh Y. (1992). Rates of experimental microbiological contamination of fish exposed to polluted water. *Water Research*, **26**, 1621–1627.

Fawell J., Fewtrell L., Hydes O., Watkins J. and Wyn-Jones P. (2005). A Protocol for Developing Water Reuse Criteria with Reference to Drinking Water Supplies. UKWIR Report Ref No. 05/WR/29/1. UK Water Industry Research Limited, London, UK.

Fielding K. S., Dolnicar S. and Schultz T. (2019). Public acceptance of recycled water. *International Journal of Water Resources Development*, **35**(4), 551–586.

Fonner R. and Sylvia G. (2015). Willingness to pay for multiple seafood labels in a Niche Market. *Marine Resource Economics*, **30**, 51–70.

Friedler E. and Lahav O. (2006). Centralised urban wastewater reuse: What is the public attitude? *Water Science & Technology*, **54**, 423–430.

Friedler E., Lahav O., Jizhaki H. and Lahav T. (2006). Study of urban population attitudes towards various wastewater reuse options: Israel as a case study. *Journal of Environmental Management*, **81**, 360–370.

Gaterell M. R., Gay R., Wilson R., Gochin R. J. and Lester J. N. (2000). An economic and environmental evaluation of the opportunities for substituting phosphorus recovered from wastewater treatment works in existing UK fertiliser markets. *Environmental Technology*, **21**, 1067–1084.

Gbewonyo K. (2007). Wastewater Irrigation and the Farmer: Investigating the Relation between Irrigation Water Source, Farming Practices, and Farmer Health in Accra, Ghana. Unpublished thesis, Harvard College, Cambridge, Massachusetts.

Gerstl S. (2001). The Economic Costs and Impact of Home Gardening in Ouagadougou, Burkina Faso. PhD dissertation, University of Basel, Switzerland.

Gibson F. L. and Burton M. (2014). Salt or sludge? Exploring preferences for potable water sources. *Environmental and Resource Economics*, **57**, 453–476.

GTZ (2009). The Chinese Waste Water Treatment Sector. CDM Perspectives in China. Opportunities for German Know-How and CDM Application. GTZ, Beijing Office, Beijing, China.

GWI (Global Water Intelligence) (2010). Municipal Water Reuse Markets 2010. Media Analytics Ltd., Oxford, UK.

Gustafsod P. E. (1998). Gender differences in risk perception: Theoretical and methodological perspectives. *Risk Analysis*, **18**, 805–811.

Haddad B., Rozin P., Nemeroff C. and Slovic P. (2009). The Psychology of Water Reclamation and Reuse: Survey Findings and Research Road Map. The WaterReuse Foundation, Alexandria, VA.

Hamilton A. J., Stagnitti F., Xiong X., Kreidl S. L., Benke K. K. and Maher P. (2007). Wastewater irrigation the state of play. *Vadose Zone*, **6**(4), 823–840.

Hanjra M. A., Drechsel P., Mateo-Sagasta J., Otoo M. and Hernandez-Sancho F. (2015). Assessing the finance and economics of resource recovery and reuse solutions across scales. In: Wastewater: Economic Asset in an Urbanizing World, P. Drechsel, M. Qadir and D. Wichelns (eds.), Springer, London, UK, 113–136.

Harrison J. and Wilson D. (2012). Towards sustainable pit latrine management through LaDePa. *Sustainable Sanitation Practice*, **13**, 25–32.

Hartley T. W. (2006). Public perception and participation in water reuse. *Desalination*, **187**, 115–126.

Helmer, R. and Hespanhol, I. (eds.). (1997). Water Pollution Control – A Guide to the Use of Water Quality Management Principles. Published by E and FN Spon on behalf of the United Nations Environment Programme, Water Supply and Sanitation Collaborative Council and the World Health Organization, London, UK, 510.

Higgins J., Warnken J., Sherman P. P. and Teasdale P. R. (2002). Surveys of users and providers of recycled water: Quality concerns and directions for applied research. *Water Research*, **36**, 5045–5056.

Hope L., Keraita B. and Akple M. S. K. (2008). Washing of vegetables grown in urban farms in irrigation water in Kumasi, Ghana. *Urban Agriculture Magazine*, **20**, 29–30.

Hurlimann A. C. (2008). Community Attitudes to Recycled Water Use: An Urban Australian Case Study – Part 2 Research Report 56. The Cooperative Research Centre for Water Quality and Treatment, Australia.

Hurlimann A. C. (2009). Water supply in regional Victoria, Australia: A review of the water cartage industry and willingness to pay for recycled water. *Resources, Conservation and Recycling*, **53**, 262–268.

Hurlimann A. and Dolnicar S. (2010). When public opposition defeats alternative water projects – The case of Toowoomba Australia. *Water Research*, **44**, 287–297.

Hurlimann A. and Dolnicar S. (2016). Public acceptance and perceptions of alternative water sources: A comparative study in nine locations. *International Journal of Water Resources Development*, **32**, 650–673.

Hurlimann A. and McKay J. M. (2006). What attributes of recycled water make it fit for residential purposes? The Mawson Lakes experience. *Desalination*, **187**, 167–177.

Jimenez B. (2005). Treatment technology and standards for agricultural wastewater reuse: A case study in Mexico. *Irrigation and Drainage Journal*, **54**(1), 23–35.

Keraita B. and Drechsel P. (2015). Consumer Perceptions of Fruit and Vegetable Quality: Certification and Other Options for Safeguarding Public Health in West Africa. (IWMI Working Paper 164). International Water Management Institute (IWMI), Colombo, Sri Lanka, 32pp. http://doi.org/10.5337/2015.215

Keraita B., Drechsel P. and Konradsen F. (2008). Perceptions of farmers on health risks and risk reduction measures in wastewater-irrigated urban vegetable farming in Ghana. *Journal of Risk Research*, **11**(8), 1047–1061.

Keraita B., Drechsel P., Seidu R., Amerasinghe P., Cofie O. and Konradsen F. (2010). Harnessing farmers' knowledge and perceptions for health risk reduction in wastewater irrigated agriculture. In: Wastewater Irrigation and Health: Assessing and Mitigation Risks in Low-Income Countries, P. Drechsel, C. A. Scott, L. Raschid-Sally, M. Redwood and A. Bahri (eds.), Earthscan-IDRC-IWMI, UK, pp. 337–354.

Keremane G. B. (2007). Urban Wastewater Reuse for Agriculture: Governance Paradigms and Institutional Arrangements in Australia and India. PhD thesis, University of South Australia. https://www.irrigationaustralia.com.au/documents/item/271 (accessed 22 January 2022).

Kilelu C. (2004). Wastewater Irrigation and Farmer's Perception of Health Risks and Institutional Perspectives: A Case Study from Maili Saba, Nairobi. CFP Report 38. IDRC, Ottawa.

Knudsen L. G., Phuc P. D., Hiep N. T., Samuelsen H., Jensen P. K., Dalsgaard A., Raschid-Sally L. and Konradsen F. (2008). The fear of awful smell: Risk perceptions among farmers in Vietnam using waster and huma excreta in agriculture. *Southeast Asian Journal of Tropical Medicine and Public Health*, **39**(2), 341–352.

Krogmann U., Gibson V. and Chess C. (2001). Land application of sewage sludge: Perceptions of New Jersey vegetable farmers. *Waste Management and Research*, **19**, 115–125.

Lagerkvist C., Hess S., Okello J. and Karanja N. (2013). Consumer willingness to pay for safer vegetables in urban markets of a developing country: The case of Kale in Nairobi, Kenya. *Journal of Development Studies*, **49**, 365–382.

Lahnsteiner J., Du Pisani P., Menge J. and Esterhuizen J. (2012). More than 40 years of direct potable reuse experience in Windhoek. In: Milestones in Water Reuse. The Best Success Stories, V. Lazarova, T. Asano, A. Bahri and J. Anderson (eds.), IWA Publishing, London, pp. 351–364.

Lan N. T., Dalsgaard A., Cam P. D. and Mara D. (2007). Microbiological quality of fish grown in wastewater-raised and non-wastewater fed fishponds in Hanoi, Vietnam: Influence of hygiene practices in local retail markets. *Journal of Water Health*, **5**, 209–218.

Lazarova V., Takashi A. T., Akiça Bahri A. and Anderson J. (2013). Milestones in Water Reuse: The Best Success Stories. IWA Publishing, London, UK.

Lim M-H. and Seah H. (2012). NEWater: A key element of Singapore's water sustainability. In: Milestones in Water Reuse. The Best Success Stories, V. Lazarova, T. Asano, A Bahri and J. Anderson (eds.), IWA Publishing, London, pp. 53–62.

Macpherson L. and Slovic P. (2012). Stigma and fear – changing mental models about reuse. *WE&T Magazine*, **20**(7).

Madany I. M., Al-Shiryan A., Lori I. and Al-Khalifa H. (1992). Public awareness and attitudes toward various uses of renovated water. *Environment International*, **18**, 489–495.

Mahjoub O. (2013). Awareness-raising workshops for farmers and rural women on the risks associated with the reuse of wastewater in agriculture: Application to the region of d'Oued Souhil, Nabeul, Tunisia. UN-Water. Proceedings of the Safe Use of Wastewater in Agriculture, International wrap-up event, Teheran, Iran.

Mancy K. H., Fattal B. and Kelada S. (2000). Cultural implications of wastewater reuse in fish farming in the Middle East. *Water Science and Technology*, **42**, 235–239.

Marks J. S. (2004). Advancing community acceptance of reclaimed water. *Water, Journal of the Australian Water Association*, **31**(5), 46–51.

Marks J. S., Martin B. and Zadoroznyj M. (2006). Acceptance of water recycling in Australia: National baseline data. *Water, Journal of the Australian Water Association*, **33**(2), 151–157.

Martinsen C. (2008). Social marketing in sanitation. More than selling toilets. *Stockholm Water Front*, **1**, 14–16.

Mayilla W., Keraita B., Ngowi H., Konradsen F. and Magayane F. (2017). Perceptions of using low-quality irrigation water in vegetable production in Morogoro, Tanzania. *Environment, Development and Sustainability*, **19**, 165–183.

McKay J. M. and Hurlimann A. (2003). Attitudes to reclaimed water for domestic use: Part 1. *Water*, **30**(5), 45–49.

Menegaki A. N., Mellon R. C., Vrentzou A., Koumakis G. and Tsagarakis K. P. (2009). What's in a name: Framing treated wastewater as recycled water increases willingness to use and willingness to pay. *Journal of Economic Psychology*, **30**, 285–292.

Mojid M. A., Wyseure C. L., Biswas S. K. and Hossain A. B. M. Z. (2010). Farmers' perceptions and knowledge in using wastewater for irrigation at twelve peri-urban areas and two sugar mill areas in Bangladesh. *Agricultural Water Management*, **98**, 79–86.

Molinos-Senante M., Hernandez-Sancho F. and Sala-Garrido R. (2010). Economic feasibility study for wastewater treatment: A cost–benefit analysis. *Science of the Total Environment*, **408**, 4396–4402.

Morf L. and Koch M. (2009). Synthesis Report for Interested Specialists: Zurich Sewage Sludge Disposal with Special Consideration of the Resource Aspects. Baudirection Kanton, Zurich.

Murphy S. (2018). Biosolids: Sydney Sewage Turned Organic Fertiliser Improves NSW Drought-hit Soil. Available at: https://www.abc.net.au/news/2018-07-07/sydney-waste-turned-biosolids-fertiliser-helps-nsw-drought-soil/9931664 (accessed 22 January 2022).

Murray A., Cofie O. and Drechsel P. (2011). Efficiency indicators for waste-based business models: Fostering private-sector participation in wastewater and faecal sludge management. *Water International*, **36**, 505–521.

Ngigi M. W., Okello J. J., Lagerkvist C. L., Karanja N. K. and Mburu J. (2011). Urban consumers' willingness to pay for quality of leafy vegetables along the value chain: The case of Nairobi Kale Consumers, Kenya. *International Journal of Business and Social Science*, **2**(7), 208–216.

Nguyen Q. and Leung P. (2009). Do fishermen have different attitudes towards risk? An application of prospect theory to the study of Vietnamese fishermen. *Journal of Agricultural and Resource Economics*, **34**, 518–538.

Niang S. (2002). Risk management in the recovery of wastewater in urban agriculture. In: Advances in Crop-Livestock Integration in West African Cities, O. O. Akinbamijo, S. T. Fall and O. B. Smith (eds.), International Trypanotolerance Centre, ISRA, IDRC, Ottawa, Canada, 101–116.

Nikiema J., Cofie O., Impraim R. and Drechsel P. (2012). Fortified excreta pellets for agriculture. Paper presented at the Second International Faecal Sludge Management Conference in Durban, 29–31st October. Available at: https://www.susana.org/_resources/documents/default/2-1781-2-1624-nikiema.pdf (accessed 22 January 2022).

ONAS (2012). Annual Report. Office National de l'Assainissement, Tunis, Tunisia.

Ortega D. L., Wang H. L., Wu L. and Olynk N. (2011). Modeling heterogeneity in consumer preferences for select food safety attributes in China. *Food Policy*, **36**, 318–324.

Otoo M. and Drechsel, P. (eds.), (2018). Resource Recovery From Waste: Business Models for Energy, Nutrient and Water Reuse in Low- and Middle-Income Countries. Routledge, London, 816 pp.

Otoo M., Drechsel P. and Hanjra M. A. (2015). Business models and economic approaches for nutrient recovery from wastewater and fecal sludge. In: *Wastewater: Economic Asset in an Urbanizing World*, P. Drechsel, M. Qadir and D. Wichelns (eds.), Springer, Dordrecht, Netherlands, pp. 247–268.

Ouedraogo B. (2002). Perceptions of Ouagadougou market gardeners on water, hygiene and disease. *Urban Agriculture Magazine*, **8**, 24–25.

Ozengin N. and Elmaci A. (2007). Performance of duckweed (Lemna minor L.) on different types of wastewater treatment. *Journal of Environmental Biology*, **28**(2), 307–314.

Pal D. and Das Gupta C. (1992). Microbial pollution in water and its effect on fish. *Journal of Aquatic Animal Health*, **4**, 32–39.

Po M., Kaercher J. D. and Nancarrow B. E. (2004). Literature Review of Factors Influencing Public Perceptions of Water Reuse. Australian Water Conservation and Reuse Research Program, CSIRO Land and Water, Canberra, Australia.

Po M., Nancarrow B. E., Leviston Z., Poter N. B., Syme G. J. and Kaercher J. D. (2005). Predicting Community Behaviour in Relation to Wastewater Reuse: What Drives Decisions to Accept or Reject? CSIRO, Perth.

P-REX (2013). Market and regulatory aspects of P recycling. *SCOPE Newsletter*, **94**, 3–5.

Price J., Fielding K. and Leviston Z. (2012). Supporters and opponents of potable recycled water: Culture and cognition in the toowoomba referendum. *Society & Natural Resources*, **25**, 980–995.

Prouty C., Mohebbi S. and Zhang Q. (2018). Socio-technical strategies and behavior change to increase the adoption and sustainability of wastewater resource recovery systems. *Water Research*, **137**, 107–119.

Rahman M. M., Salleh M. A. M., Rashid U., Ahsan A., Hossain M. M. and Ra C. S. (2014). Production of slow release crystal fertilizer from wastewaters through struvite crystallization – A review. *Arabian Journal of Chemistry*, **7**(1), 139–155.

Rashid M. M., Kattou'a M. G., Al-Khatib I. A. and Sato C. (2017). Farmers' attitude toward treated sludge use in the villages of West Bank, Palestine. *Environmental Monitoring and Assessment*, **189**(7), 1–14.

Ravishankar C., Nautiyal S. and Seshaiah M. (2018). Social acceptance for reclaimed water use: A case study in Bengaluru. *Recycling*, **3**, 1–12.

Robinson K. G., Robinson C. H. and Hawkins S. A. (2005). Assessment of public perception regarding wastewater reuse. *Water Science and Technology: Water Supply*, **5**(1), 59–65.

Rock C., Solop F. I. and Gerrity D. (2012). Survey of statewide public perceptions regarding water reuse in arizona. *Journal of Water Supply: Research and Technology – Aqua*, **61**, 506–517.

Ross V. L., Fielding K. S. and Louis W. R. (2014). Social trust, risk perceptions and public acceptance of recycled water: Testing a social-psychological model. *Journal of Environmental Management*, **137**, 61–68.

Rozin P., Haddad B., Nemeroff C. and Slovic P. (2015). Psychological aspects of the rejection of recycled water: Contamination, purification and disgust. *Judgment and Decision Making*, **10**, 50–63.

Saliba R., Callieris R., D'Agostino D., Roma R. and Scardigno A. (2018). Stakeholders' attitude towards the reuse of treated wastewater for irrigation in Mediterranean agriculture. *Agricultural Water Management*, **204**, 60–68.

Scott C. A., Zarazua J. A. and Levine G. (2000). Urban-Wastewater Reuse for Crop Production in the Water-Short Guanajuato River Basin, Mexico. Research Report 41. International Water Management Institute, Colombo, Sri Lanka.

Scott B., Curtis V., Rabie T. and Garbrah-Aidoo N. (2007). Health in our hands, but not in our heads: Understanding hygiene motivation in Ghana. *Health Policy and Planning*, **22**(4), 225–233.

Scott C., Drechsel P., Raschid-Sally L., Bahri A., Mara D., Redwood M. and Jiménez B. (2010). Wastewater irrigation and health: Challenges and outlook for mitigating risks in low-income countries. In: Wastewater Irrigation and Health: Assessing and Mitigation Risks in Low-Income Countries, P. Drechsel, C. A. Scott, L. Raschid-Sally, M. Redwood and A. Bahri (eds.), Earthscan-IDRC-IWMI, UK, pp. 381–394.

Seidu R., Heistad A., Jenssen P. D., Drechsel P. and Stenström T. A. (2008). Quantification of the health risks associated with wastewater reuse in Accra, Ghana: A contribution toward local guidelines. *Journal of Water and Health*, **6**(4), 461–471.

Shu L., Schneider P., Jegatheesan V. and Johnson J. (2006). An economic evaluation of phosphorus recovery as struvite from digester supernatant. *Bioresource Technology*, **97**(17), 2211–2216.

Silva-Ochoa P. and Scott C. A. (2004). Treatment plant effects on wastewater irrigation benefits: Revisiting a case study in the guanajuato river basin, Mexico. In: Wastewater Use in Irrigated Agriculture: Confronting the Livelihood and Environmental Realities, C. A. Scott, N. I. Faruqui and L. Raschid-Sally (eds.), Commonwealth Agricultural Bureau International, Orient-Longman, and International Development Research Centre, Ottawa, Canada, 145–152.

Smit J. and Nasr J. (1992). Urban agriculture for sustainable cities: Using wastes and idle land and water bodies as resources. *Environment and Urbanization*, **4**, 141–152.

Strauss M. (2006). Low-cost options for treating faecal sludges (FS) in developing countries. Paper presented at the International symposium on Faecal Sludge Management (FSM) Policy. May 9–12th, Dakar.

Sturm B. S. M. and Lamer S. L. (2011). An Energy Evaluation of Coupling Nutrient Removal from Wastewater with Algal Biomass Production. Applied Energy, **88**(10), 3499–3506. https://doi.org/10.1016/j.apenergy.2010.12.056.

Thebo A. L., Drechsel P., Lambin E. F. and Nelson K. L. (2017). A global, spatially-explicit assessment of irrigated croplands influenced by urban wastewater flows. *Environmental Research Letters*, **12**, 1–12.

Tilley E., Lüthi C., Morel A., Zurbrügg C. and Schertenleib R. (2008). Compendium of Sanitation Systems and Technologies. Sandec, the Department of Water and Sanitation in Developing Countries of Eawag, the Swiss Federal Institute of Aquatic Science and Technology, Dübendorf, Switzerland.

Tsagarakis K. P., Mellon R. C., Stamataki E. and Kounalaki E. (2007). Identification of recycled water with an empirically derived symbol increases its probability of use. *Environmental Science & Technology*, **41**, 6901–6908.

U.S. Environmental Protection Agency (USEPA) (2012). Guidelines for Water Reuse, Chapter 8. EPA. EPA/600/R-12/618. USEPA/USAID, Washington, D.C.

Van der Hoek W., Ul-Hassan M., Ensink J., Feenstra S., Raschid-Sally L. and Munir S. (2002). Urban Wastewater: A Valuable Resource for Agriculture. International Water Management Institute Research Report 63. Colombo, Sri Lanka.

Viaene J., Lancker J. V., Vandecasteele B., Willekens K., Bijttebier J., Ruysschaert G., De Neve S. and Reubens B. (2015). Opportunities and barriers to on-farm composting and compost application: A case study from northwestern Europe. *Waste Management*, **48**, 181–192.

Vo Q. H. (1996). Wastewater Reuse through Aquaculture in Hanoi: Status and Prospects. Master's Thesis, Asian Institute of Technology, Bangkok, Thailand.

Wallis-Lage C. (2013). Resource Recovery: The Next Wave. Turning wastewater treatment plants into recovery centers for valuable resources, May 13, 2013, Water and Wastes Digest. Available at: http://www.wwdmag.com/resource-recovery-next-wave (accessed 22 June 2014).

Wegner-Gwidt J. (1991). Winning support for reclamation projects through pro-active communication programs. *Water Science and Technology*, **24**, 313–322.

Weldesilassie A. B., Boelee E., Drechsel P. and Dabbert S. (2010). Wastewater use in crop production in peri-urban areas of Addis Ababa: Impacts on health in farm households. *Environment and Development Economics*, **16**, 25–49.

Wester J., Timpano K. R., Çek D. and Broad K. (2016). The psychology of recycled water: Factors predicting disgust and willingness to use. *Water Resources Research*, **52**, 3212–3226.

Whittington D., Sadoff C. and Allaire M. (2013). The Economic Value of Moving Toward a More Water Secure World. Global Water Partnership Technical Committee (TEC) Background Papers 18. GWP, Stockholm.

WHO (2006). Guidelines of the Safe Use of Wastewater, Excreta and Grey Water, vol. 2. Wastewater Use in Agriculture. World Health Organization, Geneva, Switzerland.

WHO (2015). Sanitation Safety Planning Manual. World Health Organization, Geneva, Switzerland.

WHO (2018). Guidelines on Sanitation and Health. World Health Organization, Geneva, Switzerland.

Wilson Z. and Pfaff B. (2008). Religious, philosophical and environmentalist perspectives on potable wastewater reuse in durban, South Africa. *Desalination*, **228**, 1–9.

Winkler M.-K. H., van Rossum F. and Wicherink B. (2013). Approaches to urban mining: Recovery of ammonium and phosphate from human urine. *G.I.T. Laboratory Journal*, Nov 11.

Woldetsadik D., Drechsel P., Keraita B., Itanna F. and Gebrekidan H. (2018). Farmers' perceptions on irrigation water contamination, health risks and risk management measures in prominent wastewater-irrigated vegetable farming sites of Addis Ababa, Ethiopia. *Environment Systems and Decisions*, **38**, 52–64.

Woltersdorf L., Scheidegger R., Liehr S. and Döll P. (2016). Municipal water reuse for urban agriculture in Namibia: Modeling nutrient and salt flows as impacted by sanitation user behavior. *Journal of Environmental Management*, **169**, 272–284.

WRF (2022). Biosolids. Water Research Foundation, USA. Available at: https://www.waterrf.org/research/topics/biosolids (accessed 25 January 2022).

Zimmerman R., Bord R., Mathiowetz N. and Slovic P. (1991). Survey of Public Perceptions of Environmental Risk: New York City Sludge Management Plan. Report for the New York City Sludge Management Program. New York University, New York.

doi: 10.2166/9781780409566

Chapter 16

The environmental impacts of resource recovery

Tessa van den Brand[1] *and Aalke Lida de Jong*[2]

[1]*KWR Water Research Institute, Groningenhaven 7, 3433 PE Nieuwegein, The Netherlands*
[2]*AquaMinerals, Groningenhaven 7, 3433 PE Nieuwegein, The Netherlands*

16.1 ENVIRONMENTAL IMPACTS BEYOND LOCAL WATER QUALITY

Resource recovery has significant potential to increase the sustainability in the water sector and reduce its direct and indirect environmental impacts. Nonetheless, pursuing resource recovery does not guarantee a reduction in life cycle environmental impacts, as the inputs required for recovery may have negative environmental consequences that exceed any benefits. For example, the equipment, reactors, chemicals, and energy required to achieve the anaerobic digestion of wastewater sludge may result in net negative outcomes if the adverse environmental impacts associated with the additional consumables and infrastructure outweigh the benefits of the recovered energy. Therefore, robust evaluations of novel processes and wastewater management pathways are required to determine if environmental impacts are actually reduced, to identify opportunities and short-comings of the different options, and to reveal the potential for optimization. Life cycle assessment (LCA) is an example of a tool that helps to facilitate this analysis and discussion.

Generally, evaluations of the environmental impacts of resource recovery should take into account:

(1) The impact of the recovery process itself (e.g., energy and material consumption);
(2) The impact of the recovery on the overall treatment system (e.g., struvite recovery can reduce a wastewater treatment plant's energy demand associated with sludge processing);
(3) The avoided emissions due to the offset of other resources (e.g., recovered ammonia from wastewater as a substitute for nitrogen fertilizer);
(4) Any other emissions, avoided emissions, or changes in the wider system across other sectors (e.g., decreased use of household chemicals as a result of centralized softening of drinking water).

Although this chapter will discuss the characterization of environmental impacts, it is important to note that a reduction in environmental impacts is not inherently more sustainable. This distinction has to do with the definition of sustainability, which does not only include the environmental impacts, but also an evaluation of social and economic aspects (as discussed in Chapters 13–15 and section 16.5 in this chapter).

16.2 LEARNING OBJECTIVES

At the completion of this chapter you should be able to:

- Explain the difference between a CO_2-footprint and an LCA and the value of environmental impact analyses.
- Perform the key steps required to conduct a well-documented LCA study, and identify the accompanying decisions and challenges that must be overcome.
- Describe the importance of a comprehensive data inventory to enable the characterization of environmental impacts.
- Demonstrate an understanding of things to keep in mind when interpreting LCA results.

16.3 CONCEPTUAL OVERVIEW

Although evaluating the environmental impact of a resource recovery method is crucial, it is not easy to characterize global environmental impacts or compare the impacts of two or more scenarios. Different tools are available to characterize environmental impacts, each with its own advantages and disadvantages. For resource recovery processes, relevant approaches include CO_2-footprint (a.k.a. carbon footprint), water footprint, material flow analyses (MFA), and life cycle assessment (LCA). This chapter will focus on LCA methods, which can also be applied to calculate water and carbon footprints. The water footprint and carbon footprint characterize water use and the greenhouse gas emissions (normalized to carbon dioxide), respectively, associated with a given process (e.g., through transport, energy consumption, or the production of required chemicals). LCA methods, however, can include a broader range of metrics including human health impacts, resource depletion, effects on ecosystem health, and other impact categories.

16.3.1 Life cycle assessment

Environmental issues are increasingly recognized as important factors in decision making. However, the existence of many different categories of environmental impacts makes direct, holistic comparisons across multiple resource recovery alternatives difficult. For instance, one scenario may result in reduced use of chemicals, while another scenario may significantly reduce energy demands: the tradeoffs across alternatives require robust and transparent assessment methodologies.

Upstream and downstream effects, which can relate to the inputs required to operate a treatment process and the potential uses of any recovered products, also complicate environmental impact evaluations. This requires a cradle-to-grave-approach. A higher quality and purity of a recovered product usually results in greater environmental benefits from its use, because such a product can replace more refined primary materials which are often associated with larger environmental impacts. For instance, using calcium carbonate recovered from softening processes to replace milled limestone in glass can result in greater environmental benefits than its use as a substitute for sand in concrete (Palmen *et al.*, 2015). However, the environmental impacts associated with resource recovery (and refinement) must also be considered. For instance, reprocessing a recovered material may consume substantial amounts of additional energy and chemicals, or an impure recovered substance may be sufficient to replace a high-grade chemical without further processing. Therefore, it is crucial to use a cradle-to-grave-approach to compare equally between the proposed and conventional scenario, accounting for these upstream and downstream impacts. LCA is a tool that helps to compare the sustainability of different scenarios in this way, facilitating the discussion of tradeoffs across types of environmental impacts.

16.3.2 Typical LCA steps

An LCA study includes the following phases: goal and scope definition, inventory analyses, impact assessment, and interpretation (ISO 14040:2006). Goal and scope definition is completely determined

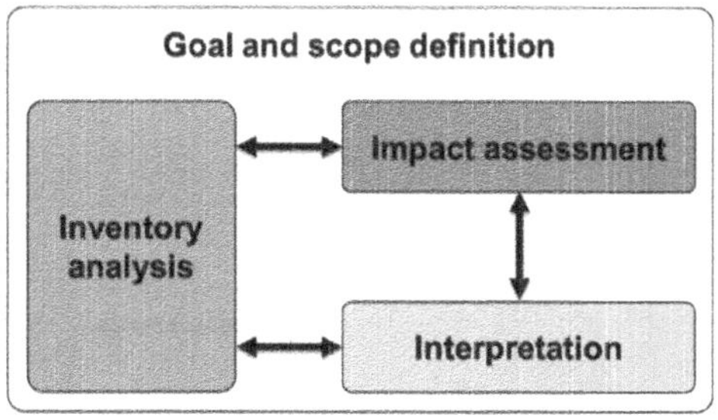

Figure 16.1 Visualization of typical LCA phases. Adapted from ISO 14040.

by the research question. Goals may include determining the environmental impacts of a product or process, comparing the impact of different products or processes, or giving direction to scientific research and development. The case studies in this chapter provide examples of these different applications in the context of resource recovery. Scope definition involves setting the boundaries of the system that will be studied and specifying the functional unit. The functional unit is required to fairly compare across different scenarios. For instance, when comparing different processes to treat wastewater, the functional unit can be the treatment of 1 m^3 of wastewater to a certain effluent quality (Byrne *et al.*, 2017). Ultimately, all environmental impacts must be normalized to the specified functional unit.

The inventory analysis compiles and documents the emissions and resources consumed. An impact assessment is then performed, considering the effects on different impact categories. The emissions and resources are classified according to their relevant impact categories (i.e., those categories on which the emissions and resources exert an impact), and then each emission and resource consumed is characterized (i.e., converted into normalized units of indicators) using impact assessment models (see section 16.3.3).

These steps interact which each other (Figure 16.1). For example, when a certain factor is identified as a hot spot of environmental impacts, it is important to intensify the inventory analysis of this particular assumption. Usually, the inventory analysis – and the determination of its quality – consumes the most time in the execution of an LCA. In all cases, however, it is important to start with the goal and scope definition.

16.3.3 Impact assessment

There are several life cycle impact assessment (LCIA) methodologies available for the translation of emissions and resources into impacts, and these can vary widely in the impact categories they cover. Examples of the most commonly applied methodologies are: CML 2002, Eco-indicator 99, Ecopoints 2006, EDIP 97, EDIP2003, |EPS2000, IMPACT 2000+, LIME, LUCAS, ReCiPe, TRACI, and MEEup (Byrne *et al.*, 2017; ILCD Handbook, 2010). The most suitable methodology will depend on the goal and scope of the LCA, which determines the impact categories that should be included in the analysis and the level at which results should be presented (i.e., endpoint or midpoint indicators). An endpoint method looks at environmental impact at the end of the cause-effect chain (e.g., extinction of species). A midpoint method looks at the impact earlier along the cause-effect chain, before reaching the endpoint (e.g., ecotoxicity). A midpoint indicator focuses on a single environmental problem, such as human toxicity, metal depletion, or particulate formation. The water footprint and carbon footprint are considered midpoint indicators. Endpoints aggregate across multiple environmental issues to make decision making easier, but uncertainty increases with each aggregation step.

For the examples in this chapter, we will use the ReCiPe method. The ReCiPe method analyses three different types of damage at the endpoint level: human health, ecosystem quality, and resources

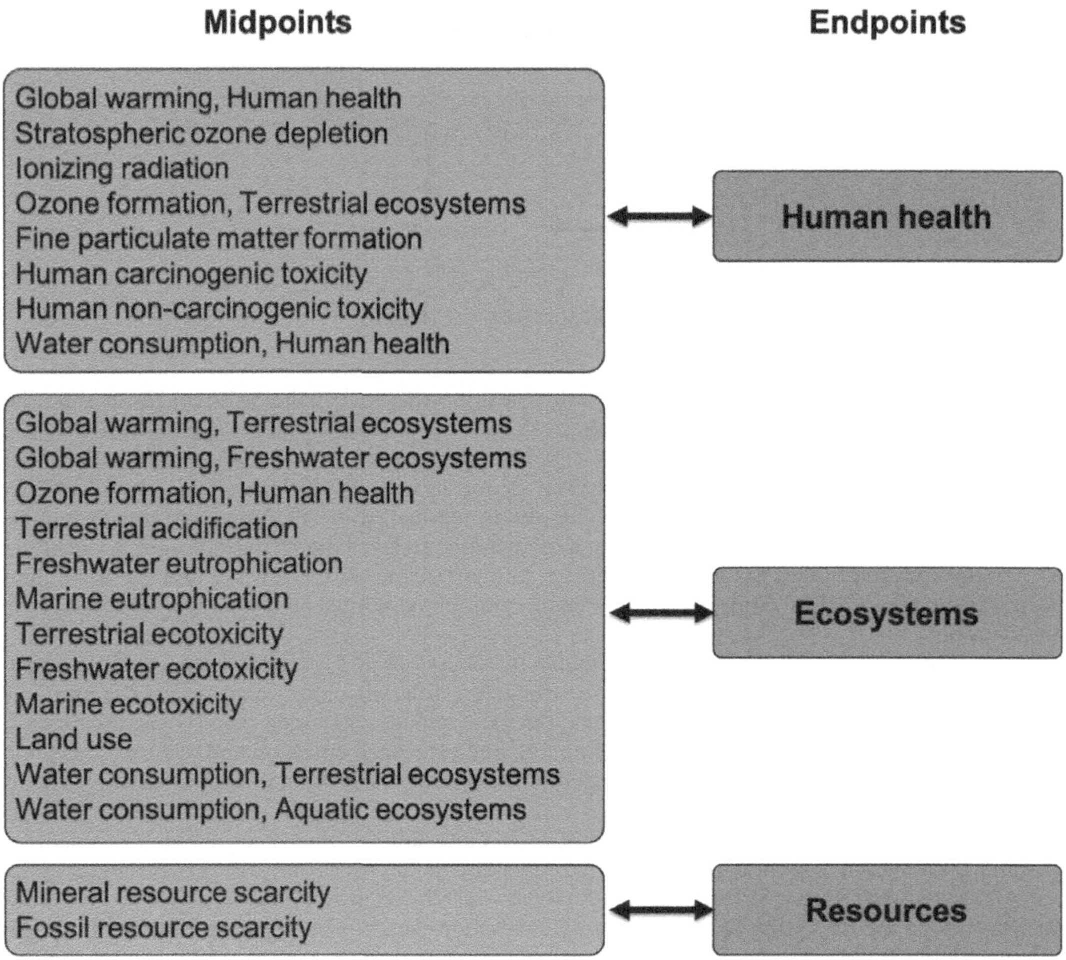

Figure 16.2 Overview of midpoints and endpoints in the ReCiPe method. Adapted from Huijbregts *et al.* (2016).

(Figure 16.2). Each endpoint has a different indicator and unit (Table 16.1). The endpoint scores are calculated from the midpoint scores using weighting and normalization factors.

Analyzing the environmental impacts associated with the ReCiPe method's three endpoints – human health, ecosystems, and resources – still does not provide a single value representing the overall environmental impact. Evaluating a scenario's total environmental impact requires a method

Table 16.1 Overview of endpoint categories and indicators for the ReCiPe method.

Impact Category/Endpoint	Indicator	Unit
Damage to human health	Disability-adjusted loss of life year	DALY
Damage to ecosystem diversity	Number of species lost per year	Yr^{-1}
Damage to resource availability	Increased cost	$

Adapted from Goedkoop *et al.* (2009).

to either navigate tradeoffs across impact categories or to combine all of the endpoint impacts through normalization and weighting. In the ReCiPe method, each endpoint impact can be assigned a certain number of 'ecopoints'. One thousand ecopoints correspond to the total annual environmental impact of one person in Western Europe (https://www.pre-sustainability.com/download/EI99_Manual.pdf).

16.3.4 LCA standards (ISO standards)

The quality and relevance of LCA studies, and the extent to which they can be applied and interpreted, depends upon the methodology used. Transparent documentation is crucial. The International Organization for Standardization (ISO) published standards to provide guidance on methodological choices, transparency, and rules for reporting. Examples of LCA-related ISO standards are ISO 14040 (Environmental management – Life cycle assessment – Principles and framework) and ISO 14044 (Environmental management – Life cycle assessment – Requirements and guidelines).

At a minimum, ISO standards state that a well-documented LCA should contain the following content and details: approach and method, goal definition, explanation of the scope, description of the evaluated scenario(s), selection of functional unit, description of system boundaries, and details on data collection and methodology (software, database, method, etc.). More specific details can be found in the ISO standards and, as they relate to wastewater management, in a recent review by Corominas *et al.* (2020).

16.3.5 (Avoided) emissions

When resource recovery is applied in the water cycle, it could also avoid environmental impact in other processes and/or sectors. The power to protein concept, in which proteins are made from wastewater resources (Powertoprotein.eu), provides a helpful illustration. The effort to produce protein from wastewater resources occurs within the water cycle, but the environmental benefits would be seen in the food industry, which can reduce its protein production (from cattle or fish, e.g.). Therefore, it is crucial to ensure that the system boundary includes not only the water cycle (where the effort occurs) but also other sectors where the benefits (or costs) may be realized.

In contrast to avoided emissions, resource recovery can also cause additional environmental impacts in other processes. For example, if a WWTP that currently digests sludge to produce biogas installs a process to recover cellulose before the digester, the organic input into the digester will decrease, resulting in lower biogas production (Visser *et al.*, 2016). This reduction in biogas production could result in less on-site electricity and heat production, for instance, potentially requiring pipeline methane or grid electricity to meet the plant's operational needs. In all cases, any changes in resource consumption must be quantified.

16.3.6 Interpretation

The goal of the LCA study influences interpretation of the results. For example, using a tree structure (Figure 16.5) to perform a hotspot analysis can show which individual process has the largest environmental impact. A mistake often made while interpreting LCA results is to consider only differences in the single aggregated score (e.g., ecopoints in ReCiPe). Only focusing on ecopoints will not indicate which midpoint indicators cause the aggregate environmental impact to change, or which indicator contributes the most to the aggregated score.

When scenarios are compared, one should consider the uncertainty in results stemming from assumptions throughout the execution of the LCA, avoiding overinterpretation of minor differences between scenarios without rigorous characterization of uncertainties. Furthermore, the level of detail can be important in interpretation. When evaluating an overall environmental impact, a value presented in ecopoints may be sufficient, while reporting midpoint level indicators may be required to achieve improvements in specific environmental outcomes (e.g., greenhouse gas emissions). When reporting any LCA findings, transparency about assumptions, midpoint values, uncertainty and sensitivity analyses, and method of weighting (if weighting used) is crucial.

Sensitivity analyses are usually important for interpreting LCA results. Although there are more robust methods for global and local sensitivity analyses, the most commonly employed sensitivity analyses in the literature are one-at-a-time approaches. For instance, a sensitivity analysis may assess how an individual parameter affects the metrics of interest by varying the value of this parameter while holding all others constant. Another option is to investigate best and worst case scenarios. For instance, wastewater collection can occur through a sewer network or through truck transport. While the overall analysis may suggest that the first scenario has lower impacts and is preferred over the second, a sensitivity analyses may reveal that energy is the main contributor within second scenario. If the environmental impact of energy production is expected to decrease in the (near) future, the second scenario may be favorable. The same could be true if an increase in efficiency is expected. As a result, performing robust sensitivity analyses is advisable.

16.4 CASE STUDIES

Depending on the goal of the study, an LCA can be performed at different levels of detail. Therefore, it is important to clearly formulate the goal and scope of the study. Examples of reasons to complete an LCA study include:

- to determine the potential of a certain (resource recovery) concept;
- to direct and focus scientific research of a (resource recovery) concept;
- to calculate the environmental impact of a certain process or product;
- to compare alternatives for a certain scenario.

The first two goals can be achieved by performing a quick scan analysis using LCA, which will be explained in section 16.4.1. The calculation of the environmental impact of a certain process and the comparison between an alternative and a conventional scenario (the final two goals) require an intensive LCA and will be explained in section 16.4.2. As explained in the second example, new insight after the start of the project may have resulted in other choices if this information were known from the beginning.

16.4.1 LCA as quick scan analysis

As mentioned above, the reasons to perform a quick scan analysis could be to investigate the potential of a resource recovery technology or concept, or to further direct and focus scientific research along a specific pathway to reduce environmental impacts. This section demonstrates a quick scan analysis with an example regarding iron flocculants.

As illustrated in Chapter 3 (Table 16.1), iron sludge is formed during the coagulation/flocculation treatment step in drinking water production (Hofman-Caris *et al.*, 2019). If the iron can be recovered, it can be reused as a flocculant in drinking water production or wastewater treatment. Before intensive scientific research on the recovery of iron is initiated, a quick scan analysis can show whether this possibility may decrease the environmental impacts of the process. To this end, only the chemicals required (and their transport) to recover flocculants from iron sludge ('$FeCl_3$ recovered' scenario) are roughly compared with the purchase of a commercial flocculant, including its manufacturing and transport ('$FeCl_3$ commercial' scenario). Also, a third scenario, employing a recycled acid, is investigated as the '$FeCl_3$ recovered green' scenario. The inventory data for these three scenarios is presented in Table 16.2. This analysis will not only help characterize the potential of the suggested concept, but may also identify hot spots for continuation of scientific research. The quick scan LCA analysis used the SimaPro 8.0 software, EcoInvent 3.0 database, and ReCiPe Endpoint E 1.12 method. The functional unit (FU) is the production of 1 kg of flocculant, as the function of both scenarios is to deliver high-quality flocculants.

Figure 16.3 demonstrates that iron flocculants from recovered iron sludge have an environmental impact that is approximately one-third lower than commercial flocculants. In this concept, a virgin

Table 16.2 Inventory data for a quick scan analysis of flocculant recovery from iron sludge with a functional unit of 1 kg flocculant.

$FeCl_3$ commercial	
Iron (III) chloride usage	2.9 kg
Iron (III) chloride transport	0.363 tkm
$FeCl_3$ recovered	
HCl usage	2.61 kg
HCl transport	0.131 tkm
$FeCl_3$ recovered green	
HCl transport	0.131 tkm

hydrochloric acid was used, but if a recycled acid could be applied, the environmental impacts are likely to decrease further. However, the $FeCl_3$ recovered 'green' scenario does not include any impacts from the recycled acid used (with the chosen allocation method). Ultimately, all activities required to isolate HCl from an industrial waste material should be included in the data inventory. At present, however, information about the additional effort required to produce and use a recycled acid are unknown and therefore not included. The actual environmental impact of a resource recovery concept that uses recycled acid will likely fall between the $FeCl_3$ recovered scenario and $FeCl_3$ recovered green scenario. Based on this result, the recovery of iron flocculants from iron sludge is promising, and continued scientific research to better understand the technical aspects appears to be worthwhile. Various research questions could be investigated, such as: What is the safety and efficiency associated

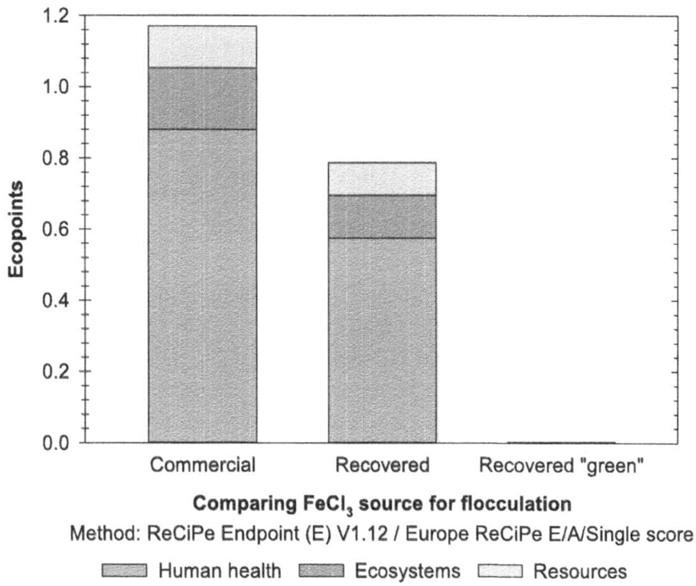

Figure 16.3 A quickscan analysis regarding the reuse of iron sludge as flocculant. Midpoints are represented as: human health (blue), ecosystems (orange) and resources (grey). Note, for the recovered 'green' scenario, any impacts from acid recovery are not included.

with recovered flocculants? Which technical problems may occur, and how can they be solved? The scenario including a recycled acid shows that the impact of the acid represents a hot spot in the total environmental impact. Therefore, efforts to use less acid or a recycled acid should be a focus of continued research. Of course, a more detailed LCA should reevaluate the concept after more comprehensive research has been completed.

16.4.2 An intensive LCA study: Calcite

An intensive LCA study can be conducted to calculate the environmental impact of a certain process/product or to perform a detailed comparison of multiple scenarios. An example regarding calcite (calcium carbonate) will be discussed to demonstrate the function of an intensive LCA study, specifically comparing a conventional system to a proposed future scenario.

In this example, the conventional pellet softening process of a hypothetical drinking water treatment plant is evaluated and compared to a proposed innovative concept. During conventional pellet softening, calcium from the raw water precipitates on a garnet or river sand seeding material. The residual of this process consists of calcite pellets with a garnet kernel. These pellets are used as a substitute for primary materials like sand and low grade crushed or milled limestone. They can also serve as a raw material for limestone products.

In the innovative concept, the garnet seeds are replaced by calcite seeds, resulting in residual pellets that contain 100% calcite. The pure calcite pellets are dried, milled, and sieved at an external location to produce both the calcite seeds and a commercial-grade fine milled limestone product, which is sold to third parties for high-end applications.

The data inventory of the conventional scenario (using garnet as seed) and the innovative concept (using calcite as seed) are presented in Figure 16.4 and Table 16.3, reflecting all process changes that will be evaluated in the LCA.

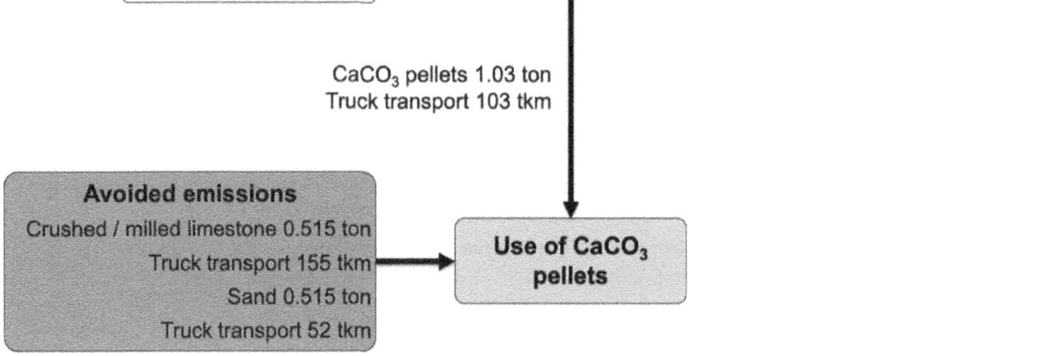

Figure 16.4 Data inventory for softening by the conventional scenario.

Table 16.3 Data inventory for all three scenarios.

Process	Conventional	Innovative Original	Innovative Revised
Inoculum	0.06 ton garnet seed 12 tkm truck transport 1.320 tkm ship transport	0.06 ton limestone residue 12 tkm truck transport	0.111 ton calcium carbonate 122 tkm truck transport 22.2 MJ energy
Lost to sewer	0.03 ton $CaCO_3$	0.03 ton $CaCO_3$	0.056 ton $CaCO_3$
External transport	103 tkm truck transport	103 tkm truck transport	106 tkm truck transport
Avoided emissions	0.515 ton $CaCO_3$ 0.515 ton sand 207 tkm truck transport	1.03 ton $CaCO_3$ 309 tkm truck transport	1.06 ton $CaCO_3$ 317 tkm truck transport

In short, this analysis employed SimaPro 8.0 software, the EcoInvent 3.0 database, and the ReCiPe Endpoint (E) 1.12 method. The functional unit is the removal of 1 ton of calcium carbonate from water, so that the comparison is based on the production of this recovered resource. A detailed description of the data inventory and assumptions is given in Palmen *et al.* (2015). However, new insights have been obtained after that first LCA study. More specific information became available about calcite, resulting in a revised inventory data set (Table 16.3). This third scenario is referred to as the revised innovative concept. The revision resulted from using a more precise method to model the calcite inoculum within the LCA. This illustrates that while performing LCA research, it is important to realize that the outcome is strongly dependent on the data inventory. As a result, new insights can drastically affect the environmental impact calculation, potentially leading to new conclusions. Accordingly, this analysis includes not only the comparison between the conventional and innovative scenario, but also illustrates more robust outcomes of LCA analyses (Figure 16.6).

To indicate the importance of iteratively improving a LCA study, Figure 16.6 shows that the revised assumptions further reduce the environmental impacts of the innovative scenario, improving from −2.8 to −5.3 ecopoints per ton of calcium carbonate removed from water. Note that a negative value signifies a positive contribution to the environment – in this case, avoided emissions. The changes in the data inventory are not only a result of new insights (more detailed information regarding calcite production), but also of improved information in the inventory database used (an update to the EcoInvent database) and impact assesment method (Huijbregts *et al.*, 2016). Given that these new assumptions lead to an environmental impact that improved almost twice, it is very important to occasionally revise an LCA study and reevaluate the assumptions used. Also, it should always be remembered that LCA results should not be the sole contributor to decision making, but they can be used as a tool to facilitate discussions.

Based on the revised assumptions, a comparison between the conventional and innovative scenarios (Figure 16.6) illustrates that the environmental impact of the innovative concept (using calcite as seeds/seeding material) has a substantially lower environmental impact (−5.3 Pt) than the conventional situation with garnet as seed (−3.8 Pt). A detailed evaluation of the innovative revised calcite scenario, based on the tree structure (Figure 16.5), specifies that the improved environmental impact is a direct result of the increase in avoided emissions for crushed/milled limestone (from 1.03 to 1.06 tons). This increase in avoided limestone is caused by the higher quality of the calcium carbonate recovered in this scenario. The tree structure shows which processes (transport, chemicals, etc.) are included in the LCA analyses. Furthermore, red arrows show a negative impact, while green arrows show the positive environmental impacts (avoided emissions). The size of each arrow shows the degree to which it contributes to the total environmental impact. Therefore, the tree structure is a useful tool to identify hotspots in a scenario, and it also provides guidance on which parameters to consider in sensitivity analyses.

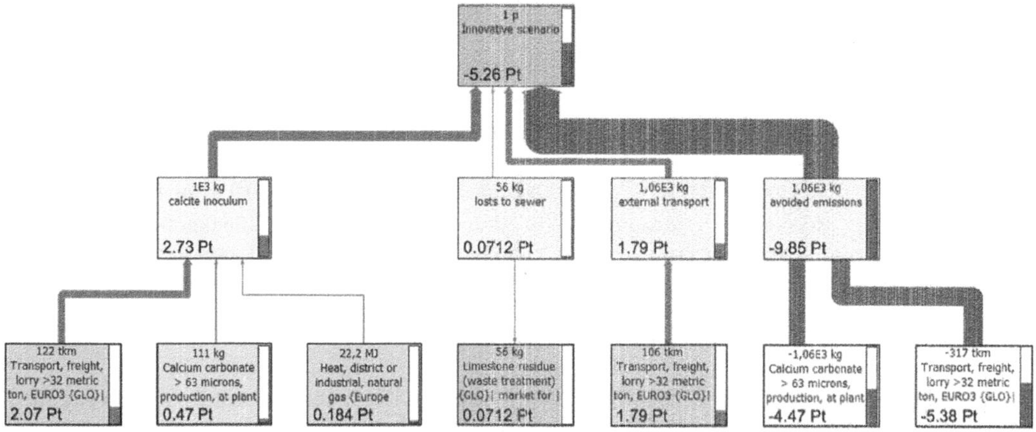

Figure 16.5 Tree structure of the innovative revised scenario (calcite as inoculum), based on ReCiPe Endpoint (E) V1.13.

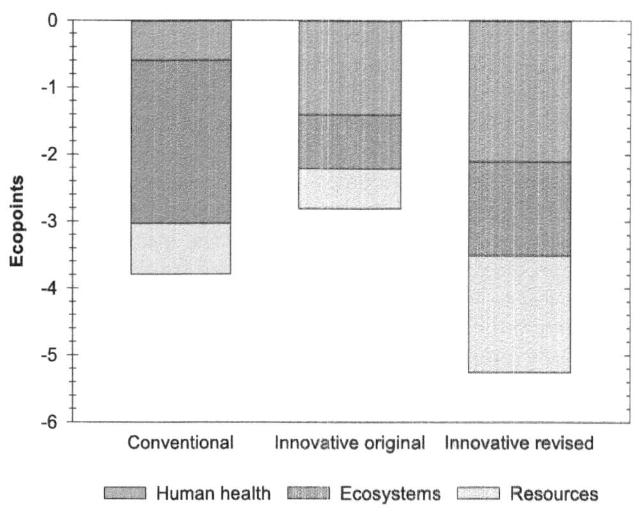

Figure 16.6 Comparison between the conventional (garnet) scenario and two versions of the innovative (calcite) scenario, based on original (center) and revised (right) assumptions. Method: ReCiPe Endpoint (E) V1.13/Europe ReCipE E/A/single score.

16.5 CHALLENGES MOVING FORWARD

The process of measuring, calculating, and discussing environmental impact faces several challenges. A common problem with LCA and other environmental impact studies is its use to determine whether one scenario is better than another. However, the results cannot be considered as a definitive indication of superiority/inferiority of a scenario. There are too many assumptions and issues related to data availability, time-framing, midpoint and endpoint metric selection, and other challenges and considerations (including economics, etc.), all of which must be considered. Several of the most crucial points and challenges related to calculating and measuring environmental impact are discussed below.

16.5.1 Environmental impact terms

An important issue within this discipline is the use of different definitions for environmental impact, sustainability, and circular economy. It is essential to be aware of the different definitions and assure that all parties understand these distinctions when collaborating. Sustainability is not the same as environmental impact. Sustainability goes beyond environmental impact, as it also includes effects on people and the economy. A sustainable scenario meets the needs of the present without compromising the ability of future generations to meet their needs (World Commission on Environment and Development (WCED), 1987), while an assessment of environmental impacts may only consider aggregated global impacts and the present. Sustainability is concerned with the 'three P's': people, profit, and planet. We should pursue favorable outcomes in all three categories to continue to move toward more sustainable systems. Environmental impact calculations, such as those in LCA, do not include any social or economic evaluations, but focus on elements like emissions (e.g., CO_2, CH_4, H_2S), use of resources (e.g., energy, metals, water), and the potential for human, environmental, and ecological consequences from these emissions and resource demands. The circular economy concept represents another way to talk about environmental impacts. The idea of the circular economy is to regenerate and reuse materials and resources (onsite, preferably), avoiding the use of virgin material and the production of waste. These efforts do not always result in a lower LCA environmental impact calculation; for example, the regeneration of a material may require large quantities of energy and chemicals. Ultimately, however, a circular economy would progress toward greater use of renewable energy and material sources, converging to environmental sustainability.

16.5.2 Data inventory

There are several issues related to the accuracy and precision of the data inventory. Examples of data inventories are given in Tables 16.2 and 16.3, and Figure 16.4. First, completing a data inventory often requires numerous assumptions. Examples of cases where assumptions may be required include: (1) when a future scenario is analyzed; (2) the actual consumption data are only known for pilot studies; or (3) when values (energy for instance) are known for the entire facility but not for one particular process. It is important to discuss and quantify the effect and importance of any assumptions. The best approach for evaluating assumptions is to perform uncertainty and sensitivity analyses, the latter of which will indicate whether a given assumption has drastic effects on the total calculated environmental impact. When the effect of your assumption is significant, additional research may be required to improve the data inventory. Ultimately, new insights can alter the data inventory considerably. It is therefore advisable to revise environmental impact studies frequently.

16.5.3 Completeness and representativeness of database information

Conducting an accurate environmental impact study is strongly dependent on the quality of the database. It is critical to use a database that is trustworthy – for example, one that goes through regular and independent review processes. The database should be transparent and well documented, and the origin of the data should be clear.

The data in commercially available databases may not always be representative for a given case. Examine the completeness of the process with regard to issues such as whether transport is included, whether the material is packaged, or whether the chemical production method aligns with the specifics of the application. For example, sodium hydroxide can be produced by different methods, resulting in different environmental impacts. Furthermore, sometimes processes in a database represent the mass of the bulk solution, but at other times the database corrects for the quantity of active substance contained within the solution. Always check if the processes behind the indicator apply to the case at hand, and if not, try to obtain data from the parties in the value chain.

Another issue relates to whether all the data required for the impact analysis are available in the database. For example, the ReCiPe method does not include characterization factors for some emissions (i.e., there is not a default value to normalize an emission to the constituent units, such as kg-CO_2 eq. for global warming potential). According to the ReCiPe method, chloride, sodium, and bromide emissions to water have no environmental impact (Huijbregts *et al.*, 2016), although impacts do exist in some cases. Be aware of the fact that some emissions are not included due to a lack of impact information. Even when they are included as components in the model, the model will assign them an impact value of zero. These instances should be documented and reported as part of the limitations of a study.

16.5.4 The representative case study

To learn from a certain case study or to draw conclusions from an evaluation of a resource recovery concept, a representative case study can be very helpful. However, the development of a representative case study is rather difficult and perhaps even impossible. Not only is the current conventional scenario different in each country (or region), but market issues also play a key role in the environmental impact. For instance, case studies related to energy supply can be severely affected by geographic differences and variances in markets (International Energy Agency, 2014). The environmental impact of the electricity market mix in the Netherlands is more than 10 times higher than in Norway, because Norway's electricity supply includes a greater fraction of renewable sources (EcoInvent 3.0 database information of market conform electricity database in 2019). As a result, for a resource recovery method that is highly reliant on electricity, its environmental impacts may vary significantly when estimated for different countries.

Another essential parameter that can differ among case studies is the quality of the resource that is recovered. If calcium carbonate is recovered within the drinking water treatment process, the quality of the lime may vary. This variation is a direct result of differences in influent water quality and composition. As a result, one should be conscious of these challenges and avoid drawing general conclusions from specific case studies.

16.5.5 Evaluating future improvements to resource recovery processes

Resource recovery often involves new technologies that have not been fully developed, resulting in relatively high energy or material requirements. As a result, an LCA assessing whether a new resource recovery process can reduce environmental impacts relative to the current situation may have a negative outcome. If improvements are feasible in the near future, one could also assess the scenario with those improvements to provide a realistic outlook on the future situation.

16.5.6 Market information

Recovered resources that are placed on the market have the potential to replace other (mainly primary) resources. When calculating the avoided emissions, the choice of the substituted material and the process in which the material is applied both matter. For example, $CaCO_3$ pellets from drinking water softening could be used directly in the glass industry without further processing to replace milled limestone. However, another quality (composition) or status (e.g., milled) may differ depending on the market the recovered resource is applied. The avoided emission should be corrected for this as much as possible. This leads to a substantial difference in the avoided impacts.

Challenges arise when there is no information available on the current market mix or when a new market is created. In the first case, for example, a small survey could provide the necessary information. In the latter case, the avoided product could be modeled by choosing a surrogate product that fulfills the same function.

16.5.7 Allocation and system expansion

In general, allocation plays a key role in LCA. Allocation is applied to divide a certain environmental impact across the processes involved. Especially in resource recovery allocation, questions arise. For example, which part of the impact of drinking water or wastewater treatment should be allocated to the recovered residuals? There is no perfect approach. Some common methods include the following:

(1) Allocation of the impacts based on economic value, mass, or energy. For instance, when a tree is used to build a shelter, suppose only half of the wood is useful in building the shelter. The other half could be used to make paper. In this case, when using mass allocation, only the consumption of half of the tree is included in the LCA of the shelter, as the other half is used for an end product (paper) that is outside the scope of the shelter study.
(2) The 'cut-off' or 'recycled content' method, in which a recovered or recycled material is assigned no environmental impact. In contrast, all impacts are allocated to primary materials. In the example in section 16.4.1, we used the cut-off method, meaning that the recycled acid used to produce the green coagulant was assigned zero environmental burden.
(3) 'System expansion' or 'substitution' (favored by the ISO 14040). In system expansion, co-products are considered alternatives to other products on the global market. For instance, we want to evaluate the impact of a wastewater treatment plant, taking into account the struvite that is recovered from the wastewater. The struvite replaces fertilizer on the market (avoided production), which allocates a negative contribution (less impact) to the environmental impact of the wastewater treatment plant.

16.5.8 Process design and its influence on resource recovery (and vice versa)

As described in Chapter 3, drinking water production sites are designed to achieve a certain goal, such as the production of hygienic, safe, colorless, and odor-free water. When new scenarios are proposed to maximize the recovery of certain resources, they should not diminish the efficacy of drinking water production. However, simple changes in process design can occasionally lead to better quality of the recovered resource - for instance, the use of calcite for the softening process in the Netherlands (Chapter 3).

When an environmental impact of an innovative process design is calculated, all effects on the total water cycle should be included. For example, when more water is lost during the last processing step in drinking water treatment, more water must be treated in the preceding steps, increasing the total environmental impact of the drinking water process. Other examples include a change in the type of chemical used (thereby changing required chemical transport), altered recovered resource quality, and modifications in the market where the resource is sold.

16.5.9 Comparisons with previous studies

During research, comparisons with previous studies are both desired and necessary. However, it is crucial to understand that new results cannot always be directly compared to previous studies. For example, differences in the functional unit require that each set of results be reframed using a consistent functional unit. Also, different databases or impact assessment methods may have been used. Furthermore, in the case of ecopoints, it is vital to know that the total annual impact of one person living in Western Europe is defined as 1000 ecopoints (Baayen, 2000). As per capita environmental impacts change, the values from different years cannot be directly compared.

16.6 CHAPTER SUMMARY

In this chapter we have discussed the crucial steps in performing a comprehensive environmental impact study. In particular, LCA represents an established methodology to transparently characterize the global environmental impacts across the life cycle of a product or process. Given that many assumptions must be made to execute an LCA, the use of well-studied and documented inventory data is critical. These data sets can change over time, which is why it is advisable to update previous LCA studies before leveraging them for decision-making.

When executing an LCA, quick scan analyses or intensive LCAs can offer complementary insight. Reasons to perform a quick scan analysis could be to investigate the potential of a resource recovery technology or concept, or to further direct and focus scientific research along a specific pathway to reduce environmental impacts.

16.7 EXERCISES

Exercise 16.1: Suppose a new vendor is found for the calcite-seed pellets (from the revised innovative scenario described in Figures 16.5 and 16.6), which would reduce transport requirements to the treatment plant by 50%. Calculate the overall environmental impact of revised innovative scenario before and after this transportation savings in units of Ecopoints (Pts).

Exercise 16.2: Suppose additional research showed that the innovative revised process described in section 16.4.2 eliminates losses to the sewer. Based on the data in Table 16.3 and Figure 16.5, what is the new single score resulting from the LCA of the 'innovative revised' scenario with no calcium carbonate loss to the sewer?

Exercise 16.3: A drinking water production site generates 150,000 m^3 of water and 680 tons of $CaCO_3$ per day. The calcium carbonate recovery offsets conventional calcium carbonate production. The environmental impacts of calcium carbonate production can be extracted from Figure 16.5. Determine the unit impacts of conventional calcium carbonate production (in units of Pts per ton $CaCO_3$) and the environmental benefits of its recovery (in units of Pts per year).

Exercise 16.4: You are performing a quick scan analysis of a process that requires the following materials and processes for construction and across its lifetime: 3.9 tons of reinforcing steel, 179 m^3 of concrete, and 16,600 Wh of electricity. Over the same lifetime, nutrient recovery and distribution to agriculture offset 139 tons of nitrogen fertilizer and 55 tons of phosphorus fertilizer. Given the unit impacts below: (a) determine the relative global warming impact of steel, concrete, and electricity (determine the total impacts of the three, and each's contribution to that total as a percentage); and (b) determine what percentage of impacts are offset by both nitrogen fertilizer offsets and phosphorus fertilizer offsets.

- Nitrogen fertilizer: 11.5 kg-CO_2 eq. per kg N
- Phosphorus fertilizer: 2.1 kg-CO_2 eq. per kg P
- Reinforcing steel: 2.5 kg-CO_2 eq. per kg steel
- Concrete: 398 kg-CO_2 eq. per m^3 concrete
- Electricity: 0.64 kg-CO_2 eq. per kWh

Exercise 16.5: A resource recovery and treatment process results in fugitive emissions of nitrous oxide (N_2O) and methane (CH_4). Ultimately, 0.5% of influent nitrogen is released as N_2O and 2 g of CH_4 is released per m^3 treated. If the characterization factors of N_2O and CH_4 are 298 g-CO_2 eq. per g N_2O and 28 g-CO_2 eq. per g CH_4, what is the total global warming potential resulting from fugitive emissions (in units of g-CO_2 per m^3 treated) for a wastewater with 35 mg-N·L^{-1} in the influent?

16.8 DISCUSSION QUESTIONS

Question 16.1: What is the value offered by conducting a LCA study, and what is the practical difference between a quick-scan and an intensive LCA study? When might you apply one vs. the other?

Question 16.2: Why should you be cautious about drawing generic conclusions from specific case studies?

Question 16.3: What are three examples in which a quick scan environmental impact assessment could provide added value to a resource recovery decision?

Question 16.4: What challenge in the interpretation phase of LCA do you think has the highest impact on results? Why?

Question 16.5: What is an alternative functional unit for the intensive LCA calcite study (instead of the removal of 1 ton of calcium carbonate from water)? Can you explain why this alternative functional unit may not have been chosen?

Question 16.6: A research group is deciding whether to focus on a particular treatment process and whether to modify a specific component, as it may reduce the environmental impact. Do you advise performing a quick scan analysis or an intensive LCA study? Why?

Question 16.7: After five years, the LCA concerning calcite (described in section 4.2) is repeated. How can we compare the results?

Question 16.8: Choose a correct functional unit for a LCA that compares a wastewater treatment system with and without anaerobic digestion of sludge (leading to biogas production). Discuss why the selected functional unit is appropriate.

Question 16.9: You are asked to perform an LCA of an entire drinking water production plant. List what you would include in your data inventory. Hypothesize which of these items may contribute the most to the plant's environmental impacts (representing a hotspot).

Question 16.10: A drinking water production plant improves its softening process, which is the sixth treatment step in its process sequence. This improvement decreases water loss. Previously, the environmental impact of the plant has been expressed relative to a functional unit representing the production of 1 m^3 of water. When revising the estimate of the plant's environmental impact, what items will be important to include in the data inventory?

ACKNOWLEDGEMENT

The authors would like to thank C. H. M. Hofman-Caris for her contribution to this chapter.

FURTHER READING

Betaprocess Bioenergy. (2016). The Value Pyramid. Available at: http://www.betaprocess.eu/the-value-pyramid.php (accessed 24 April 2018).

European Commission – Joint Research Centre – Institute for Environment and Sustainability: International Reference Life Cycle Data System (ILCD). (2010). Handbook – General Guide for Life Cycle Assessment – Detailed Guidance, 1st edn. EUR 24708 EN. Publications Office of the European Union, Luxembourg.

Fu F. and Wang Q. (2011). Removal of heavy metals ions from wastewaters: a review. *Journal of Environmental Management*, **92**(3), 407–418.

Reijnders L. (2014). Phosphorus resources, their depletion and conservation, a review. *Resources, Conservation and Recycling*, **93**, 32–49.

Renou S., Thomas J. S., Aoustin E. and Pons M. N. (2008). Influence of impact assessment methods in wastewater treatment LCA. *Journal of Cleaner Production*, **16**, 1098–1105. https://www.pre-sustainability.com/download/EI99_Manual.pdf

REFERENCES

Baayen H. (2000). Eco-indicator 99; Manual for Designers; A Damage Oriented Method for Life Cycle Impact Assessment. Ministry of Housing, Spatial Planning and the Environment, The Hague, The Netherlands.

Byrne D. M., Lohman H. A. C., Cook S. M., Peters G. M. and Guest J. S. (2017). Life cycle assessment (LCA) of urban water infrastructure: emerging approaches to balance objectives and inform comprehensive decision-making. *Environmental Science: Water Research & Technology*, **3**, 1002–1014.

Corominas L., Byrne D. M., Guest J. S., Hospido A., Roux P., Shaw A. and Short M. (2020). The application of life cycle assessment (LCA) to wastewater treatment: a best practice guide and critical review. *Water Research*, **184**, 1–18.

Goedkoop M., Heijungs R., Huijbregts M., de Schrijve A., Struijs J. and van Zelm R. (2009). ReCiPe 2008: A Life Cycle Impact Assessment Method Which Comprises Harmonized Category Indicators at the Midpoint and the Endpoint Level: Report I Characterization. Ministry of Housing, Spatial Planning and the Environment, The Netherlands.

Hofman-Caris C. H. M., Harmsen D. J. H., Siegers W. G. and van den Brand T. P. H. KWR rapport 2019.086 'TKI project HerCauwer – Reuse of Coagulant From Aquatic Iron'. KWR, Nieuwegein, The Netherlands.

Huijbregts M. A. J., Steinmann Z. J. N., Elshout P. M. F., Stam G., Verones F., Vieira M. D. M., Hollander A. and Van Zelm R. (2016). ReCiPe2016: A Harmonized Life Cycle Impact Assessment Method at Midpoint and Endpoint Level. RIVM Report 2016-0104, Bilthoven, The Netherlands.

International Energy Agency. (2014). Energy Policies of IEA Countries; European Union 2014 review. Available at: https://www.iea.org/publications/countryreviews/ (accessed 7 May 2018).

ISO (14040 2006). Environmental Management – Life Cycle Assessment – Principles and Framework. International Organization for Standardization, Geneva, Switzerland.

ILCD Handbook. (2010). European Commission – Joint Research Centre – Institute for Environment and Sustainability: International Reference Life Cycle Data System (ILCD) Handbook – General Guide for Life Cycle Assessment – Detailed Guidance, 1st edn. EUR 24708 EN. Publications Office of the European Union, Luxembourg.

Palmen L., Hofs B., Hijnen W., van den Brand T. P. H., Borsboom M. and van der Kolk O. (2015). Softening with calcite: On the market with Dutch lime pellets. Rapport number: KWR 2015.107. KWR, Nieuswegein, The Netherlands.

Powertoprotein.eu (accessed 14th September 2018).

Visser C., Odegaard I. Y. R., Naber N. R., Bergsma G. C. and van Nieuwenhuijzen A. F. (2016). Life Cycle Assessment of Raw Materials from Sewage Water. STOWA report number 2016-22. STOWA, Amersfoot, The Netherlands.

World Commission on Environment and Development (WCED). (1987). Our Common Future. United Nations, Geneva.

doi: 10.2166/9781780409566

Chapter 17

Resource recovery 2050 and beyond: water in a climate-neutral and cyclic economy

Willy Verstraete

Center for Microbial Ecology and Technology (CMET), Ghent University, Coupure Links 653, 9000 Ghent, Belgium

It is hard to believe but in my youth in the 1960s, in the villages and the cities in Flanders, Belgium, sanitation was simply and only a matter of direct drainage of the sewage to the rivers and simple transporting (sometimes still by horse wagon) of the night soil to the fields. Gradually, we have seen the evolvement of technologies to make sewage and night soil 'less offensive' to the citizen. In the past decades, the notion that not only humans, but also the whole environmental system, needed to be respected has become part of the mindset. Thus, advanced technology to treat more and more thoroughly has been installed. The counterpart of this evolution is that wastewater and wastes have become stigmatized as 'total disorder and absolute danger'. Hence, the line of full 'disintegration of wastes' excluding any type of recovery has been predominant in urban planning and management.

Today, we are experiencing the indications that some tipping points of the carrying capacity of the planet have been passed. We have become aware that we all, as citizens and consumers, have major interest in approaches which deal with the recycling of urban resources. Yet, we have also experienced that biological agents, even in their very rudimentary form of viruses, can – even in our times of technological supremacy – cause pandemics. Hence, all recycling will require careful attention in terms of safety. Safe recycling should have positive impacts on urban sustainability and economy and help to abate climate change. In this context, we will have to rethink a great many of our current urban infrastructures and processes. Although predicting the future is risky, some dreams can be formulated for the wastewater in the sense that they may help to set research and development in that direction.

First of all, the water present in sewage will be made fit, not to just discharge as now is most often the case, but to re-use at a very substantial level for process and even drinking water purposes. A perfect example that this is possible, and with the knowledge and the consent of the consumer, can be found in Wulpen, Belgium. Indeed, for several decades, the drinking water produced by the local water company contains up to a level of 50% recycled water, and this without any problem in terms of quality (Dewettinck *et al.*, 2001). Of course, there are still many aspects of grey pollutants and pathogens that need to be monitored and if possible eliminated, but the overall pressure on the utilities to bring forward the effective closure of the water cycle will continue to increase. We will see more and more full treatment, followed by reverse osmosis to capture the 'H20' and transport of the brines to proper deposits or to the sea.

Secondly, the current line of which the microbes are predominantly set up to 'destroy' and to make unwanted molecules 'disintegrate' will become reversed to a system which will focus on the unique biosynthetic capacity of the microbes. Indeed, from urban wastes and wastewaters, a number of building blocks such as ammonia, CO_2, and minerals can be captured in a way that makes them adequate for reuse. Of course, the overall process system will right from the beginning have to be focused on the aspect of 'clean' recovery. Anaerobic digestion, ammonia stripping, and phosphate precipitation will be the upfront key processes; activated sludge, nitrification, denitrification, and anammox will become of secondary use. The generated clean resources will be upgraded using microbial processes, but the special feature will be that the latter will represent so much demand that value will be supplemented to them. To be more specific, by adding quality organic carbon or quality energy in the form of hydrogen gas, one can upgrade the aforementioned building blocks into microbial biomass (De Vrieze *et al.*, 2020). The quality organic carbon can be a carbohydrate generated by a crop grown to produce such carbohydrates for wastewater upgrading (for instance a starchy crop not demanding heavy fertilization nor pest control). The quality hydrogen can be generated by electrolysis of water; the hydrogen being part of the H_2-based economy as depicted for the near future by Van Wijk *et al.* (2017). The microbial biomass thus harvested will contain the nutrients delivered via the wastes. Yet, the harvested microbial biomass will qualify to enter cleantech industrial process and value chains (for instance, having gasification as a corner stone) and thus become part of the production of food, feed, organic fertilizer, and biodegradable polymers.

The treatment of what we currently call 'wastes' will be fully revised: the first step will be dedicated to capturing and generating clean and safe building blocks; the second step will deal with upgrading the latter to new products accepted by the consumer who will request a sustainable urban society.

REFERENCES

De Vrieze J., Verbeeck K., Pikaar I., Boere J., Van Wijk A., Rabaey K. and Verstraete W. (2020). The hydrogen gas bio-based economy and the production of renewable building block chemicals, food and energy. *New Biotechnology*, **55**, 12–18.

Dewettinck T., Van Houtte E., Geenens D., Van Hege K. and Verstraete W. (2001). HACCP to guarantee safe water reuse and drinking water production: a case study. *Water Science and Technology*, **43**, 31–38.

Van Wijk A., Van Der Roest E. and Boere J. (2017). Solar Power to the People. IOS Press BV, Amsterdam, The Netherlands.

doi: 10.2166/9781780409566

Outlook for the carbon-negative circular water economy

Glen T. Daigger

Department of Civil and Environmental Engineering, University of Michigan, 177 EWRE Building, 1351 Beal Street, Ann Arbor, MI 48109, USA

My first public statement concerning the recovery of resources from the used (waste) water stream was an editorial I published in *Water Environment Research* in 1998 [1]. My first set of formal presentations was in 2001 when I was the American Academy of Environmental Engineers (now American Academy of Environmental Engineers and Scientists, AAEES) Kappe Lecturer. I visited about a dozen universities during the year of service to this lectureship, and one of the lectures I offered was titled 'The Future of Wastewater Treatment'. In it I presented a vision of wastewater treatment plants as 'factories' harvesting water, energy, nutrients, and other material from the used water stream. Examples of treatment systems where this was already happening were also offered. I relate this, not to suggest that I am prescient, but to simply indicate that I have been thinking this way for a long time. I was at least partially helped along by the fact that I grew up on a farm. On a farm the total focus is on creating useful products out of nature. By the way, it was a pig farm, so I had quite an early introduction to the business end of waste management!

So, let me suggest that, rather than asking 'why should we extract resources from the used water stream?', the truly relevant question is 'since we can extract useful products out of the used water stream, why wouldn't we?'. Assuming we can extract them practically, and at a cost that is at least partially compensated by the value of the product, why not produce the inherent value created by their production? I have never seen a reason why we would not create as much value for society as we reasonably can. We have actually been doing this as a profession for decades. Water reuse is a long-standing practice, with a history that goes back into at least the 1950s and much earlier if one considers historic practices in some cultures. Anaerobic digestion and biogas utilization to produce heat and electricity is also quite a long-standing practice. In fact, through the 1960s there were quite a number of plants that were energy self-sufficient based on this practice. This was possible because effluent discharge standards were not as strict as they are today. Consequently, liquid stream treatment often progressed to the point that electricity produced from biogas could sustain it. Recycling of nutrients and organic matter through biosolids reuse is also quite a long-standing practice. When I was new to the profession I heard plenty of stories about how people lined up to take digested and dried biosolids from sand drying beds because it was such an excellent fertilizer.

So, what is different today? For one thing, effluent discharge standards have become much stricter, and compliance is more strictly enforced. Financial support for treatment used to be much more problematic than today, so the priority was to keep operating costs low (e.g. by producing all the energy needed for treatment), and plants did the best they could in terms of effluent quality given these financial constraints. The establishment of specific discharge standards and their strict enforcement (resulting from passage and enforcement of the Clean Water Act in the U.S. and similar legislation elsewhere) reversed this. The first priority today is compliance with effluent discharge standards, mandating the provision of sufficient resources to do the job. A combination of factors in the 1970s and 1980s also led to a decline in the practice of anaerobic digestion (which was quite popular before that), including enthusiasm for new technologies, the procedures established for economic analysis to receive Clean Water Act funding, and poor industrial waste control practices at the time. Sometimes the old practices return, and this has happened in the U.S. where anaerobic digestion has returned as a preferred biosolids stabilization technology.

Another difference between the time prior to the 1970s and today is that, at least in the developed countries, we are financially much wealthier than we were in the 1960s and earlier. It may not 'feel' this way to many, but by every objective measure we are wealthier. While we have greater financial wealth, and corresponding freedom in terms of the actions we can take, we are increasingly resource constrained. Our use of renewable resources (including water) exceeds the rate at which these resources can be renewed naturally, and we are consuming non-renewable resources at alarming rates. There is also the issue of climate change. While the population of the developed countries has roughly stabilized, population growth is continuing in the developing countries and living standards are increasing in many of these countries. Thus, demand continues to grow while supplies shrink, as long as we continue to live the way we do. This shortfall between resource availability and our current rate of consumption means that we have no choice but to change. We have to use our available resources much more efficiently (become much more resource-efficient), and it will be better if we do so by choice rather than be forced to do so by scarcity. I am convinced that we have sufficient resources to allow all humans on earth to lead enriching and productive lives. To do so we will have to use our available resources (including water) much more efficiently, including a dramatically increased rate of recycling and reuse. The good news is that, since our current and growing resource constraints are human-made, we have the means to correct the situation.

This brings us back to the water cycle. The planet's renewable fresh water supply is already fully allocated (and in some instances over-allocated), while water demands continue to increase. The only available means to meet the growing demand are conservation and reuse. The most valuable resource in used water is the water itself. The practice of water reuse is already well established, and it is increasing rapidly in water-short locations around the planet. There are few constraints on dramatic acceleration of this practice, except for the limiting quantity of used water in some locations. Water reuse is already the norm (as compared to treatment and discharge) in many water-short locations, and this will become quite common as more than half the human population faces water scarcity in the not too distant future.

What about other resources in the used water stream? Practices for recycling some resources are already well-established, and new technologies are developing. We can expect resource recovery to increase, not only in quantity but also in the specific resources recovered as scarcity created by resource constraints drives increased prices and opens up opportunities for alternate sources of certain materials. Resource recovery is also a strategic issue for utilities as, unless the residual resources in the used water stream are recovered as valuable products, they remain as wastes that must be disposed of. Utilities are better positioned to continue operation if they are producing products that have a market, rather than wastes that many will oppose. Thus, the strategic advantage of resource recovery is that it can garner public support for the utility to continue necessary operations. This offers intrinsic value to the utility, even if revenue from reuse of the resources does not fully cover the cost of production.

Returning to my farming days, the old saying about a slaughter-house was that 'all parts of the pig are used, except the squeal!'. Why would we not do the same for used water?

Resource recovery is achieved by implementing a combination of separations and conversions. In this regard, one wonders why we begin by combining all of the sources of used water into a single stream. The historic answer, of course, is to minimize the piping network given that we were transporting all of the used water from the metropolitan area to a single or a small number of quite large processing facilities (which we historically called treatment plants). So, we save on infrastructure in a centralized configuration, but at the cost of making separation and recovery more difficult. What if we distribute processing more throughout the service area and, thereby, make the piping infrastructure needed to keep individual streams separate less burdensome? At the household and commercial building scale, we can easily think about keeping blackwater, yellow water, and greywater separate. Blackwater (containing feces and kitchen waste) accounts for much of the carbon in used water, yellow water (urine) contains much of the nutrients, and greywater (water used for washing clothes, ourselves, and other things) represents the stream from which much of the actual water content of used water originates. This approach is typically referred to as source separation, which is a misnomer in that the sources originate separately. We could, more accurately, refer to this as 'maintaining separation'.

Resource recovery is inherently enabled with an approach like this, as we avoid difficult separations required when we mix these sources. This approach also allows us to optimize the scale where we apply various separations and conversions, minimizing the transport of the heavier components of used water and optimizing the size of individual treatment facilities relative to their economies of scale. Water represents the largest mass in the used water stream. By separating greywater this less contaminated water can be aggregated from a small subset of the urban population and treated economically using less chemical- and energy-intensive approaches to produce fit-for-purpose product water closer to locations of demand. Water reuse is thereby enabled by a distributed system of greywater treatment and fit-for-purpose water production facilities. Because of their much lower mass, blackwater and yellowwater can be conveyed to more centralized facilities for processing. Greywater treatment technologies are well established and their economies of scale are well known. For example, the attractive size from an economy of scale perspective for greywater treatment may be around 20 000–50 000 PE, while effective sizes are likely different for for blackwater and yellow water processing. Hybrid combined distributed and centralized systems of this type represents one of my visions of the future.

Current practice is to extract nutrients, energy, and water (NEW) from the used water stream. Water is, and will likely continue to be, the most valuable resource in the used water stream and will continue to be recovered. The recovery of nutrients (nitrogen, phosphorus, and also ideally potassium) is also likely to continue. The changing energy picture may refocus us from converting carbon into energy and, instead, recover it as feedstock into the production of carbon-based materials. Declining costs for renewables (solar, wind) are making policy discussions about continued use of fossil fuels for energy production moot. We currently need carbon-based energy production as a component of the energy system because of its ability to turn on and turn off in response to demand (meet peak demands). How long will this persist, and will this function be replaced by dramatically improved battery technology? We shall see. Even if carbon-based energy production remains as part of the energy mix, the market may yet intercede as the higher value of some carbon-based products may make conversion of carbon collected from used water into such products economically more attractive than conversion to energy. Again, we shall see. The key point is that we must be more flexible in the products we produce to adjust to varying market conditions, not only in the future but even today.

Producing commercially viable products requires meeting product specifications, in sufficient quantities, according to the scheduled needs of the consumer, and with necessary marketing and distribution channels. The water profession knows how to do this for water (our core business), but we are not particularly good at this for other products. The products that can be produced from used water also depend on the quantity of raw materials (what we have traditionally called COD, BOD_5, TN, TP, etc.). Thus, while product quality is under our control, quantity and timing is not. While individual utilities can produce product, there may be the need for commercial entities (public or private, for-profit or non-profit) that aggregate products from several utilities and provide logistics, marketing, and distribution to the customer. There are actually many examples in the commercial world. My brother is a family farmer running a dairy farm and is in business because entities like this make the milk that he produces commercially viable.

When might all this come to pass? The answer is that it is already happening. Thus, it is a matter of scaling up and continuing to evolve. I have always been impatient when I see solutions, especially those needed as badly as these. This will all certainly be 'old hat' by the end of the 21st century, but I am hopeful that the transition will be well underway by the middle of the century. This represents the first major re-invention of our profession in the past hundred years or so, and I am hopeful that it will occur in my lifetime. I am also envious of those coming into the profession at this time. I know that they will find inventing this new future to be an exciting and enriching endeavor. I do hope, however, that they will come to know the past as I have so that we can build on what we have learned and avoid the mistakes of the past. But, what a glorious future we have in front of us!

REFERENCE

1. Daigger G. T. (1998). Waste or resource? *Water Environment Research*, **70**(5), 979.

doi: 10.2166/9781780409566

Hastening the arrival of the resource recovering water future

David L. Sedlak

Department of Civil and Environmental Engineering, University of California, Berkeley, CA, 94720, USA

Over the past decade, water technologists have put forward the idea that wastewater is an outmoded concept [1–6]. Following up on the observation that sewage contains valuable resources that are not being recovered, these pioneers have been building bench-top and pilot-scale prototypes of treatment plants capable of obtaining energy, clean water, nutrients and minerals from municipal and industrial wastewater. In several cases, resource recovery has been realized in full-scale resource recovery plants that produce recycled water, fertilizer and electricity from wastewater [7]. Despite these achievements and the strong advocacy of leading members of the water research community, design engineers and decision-makers in most parts of the world have been slow to embrace resource recovery by investing in this new concept [4, 8].

The apparent disconnect between the advocates of resource recovery and those responsible for designing, building and operating urban water infrastructure raises questions about whether or not the resource recovery future will ever arrive. By considering the forces that shape current investments in urban water infrastructure and the ways in which water institutions have changed in the past in response to various challenges and opportunities, we can gain insight into the best ways to assure that the resource recovering future will arrive.

When attempting to predict future changes in any type of infrastructure, it is important to recognize that the adoption of new technologies usually depends upon forces unrelated to engineering design and process performance. Factors such as the rate at which equipment is normally replaced, the size and competitiveness of the market for the new technology and the nature of the institutions responsible for managing and regulating the infrastructure are critical to change in every technological endeavor [9]. One of the most important factors influencing technology adoption is the nature of the industrial sector.

At one extreme, the telecommunications sector undergoes rapid change. In this sector, innovative technologies are rapidly adopted because most of the necessary equipment has a lifespan of less than a decade and equipment manufacturers and service providers gain advantages that translate into large profits from modest advances. Change comes more slowly to the infrastructure involved in electricity production because existing infrastructure (e.g. oil refineries, pipelines, power plants) has lifetimes of several decades and the market is subject to considerable government regulation. Wastewater

Table 17.1 Representative technologies in telecommunications, energy and wastewater treatment from 1990 to 2050.

Sector	Sectorial Nature	1990s	2020	2050
Telecommunications	Rapid replacement; competition	Landlines, modem	Wireless, fiber optic	Internet-of-things
Electricity production	Longer lifetimes; private and public	Coal, oil, nuclear	Distributed solar, windfarms	Decarbonization, battery storage
Wastewater treatment (nutrients)	Slow renewal; public ownership	Biological nutrient removal, biosolids	Annamox, struvite recovery	Anaerobic N&P recovery
Wastewater treatment (water reuse)	Slow renewal; public ownership	Non-potable reuse; agric., industry	Centralized potable reuse	Building-scale potable reuse

infrastructure is even slower when it comes to embracing new technologies because the useful lifetime for this infrastructure is often half a century or more [10]. Furthermore, the potential risks to public health and the environment from mismanagement of water systems have led to public ownership and extensive government regulation.

To illustrate the importance of the industry sector to the rate of adoption of new technologies, consider the period from 1990 to 2050 (Table 17.1). Since 1990, the telecommunications industry evolved from wired to wireless communication. On the basis of current investments, it seems likely that the wireless revolution will soon see more innovation and investment as it transitions to the internet-of-things [11]. Since 1990, electricity generation infrastructure also has undergone substantial change in response to efforts related to public policies to curb the release of greenhouse gases. As a result, we can expect continued expansion of wind and solar power generation along with greater availability of energy storage systems (e.g. batteries) over the next three decades. In contrast, progress towards recovery of resources from wastewater since 1990 has been modest, with advances largely confined to a small number of locations populated by early adopters of new technologies. Using current trajectories, we can expect that a handful of new resource recovery technologies will progress to the full scale over the next three decades and that more efficient processes will be invented, but few wastewater facilities will look substantially different from the systems that existed in 1990. Thus, without more effective coordination and collaboration, the resource recovering water future may not arrive in many places prior to 2050.

Environmental engineers interested in seeing the widespread adoption of resource recovery from wastewater during their professional careers need to understand the ways in which infrastructure investments are made, the attributes that appeal to decision-makers use this knowledge to develop their technologies and introduce them into practice.

Under most conditions, change comes to urban water systems because design engineers and their clients choose technologies that offer perceived social, environmental or financial benefits. But these advantages alone may not be enough. The people responsible for making decisions about the use of public funds are hesitant to take risks on unproven technologies. To overcome this barrier and achieve acceptance by the practitioner community, research and development needs to demonstrate reliable performance under realistic operating conditions (e.g. at full scale). Such success is even more potent if the technology can be adopted into design manuals [12].

Technologists often struggle to make the transition from the bench-top to the demonstration scale without help from the community of practicing engineers and equipment manufacturers. Resource recovery technologies are more likely to be accepted if they are supported by a healthy innovation ecosystem (i.e. professionals who work for equipment manufacturers, water service providers and regulators who interact with each other and develop mutual trust) and a critical mass of potential users of the technology. For example, over the past 30 years, on-site wastewater treatment has made considerably more progress in Beijing than in other Chinese cities because the city had a vigorous innovation ecosystem that lobbied for government mandates to install treatment systems [13]. The innovation ecosystem that built up around an initial program designed to serve international hotels grew over time and adapted to changing needs of the city for wastewater treatment in low density developments on the perimeter of the city. A similar process is likely to apply to the diffusion of nutrient recovery technologies: as more wastewater treatment plants build facilities to recover nutrients from wastewater, opportunities for vendors who market fertilizers produced from the recovered materials will increase, which will, in turn, create more incentives for designers to include nutrient recovery in new projects.

In addition to the incremental pathway described above, change also comes to urban water infrastructure as a response to a real or perceived crisis. In situations in which existing practices are unable to solve a problem that is a high priority for members of the public or government officials, the traditional reluctance of environmental engineers and their clients to take risks diminishes. This mode of change has been crucial to the two periods of the 20th century when many of the largest investments were made to urban water systems: the construction of drinking water treatment plants in the early part of the 20th century and the upgrading of municipal wastewater treatment plants that occurred between 1970 and 1990 [14]. In both cases, the change only occurred after the public's attention was focused on the problem and the engineers tasked with coming up with solutions recognized the inadequacy of their existing technology toolbox.

The common threads unifying the incremental and the rapid modes of change that will determine the timing of the resource recovery future are the development of new technologies that offer substantial benefits relative to existing practices and the existence of external forces (i.e. crises) that create opportunities for the new technologies to gain a foothold in the market although it is difficult to accurately predict the specific technologies that will dominate the resource recovery future, it is possible to anticipate where and when the greatest opportunities for resource recovery are most likely to occur. With this knowledge, environmental engineers should have a better idea about how to speed the transition to a resource recovery future throughout the water sector. Some of the most prominent opportunities for hastening the arrival of the resource recovering future are briefly described below.

The Fourth Industrial Revolution: Many industries are currently undergoing rapid change driven by advances in robotics, additive manufacturing, electronics and materials science [11, 15]. As these changes start to affect the water industry, modular wastewater treatment equipment will become less expensive to build, easier to control and more reliable. These modules will affect the ways in which centralized (i.e. large) wastewater treatment plants are designed and operated, favoring standardized treatment trains instead of the current approach to design in which each element of the treatment train is based upon a customized design. Early development of modules that enable resource recovery may create opportunities for the technologies to be adopted. Modularization and autonomous control may also create opportunities for distributed treatment and/or resource recovery at the household or building scale [16, 17].

Decarbonization of Electricity Production: In response to concerns about climate change, the rate of transformation of electricity generating systems is likely to accelerate over the next two decades. One possible outcome of these investments is significant reductions in the cost of electricity (e.g. a wind farm was recently contracted with a promise to deliver electricity at \$0.02 per kWh [18], which is considerably lower than the present cost of generating electricity from coal, oil or natural gas). Although the potential for replicating this type of low-cost renewable energy project in densely populated areas around the world is unclear, current trends in development of renewable energy and natural gas extraction suggest that renewable electricity will be less expensive than fossil fuels in the near future [19]. If such changes materialize, resource recovery technologies that are energy intensive (e.g. reverse osmosis) may become more attractive. Alternatively, government policies (e.g. taxes on fossil fuels) could make energy recovery from wastewater more attractive relative to energy sourced from traditional, non-renewable sources. Either way, the rate of adoption of resource recovery systems at wastewater treatment plants will be affected by changes in the ways in which electricity is produced.

Increasing Water Scarcity: Changes in precipitation patterns, higher temperatures and increased competition for water resources are likely to result in more water scarcity in the future, especially in large cities in arid regions [20]. In locations where water scarcity is already being experienced, such as Southern California, Western Australia and Israel, large investments have already been made in infrastructure for water reuse (i.e. the recovery of drinking water or water suitable for irrigation, cooling or industrial uses from municipal wastewater). Such projects have created opportunities for new technologies to be developed and implemented at scale, despite their relatively high costs and the risks associated with their adoption [7, 21]. The construction of water recycling facilities may also provide opportunities to integrate resource recovery into the overall treatment train (e.g. reuse of wastewater for agriculture or landscaping is a form of nutrient recovery whereas nutrients are sometimes a problem when water is reused for cooling or toilet flushing).

Urbanization of the Global South: Population projections for Africa, South America and Southeast Asia indicate rapid increases in urban populations over the next 50 years. In many cases, this urbanization will not be accompanied by industrialization. For example, Lagos, Nigeria is expected to grow from a population of around 15 million today to approximately 90 million people by 2100 [22] without the kind of industrialization and increase in wealth that accompanied the rapid growth of cities in China, Korea and Singapore. Irrespective of the per capita income in these emerging megacities, large investments in wastewater infrastructure are going to be needed for protection of public health and water security in the coming decades. As illustrated by many of the proposed designs of distributed sanitation systems [6], resource recovery may prove beneficial to the overall business model for these systems.

The difference between a waste disposal problem and a resource recovery opportunity is the value that society places on the social, environmental and financial costs and benefits associated with resource recovery. Advances in technologies for resource recovery will eventually reach a point at which some form of resource recovery will be implemented whenever wastewater is treated. However, a variety of factors related to the ways in which wastewater treatment plants are designed and operated is currently slowing or preventing the adoption of resource recovery technologies. Environmental engineers interested in accelerating the rate of technology adoption need to expand their perspective beyond the technological framework described in engineering textbooks. By designing technologies that are compatible with the opportunities and challenges facing the water sector by understanding the needs of decision-makers, regulators and investors, a resource recovering future may arrive in time to make a difference.

REFERENCES

1. Matassa S., Matassa S., Batstone D. J., Hülsen T., Schnoor J. and Verstraete W. (2015). Can direct conversion of used nitrogen to new feed and protein help feed the world? *Environmental Science & Technology*, **49**(9), 5247–5254.
2. Mayer B. K., Baker L. A., Boyer T. H., Drechsel P., Gifford M., Hanjra M. A., Parameswaran P., Stoltzfus J., Westerhoff P. and Rittmann B. E. (2016). Total value of phosphorus recovery. *Environmental Science and Technology*, **50**(13), 6606–6620.
3. Puyol D., Batstone D. J., Hülsen T., Astals S., Peces M. and Krömer J. O. (2017). Resource recovery from wastewater by biological technologies: opportunities, challenges, and prospects. *Frontiers in Microbiology*, **7**, 1–23.
4. Marlow D. R., Moglia M., Cook S. and Beale D. J. (2013). Towards sustainable urban water management: A critical reassessment. *Water Research*, **47**(20), 7150–7161.
5. Daigger G. T. (2009). Evolving urban water and residuals management paradigms: water reclamation and reuse, decentralization, and resource recovery. *Water Environment Research*, **81**(8), 809–823.

6. Larsen T. A., Hoffmann S., Lüthi C., Truffer B. and Maurer M. (2016). Emerging solutions to the water challenges of an urbanizing world. *Science (New York, N.Y.)*, **352**(6288), 928–933.
7. Hering J. G., Waite T. D., Luthy R. G., Drewes J. E. and Sedlak D. L. (2013). A changing framework for urban water systems. *Environmental Science and Technology*, **47**(19), 10721–10726.
8. Porter J. J. and Birdi K. (2018). 22 reasons why collaborations fail: lessons from water innovation research. *Environmental Science and Policy*, **89**, 100–108.
9. Markard J. (2011). Transformation of infrastructures: sector characteristics and implications for fundamental change. *Journal of Infrastructure Systems*, **17**(3), 107–117.
10. Kiparsky M., Sedlak D. L., Thompson Jr., B. H. and Truffer B. (2013). The innovation deficit in urban water: The need for an integrated perspective on institutions, organizations, and technology. *Environmental Engineering Science*, **30**(8), 395–408.
11. Zhong R. Y., Xu X., Klotz E. and Newman S. T. (2017). Intelligent manufacturing in the context of industry 4.0: A review. *Engineering*, **3**(5), 616–630.
12. WEF (2018). *Design of Water Resource Recovery Facilities*, 6th edn. WEF & ASCE (eds). McGraw-Hill Education, New York.
13. Binz C., Truffer B., Li L., Shi Y. and Lu Y. (2012). Conceptualizing leapfrogging with spatially coupled innovation systems: The case of onsite wastewater treatment in China. *Technological Forecasting and Social Change*, **79**(1), 155–171.
14. Sedlak D. L. (2014). Water 4.0: The Past, Present and Future of the World's Most Vital Resource. Yale University Press, New Haven.
15. Schwab K. (2017). The Fourth Industrial Revolution. Crown Business, New York.
16. Kavvada O., Nelson K. L. and Horvath A. (2018). Spatial optimization for decentralized non-potable water reuse. *Environmental Research Letters*, **13**(6), 1–12.
17. Kavvada O., Tarpeh W. A., Horvath A. and Nelson K. L. (2017). Life-cycle cost and environmental assessment of decentralized nitrogen recovery using ion exchange from source-separated urine through spatial modeling. *Environmental Science and Technology*, **51**(21), 12061–12071.
18. Jackson F. (2019). Wind Power Breaks New Record At $0.02 Per KWhr, in Forbes.
19. Lazard (2018). Lazard's Levelized Cost of Energy Analysis--Version 12.0. Available from: https://www.lazard.com/media/450784/lazards-levelized-cost-of-energy-version-120-vfinal.pdf (accessed 26 January 2022).
20. Florke M., Schneider C. and McDonald RI. . (2018) Water competition between cities and agriculture driven by climate change and urban growth. *Nature Sustainability*, **1**(1), 51–58.
21. Harris-Lovett S. R., Binz C., Sedlak D. L., Kiparsky M. and Truffer B. (2015). Beyond user acceptance: A legitimacy framework for potable water reuse in California. *Environmental Science & Technology*, **49**, 7552–7561.
22. Hoornweg D. and Pope K. (2017). Population predictions for the world's largest cities in the 21st century. *Environment and Urbanization*, **29**(1), 195–216.

doi: 10.2166/9781780409566

One water/one health: Used water management in 2050 and beyond

Julian Sandino

Global Wastewater Solutions Director, Jacobs Engineering Group, USA

I appreciate the authors asking me to share my thoughts on the challenges and opportunities related to the future of used water management. I want to commend them for this publication, because although there are plenty of textbooks and reference materials on this topic that review today's needs and state of the practice, very few address the importance of changing our future approach.

Certainly, our industry's recent emphasis on adopting a Circular Economy paradigm is a welcome departure from the traditional approach to used water management over the past century. However, the concept of recovering resources from our environmental health protection efforts is not a new one. Buckminster 'Bucky' Fuller (1895–1983, American philosopher, technology icon, and visionary), known for inventing the geodesic dome and coining the term Spaceship Earth, could not have anticipated this better with his 1971 quote 'Pollution is nothing but the resources we are not harvesting. We allow them to disperse because we've been ignorant of their value.'

My perspective on used water management is empirically derived, the result of more than 30 years' experience as a practitioner in the consulting engineering sector, mostly as a technologist involved in process design of treatment facilities, and more recently, in long range planning efforts for several large cities worldwide. In sharing my vision of how we must confront the future in our industry, let us begin by looking at what got us here.

Today's typical developed-world used water management urban solutions can be traced back to early 20th century planning efforts. They are usually comprised of a highly-articulated collection system connected to treatment facilities near a receiving water body. Our current infrastructure has evolved from these early planning concepts into implementation (and subsequent expansions) almost unchanged. Initially driven by the goal of protecting public health, then evolving to embrace environmental health goals to mitigate impacts on nature from discharges, the solutions we currently depend on were developed within the context of a very different world. That 'world' had a highly predictable climate, resources to spare, and only a few billion people, mostly in rural areas. Utilities provided wastewater management services by developing long-term master plans focused on the needs of what they thought (or wanted to believe) was a discernable future. Treatment options were based on technologies that arguably had evolved little in almost a century. The adopted solutions were based strictly on a consumer perspective, one that determined the best option to be the one requiring the least financial, energy, chemical, and manpower resources to meet demands defined by a certain vision of the future.

However, all recent predictions of what the world will look in the year 2050 and beyond paint a drastically different picture to that under which we currently operate – altered climate, constrained resources, and populations in excess of 10 billion living mostly in mega-urban centers, which makes us even more susceptible to the ravages of a pandemic as demonstrated recently by COVID-19. Therefore, it is highly unlikely that our current used water management approaches – and corresponding enabling technologies – will adequately meet tomorrow's radically different environmental, social, and financial requirements. So, in simple language, counting on current approaches for future needs seems downright stupid! Here are three general concepts I suggest should be part of a better approach to develop used water solutions aligned with the circular economy precept of this book, and still be relevant in 2050…and hopefully a bit beyond.

17.1 ONE WATER, ONE HEALTH

Like the now widely-accepted concept of *One Water* (one that recognizes the integrity of the water cycle and water as a single resource), we must adopt a *One Health* triple-component goal with social (public), environmental, and economic (financial) components, in the same way we have embraced the triple bottom line concept to properly evaluate sustainability in infrastructure decisions. Adding the financial health component incorporates the basic tenets of the Circular Economy concept, involving the necessary shift of the water industry's role from one of being a consumer of resources to a producer of them.

However, successfully engaging the economy's value chain must be based on offering products that the market needs (what the market 'pulls') and not by 'pushing' recovered products with no demand or with a substantial competitive disadvantage. This approach will require our industry to learn how to assess (and effectively compete) in markets using recoverable resources, but not at the expense of non-transferable public and environmental health protection responsibilities. Of equal importance, of course, will be the task of educating the public that we serve about the urgency of adopting this more holistic perspective.

17.2 GIVE UP THE CRYSTAL BALL

We must acknowledge that we cannot futureproof today's decisions by hoping to predict the future with certainty. To position for the future, utilities must abandon their traditional long-term master planning exercises, replacing them with more dynamic adaptive and strategic positioning approaches based on shorter planning horizons and multiple likely future end-point scenarios. This approach allows for short-term decision-making (by discarding least-adaptable alternatives) while still offering other plausible, longer-term opportunities. For the practitioner, process flow sheets and corresponding facility recommendations will need to be highly flexible to enable course corrections if assumptions regarding externalities should change. External factors may include climate, regulations, service area needs, technology advances, financial conditions, and (in the Circular Economy) viable markets for recoverable resources. The challenge is that flexible alternatives are inherently more expensive (at least in the short term) than highly-optimized, rigid options, so it is important that we also develop new evaluation methodologies and gain acceptance and support for them from those in decision making circles.

17.3 DARE TO DISRUPT

Expanding or upgrading existing facilities or developing greenfield solutions under the Circular Economy concept will require changing how we configure unit processes. In addition to the traditional 'how much I must remove' point of view (satisfying the traditional public and environmental goals we have relied upon) we must now also think about 'how much can we recover' (the proposed new financial aspect of *One Health*). So, it is not just about considering facility configurations on their 'percent removal' merits any more, but also on their 'percent recovery' potential. Apart from reclaimed water, energy generation (from recovered biogas), and agricultural and landscape use of biosolids, there are few examples in our industry of successfully applying this way of thinking. Most of our 'proven' unit processes (and their arrangement) are still based on removing pollution, not recovering a resource. Adopting 'disruptive' technologies, rather than continued tweaking of existing solutions, is probably the answer to our industry's current needs, because it provides necessary flexibility and adaptability to plan for an uncertain future in this new Circular Economy paradigm.

Developing these new (and somewhat unproven) technologies may be time and resource consuming for a utility. Therefore, it is essential to adopt collaborative applied research and product development strategies, including participating in subscriber-based research and marketing programs, so that we can share knowledge and resources across the industry.

So, I fare you well, users of this book, as you embark on one of mankind's most importance quests – the proper management of the limited resources we depend on for our survival, now adding to your toolbox the sound principles brought forward by this book. Saludos!

IWA Publishing's authorised EU representative for General Product Safety Regulations is Diane D'Arras, 15 rue Duret, 75116 Paris, France, e-mail: safety@iwap.co.uk.

Printed and bound by CPI Group (UK) Ltd, Croydon, CR0 4YY

12/05/2026

02108870-0015